# Advances in Sustainable Cultivation of Horticultural Crops

# Advances in Sustainable Cultivation of Horticultural Crops

Guest Editors

**Josinaldo Lopes Araújo Rocha**
**Márcia Michelle Queiroz Ambrósio**
**Railene Hérica Carlos Rocha Araújo**

MDPI

Basel • Beijing • Wuhan • Barcelona • Belgrade • Novi Sad • Cluj • Manchester

*Guest Editors*
Josinaldo Lopes Araújo Rocha
Department of Agricultural Sciences
Federal University of Campina Grande
Campina Grande
Brazil

Márcia Michelle Queiroz Ambrósio
Department of Agronomic and Forest Sciences
Universidade Federal Rural do Semi-Árido
Mossoró
Brazil

Railene Hérica Carlos Rocha Araújo
Department of Food Engineering
Federal University of Campina Grande
Campina Grande
Brazil

*Editorial Office*
MDPI AG
Grosspeteranlage 5
4052 Basel, Switzerland

This is a reprint of the Special Issue, published open access by the journal *Horticulturae* (ISSN 2311-7524), freely accessible at: https://www.mdpi.com/journal/horticulturae/special_issues/13AEB9K5TT.

For citation purposes, cite each article independently as indicated on the article page online and as indicated below:

| Lastname, A.A.; Lastname, B.B. Article Title. *Journal Name* **Year**, *Volume Number*, Page Range. |
|---|

**ISBN 978-3-7258-7629-7 (Hbk)**
**ISBN 978-3-7258-7630-3 (PDF)**
**https://doi.org/10.3390/books978-3-7258-7630-3**

Cover image courtesy of Márcia Michelle Queiroz Ambrósio

# Contents

# About the Editors

**Josinaldo Lopes Araújo Rocha**

Josinaldo Lopes Araújo Rocha is a Full Professor at the Department of Agricultural Engineering, Federal University of Campina Grande (UFCG), and a permanent faculty member of the Graduate Program in Soil Science at the Federal University of Paraíba (UFPB). His work focuses on plant mineral nutrition and its interrelationship with tolerance to water and salt stresses. Additionally, he has developed projects involving the use of nanofertilizers and the utilization of mining waste as a source of plant nutrients, as well as the biofortification of vegetables with iron and zinc.

**Márcia Michelle Queiroz Ambrósio**

Márcia Michelle de Queiroz Ambrósio is a full professor in the Department of Agronomic and Forestry Sciences at the Federal Rural University of the Semi-Arid Region (UFERSA) and a permanent member of the faculty of the Postgraduate Program in Phytotechnics at UFERSA. Her work focuses on the alternative management of plant diseases, especially those caused by soil-borne pathogens, using physical, cultural, and biological control under field conditions and during the post-harvest of fruit crops.

**Railene Hérica Carlos Rocha Araújo**

Railene Hérica Carlos Rocha Araújo holds a Doctor of Science in Crop Science (Fitotecnia) from the Federal University of Viçosa (UFV), a Master's degree in Crop Science from the Federal Rural University of the Semi-Arid Region (UFERSA), and an undergraduate degree in Agronomic Engineering (UFERSA). She is a Full Professor at the Department of Food Engineering, Federal University of Campina Grande (UFCG), and a permanent faculty member of the Graduate Programs in Food Engineering and Process Engineering (UFCG), working in the areas of Experimental Statistics and Postharvest Physiology of Fruits, Vegetables, and Edible Flowers.

# Preface

This reprint, "Advances in Sustainable Cultivation of Horticultural Crops," presents cutting-edge research on key topics such as remote sensing, hydroponics, environmental stresses, seed and plant physiology, irrigation, nanofertilizers, organic fertilizers, and their interactions. The objective is to support the transition to more resilient, low-impact horticulture. Motivated by climate change, water scarcity, and soil degradation, this work offers evidence-based solutions. It is intended for students, researchers, and professors in horticulture and related fields, as well as professionals seeking sustainable practices.

**Josinaldo Lopes Araújo Rocha, Márcia Michelle Queiroz Ambrósio, and Railene Hérica Carlos Rocha Araújo**
*Guest Editors*

*horticulturae*

MDPI

*Editorial*

# Advances in Sustainable Cultivation of Horticultural Crops

**Josinaldo Lopes Araújo Rocha [1,*], Márcia Michelle de Queiroz Ambrósio [2] and Railene Hérica Carlos Rocha Araújo [3]**

[1] Academic Unit of Agricultural Engineering, Center for Technology and Natural Resources, Federal University of Campina Grande, Campina Grande 58429-900, PB, Brazil
[2] Department of Agronomic and Forest Sciences, Federal University of Semi-Árido, Mossoró 59625-900, RN, Brazil; marciamichelle@ufersa.edu.br
[3] Academic Unit of Food Engineering, Center for Technology and Natural Resources, Federal University of Campina Grande, Campina Grande 58429-900, PB, Brazil; railene.herica@professor.ufcg.edu.br
* Correspondence: josinaldo.lopes@professor.ufcg.edu.br

## 1. Introduction

Horticulture is an activity with a colossal economic and social impact worldwide due to its potential to generate employment and income—particularly due to the significant labor demand throughout its production chain—and its contribution to food security and the fight against hunger, challenges intensified by population growth and climate change [1]. In 2024, global production of fruits and vegetables reached 2.13 billion tons, directly generating around 732 billion dollars and even approximately 1.3 trillion dollars when considering the entire production chain [2,3].

This sector is so economically important largely because of research efforts in various areas of agriculture. In the last decade, the field has seen important research innovations ranging from controlled-environment cultivation and process automation to the use of nanotechnology, genetic improvement, biostimulants, biological control, and the expansion of urban horticulture, including the production of microgreens [4–7].

However, all this technical progress is only rational if it is guided by the pursuit of truly sustainable horticulture, which, according to the FAO [3], must combine increasing food production with environmental preservation, actively confronting ecosystem degradation and the effects of climate change, within a balance that encompasses social, economic, and ecological dimensions.

Accordingly, there are still considerable challenges related to the rational use of inputs such as fertilizers and pesticides; innovations in more efficient irrigation systems and mitigation of water deficits in arid and semi-arid environments; the improvement of protected cultivation techniques; the integration of new technologies into sustainable, large-scale production systems and the utilization of organic waste generated by this very activity; and the recovery of degraded areas [8–12]. Therefore, the scientific and technological advances required for the sustainable cultivation of horticultural crops are primarily characterized by interdisciplinarity in the field of plant production, as reflected in the 16 articles published in this Special Issue, which will be briefly discussed below.

## 2. Overview of Published Articles

The 16 published articles in this Special Issue cover topics such as remote sensing, seed physiology, plant physiology, irrigation, nanofertilizers, and the interaction between these topics (Table 1), reinforcing the fact that the sustainable cultivation of horticultural crops is an interdisciplinary endeavor involving different subsections of the agricultural sciences. These themes converge towards the search for solutions for sustainable cultivation in the face of challenges posed by climate change, particularly those arising from the inefficient use of resources such as water and soil in horticultural production.

**Table 1.** A list of the contributions in this Special Issue (categorized by agricultural science subfield) and their main findings.

| Contributions | Plant | Technology or Research Topic | Main Findings |
|---|---|---|---|
| 1 | Raspberries (*Rubus idaeus*) and Blackberries (*Rubus fruticosus*) | Evaluation of growth and production of varieties | Varieties demonstrated consistent vegetative growth and yield, with distinct seasonal fruiting patterns. |
| 2 | Cauliflower (*Brassica oleracea*) | Defining planting dates | Early strategic planting, with less environmental stress, maximizes cauliflower resilience and yield, with 'Flame Star' being the top-performing cultivar. |
| 3 | Tomato (*Solanum lycopersicum*) | Introducing nanoparticles and conventional sources of iron and zinc via foliar vs. deficit irrigation | Iron nanoparticles (200 mg $L^{-1}$) increased fruit iron content under both full-irrigation and water deficit conditions. A conventional zinc source, under full irrigation (4.5 g $L^{-1}$), increased both fruit production and fruit zinc content. |
| 4 | Green peppers (*Capsicum annuum*) | Introducing zinc nanoparticles (ZnONPs) via foliar and bioinoculants via soil vs. deficit irrigation | Adding ZnONPs plus *Bacillus subtilis* mitigates deficit irrigation, improving vitamin C and mineral nutrient (Ca, P, Mg, and Fe) content in fruits, thereby enhancing their nutritional quality index values. |
| 5 | Pomegranate (*Punica granatum*) | Synthesis and seed priming of silver (Ag) nanoparticles (NPs) | Direct imaging (SEM/TEM) and elemental chemical analysis (EDX/XRF/ICP-OES) provide confirmation that the AgNPs adhered to the surface and were also taken up into the seed. |
| 6 | Potato (*Solanum tuberosum*) | Deficit irrigation vs. application of nitrogen fertilizer | Deficit irrigation ($\leq$20%) and nitrogen reduction are viable without yield loss, provided cultivar-specific tolerances for drought and nitrogen efficiency are accounted for. |
| 7 | Lettuce (*Lactuca sativa*) | Irrigation methods and light intensity | Light intensity of 400 µmol $m^{-2}$ $s^{-1}$ combined with non-circulating irrigation during the final 5 days before harvest improved anthocyanin content, deepened red pigmentation, and increased antioxidant activity. |
| 8 | Bell pepper (*Capsicum annuum*) | Irrigation methods | The optimal irrigation method consists of supplying water at a moderate rate (with 3 drippers per plant) and adjusting the irrigation end time according to the specific variety and crop vigor stage. |
| 9 | Strawberry (*Fragaria* × *ananassa*) | Hydroponics system monitoring | Closed-loop hydroponic systems with biweekly nutrient solution correction achieved strawberry yields equivalent to conventional open systems while nearly doubling nutrient use efficiency. |
| 10 | Lemon balm (*Melissa officinalis*) | Organic manure (poultry, sheep, and cattle) and consecutive harvests | Sheep manure improved essential oil production and quality, with the second and third harvests offering optimal compositions for industrial applications. |
| 11 | Caraway plant (*Carum carvi*) | Organic manure and foliar application of vitamins and yeast | Organic manure (15 ton $ha^{-1}$), along with foliar vitamin $B_1$ (100 mg $L^{-1}$), enhanced plant growth, seed yield, essential oil content, and nutrient uptake. |
| 12 | Potato (*Solanum tuberosum*) | Adding *Trichoderma asperellum* to compost and vermicompost | Vermicompost, when heat-treated and enriched with *T. asperellum*, improved both the quantity and weight of mini tubers per plant. |
| 13 | Artichoke (*Cynara cardunculus*) | Using proximal and remote monitoring for distinguishing between organic and conventional plant management | An MFA fluorometer distinguished physiological differences between organic and conventional artichoke management. Remote sensing via UAS was less effective. |
| 14 | Genus *Phoenix* (*Arecaceae*) | The difference in germination between orthodox and recalcitrant varieties | Seed storage and germination in Phoenix palms are influenced more by species identity and geographic origin than by seed age. |

https://doi.org/10.3390/horticulturae12040476

**Table 1.** *Cont.*

| Contributions | Plant | Technology or Research Topic | Main Findings |
|---|---|---|---|
| 15 | Six halophytic species | Factors that affect seed germination | While salinity consistently hinders germination, targeted scarification can counteract this for species like *Medicago marina* and *Ammophila arenaria*, emphasizing the need for customized cultivation protocols. |
| 16 | Hops (*Humulus lupulus*) | Biological control of *Verticillium* spp. | *Pseudomonas* sp. HX1, *Streptomyces luteogriseus* HR40, and *Streptomyces flavofungini* HR77 reduced the disease severity index by 32.56%. |

In Contribution 1 (Pruteanu et al., 2025), the authors evaluate the influence of growth and production parameters on raspberries (*Rubus idaeus*) and blackberries (*Rubus fruticosus*) cultivated in Romania during different seasons, aiming to identify the most adapted and productive cultivars for the region. The authors found that raspberries showed consistent fruit production, with a peak in June–July and a slight resurgence in September, while blackberries exhibited high production in early summer, followed by a significant drop in August. In the same vein, in Dhukhuchhu et al.'s work (2025) (Contribution 2), the authors investigate planting strategies for cauliflower under variable climatic conditions. They identify the best planting times and densities to ensure sustainable and resilient production, offering practical guidelines for farmers in regions with harsh winters and short growing seasons.

The authors of Contributions 3 (Almeida et al., 2025) and 4 (Martins et al., 2024) study the effectiveness of nanoparticles containing zinc and iron and/or bio-inoculants as a method of mitigating water deficit in tomatoes and bell peppers relative to their respective conventional sources. In Contribution 3, the authors show that foliar application of iron oxide nanoparticles ($NPFe_2O_3$) significantly improves the physiological performance, productivity, and iron content of tomato plants. Under deficit irrigation (50% ETc), a conventional zinc source ($ZnSO_4.7H_2O$) promoted better crop performance. In turn, in contribution 4, the authors reveal that the combination of zinc oxide nanoparticles (nano ZnO) with bio-inoculants (plant-growth-promoting rhizobacteria) successfully mitigated the negative effects of water deficit on the nutritional quality of green bell peppers, increasing vitamin C content, total phenols, and antioxidant activity. According to the authors, a challenge in these studies remains the lack of understanding regarding the mechanisms of penetration and incorporation of nanoparticles into cells, along with their modes of action. In this regard, the authors of Contribution 5 (Alymanani et al., 2025) synthesized silver nanoparticles (AgNPs) using pomegranate peel waste and evaluated their performance as a priming agent and absorption in pomegranate seeds. The authors observed, through EDX, XRF, transmission electron microscopy imaging, and ICP-OES analyses, that the nanoparticles effectively adhered to and penetrated the pomegranate seeds.

Contributions 6 (Barka et al., 2025), 7 (Ruangsangaram et al., 2025), and 8 (Yang et al., 2025) address the issue of irrigation, although with different focuses. In Contribution 6, the authors study the possibility of reducing nitrogen (N) doses (from 165 kg $ha^{-1}$ to 131 kg $ha^{-1}$) in potato cultivars (*Solanum tuberosum*) as a function of irrigation depth (70%, 80%, and 100% of ETc), observing that a 20% reduction in N dose and irrigation depth generally does not harm the yield of the varieties Mesa Russet or Russet Norkotah3, while the other varieties suffer a decline in production. For the authors, these results suggest that irrigation and nitrogen inputs can be reduced without compromising productivity, but the reductions should be determined based on each cultivar. Contribution 7 refers to a study on lettuce (*Lactuca sativa*) in which the authors tested two light intensities (300 and 400 $\mu mol\ m^{-2}\ s^{-1}$) and two irrigation systems (recirculating vs. non-recirculating),

https://doi.org/10.3390/horticulturae12040476

applied one week before harvest. Recirculating irrigation resulted in higher plant fresh and dry weights, regardless of light. In this study, shoot fresh weight was approximately 22% lower under non-recirculating irrigation. In Contribution 8, irrigation methods are studied in two varieties (Maldonado and Nagano) of bell pepper (*Capsicum annuum*), along with different cut-off times (ending 3 h before sunset and ending 4 h before sunset) and irrigation levels (two, three, and four drippers with a flow rate of 2 L $h^{-1}$). Irrigation had a greater impact on the yield of the Maldonado pepper. Water productivity (WP) increased with the reduction in irrigation in the May–June and November–December periods for both varieties, but the responses to cut-off times varied. The authors concluded that the ideal method for cultivating bell pepper in the northern region of South Korea is irrigation with three 2 L $h^{-1}$ drippers, with the cut-off time adjusted according to the variety and plant vigor.

An interesting and innovative study on hydroponic cultivation was carried out by the authors of Contribution 9 (Lim et al., 2024). In this study, two methods of correcting a nutrient solution, in a closed or open circuit, were evaluated for the cultivation of Strawberry (*Fragaria* × *ananassa*). In the first experiment, the authors controlled the system solely based on the electrical conductivity of the drained solution in an open circuit, while in experiment 2, the authors controlled the system based on the ionic composition of the drained nutrient solution, at intervals of two or four weeks, in a closed circuit. The authors concluded that there were no differences in efficiency between the systems; however, in the closed circuit, correcting the ionic composition of the solution every two and four weeks led to 94% and 88% greater nutrient efficiency, respectively, than in the open-circuit system.

Contributions 10 (Fallah et al., 2025), 11 (Hassan et al. 2025), and 12 (Peña et al. 2025) evaluate the use of organic waste, plant stimulants (Contribution 10), and beneficial microorganisms (Contributions 11 and 12). In Contribution 10, the authors study the influence of different types of manure—poultry manure (PM) (3200 kg $ha^{-1}$), sheep manure (SM) (5400 kg $ha^{-1}$), and cattle manure (CM) (6200 kg $ha^{-1}$)—and a control (CO) (without manure), equivalent to 100 kg N kg $ha^{-1}$, and harvest periods (21 June, 14 August, and 6 October 2019) on the essential oil composition of Lemon balm (*Melissa officinalis*), concluding that the manures significantly improved the quality of the essential oil in terms of neral and geranial percentages relative to the control. Notably, between mid-August and early October, there was a substantial increase in the levels of these valuable compounds. The authors of Contribution 11 tested doses of organic residue (0, 5, 10, and 15 tons $ha^{-1}$) and foliar stimulants (vitamin B1 at doses of 50 and 100 mg $L^{-1}$; vitamin E at doses of 50 and 100 mg $L^{-1}$; and active yeast at doses of 100 and 150 mL $L^{-1}$) on caraway plants (*Carum carvi*). They concluded that organic fertilization (15 tons per hectare) with foliar vitamin $B_1$ (100 mg $L^{-1}$) improved plant growth, seed production, essential oil content, and nutrient uptake. Peña et al. (2025) (Contribution 12) evaluated the effect of inoculating *Trichoderma asperellum* into organic compost and vermicompost as a strategy for producing seed potatoes (*Solanum tuberosum*). The authors concluded that vermicompost, when heat-treated and enriched with *T. asperellum*, improved both the quantity and weight of mini tubers per plant.

Deidda et al. (2025) (Contribution 13) conducted a study on artichokes (*Cynara cardunculus*), in which they used proximal and remote sensing techniques to distinguish conventional and organic farming systems in relation to plant characteristics. Remote sensing (RS) using a UAS was leveraged to assess the overall vegetative vigor of the plants through the NDVI index, capturing the entire canopy (leaves and flower heads) broadly and quickly. In contrast, MFA (proximal fluorometer) was used to analyze specific physiological parameters, such as chlorophyll, nitrogen, flavonoids, and stress, in individual leaves, allowing the detection of subtle differences between management practices. The

https://doi.org/10.3390/horticulturae12040476

authors observed significant physiological differences between organic and conventional artichoke cultivation. Remote sensing via an unmanned aerial vehicle proved less effective than the proximal sensor (MFA).

Contributions 14 (Obón et al.) and 15 (Castañeda-Loaiza et al.) explore the factors affecting seed germination. In Contribution 14, the authors study potential differences in the germination capacity of seeds from the genus *Phoenix* (Arecaceae) and their initial growth, depending on seed origin (orthodox or recalcitrant) after 10 years of storage at 5 °C. They observed that species identity and the geographic origin of the seeds influenced germination success more than seed age. Seedling development followed a conserved seasonal pattern across all species, with leaf emergence synchronized in September and between March and July, followed by winter dormancy. In Contribution 15, the effects of salinity, substrate composition, and types of dormancy breaking on the germination of six halophyte species (*Mesembryanthemum crystallinum*, *Salicornia ramosissima*, *Medicago marina*, *Ammophila arenaria*, *Portulaca oleracea*, and *Atriplex halimus*) were evaluated. The authors concluded that *P. oleracea* showed the highest germination rate (95.6%) in coconut fiber under irrigation with non-saline water and the shortest mean germination time (5.2 days). *A. halimus* did not germinate under the tested conditions. Sulfuric acid scarification improved the germination rate of M. marina by 42.2%, while ultrasound scarification improved the germination rate of A. arenaria by 35.5%.

Finally, the authors of Contribution 16 (Ghoreshizadeh et al.) tested the biological control of *Verticillium* spp., which causes dieback disease in hops (*Humulus lupulus*). They isolated bacteria from the rhizosphere and xylem of a hop plant. Six strains (three Pseudomonas from the xylem and three Streptomyces from the roots) were selected based on their in vitro antifungal activity against the pathogens *Verticillium dahliae* and/or *V. nonalfalfae*. The biocontrol potential of these strains was evaluated through plant assays in a greenhouse. The bacteria *Pseudomonas* sp. HX1, *Streptomyces luteogriseus* HR40, and Streptomyces flavofungini HR77 were the best-performing isolates, reducing the disease severity index by 32.56%.

## 3. Conclusions

The studies published in this Special Issue provide us with an important direction for advances in sustainable horticulture, with thematic approaches marked by interdisciplinarity but also convergence. The integration of precision technologies, such as sensing and nanotechnology, with biological practices, such as microbial control and the use of biostimulants, reveals a promising path toward optimizing water and nutritional resources, mitigating abiotic stresses, and improving product quality. The valorization of organic waste and efficient cultivation systems, such as closed-circuit hydroponics, reinforces the principle of efficient use and conservation of natural resources. Future perspectives point to the need to personalize these solutions, considering the specific responses of each cultivar and local conditions, and to deepen our understanding of new technologies' mechanisms of action at the physiological and molecular levels. The continued synergy between basic and applied research will be fundamental to consolidating production systems that harmonize high productivity, climate resilience, and environmental and economic sustainability, combined with new advances in the fields of automation, robotics, and artificial intelligence.

**Acknowledgments:** We sincerely thank the Fapesq (Foundation to support research in the state of Paraíba) and Fapesq/PB Edital Universal 09/2021 SEECT/Fapesq (grant number 3071/2021).

**Conflicts of Interest:** The authors declare no conflicts of interest.

https://doi.org/10.3390/horticulturae12040476

**List of Contributions**

1. Pruteanu, A.; Constantin, G.A.; Vanghele, N.A.; Vlăduț, V. Evaluation of Growth and Production Parameters of Raspberries and Blackberries Cultivated in Romania. *Horticulturae* **2025**, *11*, 369. https://doi.org/10.3390/horticulturae11040369.
2. Dhukuchhu, A.; Kaya, O.; Hatterman-Valenti, H. Planting Date and Cultivar Selection Effects on Cauliflower Growth, Physiology, and Yield Performance in North Dakota Growing Conditions. *Horticulturae* **2025**, *11*, 1314. https://doi.org/10.3390/horticulturae11111314.
3. Almeida, E.C.; Pereira, F.H.F.; Ferreira, K.N.; de Sena Rodrigues, A.C.; Araújo, R.H.C.R.; de Souza, J.E.J.; Ramos, C.S.G.; Lopes, G.; dos Santos, L.C.; Neto, F.B.; et al. Physiological, Productive, and Nutritional Performance of Tomato Plants Treated with Iron and Zinc Nanoparticles via Foliar Application Under Deficit Irrigation. *Horticulturae* **2025**, *11*, 1228. https://doi.org/10.3390/horticulturae11101228.
4. Martins, B.L.R.; Ferreira, K.N.; Rocha, J.L.A.; Araujo, R.H.C.R.; Lopes, G.; Santos, L.C.d.; Bezerra Neto, F.; Sá, F.V.d.S.; Silva, T.I.d.; da Silva, W.I.; et al. Nano ZnO and Bioinoculants Mitigate Effects of Deficit Irrigation on Nutritional Quality of Green Peppers. *Horticulturae* **2024**, *10*, 969. https://doi.org/10.3390/horticulturae10090969.
5. Alyamani, A.A.; Morsi, M.M.; Abdelmigid, H.M. Adsorption and Incorporation of AgNPs Mediated by Seed Priming in Cultivated Taify Pomegranate: Integrated Approaches. *Horticulturae* **2024**, *10*, 647. https://doi.org/10.3390/horticulturae10060647.
6. Barka, A.M.H.; Essah, S.Y.C.; Davis, J.G. Deficit Irrigation and Nitrogen Application Rate Influence Growth and Yield of Four Potato Cultivars (*Solanum tuberosum* L.). *Horticulturae* **2025**, *11*, 849. https://doi.org/10.3390/horticulturae11070849.
7. Ruangsangaram, T.; Chulaka, P.; Mosaleeyanon, K.; Chutimanukul, P.; Takagaki, M.; Lu, N. Effects of Light Intensity and Irrigation Method on Growth, Quality, and Anthocyanin Content of Red Oak Lettuce (*Lactuca sativa* var. *cripspa* L.) Cultivated in a Plant Factory with Artificial Lighting. *Horticulturae* **2025**, *11*, 75. https://doi.org/10.3390/horticulturae11010075.
8. Yang, H.; Bae, Y.; Kim, Y.; Hyeon, S.; Choi, M.; Yang, S.; Kim, D.; Jang, D. Effects of Irrigation Methods on Growth and Water Productivity in Bell Pepper Cultivation in Northern South Korea. *Horticulturae* **2024**, *10*, 1353. https://doi.org/10.3390/horticulturae10121353.
9. Lim, M.Y.; Kim, S.H.; Roh, M.Y.; Choi, G.L.; Kim, D. Nutrient Dynamics and Resource-Use Efficiency in Greenhouse Strawberries: Effects of Control Variables in Closed-Loop Hydroponics. *Horticulturae* **2024**, *10*, 851. https://doi.org/10.3390/horticulturae10080851.
10. Fallah, S.; Maggi, F.; Ghanbari-Odivi, A.; Rostaei, M. Optimizing Lemon Balm (*Melissa Officinalis* L.) Cultivation: Effects of Different Manures on Plant Growth and Essential Oil Production During Consecutive Harvests. *Horticulturae* **2024**, *10*, 1105. https://doi.org/10.3390/horticulturae10101105.
11. Hassan, A.A.; Abdel-Rahim, A.F.A.; Al Hawas, G.H.; Alshammari, W.K.; Zewail, R.M.Y.; Badawy, A.A.; El-Desouky, H.S. Optimizing Caraway Growth, Yield and Phytochemical Quality Under Drip Irrigation: Synergistic Effects of Organic Manure and Foliar Application with Vitamins B1 and E and Active Yeast. *Horticulturae* **2025**, *11*, 977. https://doi.org/10.3390/horticulturae11080977.
12. Peña, H.; Diánez, F.; Ramírez, B.; Sulbarán, J.; Arias, K.; Huertas, V.; Santos, M. Compost and Vermicompost as Substrates Enriched with *Trichoderma asperellum* for the Production of Basic Potato Seed in the Venezuelan Andes. *Horticulturae* **2025**, *11*, 124. https://doi.org/10.3390/horticulturae11020124.
13. Deidda, A.; Sassu, A.; Ghiani, L.; Tiloca, M.T.; Ledda, L.; Cossu, M.; Deligios, P.A.; Gambella, F. Proximal and Remote Sensing Monitoring of the 'Spinoso sardo' Artichoke Cultivar on Organic and Conventional Management. *Horticulturae* **2025**, *11*, 961. https://doi.org/10.3390/horticulturae11080961.
14. Obón, C.; Pardo-Pina, S.; Johnson, D.; Rivera, D. Orthodox vs. Recalcitrant? Germination and Early Growth of *Phoenix* Species (Arecaceae) Stored for up to Ten Years. *Horticulturae* **2025**, *11*, 537. https://doi.org/10.3390/horticulturae11050537.
15. Castañeda-Loaiza, V.; Rodrigues, M.J.; Fernandes, E.; Custódio, L. A Comparative Study of the Influence of Soil and Non-Soil Factors on Seed Germination of Edible Salt-Tolerant Species. *Horticulturae* **2024**, *10*, 872. https://doi.org/10.3390/horticulturae10080872.

https://doi.org/10.3390/horticulturae12040476

16. Ghoreshizadeh, S.; Calvo-Peña, C.; Ruiz-Muñoz, M.; Dobrajc, M.; Radišek, S.; Coque, J.J.R.; Cobos, R. Isolation and Characterization of *Pseudomonas* sp. HX1, *Streptomyces luteogriseus* HR40, and *Streptomyces flavofungini* HR77 as Promising Biocontrol Agents Against Verticillium Wilt in Hops Affected by *Verticillium nonalfalfae*. *Horticulturae* **2025**, *11*, 459. https://doi.org/10.3390/horticulturae11050459.

## References

1. Ali, A.; Yanwen Tan, Y.; Medani, K.; Xia, C.; Abdullahi, N.M.; Mahmood, I.; Yang, S. Horticultural postharvest loss' and its socio-economic and environmental impacts. *J. Environ. Manag.* **2025**, *373*, 123458. [CrossRef] [PubMed]
2. FAO. Food and Agriculture Organization of the United Nations. Crops and Livestock Products. 2026. Available online: https://www.fao.org/faostat/en/#data/QCL (accessed on 25 March 2026).
3. FAO. Food and Agriculture Organization of the United Nations. Value of Agricultural Production. 2024. Available online: https://www.fao.org/statistics/highlights-archive/highlights-detail/agricultural-production-statistics-2010-2023/en (accessed on 25 March 2026).
4. Seth, T.; Mishra, G.P.; Chattopadhyay, A.; Deb Roy, P.; Devi, M.; Sahu, A.; Sarangi, S.K.; Mhatre, C.S.; Lyngdoh, Y.A.; Chandra, V.; et al. Microgreens: Functional Food for Nutrition and Dietary Diversification. *Plants* **2025**, *14*, 526. [CrossRef]
5. Loeza-Lara, P.; Jiménez-Mejía, R.; Santoyo, G. The Effect of Biostimulants on Horticultural Crops. *Horticulturae* **2024**, *10*, 1086. [CrossRef]
6. Lim, M.Y.; Kim, S.H.; Roh, M.Y.; Choi, G.L.; Kim, D. Nutrient Dynamics and Resource-Use Efficiency in Greenhouse Strawberries: Effects of Control Variables in Closed-Loop Hydroponics. *Horticulturae* **2024**, *10*, 851. [CrossRef]
7. Narayanan, M.; Ma, Y. Recent progress on conservation and restoration of soil fertility for horticulture. *Chemosphere* **2024**, *362*, 142599. [CrossRef] [PubMed]
8. Ramamoorthy, K.; Dhanraj, R.; Vijayakumar, N.; Ma, Y.; Al Obaid, S.A.; Narayanan, M. Vegetable and fruit wastes: Valuable source for organic fertilizer for effective growth of short-term crops: *Solanum lycopersicum* and *Capsicum annum*. *Environ. Res.* **2024**, *251*, 118727. [CrossRef] [PubMed]
9. Domene, M.Á.; Gómez, F.; Soria, R.; Villafuerte, A.B.; Miralles, I.; Ortega, R. Comparative Analysis of Horticultural and Animal Waste Compost: Physicochemical Properties and Impact on Plant Growth. *Agronomy* **2025**, *15*, 516. [CrossRef]
10. Teshome, F.T.; Bayabil, H.K.; Schaffer, B.; Ampatzidis, T.; Hoogenboom, G.; Singh, A. Exploring deficit irrigation as a water conservation strategy: Insights from field experiments and model simulation. *Agric. Water Manag.* **2023**, *289*, 108490. [CrossRef]
11. FAO. Food and Agriculture Organization of the United Nations. Restoration of Degraded Agricultural Lands: An Urgent Need for Agrifood System Transformation and Land Degradation Neutrality. Available online: https://openknowledge.fao.org/items/183921be-218f-4e89-b8db-b9e68d702513 (accessed on 30 January 2026).
12. Poenaru, M.M.; Manta, L.F.; Gherțescu, C.; Manta, A.G. Shaping the Future of Horticulture: Innovative Technologies, Artificial Intelligence, and Robotic Automation Through a Bibliometric Lens. *Horticulturae* **2025**, *11*, 449. [CrossRef]



https://doi.org/10.3390/horticulturae12040476

*horticulturae*

MDPI

*Article*

# Adsorption and Incorporation of AgNPs Mediated by Seed Priming in Cultivated Taify Pomegranate: Integrated Approaches

**Amal Ahmed Alyamani [1], Maissa M. Morsi [2] and Hala M. Abdelmigid [1,*]**

[1] Department of Biotechnology, College of Science, Taif University, P.O. Box 11099, Taif 21944, Saudi Arabia; a.yamani@tu.edu.sa

[2] Department of Biology, College of Science, Taif University, P.O. Box 11099, Taif 21944, Saudi Arabia; m.moasa@tu.edu.sa

* Correspondence: h.majed@tu.edu.sa

**Abstract:** A novel approach known as seed priming has been developed to improve seed germination and, ultimately, increase growth and yield. For high-value crops like pomegranates (*Punica granatum*) in high-altitude regions like the Taif area, affordable, sustainable, and efficient seed treatments are yet to be discovered. In this study, we employed a green synthesis method using pomegranate peel and coffee ground extracts to synthesize silver nanoparticles (AgNPs) at a concentration of 80 mg/mL. These AgNPs were then utilized to prime pomegranate seeds for 24 h. Furthermore, a chemical reduction method using trisodium citrate was used for comparison. The adsorption of NPs was verified using scanning electron microscopy (SEM) and energy dispersive X-ray (EDX), while their incorporation was confirmed by transmission electron microscopy (TEM). We further validated our EM results with X-ray fluorescence (XRF) and inductively coupled plasma–optical emission spectroscopy (ICP-OES). According to the findings of this study, AgNPs were observed to be present within seeds even after undergoing storage during priming. There is a possibility that the results obtained could potentially contribute to maintaining the quality of crops in a sustainable and eco-friendly manner.

**Keywords:** seed priming; silver nanoparticles; *Punica granatum*; integrated approaches

## 1. Introduction

The escalating global population is exerting substantial pressure on food security, posing a major challenge to meeting the food needs of the growing population. Fertilizers play a pivotal role in augmenting crop yields, but conventional options have certain constraints. Soluble fertilizers present a substantial environmental problem, as they can dissolve in water and easily seep into groundwater and bodies of water, leading to pollution. On the contrary, insoluble fertilizers offer a sustainable solution. Nevertheless, their restricted bioavailability hampers their effectiveness. The inability of plants to access the essential micronutrients that are locked in insoluble particles hinders their capacity to foster robust plant growth. Such challenges underscore the urgent need for cutting-edge and eco-friendly fertilization methods to support agricultural productivity while minimizing environmental impacts [1,2]. Nanotechnology offers significant promise for agricultural applications. Nanoparticles, due to their distinct qualities, possess the potential to distribute nutrients among plants more effectively. Recently, nanotechnology has garnered considerable attention, especially in the context of agriculture, as it can function as a growth stimulant, plant protection agent, additive, and nano-fertilizer [3]. Many studies have demonstrated that various nanostructures produce favorable effects on numerous crops, especially during the embryonic and seedling stages. Seeds constitute the fundamental foundation for natural crop development, as they represent the initial stage in the agricultural production process. Seed production is a vital component of the food security system [4].

Many crop seeds, including pomegranates (*Punica granatum* L.), possess a tough outer coat that regulates gas exchange, water uptake, and nutrient flow. This rigid coat, while beneficial for seed protection, can hinder water permeability and ultimately reduce germination rates [4]. This challenge is particularly pertinent for high-value crops like pomegranates, which are widely cultivated and commercially important in Saudi Arabia [4]. The fruit boasts a wealth of polyphenolic phytochemicals with potent antioxidant properties [5,6]. These properties have contributed to the use of pomegranates in traditional medicine for treating various diseases [5,6]. Pomegranate peel, which constitutes a significant portion (around 60%) of the fruit's weight, is often considered agricultural waste [7]. However, it is rich in valuable nutrients like minerals, vitamins, phenolics, and antioxidants with potential health benefits [7–9]. Finding ways to utilize this byproduct would be a sustainable practice. By overcoming the seed germination challenge with techniques like seed priming, we can ensure the continued cultivation and economic viability of this valuable crop in Saudi Arabia.

Seed priming is a technology that may provide an effective solution to overcome the challenge posed by the hard seed coat in pomegranates, which presents a significant obstacle to successful germination. This approach involves applying pre-sowing treatments with various amendments to improve the physiological state of seeds prior to germination [4,10]. Seed priming offers numerous advantages that directly address the limitations of the hard seed coat. It ensures quick and consistent germination, thereby enabling better resource management, including water, fertilizer, and labor [11,12]. Furthermore, it notably shortens the imbibition period, activates germination inducers, initiates DNA repair processes, promotes absorption, and triggers antioxidant responses linked to pre-germination metabolism [13–17]. For that, the technique of priming is applied to enhance seed germination in both optimal and unfavorable conditions [18]. By reducing the need for re-seeding and supplemental resources, seed priming contributes to sustainable pomegranate cultivation, particularly in water-scarce regions like Saudi Arabia [19].

As previously mentioned, conventional seed priming has proven to be an effective solution for addressing the difficulty posed by the hardiness of pomegranate seeds. However, recent progress in nanotechnology has opened new avenues for even more precise methods. Nanoparticles, due to their unique properties, such as a large surface area, exhibit great potential for seed priming, a technique also known as nano-priming [19]. The utilization of nanoparticles as carriers for delivering beneficial substances to seeds has the potential to improve processes such as germination and seedling development. Various studies indicate that nano-priming can stimulate the breakage of seed dormancy by activating enzymes that initiate embryonic differentiation. The outcomes of these studies suggest that nanoparticles have the potential to enhance the efficiency of seed treatment by nano-priming, thus leading to improved crop productivity [6]. This targeted delivery of specific molecules could be particularly advantageous for pomegranates, as their rigid seed coat impedes water uptake and enzyme activity. Furthermore, metal nanoparticles hold great potential as intelligent nano-fertilizers and nano-pesticides [20]. This one-step seed priming method enhances germination, supplies essential nutrients, and provides potential protection for young pomegranate seedlings. By merging the established benefits of seed priming with the precise delivery capabilities of nanoparticles, nano-priming offers a promising avenue for enhancing pomegranate cultivation.

Silver nanoparticles (AgNPs) possess a high surface area, allowing them to act as efficient carriers for delivering specific molecules directly to seeds [19]. This targeted delivery has the potential to address the limitations imposed by the rigid seed coat in pomegranates. By delivering molecules that promote water uptake, enzyme activity, or other germination processes, AgNPs could significantly enhance and synchronize germination rates. Many studies have examined the effects of AgNPs on seed germination, root elongation, and plant reactions such as cellular oxidative stress or cytotoxicity [21–25]. The biochemical and physiochemical properties of AgNPs are significant compared to bulk materials. Various applications have been developed using AgNPs, owing to their fungicidal and bactericidal properties. Moreover, these nanoparticles may affect oxidative stress in algae, bacteria, and higher plants [26]. Depending on the dosage, different sizes of AgNPs showed varying effects on the growth and inhibition of plant species [27]. Research on the influence of silver nanoparticles (AgNPs) on plant growth has produced inconsistent outcomes. While some studies have shown positive effects on seed germination and growth, others have reported negative effects. These studies encompassed a range of plant crops, including *Brassica juncea* [28], *Phaseolus vulgaris* [29], *Zea mays* [30], *Eruca sativa* [31], and *Oryza sativa* [32],

Silver nanoparticles (AgNPs) can be created for agricultural purposes through multiple methods [33–37]. Conventional approaches involve chemical reactions that employ severe chemicals. In this process, nanoparticles are generated in a liquid medium that incorporates a variety of reactants, including reducing agents such as sodium borohydride [38] and potassium bitartrate [39]. Green synthesis, which involves using plant extracts to produce AgNPs, presents a promising alternative to traditional methods that raise concerns over environmental and plant health risks [40]. This eco-friendly method has several benefits. Plant extracts function as both reducing agents, converting silver ions into metallic silver, and capping agents for stabilizing the nanoparticles [41–43]. This natural approach potentially reduces cytotoxicity compared to chemically synthesized AgNPs. The effectiveness of silver nanoparticles (AgNPs) as a seed priming agent might depend on the method used to synthesize them. AgNPs synthesized via the green synthesis method, due to their biocompatible properties, could be more effective and safer in promoting seed germination and seedling development in crops such as pomegranates.

It is crucial to verify the uptake of nano-primed AgNPs before conducting germination tests. Confirming AgNP uptake helps determine whether the seed priming process introduces nanoparticles into the seeds, which is essential for understanding how AgNPs might affect germination and subsequent growth. Without uptake verification, the results of germination tests might be inconclusive, making it unclear whether any observed effects on germination are due to the AgNPs themselves or some other factors. Demonstrating AgNP uptake strengthens the credibility of our research by showing a clear link between the priming process and the potential impact on seed germination. Techniques like SEM, EDX, TEM, XRF, and ICP-OES can be used to verify uptake, providing valuable evidence that the AgNPs are indeed entering the seeds, which improves result interpretation.

The primary objective of this study is to explore the interactions between silver nanoparticles (AgNPs) and Taify pomegranate seeds during the seed priming process. This entails examining two key aspects: adsorption, which involves analyzing the extent to which AgNPs adhere to the seed surface, and incorporation, which focuses on determining whether AgNPs can enter the seeds after priming treatment. To achieve this objective, the study employs a variety of analytical techniques. Energy-dispersive X-ray spectroscopy (EDX) and electron microscopy (EM) were used to visualize and qualitatively characterize nanomaterials on the seed surface. Additionally, XRF (X-ray fluorescence) and ICP-OES (inductively coupled plasma–optical emission spectroscopy) techniques were employed to detect nanoparticles incorporated into *Punica granatum* L. seed samples. A secondary objective of this study is to compare the effectiveness of AgNPs synthesized using green (e.g., pomegranate peel extract, ground coffee bean extract) and chemical methods in seed priming as a method for delivering AgNPs to Taify pomegranate seeds. Ultimately, by gaining a deeper understanding of the interaction between AgNPs and seeds during priming, this research aims to lay the groundwork for future studies on the potential benefits of this approach for Taify pomegranate cultivation.

## 2. Materials and Methods

### *2.1. Chemicals*

Silver nitrate ($AgNO_3$) and trisodium citrate were purchased for biological and chemical synthesis. Silver nitrate (100%, *w*/*w*) was obtained from Alfa Aesar (Ward Hill, MA, USA), and trisodium citrate dihydrate (ACS reagent, ≥99.0%) was acquired from Sigma-Aldrich (Milan, Italy). Deionized (DI) water was used to prepare all aqueous solutions.

The reagents used were all analytical grades. Glassware was washed with aqua regia (a mixture of concentrated hydrochloric acid (HCl) and nitric acid ($HNO_3$), typically in a 3:1 ratio), followed by three rinses with deionized (DI) water.

### 2.2. Plant Materials

Fifty seeds of *Punica granatum* L. (pomegranate) were acquired from commercially available, mature pomegranates purchased from a local retailer in Taif, Saudi Arabia. The pomegranates were carefully inspected visually to ensure they were free of any spots or damage. The seeds were then removed from the fleshy tissue and were washed repeatedly with deionized (DI) water to remove any surface dirt or debris that may have adhered to them.

Green synthesis materials comprised air-dried fruit peels from pomegranates, which were utilized for seed collection as a reducing agent in the synthesis of silver nanoparticles (AgNPs). Furthermore, roasted ground coffee beans purchased from commercial sources were employed as another reducing agent in the green synthesis process.

### 2.3. Preparation of Plant Extracts

To prepare plant extracts to reduce and stabilize AgNPs, we utilized the same procedure previously described by Abdelmigid et al. [44]. Freshly picked pomegranates (*Punica granatum* L.) and Arabic coffee beans (*Coffea arabica*) were purchased from a local shop in Taif governorate, Saudi Arabia. It was necessary to wash the pomegranate fruits thoroughly with distilled water and to separate the seeds before obtaining the peels. After being dehydrated at low temperature overnight to full dryness, the peels were ground into a uniform coarse powder utilizing a household blender. A room-temperature air-drying process was used to dry coffee beans after they were ground and used to prepare coffee drinks. Figure 1 illustrates the steps involved in collecting coffee ground waste and pomegranate peel (PP). Using 250 mL Milli-Q ultra-pure distilled water in a 500 mL Erlenmeyer flask, 50 g of finely powdered peels were added and boiled for 10–15 min for each waste product separately. Afterward, the aqueous plant extracts were prepared by filtering them through Whatman No. 1 filter paper.

### 2.4. Green Synthesis of Ag Nanoparticles

As a metal precursor, $AgNO_3$ was used to generate AgNPs. Silver nitrate ($AgNO_3$) solution was prepared by dissolving 0.017 g in 100 mL of deionized double-distilled water and stirring at room temperature for about 10 min. The solution was filtered with filter paper No. 4 to obtain a clear solution. Then, 15 mL of plant extract was added separately to 20 mL of aqueous 0.001 M $AgNO_3$ solution in 100 mL Erlenmeyer flasks. Gradually, the extract was poured into the solution while heating it at 70 °C for 1 h, with continuous stirring. As a result, the color of the solution changed from a pale light green to a light brown and then to a dark brown. This color change indicated the presence of AgNPs in the solution. Repeated centrifugation at 10,000 rpm for 15 min, followed by dispersion of the pellet in sterile DI water four times, was used to purify AgNPs from the solution [44]. Following 48 h of lyophilization, the nanoparticles suspended in water were analyzed for their composition and structure. The AgNPs, which were synthesized using pomegranate peel extract, have been referred to as AgNPs_PPE, while those synthesized using coffee ground beans have been distinguished as AgNPs_CE throughout the manuscript.

### 2.5. Chemical Synthesis of Ag Nanoparticles

The chemical synthesis of silver nanoparticles was carried out in this study, according to Fang et al. [45]. Distilled water (15 mL) was used to dissolve $AgNO_3$ (2 g), while trisodium citrate (1 g) was dissolved in 100 mL of distilled water. After boiling 50 mL of a $10^{-3}$ M $AgNO_3$ solution, it was gently added to trisodium citrate (1%, 5 mL, Sigma-Aldrich, Milan, Italy) for five minutes at 90 °C with vigorous stirring. Once the reduction began, 2.5 mL of trisodium citrate was added dropwise until a pale yellow appeared in the solution. To distinguish these particles throughout the manuscript, we refer to them as AgNPs_Chem. Similar to the biologically prepared AgNPs, AgNPs_Chem particles were obtained through centrifugation and subsequent drying.

**Figure 1.** Waste collection steps for pomegranate peels (**A–C**) and coffee grounds (**D–G**) (adapted from Abdelmigid et al. [44]).

### 2.6. *Characterization of Silver Nanoparticles*

#### 2.6.1. UV-Vis Spectroscopy

To characterize the prepared silver nanoparticles (AgNPs_PPE, AgNPs_CE, and AgNPs_Chem), UV-1601 spectroscopy (Shimadzu, Kyoto, Japan) was used for optical absorption measurements at wavelengths ranging from 200 to 800 nm. Autoclaved distilled water was used as a reference.

#### 2.6.2. Size Distribution and Zeta Potential

The average size and surface charge of the AgNPs synthesized at 25 °C with a detection angle of 90° were measured using a Zetasizer Nano ZS particle size analyzer (Malvern

Instruments Ltd., Malvern, Worcestershire, UK). To prevent nanoparticle aggregation, all powder samples were freshly resuspended in 0.9% saline solution and sonicated at high speed for 30 min before assessment.

#### 2.6.3. Surface Morphology

To determine the surface morphology and size of AgNPs, scanning electron microscopy (SEM, JEOL JSM-639OLA Analytical Scanning Electron Microscope, Tokyo, Japan, which is located in Taif University's Electron Microscope Unit) was performed at 20 kV ($2 \times 10^4$) V with magnifications of 500, 2000, and 3000 with scale bars of 50 μm, 10 μm, and 5 μm, respectively. Before scanning, the silver nanoparticle powders (AgNPs_PPE, AgNPs_CE, and AgNPs_Chem) were coated with gold using a Cressington Sputter Coater (108 Auto, thickness controller MTM-10, UK). X-ray diffraction (XRD) was used to assess the crystallinity of the synthesized nanoparticles (Pan Analytical, X-pert pro, The Netherlands). At 30 kV ($3 \times 10^4$ V) and 100 mA ($1 \times 10^{-1}$ A), XRD was used to document the spectrum at a wavelength of 1.5406 Å in the 2θ (from the range of 20–80°). The XRD patterns were plotted using OriginLab software® (2018) and compared with the JCPDS card number 040783.

#### 2.6.4. Fourier Transforms Infrared Spectroscopy (FTIR)

A small amount of the synthesized AgNP powder was mixed with the KBr powder to create a homogeneous mixture. The mixture was subjected to high pressure to form a thin disc-shaped pellet. The resulting KBr pellet containing AgNPs was placed in an FTIR spectrometer for analysis. KBr is advantageous in this process because it is transparent in the infrared region and does not interfere with the analysis of the AgNPs. For the detection of possible functional groups in synthesized AgNPs, FTIR spectroscopy (Agilent Technologies, Santa Clara, CA, USA) at a wavenumber between 450 and 4000 $cm^{-1}$ was used.

### 2.7. *Preparations of Nanopriming Solutions and Seed Priming*

AgNPs (AgNPs_PPE, AgNPs_CE, and AgNPs_Chem) synthesized by biological and chemical methods were freshly prepared as 80 mg/L nanopriming solutions by ultrasonic vibration (100 W (1 × 102 W), 40 kHz (4 × 104 Hz)) for 30 min [46]. The concentrations of the nanopriming solutions were selected based on the preliminary experiments. Milli-Q deionized water was used as the hydropriming solution (HP) [46]. The seeds were surface-sterilized in 1% sodium hypochlorite for 3 min and rinsed three times with deionized water. The seeds were then soaked in a AgNP-priming solution for 24 h (rotatory shaker). The ratio of seed weight to solution volume was approximately 1:4 g/mL. After applying deionized water for 3 min, the seeds were dried on Whatman filter paper. The sterile pouches were then sealed and refrigerated at a temperature of 4 °C until needed.

### 2.8. *Assaying AgNPs Adsorption on a Seed Surface by SEM Analysis*

Scanning electron microscopy (SEM) was used to evaluate the distribution and structural features of nanoparticles adsorbed on the nanoprimed seeds after washing with distilled water and air-drying. The seeds were coated with a layer of gold before scanning using a Cressington 108 Sputter Coater (auto, thickness controller MTM-10; Watford, UK).

### 2.9. *Energy Dispersive X-ray (EDX)*

A JEOL–JSM-1400 PLUS system with an EDX detector was used for identifying the chemical composition of the Ag nanoparticles and for detecting the elemental composition of the *Punica granatum* L. seeds. All samples were coated with a thin layer of gold for analysis. In the EDX spectrum, the detected signal was plotted as a function of energy. The following are the measurement conditions for the EDX detector: an acceleration voltage of 20.00 kV, a magnification of ×270, and a process time of T2. The measurement detector possesses a live time of 30.00 s, a real-time of 30.22 s, and a dead time of 1.00%. The count rate is 2528.00 CPS. The JEOL system's specific spectra processing system was operated according to the manufacturer's recommended protocols. Similarly, the calibration of the EDX detector was carried out according to the manual instructions provided.

### 2.10. Assaying AgNPs Internalization

Randomly selected nanoprimed seeds were dissected and prefixed in 4F1G in phosphate buffer solution (pH = 7.2) at 4 °C for 3 h. After that, specimens were postfixed in 2% OsO4 in the same buffer at 4 °C for 2 h. Various concentrations of acetone were then used to dehydrate samples. They were then embedded in resin for polymerization and cut into sections approximately 90 Å thick. The sections were placed on grid copper and stained with uranyl acetate for 5 min. They were then stained with 1% lead citrate for 2 min, rinsed again in pure water, and stored in grid boxes. Ultrathin sections were prepared using a Leica ultracut UCT ultramicrotome. A JEOL–JSM-1400 PLUS TEM (JEOL, Tokyo, Japan) was used to obtain TEM images.

### 2.11. Inductively Coupled Plasma Spectrometer (ICP–OES) Analysis of NP Uptake

ICP-OES was used to determine the total AgNP uptake or bioaccumulation in the nanoprimed pomegranate seeds. The following steps were followed: First, 0.1 g of finely ground pomegranate seeds were dried at 100 °C for 24 h. To dissolve all residues, 10 mL aqua regia ($HNO_3$/HCL) was added and slowly heated to 150 °C. Two replications of this procedure were conducted, each with 10 mL aqua regia added to achieve complete digestion (~2 mL). Subsequently, a Whatman filter paper was used to purify the solution, and then a 100 mL volumetric flask was used to dilute it with Millipore water. The AgNP content was quantified in triplicate using an inductively coupled plasma spectrometer (ICPE-9000, Shimadzu, Kyoto, Japan). The instrument was calibrated using a standard blank and mixed-metal calibration standard before analysis. Following the calibration plot, the most accurate linear regression correlation coefficient was calculated ($R^2 \geq 0.9999$) from the calibration plot. Multi-element standards and analytical references were purchased from AccuStandard (New Haven, CT, USA).

### 2.12. Energy Dispersive X-ray Fluorescence (EDXRF) Analysis of NP Uptake

According to Tezotto et al. [47], the examination was performed following their protocol. EDXRF analysis was conducted on the same pomegranate seeds prepared as a loose powder for ICP-OES analysis. Polypropylene film (Mylar®) was used to cover a 20 mm diameter polyethylene cup filled with one gram of ground seeds. With a Shimadzu EDX-720 energy dispersive X-ray fluorescence spectrometer, the samples were radiated in triplicate for 300 s under vacuum. X-ray tubes were used for irradiating the samples (15 kV ($15 \times 10^3$ V) for Na to Sc and 50 kV ($5 \times 10^4$ V)). There was an automatic adjustment of the current to a maximum of 1A, and a 10 mm collimator was selected. The detection was carried out with a liquid nitrogen-cooled Si (Li) detector. The same method was used to verify the precision and trueness of certified reference materials (CRMs). The intensity of element K$\alpha$ counts per second (cps/$\mu$A) was brought from the sample X-ray spectrum deconvolution using the EDX Shimadzu software package (EDX-7000/8000/8100).

## 3. Results

### 3.1. Silver Nanoparticle (AgNP) Characterization

#### 3.1.1. UV-Vis Spectroscopy

The creation of silver nanoparticles is demonstrated by the change in color from pale yellow, brown, and colorless, respectively, to dark brown in PPE, CE, and trisodium citrate (Figure 2A–C). Using UV-visible spectroscopy, the ability of the aqueous solution to produce nanoparticles was further demonstrated. Figure 2D shows the UV-vis absorption spectra of AgNPs_PPE (350 nm), AgNPs_ CE (with two distinct peaks at 320 nm and 440 nm), and AgNPs_ Chem (430 nm).

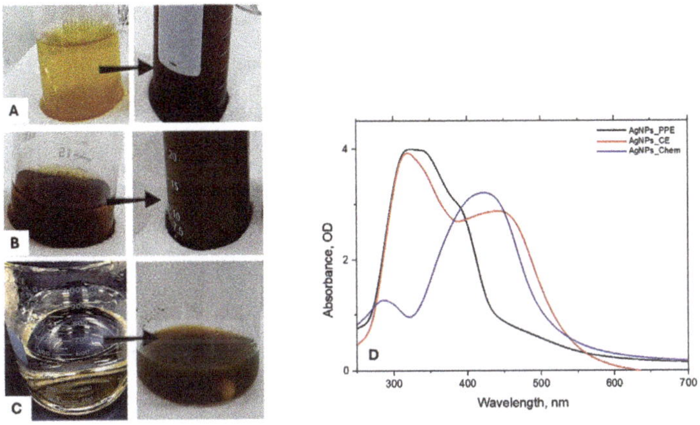


**Figure 2.** The successful creation of silver nanoparticles shown by the change in the color of (**A**) pomegranate peel extract (PPE), (**B**) coffee bean extract (CE), (**C**) trisodium citrate, and UV-vis sharp absorption spectra of AgNPs (**D**), respectively. Arrows indicate the color change.

3.1.2. Size Distribution and Zeta Potential

Zeta potential was used to estimate the stability of the produced AgNPs, where higher zeta potential values indicate more stable nanoparticles. A chemically produced AgNPs_Chem (46.7 mV) was found to be more stable than AgNP_CE (12.6 mV) and AgNPs_PPE (7.98 mV), which had the least stability. Figure 3A shows how a reduction in AgNPs stability causes particle aggregation, which leads to a size increase for AgNPs_PPE = 591.9 nm, AgNPs_CE = 273.7 nm, and AgNPs_Chem = 62.75 nm. This suggests that chemically synthesized AgNPs (AgNPs_Chem) were the most stable, followed by those from coffee extracts (AgNPs_CE), and lastly, those from pomegranate peel extracts (AgNPs_PPE).

3.1.3. Scanning Electron Microscopy (SEM) and XRD Analysis

Scanning electron microscopy (SEM) analysis, as shown in Figure 3B, provides visual evidence of the crystalline structure for both chemically and biologically synthesized AgNPs. X-ray diffraction (XRD) further confirms this crystallinity (Figure 3C). The four characteristic peaks observed in the XRD pattern correspond to the face-centered cubic (FCC) crystal structure of silver nanoparticles according to Bragg's reflection principle. These peaks appear at $2\theta$ values of 38.3°, 44.49°, 64.6°, and 77.5°, which can be assigned to Ag crystal planes (111), (200), (220), and (311), respectively. Interestingly, the XRD pattern of AgNPs_PPE reveals additional diffraction peaks at $2\theta$ of 27.90°, 32.38°, 46.43°, 55.00°, 57.00°, and 67.00°. These additional peaks warrant further investigation to identify the potential presence of other crystalline phases or impurities associated with the biological synthesis process using pomegranate peel extract.

3.1.4. FTIR Evaluation of AgNPs

Biomolecules affecting silver nanoparticle capping, reduction, and stability were identified using FTIR analysis of PPE, CE, and trisodium citrate. In Figure 3D, the differences between synthesized AgNPs and their reference are highlighted. Any change in the sharpness or position of a peak is considered a "group contribution" in the synthesis of silver nanoparticles. In comparison with PPE, the various AgNPs_PPE peaks were sharper. The sharp peak that was detected at 2927 $cm^{-1}$ was attributed to C-H stretching, 1641 $cm^{-1}$ to C-O stretching vibration in quinine, 1351 $cm^{-1}$ to the C=C aromatic ring, 1228 $cm^{-1}$ to C-O stretching in ether, ester, or phenol, and 1055 $cm^{-1}$ to C-N stretching of the aliphatic primary amine. The phenolic groups are located between 1315 and 1037 $cm^{-1}$ and 1456 and 1600 $cm^{-1}$. In comparison with AgNP extract CE, FTIR analysis revealed distinct sharp peaks. FTIR analysis of the AgNP_CE identified functional groups potentially involved in nanoparticle capping and stabilization. A broad band between 3600 and 3000 cm$^{-1}$ indicates N-H, O-H, and asymmetric C-H stretching vibrations. Additional peaks at 2922 and 2852 cm$^{-1}$ further support these functional groups. Aromatic and aliphatic C-H, C-O, and C-N vibrations were observed in the 1400–900 cm$^{-1}$ region. Peaks between 1700 and 1500 cm$^{-1}$ suggest C-C and C-N stretching vibrations. The AgNP_Chem synthesis was found to yield different peaks related to C-H stretching deformation (1282 cm$^{-1}$), nitro compound (NO) as $NO_2$ (1402 cm$^{-1}$), and C=O stretching asymmetry in $COO^-$ (1582 cm$^{-1}$), while at 3312 cm$^{-1}$, amine stretching was observed.

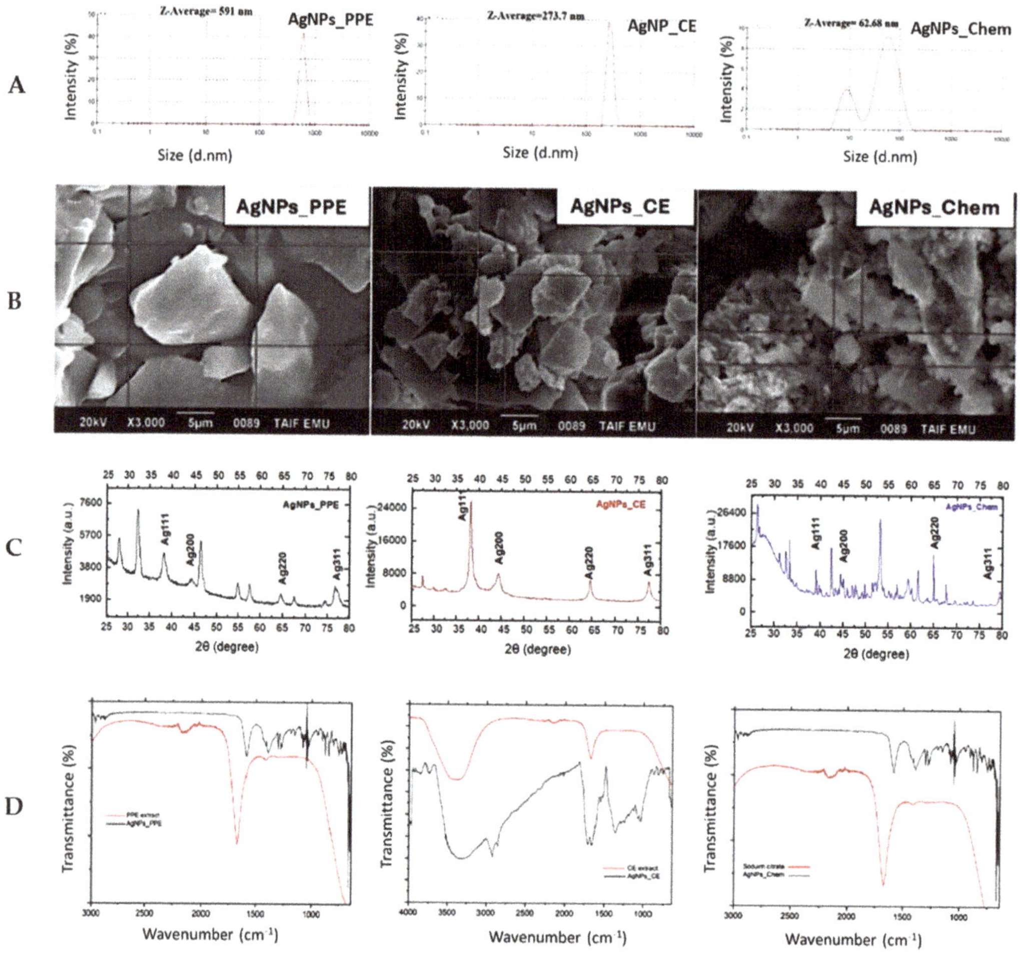


**Figure 3.** Characterization of the investigated AgNPs in this study: (**A**) zeta potential, (**B**) scanning electron microscopy, (**C**) XRD patterns, and (**D**) FTIR spectra (Adapted from Abdelmigid et al. [44]).

### 3.2. *AgNP Adsorption and Localization on Seed Coats*

SEM/EDX Analysis

Scanning electron microscopy (SEM) analysis was employed to investigate the size, shape, and distribution of AgNPs adsorbed onto the surface of Punica granatum seeds following nanopriming. Additionally, SEM was used to assess the overall morphology and integrity of pomegranate seeds (Figure 4A–D). Compared to the smooth surface texture observed in control seeds (Figure 4A), nanopriming with AgNPs resulted in significant alterations in seed coat morphology (Figure 4B–D). Notably, seeds treated with AgNPs exhibited numerous large pores with a distinctive honeycomb-like structure. The SEM images revealed that the majority of the AgNPs, regardless of their source (AgNPs_PPE, AgNPs_CE, or AgNPs_Chem), appeared to be spherical. Interestingly, the synthesized nanoparticles displayed a tendency to form both small and larger aggregates, which became entrapped within the seed coat fibers (Figure 4B–D). Differences were observed in the clustering behavior of the AgNPs depending on the synthesis method. Seeds treated with AgNPs_PPE and AgNPs_CE displayed larger clusters of nanoparticles scattered across the seed surface (Figure 4B,C). In contrast, seeds treated with AgNPs_Chem exhibited a more even distribution of individual nanoparticles, with fewer evident clusters (Figure 4D). Overall, SEM analysis provides valuable insights into the impact of AgNPs on the surface morphology of *Punica granatum* seeds. These findings suggest that nanopriming with AgNPs alters the seed coat structure and influences the distribution and clustering behavior of the nanoparticles.

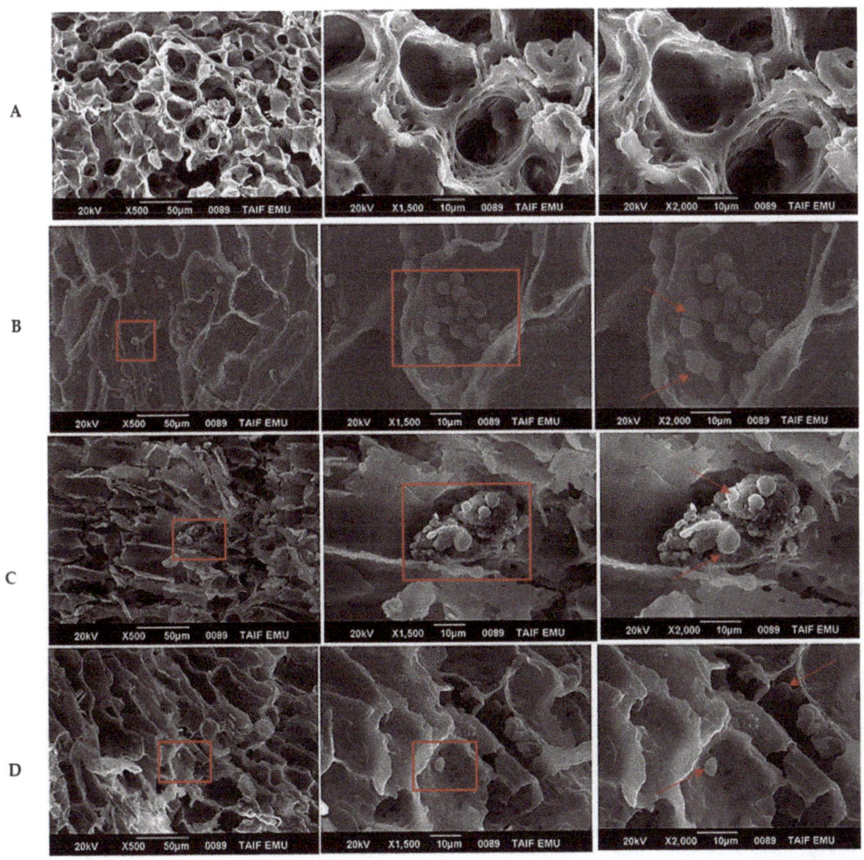


**Figure 4.** SEM images of nanoprimed *Punica granatum* seeds at various magnifications to visualize the adsorption of AgNPs. (**A**) Hydroprimed seeds (control) show no nanoparticles. (**B–D**) Seeds primed with different AgNPs: (**B**) AgNPs _PPE, (**C**) AgNPs_CE, and (**D**) AgNPs_ Chem. Red boxes and arrows highlight representative examples of adsorbed AgNPs on the seed coat surface at different magnifications (×500, ×1500, and ×2000).

Energy-dispersive X-ray spectroscopy (EDX) confirmed the presence of AgNPs on the seed coat and provided information regarding their elemental composition. EDX spectra (Figure 5) revealed a distinct Ag peak in the nanoprimed samples (AgNPs_PPE, AgNPs_CE, AgNPs_Chem), which was absent in the hydroprimed (HP) control seeds. This confirms the successful deposition of AgNPs onto the seed coat surface following the nanopriming treatments. Table 1 summarizes the quantitative elemental composition data obtained from the EDX analysis. Comparatively, the control seeds had a low mass percentage of C (47.78%) and a high mass percentage of O (52.22%) (Table 1). In the nanoprimed samples, there was a variation in carbon and oxygen contents; all nanoprimed samples displayed a decrease in oxygen content and an increase in carbon content. This suggests the adsorption of organic compounds onto the surface of the AgNPs. These organic compounds likely originated from the plant extracts (*Punica granatum* (PPE) and *Coffea arabica* (CE)) used in the synthesis process, which are known to play a role in reducing and stabilizing AgNPs. The presence of trisodium citrate, another component of the synthesis solution, also contributed to the observed carbon signal.

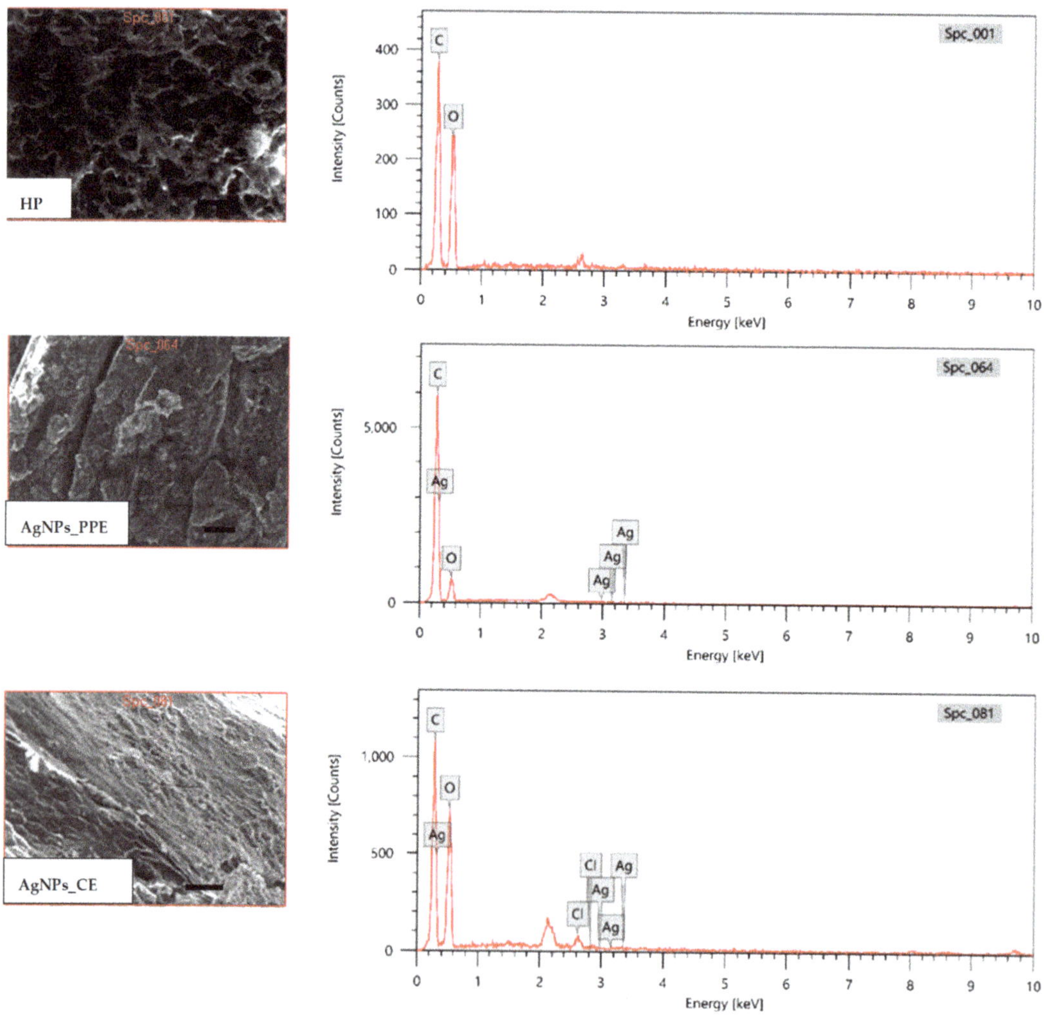


**Figure 5.** *Cont.*

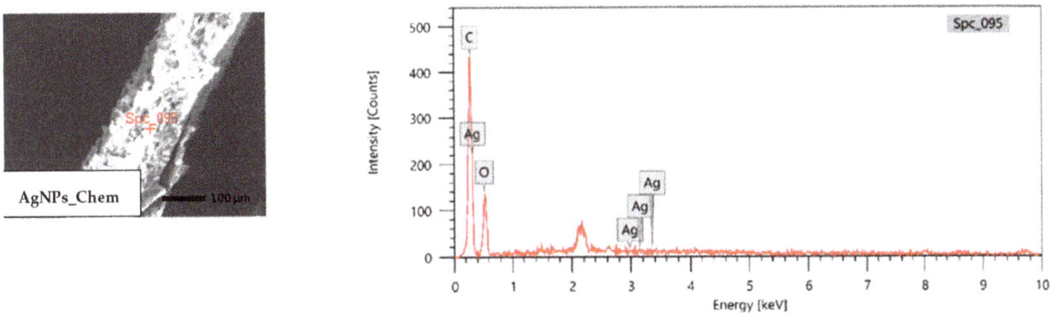


**Figure 5.** Analysis of hydroprimed (HP) and nanoprimed *Punica granatum* seeds using SEM-EDX technology. AgNPs_PPE, AgNPs_CE, and AgNPs_Chem adsorption on the surface of the seed coat is shown in panels. The element peaks (C, O, and Ag) are displayed in EDX spectra. The region of interest on the seed surface is revealed by the SEM photos (ROI).

**Table 1.** Elemental composition of control and AgNP-primed *Punica granatum* seeds determined by EDX. The amount of the adsorbed C, O, and Ag elements are expressed as mean (%) ± SE values of mass% derived from sample replications used for the study.

| Element / Treatment | Mass% | | | | |
|---|---|---|---|---|---|
| | C | O | Ag | Cl | Total |
| Control | 47.78 ± 0.54 | 52.22 ± 0.38 | 0 | 0 | 100% |
| AgNPs_PPE | 74.9 ± 0.21 | 24.9 ± 0.38 | 0.10 ± 0.04 | 0 | 100% |
| AgNPs_CE | 49.7 ± 0.33 | 49.0 ± 0.70 | 0.24 ± 0.08 | 1.07 ± 0.07 | 100% |
| AgNPs_chem | 60.4 ± 0.62 | 39.2 ± 1.32 | 0.44 ± 0.22 | 0 | 100% |

Interestingly, the nanoprimed samples exhibited variations in carbon and oxygen contents. AgNP_PPE-primed seeds had the highest carbon content (74.94%) and the lowest oxygen content (24.95%), suggesting a greater abundance of organic material on their surface than other treatments. Conversely, AgNP_CE-primed seeds displayed lower carbon content (49.69%) and higher oxygen content (49.0%), potentially indicating a different interaction with the plant extract. AgNP_Chem-primed seeds showed intermediate values (C: 60.4%, O: 39.16%). Overall, EDX analysis provided strong evidence for the successful deposition of AgNPs onto the seed coat surface and revealed differences in the surface composition of AgNPs derived from various synthesis methods.

### 3.3. *AgNP Uptake in Seeds*

#### 3.3.1. TEM Analysis

Transmission electron microscopy (TEM) was employed to investigate the potential translocation of AgNPs from the seed coat into the *Punica granatum* embryo. Figure 6A–D presents the TEM micrographs at various magnifications. In contrast to the absence of nanoparticles observed in hydroprimed control seeds (Figure 6A), TEM micrographs of nanoprimed seeds (Figure 6B–D) revealed the presence of electron-dense deposits within the embryo. These deposits appeared as individual dark particles and larger aggregates, suggesting the potential internalization of AgNPs, and higher magnification images (Figure 6C,D) provided further insight into the morphology of the internalized nanoparticles. The dark deposits exhibited a predominantly spherical or ellipsoidal shape, consistent with the expected morphology of AgNPs. These observations suggest the potential translocation of AgNPs from the seed coat to the *Punica granatum* embryo. The presence of individual particles and aggregates indicates that internalization may occur through various mechanisms.

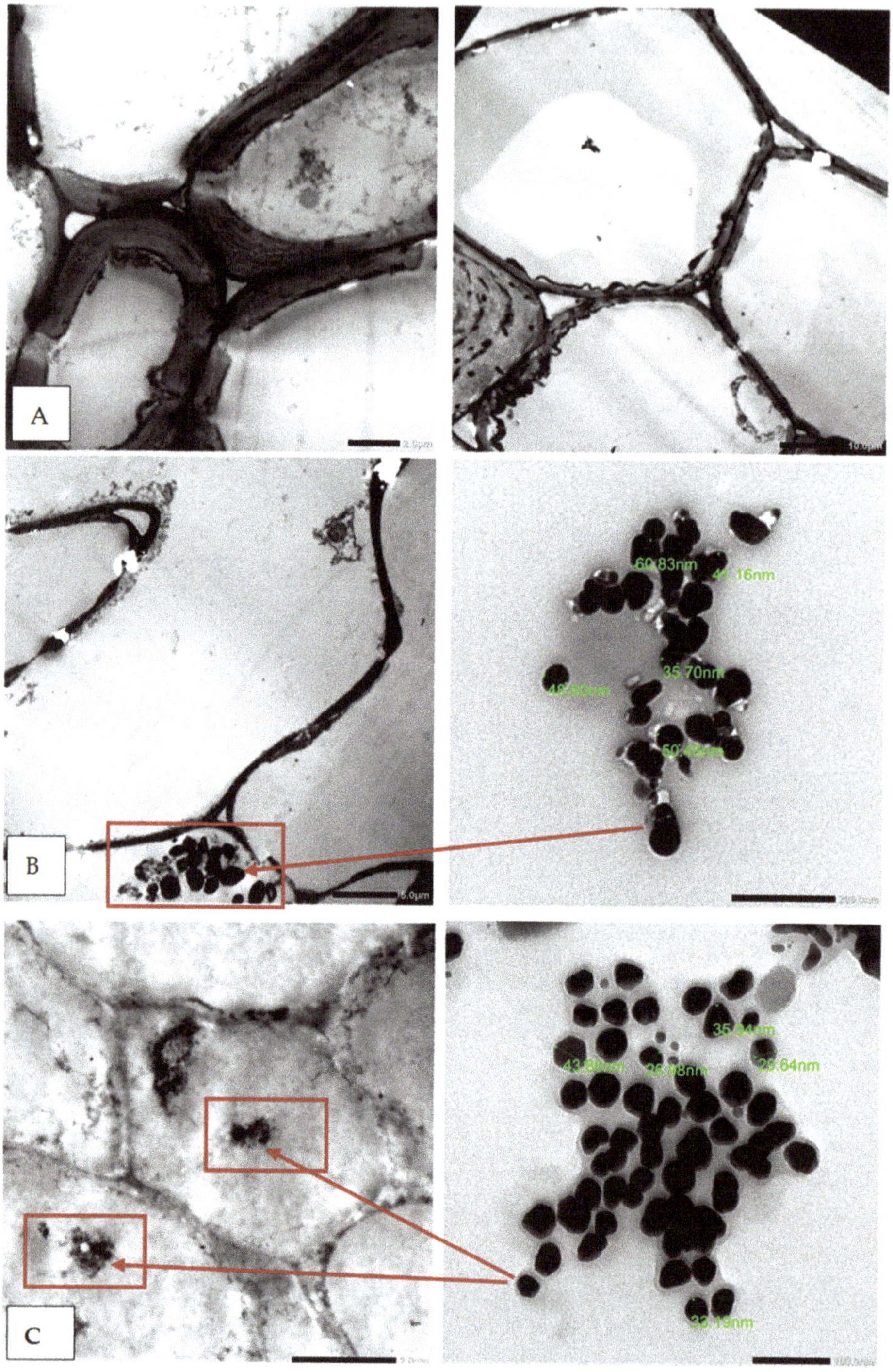


**Figure 6.** *Cont.*

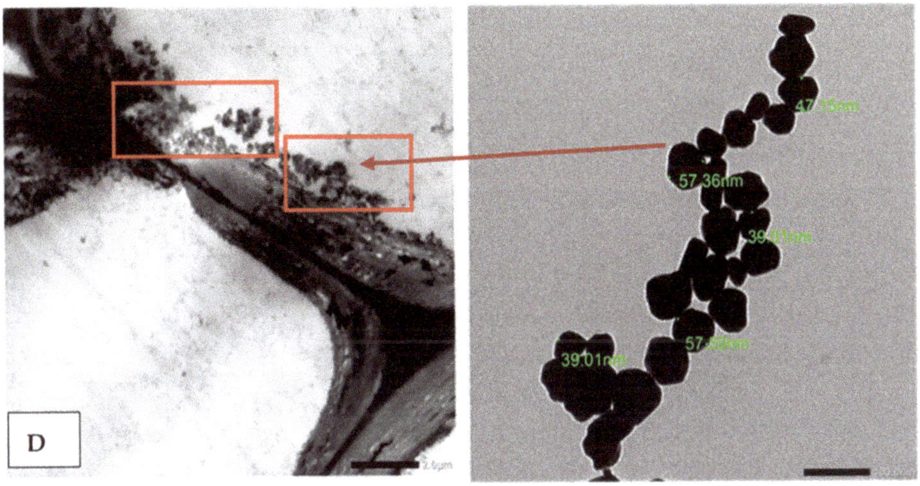


**Figure 6.** TEM images of *Punica granatum* seed cross-sections after being primed with (**A**) distilled water (hydroprimed), (**B**) AgNPs_ PPE, (**C**) AgNPs_CE, and (**D**) AgNPs_Chem. The TEM image of the manufactured NPs shows a similar pattern to the details of the Ag nanoparticle clusters found inside seed tissue. (The AgNPs are indicated by red boxes and arrows in the image).

3.3.2. X-ray Fluorescence (XRF) and ICP–OES Analyses

Although TEM provides valuable information about the morphology and size of nanoparticles within a sample, it has limitations in quantifying their total uptake. To address this, X-ray fluorescence (XRF) and inductively coupled plasma–optical emission spectroscopy (ICP-OES) techniques were employed to measure the total Ag content within Punica granatum seeds following nanopriming with AgNPs at a concentration of 80 mg/L. Both XRF and ICP-OES analyses successfully confirmed the presence of Ag in nanoprimed seed samples compared to hydroprimed controls (Figures 7 and 8). However, the techniques yielded different quantitative results for different types of AgNPs (AgNPs_PPE, AgNPs_CE, and AgNPs_Chem). ICP-OES analysis revealed a significantly higher Ag content in the AgNPs_PPE primed seeds than that in the XRF analysis. Conversely, AgNPs_CE and AgNPs_Chem displayed higher Ag values, as measured by XRF. Notably, the AgNP_ Chem-primed seeds exhibited the highest XRF value (0.083 mg/g) but the lowest ICP-OES value. XRF spectrum analysis further confirmed the presence of Ag in all nanoprimed seed samples (AgNPs_PPE, AgNPs_CE, and AgNPs_Chem), as evidenced by the distinct Ag peaks in the spectra (Figure 8). No such peaks were observed in the XRF spectra of the hydroprimed seeds.

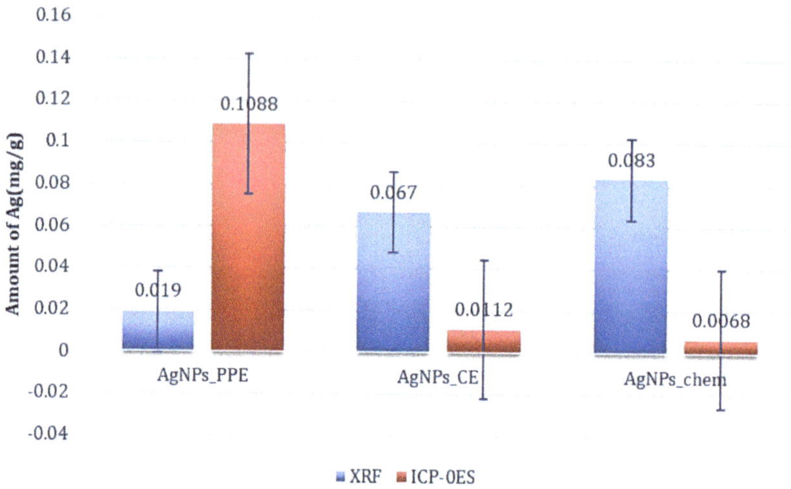


**Figure 7.** Measurements of the amount of AgNPs (mg/g) in *P. granatum* seeds subjected to AgNPs_PPE, AgNPs_ CE, and AgNPs_Chem using XRF and ICP-OES. The amount of Ag element is expressed as mean values derived from sample replications used for the investigation or mean ± SE.

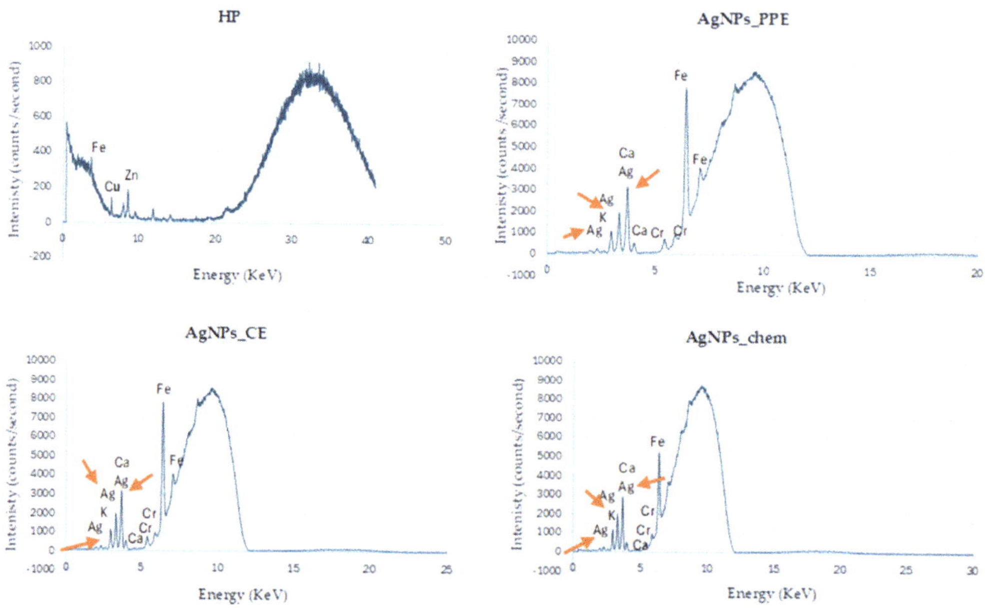


**Figure 8.** Analysis of *P. granatum* L. seeds treated with AgNPs_PPE and AgNPs_CE using X-ray fluorescence spectroscopy (XRF) compared with hydroprimed seeds (HP). Arrows indicate Ag signal in nanoprimed seeds.

## 4. Discussion

This study leverages the green synthesis of silver nanoparticles (AgNPs) using readily available plant extracts, as previously reported by our research group (Abdelmigid et al. [44]). This study leverages an eco-friendly approach for AgNP synthesis by utilizing ground coffee and aqueous pomegranate peel extracts as both reducing and stabilizing agents. This method offers advantages in terms of cost-effectiveness and reduced risk compared to traditional techniques [48]. Here, we build upon characterization data from previous studies [44] to explore the potential relationship between the size and morphology of the synthesized AgNPs and their ability to penetrate or incorporate into Punica granatum seeds. This analysis provides further evidence for the successful synthesis of functional AgNPs. For comparison, we also included chemically synthesized AgNPs_Chem using trisodium citrate as a reducing agent. Characterization techniques confirmed that all synthesized NPs were in their pure crystalline form. The observed color change during AgNP formation using plant extracts aligns with findings from similar studies. For instance, Jabir et al. [49] reported a color shift from yellow to brown in *Annona muricata* peel extract upon green synthesis of silver nanoparticles, further supporting the reduction process.

As a result of their strong surface plasmon resonances (SPRs), UV absorbance was used to verify the synthesis of AgNP_PPE, AgNP_CE, and AgNP_Chem colloidal metal nanoparticles [50–52]. Maximum absorbance was observed between 320 and 430 nanometers in the recorded UV spectrum, which indicates that silver ions have been reduced to colloidal silver. There have been reports of colloidal silver nanoparticles ranging in size from 350 to 600 nanometers [52–55]. Because of their different physical properties, including size and shape, which affect their SPRs, the AgNPs formed in this study showed different UV spectra. UV-vis spectroscopy at the SPRs 421 nm, 405–430 nm, and 420 nm showed that AgNPs were formed by reducing Ag+ to Ag0 using PPE, spent coffee ground extracts, and trisodium citrate, respectively [44,56–58]. Increasing wavelengths are associated with a decrease in the size of AgNPs, as observed in the current study [59].

Studies have demonstrated that nanoparticles with a more negative potential, such as −46.7 mV, exhibit improved repulsion forces, resulting in enhanced stability, dispersion, and colloidal properties [60]. Based on our findings, chemically synthesized AgNPs displayed smaller diameters and higher zeta potentials than PPE- and CE-synthesized AgNPs. Researchers previously found that AgNPs made from PPE (31.6 nm; 118.6–231.7 nm), CE (21–255 nm), and trisodium citrate (5–10 nm) have a smaller diameter [44,56,58–62]. The size of AgNPs can be adjusted by modifying the pH and concentration of $AgNO_3$. At alkaline pH levels, where the pH is inversely proportional to the size of the AgNPs, the reaction rate increases with successive crystallization into smaller particles. Furthermore, it is important to note that the repulsion between nanoparticles tends to increase as the pH level increases, whereas the aggregate ability tends to decrease because of the complete surface charge. However, it is worth mentioning that previous research [59] indicated that the size of silver nanoparticles (AgNPs) tends to decrease as the nitrate concentration increases from 1 M to 5 M.

X-ray diffraction (XRD) analysis confirmed the crystalline nature of the synthesized AgNPs, with characteristic peaks observed at 38.3°, 44.49°, 64.6°, and 77.5°. These peaks correspond to the face-centered cubic (FCC) structure of metallic silver, aligning with findings from previous studies utilizing trisodium citrate and other plant extracts for AgNP synthesis [4,63–66]. The presence of additional diffraction peaks in the AgNP_PPE spectrum at 27.90°, 32.38°, 46.43°, 55.00°, 57.00°, and 67.00° may be attributed to the complex phytochemical composition of pomegranate peel extract (PPE). Similar unassigned peaks have been reported in other studies [67,68].

Fourier transform infrared (FTIR) spectroscopy provided valuable insights into the functional groups potentially involved in the bioreduction and stabilization of the synthesized AgNPs. The spectra of AgNPs_PPE, AgNPs_CE, and AgNPs_Chem were compared with the reference spectra to identify the key functional groups. The presence of phenolic groups in all AgNP types is evident from the peaks observed between 1315–1037 $cm^{-1}$ and

1456–1600 $cm^{-1}$. These functional groups, particularly flavanones present in plant extracts, are well established for their role in reducing silver ions ($Ag^+$) to metallic silver (Ag0) [69]. Additionally, they can act as capping agents, stabilizing newly formed AgNPs [70]. Studies by Shankar et al. [71] support this mechanism, demonstrating the adsorption of flavanones from plant extracts onto metal nanoparticles with characteristic bands around 1074 $cm^{-1}$, potentially present in our spectra. Plant extracts, especially those rich in flavonoids, such as pomegranate peel extract (PPE), offer a multitude of bioactive compounds that contribute to AgNP synthesis and stabilization. Besides flavanols, PPE contains gallotannins, quercetin, ellagic acid derivatives, catechins, procyanidins, and kaempferol [72]. These compounds might contribute to the observed effects; the presence of free caffeine in coffee ground extract (CE) might explain the additional sharp peaks and broad bands observed in the AgNP_CE spectrum, along with the shared flavonoids with AgNPs_PPE. These findings are consistent with previous reports of caffeine vibrations in coffee extracts [4,73,74]. Finally, the FTIR spectrum of AgNPs_Chem, synthesized using trisodium citrate, displayed peaks corresponding to the functional groups associated with the reducing agent. These include C-H stretching deformation, nitro compound ($NO_2$), C=O stretching asymmetry in $COO^-$, and a possible amine stretching vibration, as reported elsewhere [75–78].

Scanning electron microscopy (SEM) revealed the presence of a porous structure on the surface of *Punica granatum* L. seeds [4,79]. This porous structure, inherent to lignocellulosic materials, likely arises due to the evaporation of volatile substances during seed coat formation [4,79]. These pores offer potential adsorption sites for mineral compounds, including AgNPs. Although SEM cannot visualize internal structures, it can be valuable for assessing particle purity and aggregation on the seed surface [80]. Previous studies have suggested that smaller nanoparticles exhibit higher surface energy, leading to increased interaction with cell walls [4,81,82]. This interaction could potentially lead to the development or expansion of existing pores, thereby enhancing the penetration of nanoparticles into the seed coat. The results of our energy-dispersive X-ray spectroscopy (EDX) analysis align with this hypothesis. Smaller AgNPs_Chem (62.68 nm) were detected in higher abundance within pomegranate seeds than larger AgNPs_CE (273.7 nm) and AgNPs_PPE (591 nm).

The higher prevalence of AgNPs_Chem within seeds aligns with observations from X-ray fluorescence (XRF) analysis. This suggests that smaller nanoparticles can readily pass through cellular membranes via existing pores or potentially promote the formation of new ones [83–85]. Additionally, their smaller size and potentially higher zeta potential might contribute to a more efficient interaction with the seed coat, facilitating pore formation, and the presence of a 3 eV optical absorbance band identified through EDX analysis confirmed the presence of pure metallic silver nanoparticles within the seeds [86,87]. Furthermore, EDX and FTIR analyses revealed the presence of organic molecules (C and O) on the AgNP surface, indicating their role as reducing and capping agents [88,89]. The broad and sharp peaks observed in the FTIR spectra suggest the presence of various functional groups associated with these capping agents [90–92]. The increased carbon (C) signal observed in the EDX analysis for AgNP-primed seeds compared to hydroprimed seeds further supports the presence of organic capping agents on the AgNP surface.

Transmission electron microscopy (TEM) analysis provided compelling evidence for the successful internalization of AgNPs into the seed embryos of *Punica granatum* L. This observation aligns with the findings of a previous study on watermelon seeds treated with AgNPs [93]. TEM images revealed the presence of AgNPs within the seed embryo, suggesting the potential for $Ag^+$ uptake and nanoparticle localization.

Interestingly, our TEM analysis did not reveal a specific or uniform distribution pattern for internalized AgNPs. This finding is consistent with observations reported by Song et al. [94], who described clusters of AgNPs surrounding plasma membranes of watermelon seed embryos. Such clustering may be attributed to the dynamic nature of the cytoplasm, where individual AgNPs can be internalized and form aggregates. Similar observations of nanoparticle agglomeration within plant cells have been reported previously. Nicolás-Álvarez et al. [95] demonstrated the presence of titanium dioxide ($TiO_2$) nanoparticles in agglomerated forms within tomato root cells, suggesting a potential shared mechanism for internalized nanoparticle behavior across plant species.

Inductively coupled plasma–optical emission spectrometry (ICP-OES) analysis confirmed the penetration of Ag ions released from AgNPs into *Punica granatum* L. seed tissues, as evidenced by the presence of detectable silver compared to hydroprimed controls. This finding aligns with the observations reported by Zhou et al. [96], who demonstrated similar Ag ion uptake in plant tissues exposed to AgNPs. While X-ray fluorescence (XRF) analysis offers a valuable tool for quantifying the elemental composition on surfaces, its penetration depth is limited compared to ICP-OES [97]. This explains the observed discrepancies between the XRF and ICP-OES results of our study. XRF analysis likely detected primarily surface-bound Ag, potentially leading to higher Ag concentrations in the AgNPs_CE and AgNPs_Chem treatments because of their larger size and potentially greater surface area. In contrast, ICP-OES requires complete sample mineralization, solubilizing both surface and internalized Ag, which may explain the lower XRF-measured Ag concentrations in AgNPs_PPE treatments, possibly due to deeper tissue penetration of these smaller nanoparticles. These observations highlight the importance of considering the limitations of the analytical technique when interpreting Ag quantification data, particularly for heterogeneous samples, such as plant tissues.

The integrated analytical approach employed in this study, utilizing SEM, EDX, TEM, XRF, and ICP-OES, successfully confirmed the absorption and incorporation of AgNPs into *Punica granatum* seeds. This approach provides a robust method for verifying the effectiveness of nanopriming before investigating its potential positive or negative effects on target crops. It is important to acknowledge that the identification and quantification of nanomaterials within complex biological matrices, such as plant tissues, requires specialized techniques, such as transmission electron microscopy (TEM) and inductively coupled plasma research [98]. Our findings regarding AgNP penetration into pomegranate seeds using TEM, XRF, and ICP-OES are consistent with the observations reported in other studies on various crops. Similar results have been documented for AgNP uptake in tomatoes [95], maize [99], and mung bean seeds [100]. Furthermore, the internalization of metal-based nanoparticles by plant tissues extends beyond AgNPs, as studies have shown comparable behavior for $Fe_3O_4$, Au, and Cu nanoparticles [101–103]. Collectively, these findings suggest a potentially widespread mechanism for nanoparticle uptake in plants.

## 5. Conclusions

This research evaluated the efficacy of synthesized silver nanoparticles (AgNPs) as seed-priming agents to minimize nanomaterial release into the environment. Our study utilized green and chemical synthesis techniques to produce AgNPs. The green synthesis technique successfully produced purified crystalline Ag nanoparticles from aqueous extracts of pomegranate fruit peel (PPE) and coffee (CE). Additionally, trisodium citrate was employed in the chemical reduction process to synthesize AgNPs. Based on the results of this work, the following key findings can be emphasized:

1. The seeds of pomegranate demonstrated successful internalization and the uptake of nanoparticles, which were validated through various analytical methods, such as UV-visible spectroscopy, zeta potential analysis, FTIR spectroscopy, XRD, SEM, and TEM.
2. The significance of the electron microscopy (EM) analysis has been emphasized in verifying the incorporation of nanoparticles in determining AgNP distribution within plants, verifying the incorporation of nanoparticles, and determining the distribution of silver nanoparticles (AgNPs) within plant tissues.
3. Elemental analysis techniques (EDX, XRF, ICP-OES) were successful in producing comprehensive outcomes that corroborated the findings obtained from the use of scanning electron microscopy (SEM) and transmission electron microscopy (TEM).

Although this research demonstrates the potential of AgNP nanopriming as a targeted approach for delivering nanoparticles to seeds, additional investigation into the optimal dosage and long-term effects of AgNPs on different plant species is necessary to ensure safe and sustainable agricultural practices.

## 6. Future Directions

Understanding the molecular mechanisms underlying plant responses to nanopriming (germination, growth, physiology, metabolism, genetics) will be essential for optimizing nanopriming protocols for improved agricultural outcomes.

**Author Contributions:** Conceptualization, H.M.A., A.A.A. and M.M.M.; methodology, H.M.A., A.A.A. and M.M.M.; software, H.M.A.; validation, H.M.A., A.A.A. and M.M.M.; formal analysis, A.A.A.; investigation, H.M.A.; resources, A.A.A.; data curation, M.M.M.; writing—original draft preparation, H.M.A.; writing, A.A.A. and M.M.M.; visualization, A.A.A.; supervision, H.M.A.; project administration, H.M.A.; funding acquisition, A.A.A. All authors have read and agreed to the published version of the manuscript.

**Funding:** The research was funded by Taif University, Saudi Arabia, Project No. (TU-DSPP-2024-213).

**Data Availability Statement:** The original contributions presented in the study are included in the article, further inquiries can be directed to the corresponding author.

**Acknowledgments:** The authors extend their appreciation to Taif University, Saudi Arabia, for supporting this work through project number (TU-DSPP-2024-213).

**Conflicts of Interest:** The authors declare no conflicts of interest.

## References


1. Singh, A.; Singh, N.Á.; Afzal, S.; Singh, T.; Hussain, I. Zinc oxide nanoparticles: A review of their biological synthesis, antimicrobial activity, uptake, translocation and biotransformation in plants. *J. Mater. Sci.* **2018**, *53*, 185–201. [CrossRef]
2. Liu, R.; Lal, R. Synthetic apatite nanoparticles as a phosphorus fertilizer for soybean (*Glycine max*). *Sci. Rep.* **2014**, *14*, 5686. [CrossRef] [PubMed]
3. Janmohammadi, M.; Sabaghnia, N. Effect of pre-sowing seed treatments with silicon nanoparticles on germinability of sunflower (*Helianthus annuus*). *Bot Lith.* **2015**, *21*, 13–21. [CrossRef] [PubMed]
4. Abdelmigid, H.M.; Alyamani, A.A.; Hussien, N.A.; Morsi, M.M.; Alhumaidi, A. Integrated Approaches for Adsorption and Incorporation Testing of Green-Synthesized $TiO_2$NPs Mediated by Seed-Priming Technology in *Punica granatum* L. *Agronomy* **2022**, *12*, 1601. [CrossRef]
5. Endo, E.H.; Ueda-Nakamura, T.; Nakamura, C.V.; Filho, B.P.D. Activity of Spray-dried Microparticles Containing Pomegranate Peel Extract against *Candida albicans*. *Molecules* **2012**, *17*, 10094–10107. [CrossRef] [PubMed]
6. Ismail, T.; Sestili, P.; Akhtar, S. Pomegranate peel and fruit extracts: A review of potential anti-inflammatory and anti-infective effects. *J. Ethnopharmacol.* **2012**, *143*, 397–405. [CrossRef] [PubMed]
7. Habibipour, R.; Moradi-Haghgou, L.; Farmany, A. Green synthesis of AgNPs@ PPE and its Pseudomonas aeruginosa biofilm formation activity compared to pomegranate peel extract. *Int. J. Nanomed.* **2019**, *14*, 6891. [CrossRef] [PubMed]
8. Leesombun, A.; Sariya, L.; Taowan, J.; Nakthong, C.; Thongjuy, O.; Boonmasawai, S. Natural Antioxidant, Antibacterial, and Antiproliferative Activities of Ethanolic Extracts from *Punica granatum* L. Tree Barks Mediated by Extracellular Signal-Regulated Kinase. *Plants* **2022**, *11*, 2258. [CrossRef]
9. Elkady, A.I. Crude alkaloid extract of Rhazya stricta inhibits cell growth and sensitizes human lung cancer cells to cisplatin through induction of apoptosis. *Genet. Mol. Biol.* **2013**, *36*, 12–21. [CrossRef]
10. Jisha, K.C.; Vijayakumari, K.; Puthur, J.T. Seed priming for abiotic stress tolerance: An overview. *Acta Physiol. Plant.* **2013**, *35*, 1381–1396. [CrossRef]
11. Bayat, M.; Zargar, M.; Murtazova, K.M.-S.; Nakhaev, M.R.; Shkurkin, S.I. Ameliorating Seed Germination and Seedling Growth of Nano-Primed Wheat and Flax Seeds Using Seven Biogenic Metal-Based Nanoparticles. *Agronomy* **2022**, *12*, 811. [CrossRef]
12. Chen, K.; Arora, R. Priming memory invokes seed stress-tolerance. *Environ. Exp. Bot.* **2013**, *94*, 33–45. [CrossRef]
13. Devika, O.S.; Singh, S.; Sarkar, D.; Barnwal, P.; Suman, J.; Rakshit, A. Seed Priming: A Potential Supplement in Integrated Resource Management Under Fragile Intensive Ecosystems. *Front. Sustain. Food Syst.* **2021**, *5*, 654001. [CrossRef]
14. Forti, C.; Ottobrino, V.; Bassolino, L.; Toppino, L.; Rotino, G.L.; Pagano, A.; Macovei, A.; Balestrazzi, A. Molecular dynamics of pre-germinative metabolism in primed eggplant (*Solanum melongena* L.) seeds. *Hortic. Res.* **2020**, *1*, 7. [CrossRef]
15. Du, B.; Luo, H.; He, L.; Zhang, L.; Liu, Y.; Mo, Z.; Pan, S.; Tian, H.; Duan, M.; Tang, X. Rice seed priming with sodium selenate: Effects on germination, seedling growth, and biochemical attributes. *Sci. Rep.* **2019**, *9*, 4311. [CrossRef] [PubMed]
16. Marthandan, V.; Geetha, R.; Kumutha, K.; Renganathan, V.G.; Karthikeyan, A.; Ramalingam, J. Seed Priming: A Feasible Strategy to Enhance Drought Tolerance in Crop Plants. *Int. J. Mol. Sci.* **2020**, *21*, 8258. [CrossRef]

17. Yan, H.; Mao, P. Comparative Time-Course Physiological Responses and Proteomic Analysis of Melatonin Priming on Promoting Germination in Aged Oat (*Avena sativa* L.) Seeds. *Int. J. Mol. Sci.* **2021**, *22*, 811. [CrossRef]
18. Ibrahim, E.A. Seed priming to alleviate salinity stress in germinating seeds. *J. Plant Physiol.* **2016**, *192*, 38–46. [CrossRef]
19. Ko, K.-S.; Koh, D.-C.; Kong, I.C. Evaluation of the Effects of Nanoparticle Mixtures on *Brassica* Seed Germination and Bacterial Bioluminescence Activity Based on the Theory of Probability. *Nanomaterials* **2017**, *7*, 344. [CrossRef]
20. Vijayaraghavan, K.; Ashokkumar, T. Plant-mediated biosynthesis of metallic nanoparticles: A review of literature, factors affecting synthesis, characterization techniques and applications. *J. Environ. Chem. Eng.* **2017**, *5*, 4866–4883. [CrossRef]
21. Cox, A.; Venkatachalam, P.; Sahi, S.; Sharma, N. Silver and titanium dioxide nanoparticle toxicity in plants: A review of current research. *Plant Physiol. Biochem.* **2016**, *107*, 147–163. [CrossRef] [PubMed]
22. Dasgupta, N.; Ranjan, S.; Mundekkad, D.; Ramalingam, C.; Shanker, R.; Kumar, A. Nanotechnology in agro-food: From field to plate. *Food Res. Int.* **2015**, *69*, 381–400. [CrossRef]
23. Takeuchi, M.T.; Kojima, M.; Luetzow, M. State of the art on the initiatives and activities relevant to risk assessment and risk management of nanotechnologies in the food and agriculture sectors. *Food Res. Int.* **2014**, *64*, 976–981. [CrossRef] [PubMed]
24. Marchiol, L.; Mattiello, A.; Pošćić, F.; Giordano, C.; Musetti, R. In vivo synthesis of nanomaterials in plants: Location of silver nanoparticles and plant metabolism. *Nanoscale Res. Lett.* **2014**, *9*, 101. [CrossRef] [PubMed]
25. Ribeiro, C.A.; Albuquerque, L.J.; de Castro, C.E.; Batista, B.L.; de Souza, A.L.; Albuquerque, B.L.; Zilse, M.S.; Bellettini, I.C.; Giacomelli, F.C. One-pot synthesis of sugar-decorated gold nanoparticles with reduced cytotoxicity and enhanced cellular uptake. *Colloids Surf. A Physicochem. Eng. Asp.* **2019**, *580*, 123690. [CrossRef]
26. Jiang, H.S.; Li, M.; Chang, F.Y.; Li, W.; Yin, L.Y. Physiological analysis of silver nanoparticles and $AgNO_3$ toxicity to *Spirodela polyrhiza*. *Environ. Toxicol. Chem.* **2012**, *31*, 1880–1886. [CrossRef] [PubMed]
27. Hojjat, S.S.; Hojjat, H. Effect of nano silver on seed germination and seedling growth in fenugreek seed. *Int. J. Food Eng.* **2015**, *1*, 106–110. [CrossRef]
28. Sharma, P.; Bhatt, D.; Zaidi, M.G.; Saradhi, P.P.; Khanna, P.K.; Arora, S. Silver nanoparticle-mediated enhancement in growth and antioxidant status of *Brassica juncea*. *Appl. Biochem. Biotechnol.* **2012**, *167*, 2225–2233. [CrossRef] [PubMed]
29. El-Naggar, N.E.; Hussein, M.H.; Shaaban-Dessuuki, S.A.; Dalal, S.R. Production, extraction and characterization of *Chlorella vulgaris* soluble polysaccharides and their applications in AgNPs biosynthesis and biostimulation of plant growth. *Sci. Rep.* **2020**, *10*, 3011. [CrossRef]
30. Mustapha, T.; Misni, N.; Ithnin, N.R.; Daskum, A.M.; Unyah, N.Z. A Review on Plants and Microorganisms Mediated Synthesis of Silver Nanoparticles, Role of Plants Metabolites and Applications. *Int. J. Environ. Res. Public Health.* **2022**, *19*, 674. [CrossRef]
31. Vannini, C.; Domingo, G.; Onelli, E.; Prinsi, B.; Marsoni, M.; Espen, L.; Bracale, M. Morphological and proteomic responses of *Eruca sativa* exposed to silver nanoparticles or silver nitrate. *PLoS ONE* **2013**, *8*, e68752. [CrossRef] [PubMed]
32. Mahakham, W.; Sarmah, A.K.; Maensiri, S.; Theerakulpisut, P. Nanopriming technology for enhancing germination and starch metabolism of aged rice seeds using phytosynthesized silver nanoparticles. *Sci. Rep.* **2017**, *7*, 8263. [CrossRef] [PubMed]
33. Li, M.; Yu, H.; Cheng, Y.; Guo, Y.; Yao, W.; Xie, Y. Simultaneous and rapid determination of polycyclic aromatic hydrocarbons by facile and green synthesis of silver nanoparticles as effective SERS substrate. *Ecotoxicol. Environ. Saf.* **2020**, *200*, 110780. [CrossRef] [PubMed]
34. Trotsiuk, L.; Antanovich, A.; Lizunova, A.; Kulakovich, O. Direct synthesis of amphiphilic polyvinylpyrrolidone-capped gold nanoparticles in chloroform. *Colloid Interface Sci. Commun.* **2020**, *37*, 100289. [CrossRef]
35. Adamo, C.B.; Junger, A.S.; Bressan, L.P.; da Silva, J.A.; Poppi, R.J.; de Jesus, D.P. Fast and straightforward in-situ synthesis of gold nanoparticles on a thread-based microfluidic device for application in surface-enhanced Raman scattering detection. *Microchem. J.* **2020**, *156*, 104985. [CrossRef]
36. Bandeira, M.; Giovanela, M.; Roesch-Ely, M.; Devine, D.M.; da Silva Crespo, J. Green synthesis of zinc oxide nanoparticles: A review of the synthesis methodology and mechanism of formation. *Sustain. Chem. Pharm.* **2020**, *15*, 100223. [CrossRef]
37. Ohara, Y.; Akazawa, K.; Shibata, K.; Hirota, T.; Kodama, Y.; Amemiya, T.; Wang, J.; Yamaguchi, T. Seed-mediated gold nanoparticle synthesis via photochemical reaction of benzoquinone. *Colloids Surf. A Physicochem. Eng. Asp.* **2020**, *586*, 124209. [CrossRef]
38. Kim, J.S.; Kuk, E.; Yu, K.N.; Kim, J.H.; Park, S.J.; Lee, H.J.; Kim, S.H.; Park, Y.K.; Park, Y.H.; Hwang, C.Y.; et al. Antimicrobial effects of silver nanoparticles. *Nanomed. Nanotechnol. Biol. Med.* **2007**, *3*, 95–101. [CrossRef] [PubMed]
39. Tan, Y.; Dai, X.; Li, Y.; Zhu, D. Preparation of gold, platinum, palladium, and silver nanoparticles by the reduction of their salts with a weak reductant–potassium bitartrate. *J. Mater. Chem.* **2003**, *13*, 1069–1075. [CrossRef]
40. Hassaan, M.A.; Hosny, S. Green synthesis of Ag and Au nanoparticles from micro and macro algae-review. *Int. J. Atmos. Ocean. Sci.* **2018**, *2*, 10–22. [CrossRef]
41. Singh, P.; Kim, Y.J.; Zhang, D.; Yang, D.C. Biological synthesis of nanoparticles from plants and microorganisms. *Trends Biotechnol.* **2016**, *34*, 588–599. [CrossRef] [PubMed]
42. Rajan, R.; Chandran, K.; Harper, S.L.; Yun, S.I.; Kalaichelvan, P.T. Plant extract synthesized silver nanoparticles: An ongoing source of novel biocompatible materials. *Ind. Crops Prod.* **2015**, *70*, 356–373. [CrossRef]
43. Khan, M.; Shaik, M.R.; Adil, S.F.; Khan, S.T.; Al-Warthan, A.; Siddiqui, M.R.; Tahir, M.N.; Tremel, W. Plant extracts as green reductants for the synthesis of silver nanoparticles: Lessons from chemical synthesis. *Dalton Trans.* **2018**, *47*, 11988–12010. [CrossRef]

44. Abdelmigid, H.M.; Morsi, M.M.; Hussien, N.A.; Alyamani, A.A.; Sufyani, A.; Moslah, N. Comparative Analysis of nanosilver Particles synthesized by different approaches and their antimicrobial efficacy. *J. Nanomater.* **2021**, *15*, 2021. [CrossRef]
45. Fang, J.; Zhong, C.; Mu, R. The study of deposited silver particulate films by simple method for efficient SERS. *Chem. Phys. Lett.* **2005**, *401*, 271–275. [CrossRef]
46. Basavegowda, N.; Lee, Y.R. Synthesis of silver nanoparticles using Satsuma mandarin (*Citrus unshiu*) peel extract: A novel approach towards waste utilization. *Mater. Lett.* **2013**, *109*, 31–33. [CrossRef]
47. Tezotto, T.; Favarin, J.L.; Paula Neto, A.; Gratão, P.L.; Azevedo, R.A.; Mazzafera, P. Simple procedure for nutrient analysis of coffee plant with energy dispersive X-ray fluorescence spectrometry (EDXRF). *Sci. Agric.* **2013**, *70*, 263–267. [CrossRef]
48. Mussatto, S.I.; Machado, E.; Martins, S.; Teixeira, J.A. Production, composition, and application of coffee and its industrial residues. *Food Bioprocess Technol.* **2011**, *5*, 661–672. [CrossRef]
49. Jabir, M.S.; Hussien, A.A.; Sulaiman, G.M. Green synthesis of silver nanoparticles from Eriobotrya japonica extract: A promising approach against cancer cells proliferation, inflammation, allergic disorders and phagocytosis induction. *Artif. Cells Nanomed. Biotechnol.* **2021**, *49*, 48–60. [CrossRef]
50. Mulvaney, P. Surface plasmon spectroscopy of nanosized metal particles. *Langmuir* **1996**, *12*, 788–800. [CrossRef]
51. Anandalakshmi, K.; Venugobal, J.; Ramasamy, V. Characterization of silver nanoparticles by green synthesis method using Pedalium murex leaf extract and their antibacterial activity. *Appl. Nanosci.* **2016**, *6*, 399–408. [CrossRef]
52. Henglein, A. Physicochemical properties of small metal particles in solution: "microelectrode" reactions, chemisorption, composite metal particles, and the atom-to-metal transition. *J. Phys. Chem.* **1993**, *97*, 5457–5471. [CrossRef]
53. Sastry, M.; Patil, V.; Sainkar, S.R. Electrostatically controlled diffusion of carboxylic acid derivatized silver colloidal particles in thermally evaporated fatty amine films. *J. Phys. Chem. B.* **1998**, *102*, 1404–1410. [CrossRef]
54. Jabir, M.S.; Saleh, Y.M.; Sulaiman, G.M. Green synthesis of silver nanoparticles using Annona muricata extract as an inducer of apoptosis in cancer cells and inhibitor for NLRP3 inflammasome via enhanced autophagy. *Nanomaterials* **2021**, *11*, 384. [CrossRef]
55. Saeb, A.; Alshammari, A.S.; Al-Brahim, H.; Al-Rubeaan, K.A. Production of silver nanoparticles with strong and stable antimicrobial activity against highly pathogenic and multidrug-resistant bacteria. *Sci. World J.* **2014**, *2014*, 704708. [CrossRef] [PubMed]
56. Saad, P.G.; Castelino, R.D.; Ravi, V.; Al-Amri, I.S.; Khan, S.A. Green synthesis of silver nanoparticles using Omani pomegranate peel extract and two polyphenolic natural products: Characterization and comparison of their antioxidant, antibacterial, and cytotoxic activities. *Beni-Suef Univ. J. Basic Appl. Sci.* **2021**, *10*, 29. [CrossRef]
57. Zuorro, A.; Iannone, A.; Miglietta, S.; Lavecchia, R. Green Synthesis of Silver Nanoparticles Using Spent Coffee Ground Extracts: Process Modelling and Optimization. *Nanomaterials* **2022**, *12*, 2597. [CrossRef] [PubMed]
58. Quintero-Quiroz, C.; Acevedo, N.; Zapata-Giraldo, J.; Botero, L.E.; Quintero, J.; Zárate-Triviño, D.; Saldarriaga, J.; Pérez, V.Z. Optimization of silver nanoparticle synthesis by chemical reduction and evaluation of its antimicrobial and toxic activity. *Biomater. Res.* **2019**, *23*, 27. [CrossRef] [PubMed]
59. Anigol, L.B.; Charantimath, J.S.; Gurubasavaraj, P.M. Effect of concentration and pH on the size of silver nanoparticles synthesized by green chemistry. *Org. Med. Chem. Int. J.* **2017**, *3*, 1–5. [CrossRef]
60. Mukherjee, S.; Chowdhury, D.; Kotcherlakota, R. Potential theranostics application of bio-synthesized silver nanoparticles (4-in-1 system). *Theranostics* **2014**, *4*, 316–335. [CrossRef]
61. Vinay, C.H.; Goudanavar, P.; Acharya, A. Development and characterization of pomegranate and orange fruit peel extract based silver nanoparticles. *J. Manmohan Meml. Inst. Health Sci.* **2018**, *22*, 72–85. [CrossRef]
62. Panzella, L.; Cerruti, P.; Aprea, P.; Paolillo, R.; Pellegrino, G.; Moccia, F.; Condorelli, G.G.; Vollaro, A.; Ambrogi, V.; Catania, M.R.; et al. Silver nanoparticles on hydrolyzed spent coffee grounds (HSCG) for green antibacterial devices. *J. Clean. Prod.* **2020**, *268*, 122352. [CrossRef]
63. Vanlalveni, C.; Lallianrawna, S.; Biswas, A.; Selvaraj, M.; Changmai, B.; Rokhum, S.L. Green synthesis of silver nanoparticles using plant extracts and their antimicrobial activities: A review of recent literature. *RSC Adv.* **2021**, *11*, 2804–2837. [CrossRef]
64. Sufyani, A.; Moslah, N.; Hussien, N.A.; Hawsawi, Y.M. Characterization and anticancer potential of silver nanoparticles biosynthesized from *Olea chrysophylla* and *Lavandula dentata* leaf extracts on HCT116 colon cancer cells. *J. Nanomater.* **2019**, *2019*, 7361695. [CrossRef]
65. Yerragopu, P.S.; Hiregoudar, S.; Nidoni, U.; Ramappa, K.T.; Sreenivas, A.G.; Doddagoudar, S.R. Chemical synthesis of silver nanoparticles using tri-sodium citrate, stability study and their characterization. *Int. Res. J. Pure Appl. Chem.* **2020**, *21*, 37–50. [CrossRef]
66. Liu, X.; Shan, K.; Shao, X. Nanotoxic effects of silver nanoparticles on normal HEK-293 cells in comparison to cancerous HeLa cell line. *Int. J. Nanomed.* **2021**, *16*, 753–761. [CrossRef] [PubMed]
67. Kumar, V.; Yadav, S.K. Plant-mediated synthesis of silver and gold nanoparticles and their applications. *J. Chem. Technol. Biotechnol.* **2009**, *84*, 151–157. [CrossRef]
68. Suvith, V.S.; Philip, D. Catalytic degradation of methylene blue using biosynthesized gold and silver nanoparticles. *Spectrochim. Acta. Part A Mol. Biomol. Spectrosc.* **2014**, *118*, 526–532. [CrossRef]
69. Jeeva, K.; Thiyagarajan, M.; Elangovan, V.; Geetha, N.; Venkatachalam, P. Caesalpinia coriaria leaf extracts mediated biosynthesis of metallic silver nanoparticles an their antibacterial activity against clinically isolated pathogens. *Ind. Crops Prod.* **2014**, *52*, 714–720. [CrossRef]

70. Dogiparthi, L.K.; Sana, S.S.; Shaik, S.Z.; Kalvapalli, M.R.; Kurupati, G.; Kumar, G.S.; Gangadhar, L. Phytochemical mediated synthesis of silver nanoparticles and their antibacterial activity. *SN Appl. Sci.* **2021**, *3*, 631. [CrossRef]
71. Shankar, S.S.; Rai, A.; Ahmad, A.; Sastry, M. Rapid synthesis of Au, Ag, and bimetallic Au core–Ag shell nanoparticles using Neem (*Azadirachta indica*) leaf broth. *J. Colloid Interface Sci.* **2004**, *275*, 496–502. [CrossRef]
72. Nasiriboroumand, M.; Montazer, M.; Barani, H. Preparation and characterization of biocompatible silver nanoparticles using pomegranate peel extract. *J. Photochem. Photobiol.* **2018**, *179*, 98–104. [CrossRef] [PubMed]
73. Al Qarni, F.; Alomair, N.A.; Mohamed, H.H. Environment-Friendly Nanoporous Titanium Dioxide with Enhanced Photocatalytic Activity. *Catalysts* **2019**, *9*, 799. [CrossRef]
74. Srivastava, S.K.; Singh, V.B. Ab initio and DFT studies of the structure and vibrational spectra of anhydrous caffeine. *Spectrochim. Acta Part A Mol. Biomol. Spectrosc.* **2013**, *115*, 45–50. [CrossRef]
75. Barbin, D.F.; Felicio, A.L.D.M.; Sun, D.W.; Nixdorf, S.L.; Hirooka, E.Y. Application of infrared spectral techniques on quality and compositional attributes of coffee: An overview. *Food Res. Int.* **2014**, *61*, 23–32. [CrossRef]
76. Chien, H.W.; Kuo, C.J.; Kao, L.H.; Lin, G.Y.; Chen, P.Y. Polysaccharidic spent coffee grounds for silver nanoparticle immobilization as a green and highly efficient biocide. *Int. J. Biol. Macromol.* **2019**, *140*, 168–176. [CrossRef]
77. Singh, J.; Bajaj, R.; Kaur, H.; Kaur, H.; Kaur, N. Chemo-bio synthesis of silver nanoparticles. *J. Nano Res.* **2016**, *4*, 00092. [CrossRef]
78. Ranoszek-Soliwoda, K.; Tomaszewska, E.; Socha, E. The role of tannic acid and sodium citrate in the synthesis of silvernanoparticles. *J. Nanopart. Res.* **2017**, *19*, 273. [CrossRef]
79. Bello, O.S.; Adegoke, K.A.; Akinyunni, O.O. Preparation and characterization of a novel adsorbent from *Moringa oleifera* leaf. *Appl. Water Sci.* **2017**, *7*, 1295–1305. [CrossRef]
80. Zhang, X.F.; Liu, Z.G.; Shen, W.; Gurunathan, S. Silver Nanoparticles: Synthesis, Characterization, Properties, Applications, and Therapeutic Approaches. *Int. J. Mol. Sci.* **2016**, *17*, 1534. [CrossRef]
81. Rawashdeh, R.Y.; Harb, A.M.; AlHasan, A.M. Biological interaction levels of zinc oxide nanoparticles; lettuce seeds as case study. *Heliyon* **2020**, *6*, e03983. [CrossRef]
82. Tang, Y.J.; Wu, S.G.; Huang, L.; Head, J.; Chen, D.R.; Kong, I.C. Phytotoxicity of Metal Oxide Nanoparticles is Related to Both Dissolved Metals Ions and Adsorption of Particles on Seed Surfaces. *J. Pet. Environ. Biotechnol.* **2012**, *3*, 126. [CrossRef]
83. Xiong, C.; Zheng, Y.; Feng, Y.; Yao, C.; Ma, C.; Zheng, X.; Jiangd, J. Preparation of a novel chloromethylated polystyrene-2-amino-1,3,4-thiadiazole chelating resin and its adsorption properties and mechanism for separation and recovery of Pt (IV) from aqueous solutions. *J. Mater. Chem. A* **2014**, *2*, 5379–5386. [CrossRef] [PubMed]
84. Kurczyńska, E.; Godel-Jędrychowska, K.; Sala, K.; Milewska-Hendel, A. Nanoparticles—Plant Interaction: WhatWe Know, Where We Are? *Appl. Sci.* **2021**, *11*, 5473. [CrossRef]
85. Xu, L.; Zhu, Z.; Sun, D.W. Bioinspired Nanomodification Strategies: Moving from Chemical-Based Agrosystems to Sustainable Agriculture. *ACS Nano* **2021**, *15*, 12655–12686. [CrossRef] [PubMed]
86. Magudapathy, P.; Gangopadhyay, P.; Panigrahi, B.K.; Nair, K.G.M.; Dhara, S. Ion beam sputtering and nanostructures of noble metals. *Phys. B.* **2001**, *299*, 142–146. [CrossRef]
87. Gowramma, B.; Keerthi, U.; Rafi, M.; Rao, D.M. Biogenic silver nanoparticles production and characterization from native stain of Corynebacterium species and its antimicrobial activity. *Biotech* **2015**, *5*, 195–201. [CrossRef] [PubMed]
88. Dada, A.O.; Adekola, F.A.; Odebunmi, E.O. A novel zerovalent manganese for removal of copper ions: Synthesis, characterization and adsorption studies. *Appl. Water Sci.* **2017**, *7*, 1409–1427. [CrossRef]
89. Femi-Adepoju, A.G.; Dada, A.O.; Otun, K.O.; Adepoju, A.O.; Fatoba, O.P. Green synthesis of silver nanoparticles using terrestrial fern (*Gleichenia Pectinata* (Willd.) C. Presl.): Characterization and antimicrobial studies. *Heliyon* **2019**, *5*, e01543. [CrossRef]
90. Corral-Bobadilla, M.; Lostado-Lorza, R.; Somovilla-Gomez, F.; Escribano-García, R. Effective use of activated carbon from olive stone waste in the biosorption removal of Fe (III) ions from aqueous solutions. *J. Clean. Prod.* **2021**, *294*, 126332. [CrossRef]
91. Rosson, E.; Garbo, F.; Marangoni, G.; Bertani, R.; Lavagnolo, M.C.; Moretti, E.; Talon, A.; Mozzon, M.; Sgarbossa, P. Activated Carbon from Spent Coffee Grounds: A Good Competitor of Commercial Carbons for Water Decontamination. *Appl. Sci.* **2020**, *10*, 5598. [CrossRef]
92. El-Azazy, M.; El-Shafie, A.S.; Morsy, H. Biochar of Spent Coffee Grounds as Per Se and Impregnated with TiO2: Promising Waste-Derived Adsorbents for Balofloxacin. *Molecules* **2021**, *26*, 2295. [CrossRef] [PubMed]
93. Acharya, P.; Jayaprakasha, G.K.; Kevin, M.; Crosby, K.M.; Jifon, J.L.; Patil, B.S. Nanoparticle-Mediated Seed Priming Improves Germination, Growth, Yield, and Quality of Watermelons (*Citrullus lanatus*) at multi-locations in Texas. *Sci. Rep.* **2020**, *10*, 5037. [CrossRef] [PubMed]
94. Cullity, B.D. *Elements of X-ray Diffraction*; Addison-Wesley Publishing: Boston, MA, USA, 1956.
95. Nicolás-Álvarez, D.E.; Andraca-Adame, J.A.; Chanona-Pérez, J.J.; Méndez-Méndez, J.V.; Borja-Urby, R.; Cayetano-Castro, N.; Martínez-Gutiérrez, H.; López-Salazar, P. Effects of TiO2 Nanoparticles Incorporation into Cells of Tomato Roots. *Nanomaterials* **2021**, *11*, 1127. [CrossRef] [PubMed]
96. Zhou, X.; Jia, X.; Zhang, Z.; Chen, K.; Wang, L.; Chen, H.; Yang, Z.; Li, C.; Zhao, L. AgNPs seed priming accelerated germination speed and altered nutritional profile of Chinese cabbage. *Sci. Total Environ.* **2022**, *808*, 151896. [CrossRef] [PubMed]
97. Yashveer, S.; Redhu, N.S.; Singh, V.; Sangwan, S.; Laxman, H.V.; Tokas, J.; Malhotra, S.; Khurana, S.; Sindhu, A. Nanoparticles in Agriculture: Characterization, Uptake and Role in Mitigating Heat Stress. *Nat. Resourses Hum. Health* **2022**, *2*, 160–181. [CrossRef]

98. Kolbert, Z.; Szőllősi, R.; Rónavári, A. Nanoforms of essential metals: From hormetic phyto effects to agricultural potential and Árpád. *J. Exp. Bot.* **2022**, *73*, 1825–1840. [CrossRef] [PubMed]
99. Tilahun Bekele, E.; Gonfa, B.A.; Sabir, F.K. Use of different natural products to control growth of titanium oxide nanoparticles in green solvent emulsion, characterization, and their photocatalytic application. *Bioinorg. Chem. Appl.* **2021**, *13*, 2021. [CrossRef] [PubMed]
100. Shah, T.; Latif, S.; Saeed, F.; Ali, I.; Ullah, S.; Alsahli, A.; Jan, S.; Ahmad, P. Seed priming with titanium dioxide nanoparticles enhances seed vigor, leaf water status, and antioxidant enzyme activities in maize (*Zea mays* L.) under salinity stress. *J. King Saud Univ. Sci.* **2021**, *33*, 101207. [CrossRef]
101. Judy, J.D.; Unrine, J.M.; Rao, W.; Wirick, S.; Bertsch, P.M. Bioavailability of Gold Nanomaterials to Plants: Importance of Particle Size and Surface Coating. *Environ. Sci. Technol.* **2012**, *46*, 8467–8474. [CrossRef]
102. Tombuloglu, H.; Slimani, Y.; Tombuloglu, G.; Korkmaz, A.D.; Baykal, A.; Almessiere, M.; Ercan, I. Impact of superparamagnetic iron oxide nanoparticles (SPIONs) and ionic iron on the physiology of summer squash (*Cucurbita pepo*): A comparative study. *Plant Physiol. Biochem.* **2019**, *139*, 56–65. [CrossRef]
103. Ibrahim, A.S.; Ali, G.A.; Hassanein, A.; Attia, A.M.; Marzouk, E.R. Toxicity and Uptake of CuO Nanoparticles: Evaluation of an Emerging Nanofertilizer on Wheat (*Triticum aestivum* L.) Plant. *Sustainability* **2022**, *9*, 4914. [CrossRef]

*horticulturae*

MDPI

*Article*

# Nutrient Dynamics and Resource-Use Efficiency in Greenhouse Strawberries: Effects of Control Variables in Closed-Loop Hydroponics

**Mi Young Lim, So Hui Kim, Mi Young Roh, Gyeong Lee Choi and Dongpil Kim ***

Protected Horticulture Research Institute, National Institute of Horticultural and Herbal Science, Rural Development Administration, Haman 52054, Republic of Korea; ericalim@korea.kr (M.Y.L.); shkim89@korea.kr (S.H.K.); rohmy17@korea.kr (M.Y.R.); chlruddl@korea.kr (G.L.C.)

* Correspondence: kimdongpil@korea.kr; Tel.: +82-555805536

**Abstract:** The importance of implementing recirculating drainage for greenhouse strawberries is often overlooked because of the low electrical conductivity (EC) of drainage and transpiration despite the large area of cultivation in Korea. In this study, we analyzed the growth of strawberry crops and their water and individual nutrient use efficiency when using closed-loop hydroponics in greenhouses. The study consisted of two parts: Experiment (Exp) 1 and Exp 2, each of which employed a different closed-loop hydroponic control method. In Exp 1, the system was controlled solely based on the EC of the drainage mixed with raw water. In Exp 2, the nutrient solution (NS) was corrected according to the ion concentrations in the drainage, with correction intervals of 2 weeks, 4 weeks, and no correction, while the control treatment was open-loop. In Exp 1, a prolonged increase in drainage EC resulted in an imbalance in NS ion composition and reduced fruit yield. In Exp 2, the NS composition was stabilized through periodic nutrient correction, and the fruit yield of the closed-loop treatments did not differ from the open-loop treatment. However, the closed-loop with 2- and 4-week correction in Exp 2 showed 94% and 88% higher nutrient use efficiency (NUE), respectively, than an open-loop system. Among the closed-loop treatments, the 2- and 4-week correction intervals had 36% and 32% higher NUE than the no-correction treatment. Furthermore, the 2-week correction interval showed a 3% improvement in NUE compared to the 4-week interval. These findings highlight the importance of considering a shorter ion correction interval for optimal NUE and normal plant growth in recirculating hydroponic systems of strawberry cultivation.

**Keywords:** recirculating hydroponics; soilless culture; nutrient uptake; nutrient management; protected horticulture

## 1. Introduction

In greenhouse horticulture, nutrient recirculation is crucial in addressing the growing global demand for freshwater and resource conservation by reducing fertilizer use. The introduction of soilless media in greenhouse cultivation provided an opportunity to transition to a closed-loop hydroponic system, as it allows for precise nutrient control and mitigates microbial risks by reducing the use cycle to one year. These systems could potentially increase water and fertilizer efficiencies by 20–30% compared to open-loop hydroponic systems that recycle drainage water [1]. Despite these advantages, most small-scale greenhouses continue to use open-loop hydroponic systems. This is mainly due to the instability of ion concentrations in the root zone, the risk of microbial contamination from recycled drainage nutrients, and the high cost of disinfection systems.

In Korea, open-loop hydroponic systems are commonly used for strawberry production, with most growers opting for drip irrigation on a mixed substrate of coir and perlite. The smaller greenhouse size of individual strawberry growers compared to tomatoes and peppers has contributed to low adoption of closed-loop hydroponics due to nutrient imbalances, microbial risks, as well as the cost of disinfection systems. The prevalence of closed-loop hydroponics in strawberry production remains scarce, although strawberries account for 15.3% of the proportion in greenhouse fruit vegetable production (5692 ha out of a total of 37,236 ha) [2]. Even if all greenhouse vegetables are included, closed-loop hydroponics currently account for only 5% of hydroponics in the Republic of Korea.

The main commercial strawberry variety in Korea is Seolhyang, a one-season flowering variety that requires low-temperature acclimatization to grow. Thus, strawberries have a relatively short cultivation period of approximately 8–9 months, from September to May [3]. This short cultivation period leads to fewer fluctuations in nutrient levels and a relatively low risk of microbial contamination in the substrate. This can be an advantage for strawberries in managing the risk of recirculation, but it can also be a disadvantage in terms of economics, as the disinfection system is not used all year round. Another feature is that strawberry nutrient solution (NS) has a low electrical conductivity (EC) range of approximately 1.0–1.3 dS $m^{-1}$. Thus, the fertilizer savings from the recirculation are relatively small compared to tomatoes and sweet peppers. On the other hand, the low EC and nutrient requirement result in a gradual change in the EC of the substrate or drainage solution [4]. Thus, while the low EC dosage can be an advantage, poor source water quality can make it more difficult to provide a balanced supply of essential nutrients in a recirculating hydroponic system. Another problem with raw water quality is that strawberries are a salt-sensitive crop. Therefore, salt ($Na^+$, $Cl^-$) accumulation in closed-loop hydroponics could be linked to reduced strawberry growth and deficiencies in essential nutrients. The specificity of these locally grown varieties and crops demonstrates the need for empirical studies of closed-loop hydroponics.

In general, most growers perform periodic analyses of substrate and drainage EC or individual ions to maintain a consistent nutrient composition in the root zone by correcting individual ion concentrations in closed-loop systems [5]. EC-based control is a simple method used in semi-closed systems, where drainage solutions are discharged when they reach pre-set minimum or maximum thresholds for EC and saline ions (such as Na and Cl) [6]. However, EC-based control has shown a tendency to decrease root-zone ions, such as $NO_3^-$, P, and K, as cultivation progresses [7] due to increased crop water and fertilizer requirements [8]. Conversely, polyvalent ions, including $Ca^{2+}$ and $SO_4^{2-}$, tend to accumulate in the substrate [9]. Therefore, it is plausible that progressive changes in crop nutrient requirements during cultivation may lead to an accumulation of root-zone nutrient imbalances, potentially increasing the physiological disorders of fruits [10]. Therefore, it is crucial to periodically analyze the drainage ions for "zero-emission" closed-loop hydroponic systems to adjust the NS composition and prevent imbalances of individual ions in the growing substrate.

Several studies have been conducted to determine the optimal nutrient balance or ion uptake characteristics of strawberry [11–13], but recirculating hydroponics have drawn less attention. In both semi-closed and closed-loop hydroponic systems, the implementation of recirculation methods may necessitate minor alterations contingent on the properties of the plant type, cultivar, growth stage, and growing media. Coir-mixed substrate consists of a certain degree of $NO_3^-$ and Na, which may result in differences in the initial root-zone nutrient composition. Therefore, this study was conducted to establish management practices for closed-loop hydroponics for strawberries in Korea. By conducting two different drainage recirculation methods, we analyzed nutrient balance in the root zone, crop growth, yield, and nutrient use efficiency according to each method. Based on these results, we aimed to determine management variables (drainage mixing EC, individual ion correction interval, etc.) for closed-loop hydroponics.

## 2. Materials and Methods

### 2.1. Closed-Loop Hydroponics Treatment of Experiments 1 and 2

In this study, two independent experiments, Experiment (Exp) 1 and Exp 2, were conducted to examine the changes in root-zone ion concentrations resulting from the implementation of closed-loop systems for strawberry cultivation. In Exp 1, the variations in ion concentrations within the root zone were evaluated under distinct drainage mixing EC levels without any initial correction in the initial nutrient solution (NS). While recirculating the drainage solution, we analyzed the composition of the nutrient and drainage solution in comparison to the initial NS and the resulting strawberry growth and fruit yield. The nutrient recirculation amount was, therefore, controlled solely based on the drain supply EC (set value of EC mixing raw water to the drainage solution) (Figure 1). The drain was once again mixed with NS from A and B tanks using a three-way valve and supplied with the pre-set EC value. The drainage was discharged if the capacity of the drainage tank was exceeded. The objective of the Exp 1 approach was to observe the long-term growth of plants and changes in underground ion composition without individually correcting for ion levels. Four treatments were performed, each with different drainage mixing EC levels (0.6, 0.9, and 1.2 dS $m^{-1}$) and a total reuse treatment (All use). The first drainage mixing occurred on 3 November 2021, and the first ion analyses were conducted on 6 December 2021.

Exp 2 was conducted to explore an appropriate nutrient correction period by analyzing the root-zone ion balance based on different ion-correction intervals. Four treatments were performed in Exp 2: open-loop, closed-loop with 2-week correction, closed-loop with 4-week correction, and closed-loop without correction. The NS was replenished with any lacking standard nutrients in the mixed drain and raw water. The corrected NS was prepared using Savvas's nutrient replenishment method [14]. Different correction periods of 2 weeks and 4 weeks were applied to analyze the effect of the ion correction period. To maintain nutrient stability in the substrate, in the 2-week correction treatment, when a new NS prescription was implemented after ion analysis, the first week was fed with the calibrated NS and the second week with the standard NS. Similarly, for the 4-week correction treatment, the first half used the corrected NS, and the second half used the standard NS (Figure 1). Yamazaki and PBG solutions were used as NSs in Exp 1 and 2, respectively (Table 1).

**Table 1.** Ion composition of the nutrient solution in experiments 1 and 2 (Exp 1, 2).

| | | | Macronutrients (mmol $L^{-1}$) | | | | | | | | | |
|---|---|---|---|---|---|---|---|---|---|---|---|---|
| | EC (dS $m^{-1}$) | pH | $NH_4$-N | K | Ca | Mg | Na | $NO_3$-N | P | S | Cl | $HCO_3$ |
| Exp 1 | | | | | | | | | | | | |
| Raw water 1 | 0.50 | 7.7 | 0.20 | 0.20 | 1.20 | 0.30 | 1.30 | 0.20 | 0.02 | 1.00 | 0.50 | 2.00 |
| Raw water 2 | 0.43 | 7.58 | - | 0.07 | 1.18 | 0.30 | 1.23 | 0.23 | - | 1.07 | 0.48 | 1.44 |
| Solution | 1.20 | 5.80 | 0.75 | 4.50 | 3.00 | 1.15 | - | 7.50 | 0.75 | 1.50 | - | - |
| Exp 2 | | | | | | | | | | | | |
| Raw water | 0.01 | 6.54 | - | - | - | - | - | - | - | 0.01 | 0.20 | 0.10 |
| Solution | 1.20 | 5.80 | 0.75 | 3.30 | 2.75 | 1.10 | - | 8.77 | 0.75 | 1.12 | - | - |

EC: electrical conductivity and Exp: experiment.

The micronutrients were 30.00 Fe, 20.00 B, 15.00 Mn, 7.00 Zn, 0.75 Cu, and 0.50 Mo (μmol $L^{-1}$) for both experiments. In Exp 2, ion correction ($I_{Cor}$) was performed using the following function:

$$I_{Cor} = I_{NS} - (I_{Dr} + I_{RW}) \quad (1)$$

where $I_{Cor}$ = corrected ion composition, $I_{NS}$ = ion composition of the NS; $I_{Dr}$ = ion composition of the drainage solution; and $I_{RW}$ = ion composition of the raw water.

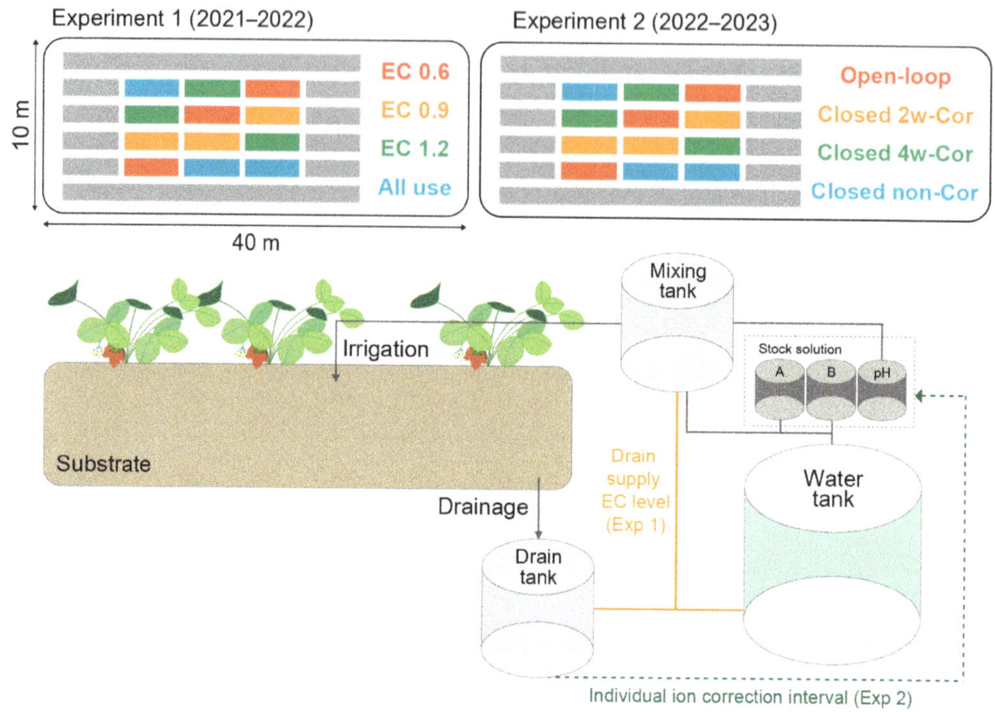


**Figure 1.** Schematic diagram of closed-loop hydroponics system and experimental set-up. Closed 2w, 4w, no cor indicate closed-loop with 2-week, 4-week interval ion correction, and no correction treatments, respectively.

Corrected NS was applied to the 3 days after ion analysis. In all experiments, the drainage solutions were filtered through sand, and 50 mg L$^{-1}$ of hydrogen peroxide was applied to the drainage tank for disinfection.

### 2.2. Preparation of Water Source

In Exp 1, groundwater was used as the source water without purification (Table 1). The results of Exp 1 showed that the salt concentration of groundwater was too high in the substrate. In Exp 2, to exclude the influence of raw water in the recirculation system and purify the groundwater, we used a membrane filter (WRO-T10, Waternix Inc., Busan, Republic of Korea) with an EC of 0.01 dS m$^{-1}$ (Table 1). In Exp 1, the raw water quality of the groundwater was measured twice at three-month intervals. The results showed that the EC was 0.5 and 0.43 dS m$^{-1}$, respectively, and contained several ions. Although the nutrient concentrations were not higher than those in the NS composition, the Na concentration was found to be 1.30 and 1.23 mmol L$^{-1}$ in the two measurements (Table 1).

### 2.3. Plant Materials and Growth Conditions

Strawberry plants (*Fragaria* × *ananassa*, cv. Seolhyang) were transplanted on 14 September 2021 and 16 September 2022 for Exp 1 and 2, respectively. The experiments were conducted in a plastic-covered greenhouse at the Protected Horticulture Research Institute, Haman, Republic of Korea (35.2° N 128.4° E). The greenhouse had a floor area of 400 m$^2$ (40 × 10 m (L, W)) and a planting density of 23 plant/m$^2$ (Figure S1A). A total of 1920 plants were grown, with 12 sectors utilized for growth analysis and ion sampling with the exclusion of the outside position in the greenhouse to consider the non-uniformity of light distribution (Figure 1). The substrate comprised a mixed substrate with a volume of 49.9% coconut coir, 35% perlite, 10% peat moss, 5% zeolite, 0.1% nutrients, pH buffer, and humectant (Greenfarm, Nongwoo Bio Co., Ltd., Suwon, Gyeonggi-do, Republic of Korea). The temperature in the greenhouse was set to 25 °C during the day and 8 °C during the night; however, due to seasonal weather changes, the minimum and maximum temperatures fluctuated between 5 °C and 37 °C. Exp 1 and 2 ended on 15 April 2022 and 29 March 2023, respectively.

Environmental data were recorded every 10 min using a CR1000 data logger (Campbell Scientific, Logan, UT, USA). The substrate was mulched with black plastic, and an irrigation pipe was placed in the center of the growing bed (Figure S1B). The nutrient solution (NS) was supplied to the plants six times per day at a rate of 33 mL/plant using a drip irrigation system. As growth progressed, the leaf area increased, and the amount of irrigation per unit time was set to 50 mL/plant. The first irrigation was conducted 1 h after sunrise, and the last irrigation was conducted 2 h before sunset. The supply EC and pH of the NS were 1.2 dS $m^{-1}$ and 6.0, respectively. The irrigation amount was adjusted to maintain drainage at 30%–40% of the water supply. The hydroponic cultivation system included supply, drainage, water, and NS tanks. Solutions A and B were mixed with raw water and supplied using a three-way valve control system with an EC meter. To collect the drainage solution, we installed a hanging gutter hydroponic system. A plastic sheet was installed on the bottom of the gutter to collect the drainage solution, and the collected drainage was designed to be moved to the back side of the greenhouse due to the difference in height of the drainage pipe (Figure S1C). Three plastic beakers were installed for each treatment for sampling drainage EC, pH analysis, and ion sampling, and the beakers were emptied daily after measuring the drainage level. EC and pH were measured with a portable EC and pH meter (HI9814N, Hanna Instruments, Smithfield, RI, USA) after the last drainage was collected at 4 pm for uniformity of the measurement.

### 2.4. Root-Zone Ion Component Analysis and Solution Sampling

All ion analyses were performed in the same manner as in Exp 1 and 2. Supply and drainage ion concentrations were measured weekly. Measurements were conducted for EC, pH, and all individual root-zone ions: $NH_4^+$-N, $NO_3^-$-N, P, K, Ca, Mg, S, Cl, Na, $HCO_3$, Fe, Mn, B, Zn, Cu, and Mo. As the concentration of the supplied NS was stable, only one repetition was conducted for the supplied NS per treatment. The anions $NO_3^-$-N, S, P, and Cl were analyzed using standard ion chromatography (IC) curves with 0, 5, 10, 20, and 40 ppm solutions. The sample and regeneration solutions were subjected to IC–mass spectrometry (MS) (ICS-5000, Dionex, Sunnyvale, CA, USA), and the baseline was stabilized (ICS-5000, Dionex) at a flow rate of 0.7 mL $min^{-1}$. The drainage solution was collected thrice per treatment at different locations on each treatment. The concentrations of metallic cations and microelements were quantified using inductively coupled plasma MS (ICP-MS; NexION 5000 Multi-Quadrupole, PerkinElmer, Waltham, MA, USA). For $NH_4^+$-N, 50 mL of a 2% boric acid solution was taken from the sample solution and connected to a collecting device. Subsequently, 25–50 mL of the blank solution was added to a Kjeldahl flask, and distilled water was added to achieve a total volume of 250~300 mL. After adding three to four drops of phenolphthalein solution, 50% NaOH or MgO was added until the colorless solution turned red. The solution was then transferred to a nitrogen evaporator and distilled. The quantification of $NO_3^-$-N was conducted via Kjeldahl (T-N) analysis, whereby a specific quantity was extracted from the blank solution to be decomposed in the $NH_4^+$-N analysis in a Kjeldahl flask and distilled into a Devarda alloy (1–2 g) [15]. All the reagents required for ion analysis were ordered from Sigma-Aldrich (St. Louis, MO, USA).

### 2.5. Analyses of Plant Growth, Yield, and Fruit Quality

Monthly analyses were conducted to assess the growth of the crops, including stem length, petiole length, leaf length, leaf width, and leaf number. Fruit yields were recorded during each harvest, and fruits were classified as marketable if they weighed $\geq$ 12 g and non-marketable if they weighed < 12 g or were deformed. Ten plants were selected from each treatment plot to examine fruit quality. Fruit quality analysis was conducted to determine fruit weight, length, width, firmness, and total soluble sugar (TSS) within eight hours of harvest. The quality analysis was performed once per week, but sometimes twice if the fruit ripened quickly and had to be harvested. Twenty plants were selected per treatment and examined in three replicates. Fruit weight and length/width were measured using weighing scales and Vernier calipers, respectively, and the TSS content was determined using a portable reflectometer (PAL-1, ATAGO Co., Ltd., Tokyo, Japan).

### 2.6. Calculation of Water and Nutrient Use Efficiency

Irrigation and drainage amounts were monitored daily via beakers placed under the cultivation bed, and water use efficiency ($WUE_{yield}$) was calculated as kg fruit yield/L by dividing the total water use by the fruit yield per treatment. Nutrient use was recorded each time by supplying the fertilizer in the NS tank, and nutrient use efficiency ($NUE_{yield}$) was calculated as kg fruit yield/kg nutrient by dividing the amount (kg) of nutrients used by the fruit yield per treatment.

### 2.7. Statistical Analyses

In the experimental setting, each treatment area was placed in the north–south direction without dividing the blocks to minimize light variation between treatments (Figure 1). Mean comparisons were performed using one-way ANOVA, and post-hoc testing was conducted using Tukey's method. All statistical analyses were conducted using the Scipy Python package (1.10.1). The pandas (1.5.3) and matplotlib (3.6.0) packages were used for data handling and visualization.

## 3. Results

### 3.1. Quality of Raw Water in Exp 1 and 2

As Na and Cl are not taken up by plants, to exclude the accumulation of Na and Cl in the substrate and the resulting nutrient imbalance, raw water was purified and used in Exp 2. As a result, most ions were not detected except for S and Cl (Table 1).

### 3.2. EC and pH Changes during Cultivation in Exp 1 and 2

Overall, EC and pH patterns were similar in all treatments in Exp 1 and 2. The EC of NS in both Exp 1 and 2 was consistently controlled (1.0–1.3 dS $cm^{-1}$) (Figure 2A,C). The drainage EC in Exp 1 started at 0.6 and remained around 1.5 until February, and finally, the drainage EC peaked at approximately 2.0 dS $cm^{-1}$ (Figure 2B). In Exp 2, the drainage EC was initially 0.4, with values of about 1.25 in the middle of January and finally 1.7 ds $cm^{-1}$ (Figure 2D). The initial pH of the NS, starting at approximately 6.5, showed a gradual decline and stabilized around 6.0. In Exp 1, the NS pH in the all-use treatment exhibited a marginally elevated value compared with the other treatments (Figure 2A). The mean pH values of the DS were 7.0 in Exp 1 and 6.6 in Exp 2 (Figure 2B,D). EC and pH were discharged from the nutrient supply system at the set values, so there were slight but not significant differences between treatments.

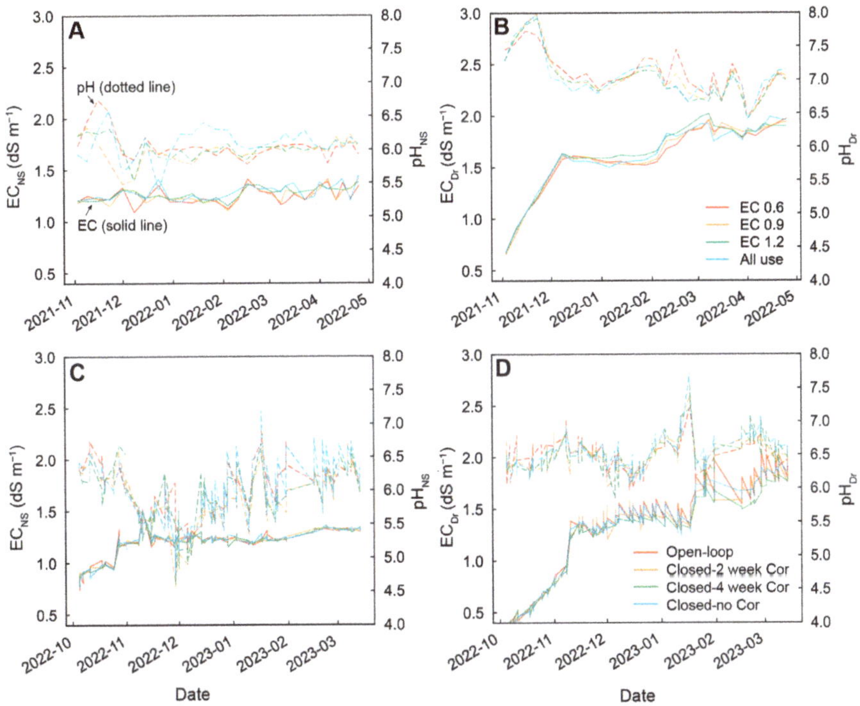


**Figure 2.** EC, pH changes of nutrient solution (NS), and drainage (Dr) during the cultivation for experiment 1 (**A**,**B**) and experiment 2 (**C**,**D**). Left and right axes indicate EC and pH and share the same ticks in all figures. Solid and dotted lines indicate EC and pH, respectively.

### *3.3. Ion Balance of NSs in Exp 1 and 2*

The equilibrium of cations and anions in the NS was analyzed by dividing the cultivation period into initial, intermediate, and final periods.

In Exp 1, the cation composition of the NS shifted towards a deficiency of K as the cultivation progressed (Figure 3A–C). Changes in Ca, Mg, and K over time (Figure S2) suggested that the K deficiency was due to an increase in Na in the NS. The anion composition of the NS rapidly shifted towards an accumulation of S and Cl and a deficiency of $NO_3^-$-N and P (Figure 3D–F). $NH_4^+$-N was found to be depleted more rapidly from the recirculation than $NO_3^-$-N (Figure S2). In the early stage, discrepancies to the initial NS were approximately 12% in $NO_3^-$-N for the EC 0.6 treatment and 15% for the other treatments. In other words, the nutrient imbalance was not as great in the initial period. However, during the middle and late stages, the EC 0.9, 1.2, and all-use treatment groups displayed depletion ranging from 31% to a maximum of 50% to the initial NS. In contrast, the EC 0.6 treatment group consistently maintained a deficiency of approximately 24% throughout the middle and late stages. Among the cations, only K exhibited a larger depletion of 14% in the EC 0.6 treatment, whereas a deficit of 18%–22% was observed in the other treatments in the late stage. The ion analysis of Exp 1 showed that monovalent ions of N, P, and K develop deficiencies in strawberries more rapidly than Ca, Mg, and S. S could accumulate in the NS because of the low uptake from the plant. In contrast to Exp 1, we continued to perform nutrient calibration in Exp 2, which resulted in all cations and anions maintaining a difference of <7% between target NS compositions, indicating a relatively balanced equilibrium in Exp 2 (Figure 3G–L).

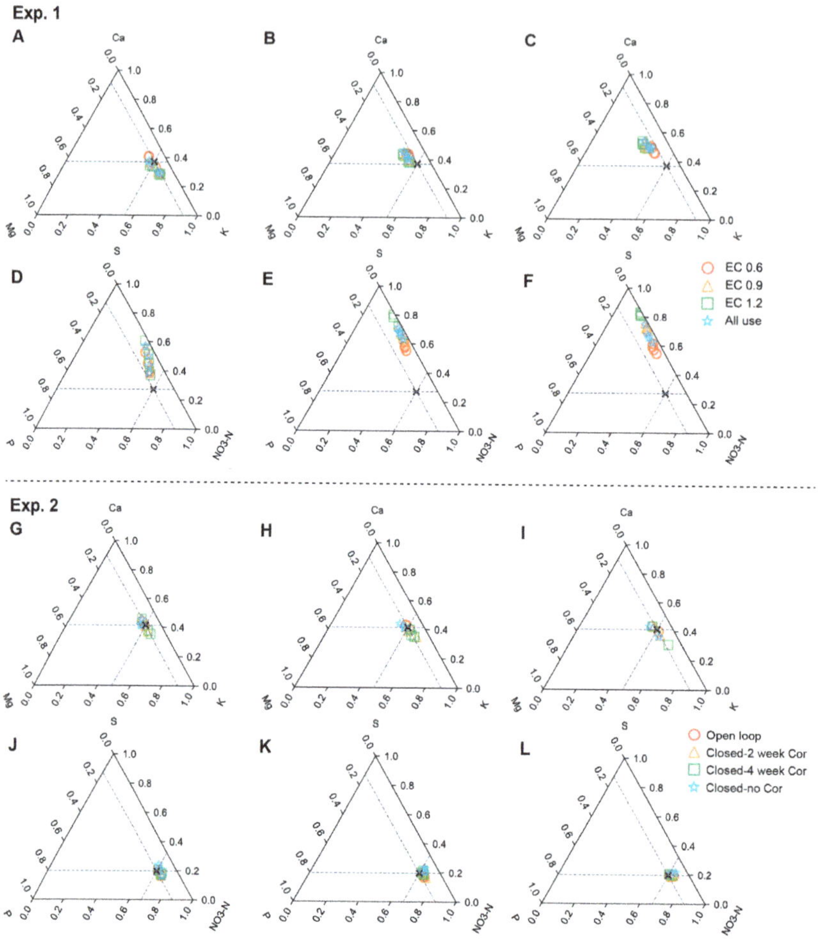


**Figure 3.** Ion balance of the nutrient solution for cations (**A–C**,**G–I**) and anions (**D–F**,**J–L**) during Experiment 1 and 2 (Exp. 1, 2). x symbol and dotted line indicate ion composition of reference nutrient solution. Panels (**A**,**D**,**G**,**J**) show the ion balance for the initial period, while panels (**B**,**E**,**H**,**K**), and (**C**,**F**,**I**,**L**) show the ion balance for intermediate and final period, respectively. Each dot represents the ion analysis results every 2 weeks.

Interestingly, even the closed-loop non-Adj processing, which does not handle correction, maintained a nearly constant nutrient equilibrium with few deficiencies in P and K (Figures 3 and S1). This suggests that there is no appreciable variation in nutrient solution and plant uptake. Therefore, the nutrient imbalance in Exp 1 suggests that the nutrient composition imbalance caused by the quality of Na and Cl in the raw water would have had a greater impact on the ion balance in the root zone.

### *3.4. Drainage Ion Changes during Cultivation in Exp 1 and 2*

Since the drainage was mixed with the NS for recirculation, the NS had a similar trend, and the ion composition of the drainage changed almost identically (Figures 4 and S1). However, since the drainage is the NS after passing through the substrate, the pattern of ions taken up by the plant was clearly visible.

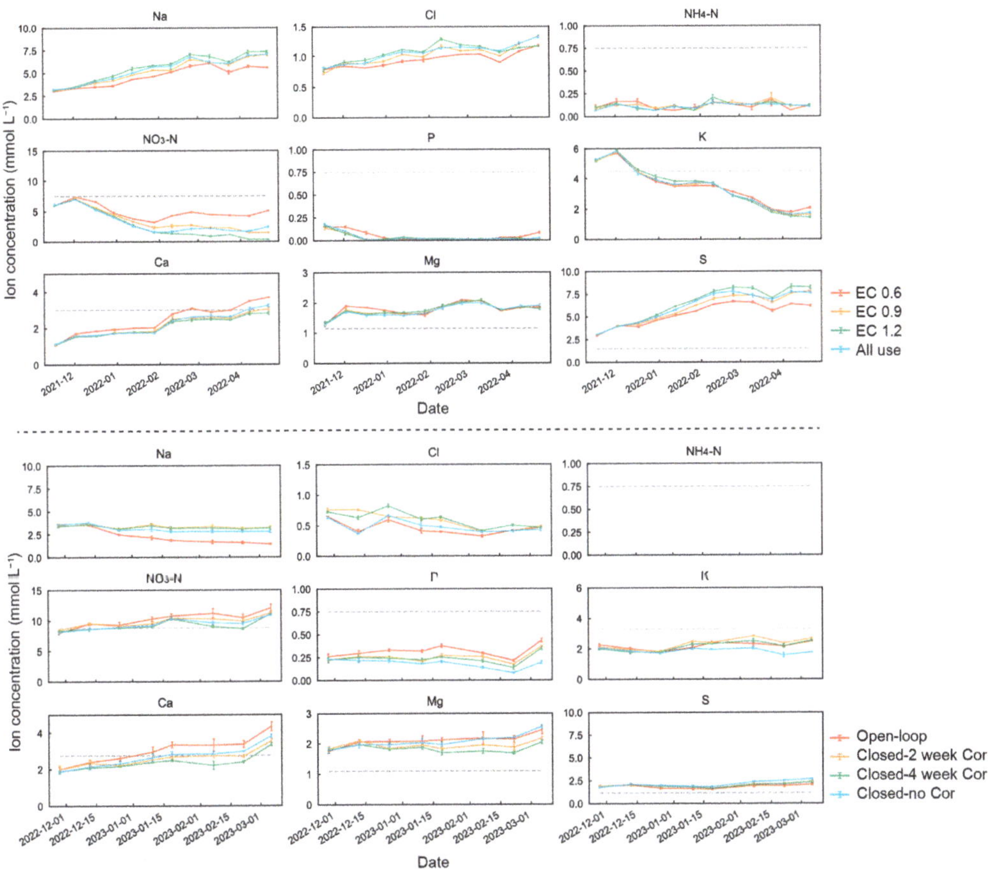


**Figure 4.** All major ion concentrations of drainage solution during cultivation period. The nine figures above and below indicate experiments 1 and 2, respectively. Vertical bars denote ± standard deviation of means (n = 3). The gray dotted lines in each figure box indicate the initial target concentration of the nutrient solution.

In Exp 1, there was a gradual accumulation of Na and Cl over the cultivation period (Figure 4). In Exp 2, because of the raw water used after reverse osmosis filtering, the levels of Na and Cl remained constant (Figure 4). Moreover, the open-loop in Exp 2 showed a lower accumulation of Na than the three closed-loop treatments, with consistent Na ion levels between 2.5 and 3.5 mmol $L^{-1}$ in the remaining treatments. This may be due to the decrease in Na eluting from the surface of the coir substrate used in the experiment, as it is increasingly replaced by Ca. In Exp 1, the imbalance of $NO_3^-$-N, P, K, and S ions was amplified with drainage reuse. $NO_3^-$-N, Ca, S, and EC 0.6 treatment, which used lower mixed EC drainage, diverged relatively less from the ion composition of the NS compared to the EC 0.9, 1.2, and all-use treatments (Figure 4). Notably, the accumulation of S showed a significant disparity, starting from 2.5 mmol $L^{-1}$ initially and escalating to 7.5 mmol $L^{-1}$ by the end of the cultivation. In Exp 2, no specific accumulation of macronutrient ions was observed in all three closed-loop treatments, including the collection treatments of 2 and 4 weeks and the non-correction treatment. Throughout the growing season, the drainage consistently had lower levels of $NH_4^+$-N, P, and K compared to the NS, whereas S and Mg levels were higher in the drainage. Moreover, $NO_3^-$-N, P, and K were higher in the drainage compared to the 2-week and 4-week correction treatments in the closed-loop treatment (Figure 4). There are no differences between the correction and non-correction treatments, maybe because the strawberries had a suitable initial NS composition and were well grown with no noticeable physiological disorder during cultivation.

### 3.5. Strawberry Fruit Yield and Quality

A higher fruit yield was observed in the EC 0.6 treatment than in the EC 1.2 and all-use treatments in Exp 1 (Figure 5A and Table 2). These results show that there may be a slight reduction in crop yield due to nutrient imbalance when All NS was reused without correction to individual ions. There is no significant difference in yield in Exp 2 (Figure 5B). In both Exp 1 and 2, fruit quality indices, including individual fruit weight, length, width, firmness, and TSS content, were not significantly different among treatments.

**Table 2.** Fruit yield and quality of strawberries grown under different hydroponic treatments.

| | Treatment | Fruit Yield (g/Plant) | Individual Fruit Weight (g/Fruit) | Fruit Length (mm/Fruit) | Fruit Width (mm/Fruit) | Firmness (kg LB/Newton) | Total Soluble Sugar (Brix°) |
|---|---|---|---|---|---|---|---|
| Exp 1 | EC 0.6 | 477 a | 31.0 a | 49.3 a | 41.2 a | 25.8 a | 10.5 a |
| | EC 0.9 | 444 ab | 30.8 a | 48.8 a | 40.8 a | 26.0 a | 10.3 a |
| | EC 1.2 | 406 b | 30.3 a | 48.5 a | 40.5 a | 26.3 a | 10.5 a |
| | All use | 422 b | 30.7 a | 49.4 a | 40.3 a | 26.0 a | 10.5 a |
| Exp 2 | Open-loop | 497 a | 36.0 a | 53.1 a | 42.6 a | 0.47 a | 10.3 a |
| | Closed 2w-Cor | 495 a | 35.3 a | 53.0 a | 42.4 a | 0.46 a | 10.2 a |
| | Closed 4w-Cor | 500 a | 35.2 a | 53.1 a | 42.6 a | 0.46 a | 10.4 a |
| | Closed non-Cor | 487 a | 35.5 a | 53.7 a | 42.8 a | 0.46 a | 10.3 a |

Different letters in the same column indicate significant differences between treatments according to Tukey's HSD test ($p < 0.05$, n = 30). Exp: experiment and EC: electrical conductivity.

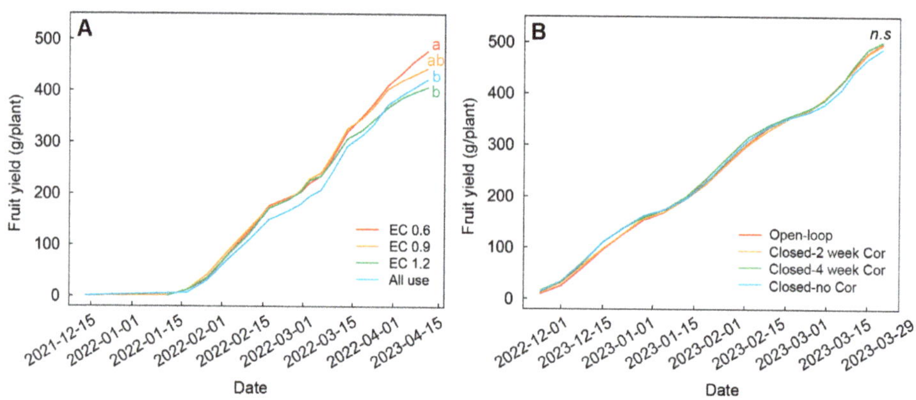


**Figure 5.** Accumulated fruit yield during cultivation period in experiment 1 (**A**) and 2 (**B**). Different letters a and b in Figure A indicate significant difference between the treatments, "*n.s*" indicates not significant ($p > 0.05$). Different letters in the figure indicate significant differences between treatments according to Tukey's HSD test ($p < 0.05$, n = 30).

### 3.6. Accumulated Water and Nutrient Use Based on Closed-Loop Hydroponics

On analyzing the water and nutrient usage in Exp 1, we observed that the volume of discharged water varied according to the drain EC level. The discharge ratios of EC 0.6, 0.9, 1.2, and all-use treatments were 14.6, 1.6, 0.7, and 0%, respectively, which resulted in different uses of total water between treatments in Exp 1 (Figure 6B). Both water-use efficiency ($WUE_{yield}$) and nutrient-use efficiency ($NUE_{yield}$) were approximately 4–6% higher in the EC 0.6 and 0.9 treatments than in the EC 1.2 and all-use treatments.

Thus, in Exp 1, the high nutrient reuse rates of EC 0.9, 1.2, and All use treatments were counteracted by the reduced yield. In Exp 2, the $NUE_{yield}$ was 3% greater in the 2-week correction treatment than in the 4-week correction treatment. However, the closed-loop 2-week and 4-week correction treatments exhibited markedly higher $NUE_{yield}$ (94% and 88%, respectively) than the open-loop treatment and showed 36% and 32% higher $NUE_{yield}$ corresponding to closed–no correction, respectively. Regarding the individual ions, Mg, $SO_4$, Ca, $NH_4^+$, $NO_3^-$, $PO_4$, and K displayed high reusability in this sequence (Figure 6A). Reusability may be achieved by decreasing the nutrient absorbency of plants within the substrate.

Accumulated nutrient use of microelements was generally higher in open-loop hydroponics but did not show a constant tendency depending on the closed-loop control method (Figure S3).

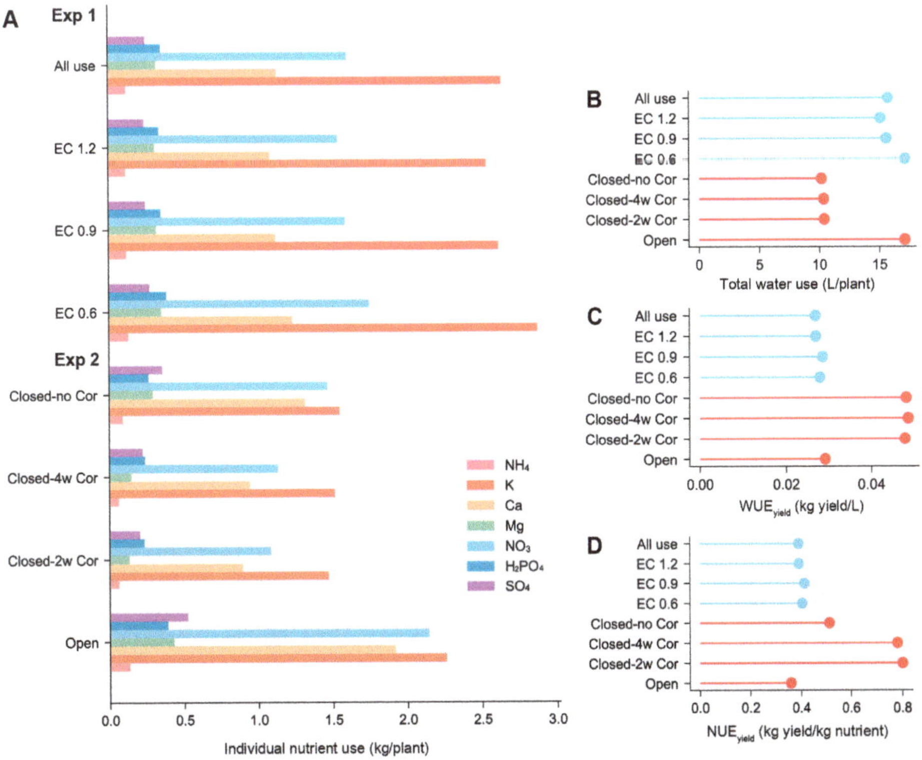


**Figure 6.** Resource use during cultivation period according to the recirculation methods in experiments 1 and 2 (Exp 1, 2). Individual nutrient use (**A**), total water use (**B**), water use efficiency ($WUE_{yield}$) (**C**), and nutrient use efficiency ($NUE_{yield}$) (**D**). The blue and red lines in (**B**–**D**) separate resource use of Exp 1 and 2, respectively. The weight of each nutrient ion was calculated based on the molar mass of the nutrient fertilizer when dissolved in water.

## 4. Discussion

In this study, we analyzed two strategies for NS management commonly used in strawberry cultivation for closed-loop hydroponics.

In Exp 1, we controlled the drainage mixing ratio, which involved adding fresh water to the drainage solution to achieve a predetermined EC [14]. This method is widely used but can lead to long-term ion depletion or accumulation in the growing substrate, causing issues for plants when there are discrepancies between nutrient uptake and the NS. The second method involved readjusting the NS composition based on the measured drainage solution's composition. This is the most used method globally [5], especially in larger greenhouses, to stabilize the nutrient composition in the root zone. The volume of nutrient modification in this method depends on various factors such as crop type, substrate type, size, and crop transpiration influenced by the greenhouse environment. As most strawberry cultivators in Asia use open-loop control, this research provides valuable data for implementing circular hydroponics specifically for strawberries. We designed Exp 1 and 2 to assess the efficacy of water and nutrient use along with evaluating the advantages of implementing closed-loop hydroponics.

### 4.1. Plant Growth According to Hydroponic Treatments

This study examined the water and NUE of strawberry crops subjected to varying water supply and drainage ion composition management methods in hydroponic cultivation. We quantitatively assessed the implications of the traditionally employed recirculation hydroponic method for crop growth and its efficacy in nutrient conservation by examining growth, yield, and fruit quality following each nutrient control regime. In Exp 1, the drainage reuse rate determined by the EC of the mixing solution was compared with that of a closed-loop system. Most agricultural practices manage the NS supply via EC due to the convenience of the EC sensor [16].

Without ion correction, the nutrient of the mixed solution can change because of the reuse of drainage water. A high set point of mixing EC could increase the mixing rate of the drainage, whereas a low drainage-mixing EC uses the solution in the drainage tank slowly. At the end of the experiment, only 14.6% of the EC 0.6 treatment had drainage discharged, so the other treatments reused almost all of the drainage solution in the tank. Thus, a higher mixing EC set point (EC 1.2, All-use treatments) encouraged most of the drainage to be used immediately as soon as it was stored in the drainage tank.

The treatments of EC 0.6, 0.9, and 1.2 were initially set up because the raw water EC was as high as 0.4~0.5 dS $m^{-1}$ (Table 1), but it seems that the venturi type of fluid supply system with a single valve opening between the tank and the 3-way-valve is not an accurate method to use on an experimental scale. As Exp 1 did not show a clear difference between EC 0.9, 1.2, and All use, the drainage dilution and NS supply method using a diaphragm pump may be considerable for more precise nutrient calibration at the experimental greenhouse scale.

The extensive use of drainage led to a decrease in strawberry yield at EC 1.2 and the All use treatment compared to EC 0.6 (Figure 5). Hence, the ion imbalance resulting from the utilization of the entire amount of drainage significantly influences crop growth. Elevated salinity levels in the drainage could contribute to the overall EC level, resulting in ion imbalance or a reduction in fruit yield (Figure 4) [17,18]. In Exp 2, there were no discernible differences in NS ion concentrations among the treatment groups (Figure 3), indicating that normal crop growth was achievable through closed-loop hydroponic cultivation by correcting the root zone nutrient composition.

Surprisingly, even the uncorrected treatment showed no difference in yield compared to the open-loop treatment. Unlike in Exp 1, where underground raw water was used, Exp 2 involved the purification process of raw water. This suggests that the quality of raw water significantly influences ion imbalance more than the initial nutrient composition. The high levels of Na, Cl, and $HCO_3$ ions in the raw water during Experiment 1 (Table 1) could have accumulated in the growing substrate, as indicated by the higher Na level in the drainage solution (Figure 4). As non-essential ions such as Na accumulate during crop growth [19], these findings emphasize the importance of drainage flushing for crop growth in EC-based hydroponic cultivation [6,20]. Additionally, purifying raw water can help alleviate the occurrence of physiological disorders or yield fluctuations in closed-loop hydroponics.

### 4.2. Root-Zone Ion Variations and Ion-Correction Intervals

In Exp 2, we expected to observe variations in the drainage ion composition patterns depending on the nutrient correction interval of 2 or 4 weeks. However, the actual drainage ion composition showed similar trends across all treatment groups (Figure 4). Crop root nutrient absorption is known to be described by transpiration, root catalytic function, or the flow-force model [21]. However, if the overall growth levels of crops are analogous, other cultivation treatments (e.g., water supply and drainage) might function as larger stochastic variables than crop absorption variables [22]. If there are no specific physiological disorders, the need for ion correction may not arise, and practical methods used by farmers, such as adjusting ion levels based on crop growth stages or weather conditions, can be considered viable alternatives.

The uptake concentrations of macronutrients such as Ca, K, and $NO_3$ in closed-loop hydroponics vary with crop type and growth stage [23,24]. Without correction, the levels of P and K slightly decreased compared to other ions (Figure 4), indicating an increased demand for these ions due to leaf area growth and fruit development. On the other hand, the relatively high levels of Mg in the drainage suggest that the initially set NS ion concentration was adequate and might have accumulated in the substrate. This trend varied slightly for each ion according to the developmental stages [25,26], highlighting the benefit of ion correction in preventing nutrient imbalances in the root zone and reducing fertilizer consumption [27]. In this study, there was no significant variability between the 2-week and 4-week calibrations, which could be attributed to the relatively low nutrient EC and low water demand of strawberry crops with small leaf areas. These findings suggest that extended ion correction periods (4 weeks or longer) may be sufficient for circular hydroponic strawberry farming, unlike taller greenhouse crops such as tomatoes and bell peppers that require increased nutrient quantities and evaporation.

### 4.3. Water and Nutrient Use Efficiency According to Each Closed-Loop Hydroponics Regime

In Exp 1, all treatments were conducted with closed-loop hydroponics, but the EC 0.6 treatment had drainage discharge when the capacity of the drainage container was exceeded, resulting in approximately 15% higher water consumption compared to the other treatment groups (Figure 6B). However, despite the higher water consumption, the EC 0.6 treatment had a relatively high yield, and there was no significant difference in $WUE_{yield}$ or $NUE_{yield}$ compared to other treatments. This indicates that for implementing a closed-loop system, crop production should not be affected by root-zone ion imbalance. However, the results of this study showed that temporary or long-term nutrient imbalance occurs when drainage solutions are simply reused, highlighting the necessity of nutrient correction during cultivation. While a system based on EC alone is simple and easy to implement, it can impact crop growth without nutrient composition correction, thus requiring correction of the drainage mixing ratio and intermittent discharge of drainage [6]. In Exp 2, the open-loop treatment discharged All drainage during the growing period, resulting in a 30–40% difference in total water use compared to the closed-loop treatments. A considerable decrease in $NUE_{yield}$ was observed in the open-loop treatment, corresponding to the amount of wasted water. Existing studies have reported NUE improvements of 35–54% in closed-loop hydroponics compared to open-loop [28], but no specific comparison between closed-loop methodologies has been conducted.

Interestingly, although water reuse was consistent across treatments, $NUE_{yield}$ was significantly better in the corrected treatments compared to the non-corrected ones, with a 30% higher efficiency observed in the corrected treatments (Figure 6). The frequency of nutrient correction through drainage composition analysis, represented by the correction interval, showed a minor 3% difference in $NUE_{yield}$ between the 2-week and 4-week correction cycles. This suggests that frequent nutrient correction based on plant absorption optimization through drainage composition analysis can enhance nutrient utilization efficiency. While the difference may seem minor, it can be effective for nutrient savings on a large scale in extensive greenhouses. In practice, crop-specific nutrient compositions are estimates based on typical fertilizer uptake, so in greenhouse cultivation, it is effective to perform chemical analysis of individual ions every fortnight to ensure that actual root zone nutrient levels are uniform [29]. Among the different nutrient types, the most frequently used ions showed a smaller nutrient-saving effect through calibration, while accumulated ions, such as Mg, $SO_4$, and Ca, yielded considerable nutrient savings through ion correction. On the other hand, $NH_4^+$, $NO_3^-$, $PO_4$, and K showed relatively less nutrient-saving effects. These findings indicate that simple reuse without nutrient correction is not sufficient for water and nutrient reuse, and nutrient correction through ion analysis enables accurate nutrient supply and approximately 30% higher nutrient use efficiency in the 2- or 4-week correction cycles compared to the non-corrected treatment. Moreover, the 2-week correction cycle demonstrated an additional nutrient-saving effect of approximately 3% compared to the 4-week cycle. This nutrient reduction effect through periodic correction is expected to have a greater impact on large-scale growers. However, this study was conducted on strawberries, which are a relatively sensitive crop to root-zone salinity (maximum root area EC to 2.0 dS $m^{-1}$), so these savings effects may be reduced if reverse osmosis purification of raw water is not conducted. This is because drainage should be discharged regularly if the raw water has a high salt concentration [30]. This study emphasizes the importance of reclaiming below-ground nutrients frequently, not only for crop growth but also for enhancing the reuse efficiency of agricultural resources through circular hydroponics. As nitrate and phosphate runoff globally reaches biologically risky levels, the study underscores the need for frequent reclamation of these nutrients [31]. When implementing a decision support system for underground ion management in commercial greenhouse horticulture, it is crucial to consider the nutrient utilization efficiency based on the root-zone ion correction cycle for each crop.

## 5. Conclusions

This study examined the growth of strawberry and the water and nutrient use efficiency when implementing the commonly used methods for greenhouse circular hydroponics systems. In the EC-based control method, the accumulation of salt affects crop production under the higher set-point of drainage mixing EC. Therefore, it was necessary to discharge drainage solutions or make nutrient corrections to ensure the normal growth of strawberries. By scrutinizing the nutrient content in the drainage every 2 weeks and correcting the nutrient solution based on these analyzed results, it was possible to maintain an equilibrium of nutrient solution. This resulted in an enhancement in nutrient use efficiency by 32–36% as opposed to cases where no nutrient corrections were implemented. Moreover, correcting the nutrient solution every two weeks showed a 3% improvement in efficiency relative to a four-week interval, suggesting that matching nutrient composition with plant absorption can lead to additional nutrient conservation. These outcomes underscore the importance of improving the decision-making processes around nutrient management in greenhouse hydroponics. Thus, we could achieve more sustainable crop production by persistently collecting data on plant nutrient consumption profiles.

**Supplementary Materials:** The following supporting information can be downloaded at: https://www.mdpi.com/article/10.3390/horticulturae10080851/s1, Figure S1: View of the greenhouse and setup for closed-loop hydroponics. Front view of the greenhouse (A), mulching plastic and irrigation lines (B), bottles for drain sampling and a return line for drainage recirculating (C); Figure S2: Biweekly measured major ion concentrations of nutrient solution during cultivation period in experiment 1 and 2 (Exp 1, 2). The gray dotted lines in each figure box indicate the target concentration of the nutrient solution; Figure S3: Accumulated micronutrients (Fe, Mn, Zn, B, Cu, Mo) consumption during cultivation period in experiment 2 (Exp 2).

**Author Contributions:** Conceptualization, M.Y.L., M.Y.R., G.L.C., S.H.K. and D.K.; methodology, D.K.; software, D.K.; validation, D.K.; formal analysis, D.K.; investigation, M.Y.L. and S.H.K.; resources, G.L.C.; data curation, D.K.; writing—original draft preparation, M.Y.L. and D.K.; writing—review and editing, D.K.; visualization, D.K.; supervision, M.Y.R.; project administration, M.Y.R.; funding acquisition, M.Y.R. All authors have read and agreed to the published version of the manuscript.

**Funding:** This work was carried out with the support of "Cooperative Research Program for Agriculture Science and Technology Development (Project No. PJ01604803)" Rural Development Administration, Republic of Korea.

**Data Availability Statement:** The original contributions presented in the study are included in the article/Supplementary Materials, further inquiries can be directed to the corresponding author.

**Conflicts of Interest:** The authors declare that they have no known competing financial interests or personal relationships that could have appeared to influence the work reported in this paper.

## References

1. Massa, D.; Magán, J.J.; Montesano, F.F.; Tzortzakis, N. Minimizing water and nutrient losses from soilless cropping in southern Europe. *Agric. Water Manag.* **2020**, *241*, 106395. [CrossRef]
2. MAFRA (Ministry of Agriculture, Food and Rural Affairs). 2022 Greenhouse Vegetable Status and Vegetable Production. 2023. Available online: https://www.mafra.go.kr/bbs/home/789/568302/artclView.do (accessed on 24 April 2024).
3. Hong, S.J.; Eum, H.L. Determination of the harvest date and ripening phase of 'Seolhyang'strawberry. *J. Bio-Environ. Control.* **2020**, *29*, 62–72. [CrossRef]
4. Lee, Y.H.; Yoon, C.S.; Park, N.I.; Yeoung, Y.R. Influence of various nutrient concentrations on the growth and yield of summer strawberry cultivars cultivated in a hydroponic system. *Hortic. Environ. Biotechnol.* **2015**, *56*, 421–426. [CrossRef]
5. Blok, C.; Voogt, W.; Barbagli, T. Reducing nutrient imbalance in recirculating drainage solution of stone wool grown tomato. *Agric. Water Manag.* **2023**, *285*, 108360. [CrossRef]
6. Massa, D.; Incrocci, L.; Maggini, R.; Bibbiani, C.; Carmassi, G.; Malorgio, F.; Pardossi, A. Simulation of crop water and mineral relations in greenhouse soilless culture. *Environ. Model. Softw.* **2011**, *26*, 711–722. [CrossRef]
7. Massa, D.; Incrocci, L.; Maggini, R.; Carmassi, G.; Campiotti, C.A.; Pardossi, A. Strategies to decrease water drainage and nitrate emission from soilless cultures of greenhouse tomato. *Agric. Water Manag.* **2010**, *97*, 971–980. [CrossRef]
8. Incrocci, L.; Massa, D.; Pardossi, A. New Trends in the Fertigation Management of Irrigated Vegetable Crops. *Horticulturae* **2017**, *3*, 37. [CrossRef]
9. Zekki, H.; Gauthier, L.; Gosselin, A. Growth, Productivity, and Mineral Composition of Hydroponically Cultivated Greenhouse Tomatoes, with or without Nutrient Solution Recycling. *J. Am. Soc. Hortic. Sci.* **1996**, *121*, 1082–1088. [CrossRef]
10. Savvas, D.; Ntatsi, G.; Passam, H.C. Plant nutrition and physiological disorders in greenhouse grown tomato, pepper and eggplant. *Eur. J. Plant Sci. Biotechnol.* **2008**, *2*, 45–61.
11. Roosta, H.R.; Afsharipoor, S. Effects of different cultivation media on vegetative growth, ecophysiological traits and nutrients concentration in strawberry under hydroponic and aquaponic cultivation systems. *Adv. Environ. Biol.* **2012**, *6*, 543–555.
12. Neocleous, D.; Savvas, D. Assessment of different strategies to balance high Mg levels in the irrigation water when preparing nutrient solution for soilless strawberry crops. *Eur. J. Hortic. Sci.* **2013**, *78*, 267–274.
13. Neocleous, D.; Savvas, D. Response of hydroponically-grown strawberry (Fragaria × ananassa Duch.) plants to different ratios of K: Ca: Mg in the nutrient solution. *J. Hortic. Sci. Biotechnol.* **2013**, *88*, 293–300. [CrossRef]
14. Savvas, D. SW—Soil and Water: Automated replenishment of recycled greenhouse effluents with individual nutrients in hydroponics by means of two alternative models. *Biosyst. Eng.* **2002**, *83*, 225–236. [CrossRef]
15. Liao, C.F.H. Devarda's Alloy Method for Total Nitrogen Determination. *Soil Sci. Soc. Am. J.* **1981**, *45*, 852–855. [CrossRef]
16. Ahn, T.I.; Son, J.E. Application of an Alternative Nutrient Replenishment Method to Electrical Conductivity-Based Closed-Loop Soilless Cultures of Sweet Peppers. *Horticulturae* **2022**, *8*, 295. [CrossRef]
17. Ehret, D.; Ho, L. The effects of salinity on dry matter partitioning and fruit growth in tomatoes grown in nutrient film culture. *J. Hortic. Sci.* **1986**, *61*, 361–367. [CrossRef]
18. Trajkova, F.; Papadantonakis, N.; Savvas, D. Comparative Effects of NaCl and $CaCl_2$ Salinity on Cucumber Grown in a Closed Hydroponic System. *HortSci* **2006**, *41*, 437–441. [CrossRef]
19. Carmassi, G.; Incrocci, L.; Maggini, R.; Malorgio, F.; Tognoni, F.; Pardossi, A. Modeling Salinity Build-Up in Recirculating Nutrient Solution Culture. *J. Plant Nut.* **2005**, *28*, 431–445. [CrossRef]
20. Giannothanasis, E.; Spanoudaki, E.; Kinnas, S.; Ntatsi, G.; Voogt, W.; Savvas, D. Development and validation of an innovative algorithm for sodium accumulation management in closed-loop soilless culture systems. *Agric. Water Manag.* **2024**, *301*, 108968. [CrossRef]
21. Le Deunff, E.; Tournier, P.H.; Malagoli, P. The Thermodynamic Flow-Force Interpretation of Root Nutrient Uptake Kinetics: A Powerful Formalism for Agronomic and Phytoplanktonic Models. *Front. Physiol.* **2016**, *7*, 243. [CrossRef]
22. Ahn, T.I.; Shin, J.H.; Son, J.E. Theoretical and Experimental Analyses of Nutrient Control in Electrical Conductivity-Based Nutrient Recycling Soilless Culture System. *Front. Plant Sci.* **2021**, *12*, 656403. [CrossRef]

23. Lee, J.Y.; Rahman, A.; Azam, H.; Kim, H.S.; Kwon, M.J. Characterizing nutrient uptake kinetics for efficient crop production during *Solanum lycopersicum* var. cerasiforme Alef. growth in a closed indoor hydroponic system. *PLoS ONE* **2017**, *12*, 0177041. [CrossRef] [PubMed]
24. Vardar, G.; Altıkatoğlu, M.; Ortaç, D.; Cemek, M.; Işıldak, İ. Measuring calcium, potassium, and nitrate in plant nutrient solutions using ion-selective electrodes in hydroponic greenhouse of some vegetables. *Biotechnol. Appl. Bioc.* **2015**, *62*, 663–668. [CrossRef]
25. Tan, X.W.; Ikeda, H.; Oda, M. The absorption, translocation, and assimilation of urea, nitrate or ammonium in tomato plants at different plant growth stages in hydroponic culture. *Sci. Hortic.* **2000**, *84*, 275–283. [CrossRef]
26. Xu, X.; Du, X.; Wang, F.; Sha, J.; Chen, Q.; Tian, G.; Zhu, Z.; Ge, S.; Jiang, Y. Effects of potassium levels on plant growth, accumulation and distribution of carbon, and nitrate metabolism in apple dwarf rootstock seedlings. *Front. Plant Sci.* **2020**, *11*, 904. [CrossRef]
27. Rufí-Salís, M.; Calvo, M.J.; Petit-Boix, A.; Villalba, G.; Gabarrell, X. Exploring nutrient recovery from hydroponics in urban agriculture: An environmental assessment. *Resour. Conserv. Recycl.* **2020**, *155*, 104683. [CrossRef]
28. Rufí-Salís, M.; Petit-Boix, A.; Villalba, G.; Sanjuan-Delmás, D.; Parada, F.; Ercilla-Montserrat, M.; Arcas-Pilz, V.; Muñoz-Liesa, J.; Rieradevall, J.; Gabarrell, X. Recirculating water and nutrients in urban agriculture: An opportunity towards environmental sustainability and water use efficiency? *J. Clean. Prod.* **2020**, *261*, 121213. [CrossRef]
29. Savvas, D.; Giannothanasis, E.; Ntanasi, T.; Karavidas, I.; Ntatsi, G. State of the Art and New Technologies to Recycle the Fertigation Effluents in Closed Soilless Cropping Systems Aiming to Maximise Water and Nutrient Use Efficiency in Greenhouse Crops. *Agronomy* **2023**, *14*, 61. [CrossRef]
30. Katsoulas, N.; Voogt, W. Recent trends in salinity control for soilless growing systems management. In Proceedings of the International Symposium on Growing Media and Soilless Cultivation, Leiden, The Netherlands, 17–21 June 2013; Volume 1034, pp. 433–442.
31. Steffen, W.; Richardson, K.; Rockström, J.; Cornell, S.E.; Fetzer, I.; Bennett, E.M.; Biggs, R.; Carpenter, S.R.; de Vries, W.; de Wit, C.A.; et al. Planetary boundaries: Guiding human development on a changing planet. *Science* **2015**, *347*, 1259855. [CrossRef]

horticulturae

MDPI

*Brief Report*

# A Comparative Study of the Influence of Soil and Non-Soil Factors on Seed Germination of Edible Salt-Tolerant Species

**Viana Castañeda-Loaiza, Maria João Rodrigues, Eliana Fernandes and Luísa Custódio** *

Centre of Marine Sciences (CCMAR/CIMAR LA), University of Algarve, Campus de Gambelas, 8005-139 Faro, Portugal; vcloaiza@ualg.pt (V.C.-L.); mjrodrigues@ualg.pt (M.J.R.); elianafernandesea@gmail.com (E.F.)

* Correspondence: lcustodio@ualg.pt

**Abstract:** Cultivating edible salt-tolerant plants (halophytes) for human consumption is increasingly important due to climate change and soil salinization, and offers sustainable agricultural solutions. Optimizing seed germination, the crucial initial stage of crop growth, is essential for enhancing crop production. This study aimed to optimize the germination of edible halophytes under greenhouse conditions, focusing on select soil (salinity and substrate) and non-soil-related factors (chemical and mechanical treatments). The target species were selected for their commercial value and included *Mesembryanthemum crystallinum* L. (crystalline iceplant), *Salicornia ramosissima* J. Woods (sea asparagus), *Medicago marina* L. (sea medick), *Ammophila arenaria* (L.) Link (European beachgrass), *Portulaca oleracea* L. (common purslane), and *Atriplex halimus* L. (Mediterranean saltbush). Salinity negatively impacted germination rates (GRs) and delayed mean germination time (MGT) across species. *P. oleracea* had the highest GR (95.6%) in coco peat under freshwater irrigation, and the shortest MGT (5.2 days). *A. halimus* did not germinate under the tested conditions. Scarification with sulfuric acid improved the GR of *M. marina* by 42.2%, while scarification with ultrasounds improved the GR of *A. arenaria* by 35.5%. Our results indicate that the choice of substrate and the application of specific treatments like scarification can significantly improve the germination of certain halophyte species under variable saline conditions.

**Keywords:** climate change; halophytes; salinization; saline agriculture; sustainability

## 1. Introduction

Freshwater is a scarce resource in most Mediterranean countries, primarily due to the agricultural sector, which accounts for 70% of global freshwater withdrawals [1,2]. Additionally, water salinization is an emerging global problem caused by anthropogenic factors such as climate change, sea-level rise, urbanization, and agriculture, as well as natural causes like tidal influences and seasonal cycles [3]. Soil salinization affects nearly 20% of cultivated land worldwide, severely limiting arable land area, reducing crop productivity, and threatening agricultural sustainability and food security [4]. Salt stress decreases the productivity of most commercial crops, restricting the necessary increase in food crop production to meet the demands of a growing population, estimated to reach an additional 2.3 billion people by 2050 [5]. To address these challenges, efforts have been made to develop more resistant crops, including of resistant cultivars, and the recruitment of salt-tolerant plant species with agronomic potential. Furthermore, achieving the United Nations Sustainable Development Goals (SDG) 2 for food security and SDG 6 for water security within the framework of the water–energy–food–ecosystem nexus necessitates a transition to diets that are both nutritious and sustainable [6,7].

Salt-tolerant plants, or halophytes, thrive in habitats with stressful climatic conditions, such as salt marshes, inland deserts, salt flats, and steppes, and can withstand soil and water salt concentrations of at least 200 $mmoL^{-1}$ [8]. Halophytes possess high biotechnological potential due to their chemical composition, which includes nutritional elements (e.g., minerals and fiber) and secondary metabolites (e.g., phenolic acids and flavonoids) with health-improving properties (e.g., antioxidant and anti-diabetic) [9,10]. Historically, halophytes have been used worldwide for food and therapeutic purposes [11], and today, several species are considered gourmet foods, including quinoa seeds (*Chenopodium quinoa* Willd.), iceplant (*Mesembryanthemum crystallinum* L.), and sea asparagus (*Sarcocornia* and *Salicornia* spp.) [12–14]. Halophytes can be cultivated in a variety of saline conditions, including marginal saline soils and greenhouse aquaponics. They hold significant economic potential as sources of food products, bioactive products with health-improving properties, animal feed, and biofuel feedstocks, while contributing to soil phytoremediation and coastal protection efforts [15]. Therefore, recruiting new species and optimizing the cultivation of halophytic species for commercial purposes is a valuable strategy to address soil and water salinization, improve growth conditions, ensure a consistent supply of nutrients, and produce high-quality, safe food products [9,16].

Seed germination is the first stage of crop growth in the plant life cycle and plays a crucial role in subsequent crop production, making it the primary stage to optimize. The germination of halophyte seeds is influenced by various environmental factors, including salinity, light, and temperature, and is species-dependent [17]. Some species exhibit reduced germination rates and significant seed dormancy, which can be overcome, for example, through the use of chemical treatments (e.g., gibberellic acid and proline) applied to *Halogeton glomeratus* (M.Bieb.) Ledeb., *Lepidium latifolium* L., and *Peganum harmala* L. [18], and physical treatments (e.g., using a 500 μm sieve and a pestle) in *Salicornia herbacea* L. [19]. Soil physical properties also affect germination, as a suitable substrate should offer optimal water availability for seed germination [20]. These properties include porosity, volume, and density of the substrate, as well as water availability for the plant [21]. There is no universal optimum substrate or mixture for all species and cultivation processes. Examples include a mixture of peat and perlite (3:1, *v*/*v*) used in the cultivation of *Limonium algarvense* Erben [22], vermicompost for *Suaeda salsa* (L.) Pall. [23], and a mixture of clay and sand for *Salicornia dolichostachya* Moss and *S. brachystachya* (GFW Meyer) König [24].

In our ongoing efforts to optimize the sustainable saline cultivation of edible halophytes, this research focused on the influence of soil-related factors (salinity and substrate) and non-soil-related factors (chemical and mechanical treatments) on the germination of seeds of six halophytic species in greenhouse conditions. The target species included *Mesembryanthemum crystallinum* L. (crystalline iceplant), *Salicornia ramosissima* J. Woods (sea asparagus), *Medicago marina* L. (sea medick), *Ammophila arenaria* (L.) Link (European beachgrass), *Portulaca oleracea* L. (common purslane), and *Atriplex halimus* L. (Mediterranean saltbush). These species were selected for their commercial interest due to their traditional uses, and nutritional and functional properties, as outlined in Table S1 (Supplementary Materials). There are few studies that have focused on the best germination conditions for the species included in this work. These studies include examining the effects of light treatments on the growth and chemical composition of *M. crystallinum* [25], investigating the impact of seed dimorphism and salinity [26], as well as bacterial inoculation [27], on the germination efficiency of *S. ramosissima*, and exploring mechanisms controlling dormancy and germination in *M. marina* [28]. Additionally, research has determined the optimal germination temperature for *A. arenaria* [29] and evaluated various factors affecting germination in *P. oleracea* [30–32] and *A. halimus* [33–35], such as site of origin, seed maturation time, seed age, salinity, and temperature. However, such studies have largely focused on germination under controlled laboratory settings, and, therefore, the practical application of these findings to commercial-scale production remains limited. Assessing the germination of edible halophytes under greenhouse conditions is critical for advancing sustainable agriculture in saline environments, since greenhouse environments offer a more realistic intermediate between laboratory and field conditions, incorporating natural variability while maintaining some control over environmental factors. Therefore, this work aims to bridge the gap by developing germination protocols that are not only scientifically robust but also scalable and applicable to commercial farming. By doing so, it addresses the urgent need for sustainable crop production methods in the face of increasing soil salinization and global food demand.

## 2. Materials and Methods

### 2.1. Seed Acquirement and Greenhouse Conditions

Seeds were purchased from seed commercial suppliers: *S. ramosissima* from Horta da Ria Lda. (Aveiro, Portugal) and *M. crystallinum*, *M. marina*, *A. arenaria* and *A. halimus*, and *P. oleracea* from Semillas Cantueso S.L., (Córdoba, Spain), stored in dry conditions in darkness at room temperature (RT, about 20 °C), and used in the assays in the same year of their collection. Experiments were conducted throughout April–June 2021 in a greenhouse located in the University of Algarve (Faro, Portugal). The average air temperature was 24.8 °C, with a minimum and maximum of 12.8 °C and 45.3 °C, respectively, and an average relative air humidity (RH) of 52.2%, with a minimum and maximum of 21.5% and 78.9%, correspondingly. Seeds were sown in styrofoam trays of 198 cavities (53 mm deep and 22.05 $cm^3$ cell capacity) filled with the corresponding substrate and water irrigation according to the experiment. All species were subjected to the same germination conditions.

### 2.2. Germination Conditions

Four replicates of 40 seeds each were sown in the corresponding substrate and irrigation water conditions. The following substrates were tested: perlite, vermiculite, coco peat, sand, coco peat and perlite (1:1 $v/v$), coco peat and vermiculite (1:1 $v/v$), perlite and vermiculite (1:1 $v/v$), and sand, organic peat and perlite (1:1:1 $v/v/v$). The salinity of the irrigation water was evaluated by using fresh water and saline water (from a well, conductivity of 20.1 mS/cm). Germination was monitored twice a week for 28 days, and seeds were considered germinated when the radicle emerged from the substrate. Two indices were used to evaluate germination: (1) the final germination percentage, expressed as germination rate (GR) and calculated as the ratio between the total number of germinated seeds and the total number of seeds, multiplied by 100, and (2) the mean germination time (MGT), calculated by Equation (1) [36]:

$$\text{MGT} = (\textstyle\sum(\text{n} \times \text{d}))/\text{N} \quad (1)$$

where n = number of seeds germinated in each interval; d = incubation period at that time point (in days); and N = total number of germinated seeds.

### 2.3. Seed Treatments

Four replicates of 20 seeds each of species with a GR lower than 50% were submitted to five treatments, as follows: (1) chemical treatment—seeds were soaked in gibberellic acid (1 g/L; Merck, Darmstadt, Germany) for 24 h at RT (approx. 20 °C); (2) chemical scarification—seeds were soaked in sulfuric acid (50%; Merck, Darmstadt, Germany) for 10 min at RT; (3) mechanical scarification—seeds were soaked in distilled water and placed for 15 min in an ultrasonic water bath (USC-TH, VWR, Darmstadt, Germany) at RT; (4) water soaking—seeds were soaked in distilled water for 24 h at RT; and (5) thermal shock—seeds were soaked in distilled water at different temperatures, namely 60 °C for 10 min and 24 h in water (W60), 75 °C for 20 min (W75), 75 °C for 20 min and at −4 °C for 20 min (W75W-4), −4 °C for 20 min and 75 °C for 20 min (W-4W75), and −4 °C for 20 min (W-4). After treatment, seeds were sown in the conditions of water salinity and substrate that allowed for the highest GR. Germination was monitored twice a week for 28 days and seeds were considered germinated when the radicle emerged from the substrate.

### 2.4. Statistical Analysis

Results were expressed as mean ± standard error of the mean (SEM). A two-way ANOVA was used to test the significance of substrate, irrigation, and their interaction on GR and MGT, and a one-way ANOVA was used to test the significance of specific treatments to improve the germination of species with a GR lower than 50%. Tukey's HSD test were performed to determine significant ($p < 0.05$) differences between individual treatments. A principal component analysis (PCA) was performed using the Pearson correlation coefficient (r) ($p < 0.05$) as an index of similarity. All the results were analyzed using the XLSTAT statistical package (v.2015.6.01.23865, Addinsoft, New York, NY, USA).

## 3. Results and Discussion

As can be seen in Table 1, GRs and MGTs varied significantly among species and cultivation conditions (one-way ANOVA, Tables S2 and S3 of the Supplementary Materials). *P. oleracea* showed the highest GR (95.6%) and the shortest MGT (5.2 days) in coco peat under freshwater irrigation, following previous findings in the same species where experiments occurred in petri dishes in controlled conditions (germination chamber) [31]. Other species showed varying GRs and MGTs depending on the substrate and salinity; for example, *M. marina* showed the best GR (11.1%) in vermiculite and *A. arenaria* (16.7%) in coco peat, both under freshwater irrigation. *M. crystallinum* reached the highest GR in a coco peat and vermiculite mix under freshwater irrigation (GR, 52.2%; MGT, 8.44 days). Similar results were obtained for *M. crystallinum* under controlled conditions in a growth chamber with different photoperiods and light intensities, cultivated in sand irrigated with freshwater and supplemented with a nutrient solution (GR, 53.75%; MGT, 3.70 days) [25]. In the case of *S. ramosissima*, the highest GR was reached in coco peat and vermiculite under freshwater irrigation (GR: 66.7%; MGT: 13.2 days). In a previous study, *S. ramosissima* seeds were germinated in petri dishes under controlled conditions and exposed to irrigation solutions with increasing salinity, which resulted in a decrease in GR of 15% for salinities higher than 342 $mmolL^{-1}$ of NaCl [26]. *M. crystallinum*, *M. marina*, and *A. arenaria* did not germinate under saltwater irrigation, which may be explained by osmotic stress, ionic toxicity, and plant adaptation to seed recollection site [37,38]. Germination response depends, to some extent, on geographic variation in environmental conditions from the original seed collection site [38]. Plants exposed to adverse salinity conditions will develop greater resilience, enabling them to thrive in extreme environments, as in the case of *C. quinoa* at 500 $mmolL^{-1}$ of NaCl with seeds from salt flats (GR about 45%) and coastal lowlands (GR about 20%) [38].

*A. halimus* did not germinate under the tested conditions, and, therefore, seed viability was checked by the tetrazolium test, an established method to score seed quality [39], revealing that not all seeds were viable. Seeds with low or nil viability are common and can be caused by different factors, including storage conditions (length and conditions of storage), the presence of incomplete embryos, and the absence of embryos [40]. *A. halimus* is a xerophytic halophyte that grows mainly in dry saline soil [41], so seeds soaked in fresh or seawater could be non-optimal conditions and the seeds would be in a dormant state, as occurs in other species of the *Atriplex* genus, such as *A. confertifolia* (Torr. & Frem.) Wats and *A. sagittata* Borkh [42,43]. Moreover, the period between seed collection and the assays (seven months) may have influenced seed germination. Research on seed storage time has shown that seeds can lose their viability after a period of 3 months for *Tamarix* species, 12 months for *Arthrocnemum indicum*, and 14 months for *Cressa cretica* [44,45]. In a previous study on the same species, a GR of 55% was achieved in seeds after 33 days in storage when germination occurred under controlled conditions under freshwater irrigation, decreasing by 10–15% after 272 days in storage [34]. ROS are continuously produced during seed storage, and, therefore, may be implicated in the loss of viability by causing oxidative damage during seed imbibition [46].

**Table 1.** Germination rates (GRs, %) and mean germination times (MGTs) of seeds of *M. crystallinum*, *S. ramosissima*, *M. marina*, *A. arenaria*, *P. oleracea*, and *A. halimus* under greenhouse cultivation.

| Substrate | Irrigation | *M. crystallinum* | | *S. ramosissima* | | *M. marina* | | *A. arenaria* | | *P. oleracea* | |
|---|---|---|---|---|---|---|---|---|---|---|---|
| | | GR | MGT | GR | MGT | GR | MGT | GR | MGT | GR | MGT |
| Perlite | Freshwater | 43.3 ± 1.5 $^{a}$ | 6.3 $^{a}$ | 52.2 ± 1.5 $^{abc}$ | 9.2 $^{b}$ | - | - | - | - | 24.4 ± 1.5 $^{d}$ | 5.0 $^{d}$ |
| | Saltwater | - | - | 5.6 ± 0.3 $^{e}$ | 9.4 $^{ab}$ | - | - | - | - | - | - |
| Vermiculite | Freshwater | 35.6 ± 3.2 $^{a}$ | 6.0 $^{a}$ | 50.0 ± 1.5 $^{abc}$ | 12.2 $^{ab}$ | 11.1 ± 0.7 $^{a}$ | 17.4 $^{ab}$ | 2.2 ± 0.3 $^{b}$ | 29.0 $^{a}$ | 62.2 ± 0.7 $^{b}$ | 5.4 $^{d}$ |
| | Saltwater | - | - | 14.4 ± 0.9 $^{de}$ | 12.6 $^{ab}$ | - | - | - | - | 1.1 ± 0.3 $^{e}$ | 13.3 $^{ab}$ |
| Coco peat | Freshwater | 40.0 ± 2.1 $^{a}$ | 6.3 $^{a}$ | 57.8 ± 1.8 $^{abc}$ | 14.2 $^{ab}$ | - | - | 17.8 ± 0.9 $^{a}$ | 20.4 $^{a}$ | 95.6 ± 0.9 $^{a}$ | 5.2 $^{d}$ |
| | Saltwater | - | - | 10.0 ± 1.2 $^{de}$ | 14.5 $^{ab}$ | - | - | - | - | 15.6 ± 0.7 $^{de}$ | 12.2 $^{bc}$ |
| Sand | Freshwater | 35.6 ± 2.2 $^{a}$ | 10.6 $^{a}$ | 57.8 ± 0.9 $^{abc}$ | 16.7 $^{ab}$ | 5.6 ± 0.3 $^{ab}$ | 15.0 $^{bc}$ | - | - | 47.8 ± 1.2 $^{bc}$ | 5.5 $^{d}$ |
| | Saltwater | - | - | 45.6 ± 2.0 $^{abc}$ | 17.1 $^{ab}$ | - | - | - | - | - | - |
| Coco peat, perlite (1:1 *v*/*v*) | Freshwater | 38.9 ± 1.5 $^{a}$ | 6.2 $^{a}$ | 35.6 ± 1.7 $^{bcd}$ | 11.5 $^{ab}$ | 6.7 ± 0 $^{ab}$ | 21.3 $^{a}$ | - | - | 83.3 ± 0.6 $^{a}$ | 5.2 $^{d}$ |
| | Saltwater | - | - | 31.1 ± 1.2 $^{cde}$ | 12.9 $^{ab}$ | - | - | - | - | - | - |
| Coco peat, vermiculite (1:1 *v*/*v*) | Freshwater | 52.2 ± 2.2 $^{a}$ | 8.4 $^{a}$ | 66.7 ± 2.6 $^{a}$ | 13.2 $^{ab}$ | 2.2 ± 0.3 $^{b}$ | 10.5 $^{cd}$ | 16.7 ± 0.6 $^{a}$ | 22.8 $^{a}$ | 64.4 ± 0.9 $^{b}$ | 5.0 $^{d}$ |
| | Saltwater | - | - | 33.3 ± 2.1 $^{cde}$ | 17.5 $^{a}$ | - | - | - | - | 1.1 ± 0.3 $^{e}$ | 19.0 $^{a}$ |
| Perlite, vermiculite (1:1 *v*/*v*) | Freshwater | 45.6 ± 1.2 $^{a}$ | 6.0 $^{a}$ | 57.8 ± 0.7 $^{abc}$ | 10.4 $^{ab}$ | 4.4 ± 0.3 $^{b}$ | 9.5 $^{d}$ | 5.6 ± 0.3 $^{b}$ | 18.3 $^{a}$ | 44.4 ± 1.8 $^{c}$ | 6.9 $^{cd}$ |
| | Saltwater | - | - | 50.0 ± 2.9 $^{abc}$ | 15.8 $^{ab}$ | - | - | - | - | - | - |
| Sand, organic peat, perlite (1:1:1 *v*/*v*/*v*) | Freshwater | 30.0 ± 1.7 $^{a}$ | 7.9 $^{a}$ | 62.2 ± 0.9 $^{ab}$ | 9.9 $^{ab}$ | 1.1 ± 0.3 $^{b}$ | 9.9 $^{d}$ | - | - | 43.3 ± 1.2 $^{c}$ | 5.4 $^{d}$ |
| | Saltwater | - | - | 55.6 ± 0.9 $^{abc}$ | 14.0 $^{ab}$ | - | - | - | - | - | - |

-: Species without germination. *A. halimus* did not germinate under the greenhouse conditions tested. Values represent the mean ± standard error of the mean (SEM) of at least three experiments performed in triplicate ($n$ = 30). In the same column, values followed by different letters are significantly different at $p < 0.05$ (Tukey HSD test).

PCA analysis revealed that overall, salinity negatively impacted germination rates and delayed MGT across the species (Figure 1), which aligns with previous studies on other halophytes, such as *Cakile maritima*, *C. quinoa*, and different *Salicornia* and *Sarcocornia* ecotypes [37,47,48]. For most halophytes, germination and seedling emergence are strongly inhibited at salt concentrations much lower than those endured by adult plants, since at the adult stage, plants develop morphological, physiological, and biochemical adaptations to salt stress in the whole plant or tissues [10]. In the wild, saline conditions can inhibit the germination of halophytic species, leading to an excessive production of reactive oxygen (ROS) and nitrogen (RNS) species [49,50]. Under normal production, ROS/RNS react as secondary messengers in a variety of cellular processes, but their excessive production in saline conditions, for example, results in oxidative damage to lipids, proteins, and DNA, ultimately resulting in cell death [51]. To regulate the excessive production of ROS/RNS, plant cells rely on efficient enzymic and nonenzymic antioxidant mechanisms [52]. Therefore, seed germination in hostile environments, including saline settings, relies on the efficiency of the antioxidant defense system and varies with species and type of stress [53]. However, seeds can remain viable for long periods and germinate when conditions are favorable, typically in early spring when rain reduces soil salinity and temperatures rise. In contrast, glycophytic seeds only survive for short periods under high salinity [45].

The PCA biplot indicates a differential response to substrates and salinity levels, with coco peat and vermiculite mixes generally showing positive correlations with higher germination rates (Figure 1). Coco peat is made from shredded coconut husks and combines the water retention of vermiculite with the air retention of perlite [54]. Considering that the most essential step during germination is the uptake of water (imbibition) by the seeds, coco peat would allow the correct hydration required for embryo growth and endosperm rupture [55].

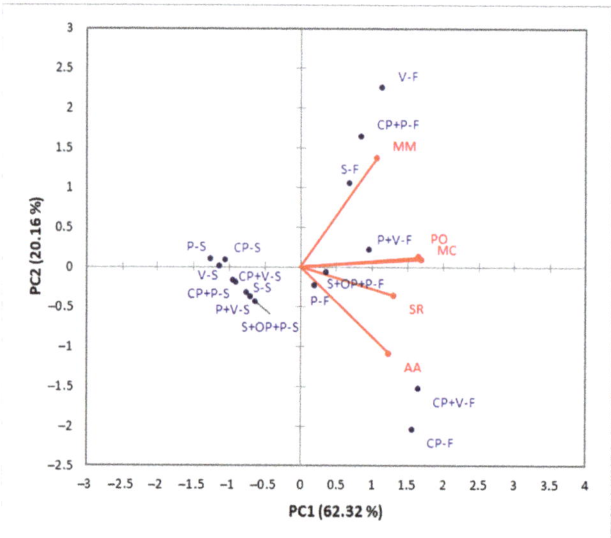


**Figure 1.** Principal components analysis (PCA) biplot of the germination of halophytes cultivated in different conditions of irrigation salinity and substrate. The graph represents the loadings on PC1 and PC2 of the parameters and the data scores of the samples. The red letters correspond to the studied species. MC: *M. crystallinum*; SR: *S. ramosissima*; MM: *M. marina*; AA: *A. Arenaria*; PO: *P. oleracea*. The blue letters correspond to the substrate followed by the irrigation water (substrate-irrigation water). Substrates: P: perlite; V: vermiculite; CP: coco peat; S: sand; CP + P: coco peat + perlite (1:1 *v*/*v*); CP + V: coco peat + vermiculite (1:1 *v*/*v*); P + V: perlite + vermiculite (1:1 *v*/*v*); S + OP + P: sand, organic peat + perlite (1:1:1 *v*/*v*/*v*). Irrigation water: F: freshwater; S: saltwater.

Seed treatments were applied to *M. marina* and *A. arenaria*, both with GRs lower than 50% (Table 2). Scarification with sulfuric acid increased the GR of *M. marina* by 42.2% (one-way ANOVA, Table S4 of the Supplementary Materials). In a previous study on the germination of *M. marina* under controlled conditions and freshwater irrigation, the highest GR was obtained after mechanical scarification (98%), followed by freezing at −20 °C (93%) and heating (25%), compared with the control (5%) [28]. Seeds from *M. marina* exhibit a strong dormancy due to the presence of a hard seed coat that prevents germination [28]. This coat-imposed dormancy, known as hardseededness, prevents seed water uptake, so that germination and the development of young seedlings occurs when evaporation and soil salinity decrease after the first rains [56,57]. Chemical scarification by sulfuric acid damages the surface of the seed coat, which increases the absorption of water and oxygen by the seeds [58,59], therefore improving germination. Scarification with ultrasounds was beneficial for *A. arenaria* (35.5% improvement). The use of ultrasounds is a simple, quick, and non-destructive method to break seed dormancy, causing fissures in the protective coating surrounding the pericarp (seed coat) and seed, increasing seedling moisture [60]. This system has been applied to break dormancy of seeds from different species, such as those from *Atriplex lentiformis* (Torr.) S. Watson, *Cuminum cyminum* L., *Zygophyllum eurypterum* Boiss. & Buhse, *Hordeum* spp (barley), *Thymus vulgaris* L. (thymes) and *Helianthus annuus* L. (sunflower) seeds [61–64].

**Table 2.** Germination rates (GRs, %) and improvement rates (IRs, %) in relation to the control group (without treatment) of seeds of *M. marina* and *A. arenaria* after the application of germination improvement treatments.

| Type of Treatment | Specific Treatment | *M. marina* | | *A. arenaria* | |
|---|---|---|---|---|---|
| | | GR | RI | GR | RI |
| Chemical treatment | Gibberellic acid 1 g/L, 24 h | 35.0 ± 1.2 [a] | +23.9 | 26.7 ± 1.9 [bcd] | +8.9 |
| Chemical scarification | Sulfuric acid 50%, 10 min | 53.3 ± 1.7 [a] | +42.2 | 26.7 ± 1.3 [abcd] | +8.9 |
| Mechanical scarification | Ultrasounds, 15 min | 25.0 ± 1.2 [a] | +13.9 | 53.3 ± 1.2 [a] | +35.5 |
| Soaking | Water at RT, 24 h | 23.3 ± 0.7 [a] | +12.2 | 30.0 ± 1.2 [abc] | +12.2 |
| | W60 | 36.7 ± 1.3 [a] | +25.6 | 20.0 ± 0.6 [d] | +2.2 |
| | W75 | 30.0 ± 1.0 [a] | +18.9 | 31.7 ± 1.7 [abc] | +13.9 |
| Thermal shock | W75W-4 | 38.3 ± 0.7 [a] | +27.2 | 53.3 ± 1.5 [ab] | +35.5 |
| | W-4W75 | 40.0 ± 1.7 [a] | +28.9 | 23.3 ± 0.9 [cd] | +5.5 |
| | W-4 | 26.7 ± 1.2 [a] | +15.6 | 1.7 ± 0.3 [d] | −16.1 |

Values represent the mean ± standard error of the mean (SEM) of at least three experiments, each performed in triplicate ($n$ = 30). In the same column, GR values followed by different letters are significantly different at $p <$ 0.05 (Tukey HSD test).

Table 3 presents a summary of the best germination conditions, including the specific treatment to improve germination if applicable, for each halophyte studied. It is important to note that achieving an optimized germination rate does not necessarily guarantee the successful growth and development of seedlings, especially under saline conditions [65]. Therefore, further optimization studies are required for subsequent cultivation phases to ensure the viability and healthy development of the plants.

**Table 3.** Best germination conditions for the studied halophyte species.

| Species | Irrigation | Substrate | Treatment |
|---|---|---|---|
| *Mesembryanthemum crystallinum* L. | Freshwater | Coco peat, vermiculite (1:1 $v/v$) | n.a. |
| *Salicornia ramosissima* J. Woods | Freshwater | Coco peat, vermiculite (1:1 $v/v$) | n.a. |
| *Medicago marina* L. | Freshwater | Vermiculite | Sulfuric acid 50%, 10 min |
| *Ammophila arenaria* (L.) Link | Freshwater | Coco peat | Ultrasounds, 15 min |
| *Portulaca oleracea* L. | Freshwater | Coco peat | n.a. |
| *Atriplex halimus* L. | No germination | No germination | No improvement |

n.a. Not applicable.

## 4. Conclusions

This study underscores the importance of optimizing germination conditions for halophytes to ensure successful saline agriculture under greenhouse conditions, making it feasible for large commercial-scale crop production. Freshwater irrigation and specific substrates, particularly coco peat combined with vermiculite for *M. crystallinum* (GR, 52.2%) and *S. ramosissima* (GR, 66.7%), significantly enhanced germination rates. Seed treatments, such as scarification with sulfuric acid for *M. marina* (GR, 53.3%) and ultrasounds for *A. arenaria* (GR, 53.3%), can further improve germination for species with low initial rates. The results obtained from this work provide a preliminary foundation for optimizing the early stages of the cultivation process. Ongoing cultivation trials are being conducted to refine the saline cultivation methods for selected species identified in this study. These trials aim to enhance biomass production and improve the biochemical profiles of the plants.

**Supplementary Materials:** The following supporting information can be downloaded at: https://www.mdpi.com/article/10.3390/horticulturae10080872/s1, Table S1: Traditional uses, and nutritional and functional properties of *Mesembryanthemum crystallinum* L., *Salicornia ramosissima* J. Woods, *Medicago marina* L, *Ammophila arenaria* (L.) Link, *Portulaca oleracea* L., and *Atriplex halimus* L.; Table S2: Two-way ANOVA of effects of irrigation, salinity, and their interactions on MGT in F- values at 0.05 level; Table S3: Two-way ANOVA of effects of irrigation, salinity, and their interactions on seed MGT in F-values at 0.05 level; Table S4: One-way ANOVA of effects of treatment to improve seed germination in F-values at 0.05 level. References [9,66–80] are cited in Supplementary Materials file.

**Author Contributions:** Conceptualization, V.C.-L., M.J.R., E.F. and L.C.; supervision, L.C.; investigation, V.C.-L.; methodology, V.C.-L.; formal analysis, V.C.-L.; writing—original draft preparation, V.C.-L.; writing—review and editing, V.C.-L., M.J.R., E.F. and L.C. All authors have read and agreed to the published version of the manuscript.

**Funding:** This study received Portuguese national funds from FCT—Foundation for Science and Technology through projects UIDB/04326/2020 (DOI:10.54499/UIDB/04326/2020), UIDP/04326/2020 (DOI:10.54499/UIDP/04326/2020), LA/P/0101/2020 (DOI:10.54499/LA/P/0101/2020) and PTDC/BAA-AGR/1391/2020. Funding was also received from Fundo Azul, by the XtremeAquaCrops project (FA_05_2017_028). This work is included in the HaloFarMs project (PRIMA/0002/2019), which is part of the Partnership on Research and Innovation in the Mediterranean Area. Luísa Custódio was supported by the FCT Scientific Employment Stimulus (CEEC-IND/00425/2017); Viana Castañeda-Loaiza by the FCT PhD grant (2020.04541.BD); Maria João Rodrigues by the FCT program contract (UIDP/04326/2020); and Eliana Fernandes by the FCT PhD grant (UI/BD/151301/2021). This work was also conducted under the framework of COST Action CA22144, titled 'Sustainable Use of Salt-Affected Land' (SUSTAIN), with support from COST (European Cooperation in Science and Technology).

**Data Availability Statement:** The original contributions presented in the study are included in the article/Supplementary Materials, further inquiries can be directed to the corresponding author.

**Conflicts of Interest:** The authors declare no conflicts of interest.

## References

1. Soligno, I. Spatial and Temporal Patterns of Global Freshwater Use for Food Production and Trade. Ph.D. Thesis, Graduate School of Politecnico di Torino, Torino, Italy, 2019.
2. Vanham, D.; Guenther, S.; Ros-Baró, M.; Bach-Faig, A. Which diet has the lower water footprint in Mediterranean countries? *Resour. Conserv. Recycl.* **2021**, *171*, 105631. [CrossRef] [PubMed]
3. Kaushal, S.S.; Likens, G.E.; Pace, M.L.; Reimer, J.E.; Maas, C.M.; Galella, J.G.; Utz, R.M.; Duan, S.; Kryger, J.R.; Yaculak, A.M. Freshwater salinization syndrome: From emerging global problem to managing risks. *Biogeochemistry* **2021**, *154*, 255–292. [CrossRef]
4. Liu, L.; Wang, B. Protection of halophytes and their uses for cultivation of saline-alkali soil in China. *Biology* **2021**, *10*, 353. [CrossRef]
5. FAO; UN. How to Feed the World in 2050. In Proceedings of the High Level Expert Forum—How to Feed the World in 2050, Rome, Italy, 12–13 October 2009.

6. Vanham, D.; Leip, A.; Galli, A.; Kastner, T.; Bruckner, M.; Uwizeye, A.; Van Dijk, K.; Ercin, E.; Dalin, C.; Brandão, M. Environmental footprint family to address local to planetary sustainability and deliver on the SDGs. *Sci. Total Environ.* **2019**, *693*, 133642. [CrossRef]
7. FAO/WHO. Sustainable Healthy Diets–Guiding Principles. In Proceedings of the Sustainable Healthy Diets–Guiding Principles, Rome, Italy, 29 October 2019.
8. Song, J.; Wang, B. Using euhalophytes to understand salt tolerance and to develop saline agriculture: Suaeda salsa as a promising model. *Ann. Bot.* **2015**, *115*, 541–553. [CrossRef] [PubMed]
9. Barreira, L.; Resek, E.; Rodrigues, M.J.; Rocha, M.I.; Pereira, H.; Bandarra, N.; da Silva, M.M.; Varela, J.; Custódio, L. Halophytes: Gourmet food with nutritional health benefits? *J. Food Compos. Anal.* **2017**, *59*, 35–42. [CrossRef]
10. Castañeda-Loaiza, V.; Oliveira, M.; Santos, T.; Schüler, L.; Lima, A.R.; Gama, F.; Salazar, M.; da Rosa Neng, N.; Nogueira, J.M.F.; Varela, J. Wild vs. cultivated halophytes: Nutritional and functional differences. *Food Chem.* **2020**, *333*, 127536. [CrossRef]
11. Petropoulos, S.A.; Karkanis, A.; Martins, N.; Ferreira, I.C.F.R. Halophytic herbs of the Mediterranean basin: An alternative approach to health. *Food Chem. Toxicol.* **2018**, *114*, 155–169. [CrossRef]
12. Custódio, L.; Rodrigues, M.J.; Pereira, C.G.; Castañeda-Loaiza, V.; Fernandes, E.; Standing, D.; Neori, A.; Shpigel, M.; Sagi, M. A review on *Sarcocornia* species: Ethnopharmacology, nutritional properties, phytochemistry, biological activities and propagation. *Foods* **2021**, *10*, 2778. [CrossRef]
13. Sánchez-Faure, A.; Calvo, M.M.; Pérez-Jiménez, J.; Martín-Diana, A.B.; Rico, D.; Montero, M.P.; del Carmen Gómez-Guillén, M.; López-Caballero, M.E.; Martínez-Alvarez, O. Exploring the potential of common iceplant, seaside arrowgrass and sea fennel as edible halophytic plants. *Food Res. Int.* **2020**, *137*, 109613. [CrossRef]
14. Sezgin, A.C.; Sanlier, N. A new generation plant for the conventional cuisine: Quinoa (*Chenopodium quinoa* Willd.). *Trends Food Sci. Technol.* **2019**, *86*, 51–58. [CrossRef]
15. Ksouri, R.; Ksouri, W.M.; Jallali, I.; Debez, A.; Magné, C.; Hiroko, I.; Abdelly, C. Medicinal halophytes: Potent source of health promoting biomolecules with medical, nutraceutical and food applications. *Crit. Rev. Biotechnol.* **2012**, *32*, 289–326. [CrossRef]
16. Hamed, K.B.; Castagna, A.; Ranieri, A.; Garcia-Caparros, P.; Santin, M.; Hernandez, J.A.; Espin, G.B. Halophyte based Mediterranean agriculture in the contexts of food insecurity and global climate change. *Environ. Exp. Bot.* **2021**, *191*, 104601. [CrossRef]
17. Gairola, S.; Hameed, A.; Rasheed, A.; Alketbi, A.; Aljasmi, M.; El-Keblawy, A. Seed germination and salinity tolerance of habitat-indifferent halophytes as associated with geographical distribution. *Seed Sci. Technol.* **2022**, *50*, 125–140. [CrossRef]
18. Ahmed, M.Z.; Gulzar, S.; Khan, M.A. Role of Dormancy Regulating Chemicals in Alleviating the Seed Germination of Three Playa Halophytes. *Ekoloji Derg.* **2014**, *23*, 1–7. [CrossRef]
19. Zhang, S.; Yu, S.; Zhang, F.; Si, L.; Yu, Y.; Zhao, X.; Zhang, D.; Wang, W. Inheritance of downy mildew resistance at different developmental stages in Chinese cabbage via the leaf disk test. *Hortic. Environ. Biotechnol.* **2012**, *53*, 397–403. [CrossRef]
20. Di Lorenzo, R.; Pisciotta, A.; Santamaria, P.; Scariot, V. From soil to soil-less in horticulture: Quality and typicity. *Ital. J. Agron.* **2013**, *8*, e30. [CrossRef]
21. Bilderback, T.E.; Warren, S.L.; Owen, J.S.; Albano, J.P. Healthy substrates need physicals too! *Horttechnology* **2005**, *15*, 747–751. [CrossRef]
22. Rodrigues, M.J.; Monteiro, I.; Castañeda-Loaiza, V.; Placines, C.; Oliveira, M.C.; Reis, C.; Caperta, A.D.; Soares, F.; Pousão-Ferreira, P.; Pereira, C. Growth performance, in vitro antioxidant properties and chemical composition of the halophyte *Limonium algarvense* Erben are strongly influenced by the irrigation salinity. *Ind. Crops Prod.* **2020**, *143*, 111930. [CrossRef]
23. Cai, J.-F.; Liu, X.-S.; Sun, K.; Jiang, F.; Zhang, M.-X.; Li, H.-L.; Yu, F.-H. Vermicompost Application Enhances Halophyte *Suaeda salsa* Performance and Improves Coastal Saline Soil Quality. *J. Soil Sci. Plant Nutr.* **2022**, *22*, 294–305. [CrossRef]
24. Rozema, J.; Van Der List, J.C.; Schat, H.; van Diggelen, J.; Broekman, R.A. Ecophysiological response of *Salicornia dolichostachya* and *Salicornia brachystachya* to seawater inundation. In *Vegetation between Land and Sea*; Springer: Berlin/Heidelberg, Germany, 1987; pp. 180–186.
25. Weeplian, T.; Yen, T.-B.; Ho, Y.-S. Growth, development, and chemical constituents of edible ice plant (*Mesembryanthemum crystallinum* L.) produced under combinations of light-emitting diode lights. *HortScience* **2018**, *53*, 865–874. [CrossRef]
26. Ameixa, O.M.C.C.; Marques, B.; Fernandes, V.S.; Soares, A.M.V.M.; Calado, R.; Lillebø, A.I. Dimorphic seeds of *Salicornia ramosissima* display contrasting germination responses under different salinities. *Ecol. Eng.* **2016**, *87*, 120–123. [CrossRef]
27. Figueira, C.; Ferreira, M.J.; Silva, H.; Cunha, A. Improved germination efficiency of *Salicornia ramosissima* seeds inoculated with Bacillus aryabhattai SP1016-20. *Ann. Appl. Biol.* **2019**, *174*, 319–328. [CrossRef]
28. Scippa, G.S.; Petrollini, E.; Trupiano, D.; Rocco, M.; Falco, G.; Di Michele, M.; Chiatante, D. Dormancy of *Medicago marina* (L.) seed. *Environ. Exp. Bot.* **2011**, *72*, 320–329. [CrossRef]
29. Bendimered, F.Z.; Mehdadi, Z.; Benhassaini, H. Étude de la germination et de la croissance foliaire de l'oyat (*Ammophila arenaria* (L.) Link) en conditions contrôlées. *Acta Bot. Gall.* **2007**, *154*, 129–140. [CrossRef]
30. El-Keblawy, A.; Al-Ansari, F. Effects of site of origin, time of seed maturation, and seed age on germination behavior of *Portulaca oleracea* from the Old and New Worlds. *Can. J. Bot.* **2000**, *78*, 279–287.
31. Chauhan, B.S.; Johnson, D.E. Seed germination ecology of *Portulaca oleracea* L.: An important weed of rice and upland crops. *Ann. Appl. Biol.* **2009**, *155*, 61–69. [CrossRef]
32. Franco, J.A.; Cros, V.; Vicente, M.J.; Martínez-Sánchez, J.J. Effects of salinity on the germination, growth, and nitrate contents of purslane (*Portulaca oleracea* L.) cultivated under different climatic conditions. *J. Hortic. Sci. Biotechnol.* **2011**, *86*, 1–6. [CrossRef]

33. Shaygan, M.; Baumgartl, T.; Arnold, S. Germination of *Atriplex halimus* seeds under salinity and water stress. *Ecol. Eng.* **2017**, *102*, 636–640. [CrossRef]
34. Osman, A.E.; Ghassali, F. Effects of storage conditions and presence of fruiting bracts on the germination of *Atriplex halimus* and *Salsola vermiculata*. *Exp. Agric.* **1997**, *33*, 149–155. [CrossRef]
35. Hamza, L.; Mederbal, K.; Regagba, Z.; Dahou, A.A.; Medjahed, M. Effect of Salinity and Temperature on Seed Germination of *Atriplex halimus* L.(*halimus* and *schweinfurthii* subspecies) Harvested in Western Algerian Region. *Indian J. Agric. Res.* **2023**, *57*, 435–441. [CrossRef]
36. Brenchley, J.L.; Probert, R.J. Seed germination responses to some environmental factors in the seagrass *Zostera capricorni* from eastern Australia. *Aquat. Bot.* **1998**, *62*, 177–188. [CrossRef]
37. Debez, A.; Ben Hamed, K.; Grignon, C.; Abdelly, C. Salinity effects on germination, growth, and seed production of the halophyte *Cakile maritima*. *Plant Soil.* **2004**, *262*, 179–189. [CrossRef]
38. Ruiz, K.B.; Aloisi, I.; Del Duca, S.; Canelo, V.; Torrigiani, P.; Silva, H.; Biondi, S. Salares versus coastal ecotypes of quinoa: Salinity responses in Chilean landraces from contrasting habitats. *Plant Physiol. Biochem.* **2016**, *101*, 1–13. [CrossRef] [PubMed]
39. Grzybowski, C.R.d.S.; Ohlson, O.d.C.; Silva, R.C.d.; Panobianco, M. Viability of barley seeds by the tetrazolium test. *Rev. Bras. Sementes* **2012**, *34*, 47–54. [CrossRef]
40. Zhang, K.; Zhang, Y.; Sun, J.; Meng, J.; Tao, J. Deterioration of orthodox seeds during ageing: Influencing factors, physiological alterations and the role of reactive oxygen species. *Plant Physiol. Biochem.* **2021**, *158*, 475–485. [CrossRef] [PubMed]
41. Arora, S.; Rao, G.G. Bio-amelioration of salt-affected soils through halophyte plant species. In *Bioremediation of Salt Affected Soils: An Indian Perspective*; Springer: Berlin/Heidelberg, Germany, 2017; pp. 71–85. ISBN 9783319482576.
42. Mandák, B. Germination requirements of invasive and non-invasive *Atriplex* species: A comparative study. *Flora-Morphol. Distrib. Funct. Ecol. Plants* **2003**, *198*, 45–54. [CrossRef]
43. Mikhiel, G.S.; Meyer, S.E.; Pendleton, R.L. Variation in germination response to temperature and salinity in shrubby *Atriplex* species. *J. Arid Environ.* **1992**, *22*, 39–49. [CrossRef]
44. Terrones, A.; Moreno, J.; Agulló, J.C.; Villar, J.L.; Vicente, A.; Alonso, M.Á.; Juan, A. Influence of salinity and storage on germination of *Tamarix* taxa with contrasted ecological requirements. *J. Arid Environ.* **2016**, *135*, 17–21. [CrossRef]
45. Gul, B.; Ansari, R.; Flowers, T.J.; Khan, M.A. Germination strategies of halophyte seeds under salinity. *Environ. Exp. Bot.* **2013**, *92*, 4–18. [CrossRef]
46. Bailly, C.; El-Maarouf-Bouteau, H.; Corbineau, F. From intracellular signaling networks to cell death: The dual role of reactive oxygen species in seed physiology. *C. R. Biol.* **2008**, *331*, 806–814. [CrossRef]
47. Panuccio, M.R.; Jacobsen, S.E.; Akhtar, S.S.; Muscolo, A. Effect of saline water on seed germination and early seedling growth of the halophyte quinoa. *AoB Plants* **2014**, *6*, plu047. [CrossRef] [PubMed]
48. Ventura, Y.; Wuddineh, W.A.; Myrzabayeva, M.; Alikulov, Z.; Khozin-Goldberg, I.; Shpigel, M.; Samocha, T.M.; Sagi, M. Effect of seawater concentration on the productivity and nutritional value of annual *Salicornia* and perennial *Sarcocornia* halophytes as leafy vegetable crops. *Sci. Hortic.* **2011**, *128*, 189–196. [CrossRef]
49. Saddhe, A.A.; Malvankar, M.R.; Karle, S.B.; Kumar, K. Reactive nitrogen species: Paradigms of cellular signaling and regulation of salt stress in plants. *Environ. Exp. Bot.* **2019**, *161*, 86–97. [CrossRef]
50. Sharma, P.; Jha, A.B.; Dubey, R.S.; Pessarakli, M. Reactive Oxygen Species, Oxidative Damage, and Antioxidative Defense Mechanism in Plants under Stressful Conditions. *J. Bot.* **2012**, *2012*, 217037. [CrossRef]
51. Hameister, R.; Kaur, C.; Dheen, S.T.; Lohmann, C.H.; Singh, G. Reactive oxygen/nitrogen species (ROS/RNS) and oxidative stress in arthroplasty. *J. Biomed. Mater. Res. Part B Appl. Biomater.* **2020**, *108*, 2073–2087. [CrossRef] [PubMed]
52. Hasanuzzaman, M.; Bhuyan, M.H.M.B.; Anee, T.I.; Parvin, K.; Nahar, K.; Mahmud, J.A.; Fujita, M. Regulation of ascorbate-glutathione pathway in mitigating oxidative damage in plants under abiotic stress. *Antioxidants* **2019**, *8*, 384. [CrossRef]
53. Hasanuzzaman, M.; Nahar, K.; Öztürk, M. *Ecophysiology, Abiotic Stress Responses and Utilization of Halophytes*; Springer: Berlin/Heidelberg, Germany, 2019; ISBN 9811337624.
54. Hussain, A.; Iqbal, K.; Aziem, S.; Mahato, P.; Negi, A.K. A review on the science of growing crops without soil (soilless culture)-a novel alternative for growing crops. *Int. J. Agric. Crop Sci.* **2014**, *7*, 833.
55. Bewley, J.D.; Bradford, K.; Hilhorst, H. Chapter 4: Germination. In *Seeds: Physiology of Development, Germination and Dormancy*; Springer Science & Business Media: Berlin/Heidelberg, Germany, 2012; pp. 133–181. ISBN 1461446929.
56. Loi, A.; Howieson, J.G.; Nutt, B.J.; Carr, S.J. A second generation of annual pasture legumes and their potential for inclusion in Mediterranean-type farming systems. *Aust. J. Exp. Agric.* **2005**, *45*, 289–299. [CrossRef]
57. Qu, X.-X.; Huang, Z.-Y.; Baskin, J.M.; Baskin, C.C. Effect of temperature, light and salinity on seed germination and radicle growth of the geographically widespread halophyte shrub *Halocnemum strobilaceum*. *Ann. Bot.* **2008**, *101*, 293–299. [CrossRef]
58. Dashti, F.; Ghahremani-Majd, H.; Esna-Ashari, M. Overcoming seed dormancy of mooseer (*Allium hirtifolium*) through cold stratification, gibberellic acid, and acid scarification. *J. For. Res.* **2012**, *23*, 707–710. [CrossRef]
59. Schelin, M.; Tigabu, M.; Eriksson, I.; Sawadogo, L.; Oden, P.C. Effects of scarification, gibberellic acid and dry heat treatments on the germination of *Balanites aegyptiaca* seeds from the Sudanian savanna in Burkina Faso. *Seed Sci. Technol.* **2003**, *31*, 605–617. [CrossRef]

60. Babaei-Ghaghelestany, A.; Alebrahim, M.T.; MacGregor, D.R.; Khatami, S.A.; Hasani Nasab Farzaneh, R. Evaluation of ultrasound technology to break seed dormancy of common lambsquarters (*Chenopodium album*). *Food Sci. Nutr.* **2020**, *8*, 2662–2669. [CrossRef] [PubMed]
61. Ashnagar, N.; Mir Mahmoodi, T.; Jalilnejhad, N. Effects of electromagnetic field and ultrasonic waves on seed germination, seedling characteristics and essence percent of thymes (*Thymus vulgaris* L.). *J. Crop Ecophysiol.* **2019**, *13*, 57–72.
62. Machikowa, T.; Kulrattanarak, T.; Wonprasaid, S. Effects of ultrasonic treatment on germination of synthetic sunflower seeds. *Int. J. Agric. Biosyst. Eng.* **2013**, *7*, 1–3.
63. Miano, A.C.; Forti, V.A.; Abud, H.F.; Gomes-Junior, F.G.; Cicero, S.M.; Augusto, P.E.D. Effect of ultrasound technology on barley seed germination and vigour. *Seed Sci. Technol.* **2015**, *43*, 297–302. [CrossRef]
64. Sharififar, A.; Nazari, M.; Asghari, H.R. Effect of ultrasonic waves on seed germination of *Atriplex lentiformis*, *Cuminum cyminum*, and *Zygophyllum eurypterum*. *J. Appl. Res. Med. Aromat. Plants* **2015**, *2*, 102–104. [CrossRef]
65. Tobe, K.; Li, X.; Omasa, K. Seed Germination and Radicle Growth of a Halophyte, Kalidium caspicum (Chenopodiaceae). *Ann. Bot.* **2000**, *85*, 391–396. [CrossRef]
66. Kang, Y.-W.; Joo, N.-M. Comparative Analysis on Phytochemical Properties, Anti-Oxidative, and Anti-Inflammatory Activities of the Different Organs of the Common Ice Plant *Mesembryanthemum crystallinum* L. *Appl. Sci.* **2023**, *13*, 2527. [CrossRef]
67. Bouftira, I.; Abdelly, C.; Sfar, S. Characterization of cosmetic cream with *Mesembryanthemum crystallinum* plant extract: Influence of formulation composition on physical stability and anti-oxidant activity. *Int. J. Cosmet. Sci.* **2008**, *30*, 443–452. [CrossRef]
68. Atzori, G.; de Vos, A.C.; van Rijsselberghe, M.; Vignolini, P.; Rozema, J.; Mancuso, S.; van Bodegom, P.M. Effects of increased seawater salinity irrigation on growth and quality of the edible halophyte *Mesembryanthemum crystallinum* L. under field conditions. *Agric. Water Manag.* **2017**, *187*, 37–46. [CrossRef]
69. Carapeto, A. Levantamento etnobotânico na reserva natural do sapal de castro marim e vila real de santo antónio. In Proceedings of the Relatório final do Projeto AGRO no 800 Rede Nacional para a Conservação e Utilização das Plantas Aromáticas e Medicinais, Castro Marim, Portugal, 29 June 2006.
70. Lima, A.R.; Castañeda-Loaiza, V.; Salazar, M.; Nunes, C.; Quintas, C.; Gama, F.; Pestana, M.; Correia, P.J.; Santos, T.; Varela, J.; et al. Influence of cultivation salinity in the nutritional composition, antioxidant capacity and microbial quality of *Salicornia ramosissima* commercially produced in soilless systems. *Food Chem.* **2020**, *333*, 127525. [CrossRef]
71. Ventura, Y.; Sagi, M. Halophyte crop cultivation: The case for *Salicornia* and *Sarcocornia*. *Environ. Exp. Bot.* **2013**, *92*, 144–153. [CrossRef]
72. Flamini, G.; Luigi Cioni, P.; Morelli, I.; Ceccarini, L.; Andolfi, L.; Macchia, M. Composition of the essential oil of *Medicago marina* L. from the coastal dunes of Tuscany, Italy. *Flavour. Fragr. J.* **2003**, *18*, 460–462. [CrossRef]
73. Wang, G.; Wang, J.; Liu, W.; Nisar, M.F.; El-Esawi, M.A.; Wan, C. Biological activities and chemistry of triterpene saponins from *Medicago* species: An update review. *Evid.-Based Complement. Altern. Med.* **2021**, *2021*, 6617916. [CrossRef] [PubMed]
74. Salem, S.; Nasri, S.; Abidi, S.; Smaoui, A.; Nasri, N.; Mutjé, P.; Ben Hamed, K. Lignocellulosic Biomass from Sabkha Native Vegetation: A New Potential Source for Fiber-Based Bioenergy and Bio-Materials. In *Sabkha Ecosystems*; Springer: Berlin/Heidelberg, Germany, 2019; pp. 407–412.
75. Merchaoui, H.; Ben Mansour, R.; Oueslati, M.; Medini, F.; Hanana, M.; Ksouri, R. A Comparative Evaluation of Total Polyphenolic Content and Antioxidant Potential of Thirty Medicinal Halophytes from the Mediterranean Region. *J. Agric. Sci. Technol.* **2019**, *21*, 1433–1446.
76. Samir, D.; Sara, C.; Widad, A. The Effects of aqueous leaf extract of *Portulaca oleracea* on haemato-biochemical and histopathological changes induced by Sub-chronic Aluminium toxicity in male wistar rats. *Pharmacol. Res. Chin. Med.* **2022**, *4*, 100101. [CrossRef]
77. Nemzer, B.; Al-Taher, F.; Abshiru, N. Phytochemical composition and nutritional value of different plant parts in two cultivated and wild purslane (*Portulaca oleracea* L.) genotypes. *Food Chem.* **2020**, *320*, 126621. [CrossRef]
78. Zeghib, K.; Boutlelis, D.A.; Menai, S.; Debouba, M. Protective effect of Atriplex halimus extract against benzene-induced haematotoxicity in rats. *Ukr. Biochem. J.* **2021**, *93*, 66–76. [CrossRef]
79. Clauser, M.; Dall'Acqua, S.; Loi, M.C.; Innocenti, G. Phytochemical investigation on *Atriplex halimus* L. from Sardinia. *Nat. Prod. Res.* **2013**, *27*, 1940–1944. [CrossRef] [PubMed]
80. Gómez-Bellot, M.J.; Lorente, B.; Ortuño, M.F.; Medina, S.; Gil-Izquierdo, Á.; Bañón, S.; Sánchez-Blanco, M.J. Recycled wastewater and reverse osmosis brine use for halophytes irrigation: Differences in physiological, nutritional and hormonal responses of *Crithmum maritimum* and *Atriplex halimus* plants. *Agronomy* **2021**, *11*, 627. [CrossRef]

*horticulturae*

MDPI

*Article*

# Nano ZnO and Bioinoculants Mitigate Effects of Deficit Irrigation on Nutritional Quality of Green Peppers

**Bruna Lorrane Rosendo Martins [1], Kaikí Nogueira Ferreira [1], Josinaldo Lopes Araujo Rocha [1,*], Railene Hérica Carlos Rocha Araujo [1], Guilherme Lopes [2], Leônidas Canuto dos Santos [2], Francisco Bezerra Neto [3], Francisco Vaniés da Silva Sá [4], Toshik Iarley da Silva [5], Whashington Idalino da Silva [1], Geovani Soares de Lima [1], Francisco Jean da Silva Paiva [1] and José Zilton Lopes Santos [6]**

1 Department of Agricultural Engineering, Federal University of Campina Grande, Campina Grande 58429-900, PB, Brazil; brunalorranerosendomartins@gmail.com (B.L.R.M.); kaicknf95@gmail.com (K.N.F.); railene.herica@professor.ufcg.edu.br (R.H.C.R.A.); whashingtonidalino@gmail.com (W.I.d.S.); geovanisoareslima@gmail.com (G.S.d.L.); francisco.jean@estudante.ufcg.edu.br (F.J.d.S.P.)

2 Department of Soil Science, Federal University of Lavras, Lavras 37200-900, MG, Brazil; guilherme.lopes@ufla.br (G.L.); leonidas.santos2@estudante.ufla.br (L.C.d.S.)

3 Agricultural Sciences Center, Federal Rural University of Semi-Arid, Mossoró 59625-900, RN, Brazil; bezerra@ufersa.edu.br

4 Agricultural Sciences Center, State University of Paraíba, Catolé do Rocha 58884-000, PB, Brazil; vanies@ufersa.edu.br

5 Center for Agricultural, Environmental and Biological Sciences, Federal University of Recôncavo da Bahia, Cruz das Almas 44380-000, BA, Brazil; iarley.toshik@gmail.com

6 Faculty of Agricultural Science, University of Amazonas, Manaus 69067-005, AM, Brazil; ziltonlopes@ufam.edu.br

* Correspondence: josinaldo.lopes@professor.ufcg.edu.br

**Abstract:** Green peppers (*Capsicum annuum* L.) are a fruit vegetable with great culinary versatility and present important nutritional properties for human health. Water deficit negatively affects the nutritional quality of green peppers' fruits. This study aimed to investigate the influence of zinc oxide nanoparticles (ZnONPs), associated with plant growth-promoting bacteria (PGPB), on the post-harvest nutritional quality of green peppers subjected to water deficit. In an open-field experiment, two irrigation levels (50 and 100% of crop evapotranspiration (Etc)), four treatments composed of a combination of ZnONPs, zinc sulfate ($ZnSO_4$), and PGPB (T1 = $ZnSO_4$ via leaves, T2 = ZnONPs via leaves, T3 = ZnONPs via leaves + PGPB via soil, T4 = $ZnSO_4$ via soil + PGPB via soil), and a control treatment (Control) were tested. Water deficit or water deficit mitigation treatments did not interfere with the physical–chemical parameters (except vitamin C content) and physical color parameters (except the lightness) of green peppers. On average, the water deficit reduced the levels of Ca (−13.2%), Mg (−8.5%), P (−8.5%), K (−8.6%), Mn (−10.5%), Fe (−12.2%), B (−12.0%), and Zn (−11.5%) in the fruits. Under the water deficit condition, ZnONPs or $ZnSO_4$ via foliar, associated or not with PGPB, increased the levels of Ca (+57% in the T2 and +69.0% in the T2), P, Mg, and Fe in the fruits. At 50% Etc, the foliar application of ZnONPs in association with PGPB increases vitamin C and mineral nutrients' contents and nutritional quality index (+12.0%) of green peppers. Applying Zn via foliar as ZnONPs or $ZnSO_4$ mitigated the negative effects of water deficit on the quality of pepper fruits that were enhanced by the *Bacillus subtilis* and *B. amyloliquefaciens* inoculation. The ZnONPs source was more efficient than the $ZnSO_4$ source. The water deficit alleviating effect of both zinc sources was enhanced by the PGPB.

**Keywords:** *Capsicum annuum*; drought stress; fruit quality; mineral nutrients

## 1. Introduction

Fruit vegetables are such important sources of vitamins and mineral nutrients in the human diet [1,2]. Due to their great versatility in cooking and their nutritional properties, green peppers in natura are one of the main vegetables consumed in Brazil [3]. In the semi-arid region of Brazil, green peppers are widely cultivated; however, water scarcity generally promotes a decrease in fruit productivity, which can also affect their nutritional quality [4].

Water deficit negatively affects the production and quality of vegetables such as tomatoes and peppers, especially during flowering, fruiting, and fruit maturation [5,6]. In some situations, water stress can increase the concentration of soluble solids, titratable acidity, and vitamin C but decreases other quality parameters, such as fruit size, firmness, and mineral elements' content [6]. The water deficit decreases fruit yield, the translocation of photoassimilates, and the content of minerals such as calcium, potassium, magnesium, and iron in the fruits [5]. The decrease in mineral nutrients in fruits also occurs due to the decrease in the diffusion process and mass flow that contribute to the largest portion of the amount of nutrients that reach the roots and, subsequently, their redistribution to the fruits [7].

The foliar application of zinc, as zinc oxide nanoparticles (ZnONPs), in some plant species, including green peppers and other fruits and vegetables, can contribute to mitigating the negative effects of environmental stresses such as water deficit and saline stress on the plants and improving some aspects of the physical–chemical quality of fruits [8–13]. Due to their high specific surface, the main mode of action of ZnONPs would be through the stimulating effect on plant metabolism, through their interaction with biomolecules, promoting an increase in enzymatic activity and the synthesis of substances that act to mitigate oxidative stress in plants [14,15].

ZnONPs can also act as an efficient source of zinc (Zn) for plants, with slower release than conventional sources, such as zinc sulfate [8,16]. In plants, zinc is involved in the synthesis of several hormones, such as gibberellin, auxin, cytokinin, and abscisic acid [17]. In addition, Zn participates in the synthesis of chlorophyll, the development of chloroplasts, and the integrity of the cell membrane, providing protection, stability, and structure [17,18]. Furthermore, zinc plays a crucial role in fruiting, seed growth, and fruit maturation, increasing resistance to environmental stress and acting as a co-enzyme in the enzymatic processes related to protein synthesis, glucose metabolism, and DNA synthesis, being also essential for the production of auxins, which regulate root growth, cell expansion, and the beginning of flowering [19–21].

Under water stress, several plant species can be favored by the presence of plant growth-promoting bacteria (PGPB) in the rhizosphere region [22–26]. Indirectly, PGPB can alleviate the effects of oxidative stress caused by drought, through the stimulation of enzymes from the antioxidant system such as catalase, superoxide dismutase, and peroxidase [27–30]. Plant inoculation with *Bacillus amyloliquefaciens* improved the survival of pepper plants under different stress conditions, resulting in a higher seedling growth rate and better physicochemical characteristics [31]. In the study conducted in [32], it was noted that the application of *B. amyloliquefaciens* to corn resulted in an increase in nutrient absorption and stimulated plant growth mechanisms, reflected in higher concentrations of amino acids, such as tryptophan, isoleucine, alanine, valine, tyrosine, and sugars, such as fructose and glucose. Similarly, for blueberry plants, inoculating the soil with different strains of *B. amyloliquefaciens and B. subtilis* increased the foliar chlorophyll content, nutrient absorption, and the levels of soluble solids, vitamin C, and anthocyanins in the fruits [33].

Although some studies already tested the influence of nano ZnO as a water stress alleviator [10,12,15], information is lacking on the effects of ZnO and PGPB, whether in combination or not, on the nutritional quality of fruiting vegetables such as green peppers exposed to water deficit, especially at low concentrations compared to the conventional zinc sulfate source.

In this present study, two main hypotheses were considered. The first hypothesis was that water deficit alters the physical–chemical quality and mineral composition of green pepper fruits. The second hypothesis is that zinc oxide nanoparticles (ZnONPs) applied via leaves, associated or not with PGPB, could mitigate the negative effects of water deficit on the physicochemical attributes of the quality and mineral composition of green pepper fruits and that ZnONPs would be more efficient than the conventional source of zinc ($ZnSO_4$*$7H_2O$).

The objective of this work was to evaluate how water deficit affects the nutritional quality of green peppers and whether the application of zinc sources could be capable of mitigating the possible negative effects of the lack of water. Therefore, the application of zinc oxide nanoparticles, associated or not with PGPB, on some physical–chemical parameters and mineral composition of green pepper fruits, from plants grown under two irrigation levels, was evaluated.

## 2. Materials and Methods

### 2.1. Characterization of the Experimental Area

The present research was carried out under field conditions, from December 2022 to April 2023, in the Experimental Farm area of the Center for Agro-food Science and Technology at the Federal University of Campina Grande (UFCG), Pombal-PB Campus. The Experimental Farm is located in the municipality of São Domingos-PB. According to the Köppen classification, the predominant climate is type Aw, hot and humid. Annual rainfall is around 800 mm, and the temperature range is always below 5 °C. The Experimental Farm is located at an altitude of 190 m, at the following coordinates: latitude 6°50′4″ S, longitude 37°53′9″ W. According to Gaussen's classification, the Mediterranean-type bioclimate, also known as medium-dry northeastern, prevails, with a dry season that lasts between 4 and 6 months.

The soil tillage consisted of plowing and harrowing with a plow harrow. One day before planting soil fertilizer, soil sampling was carried out in the area, in the 0–20 cm layer for its chemical and physical characterization (Table 1) according to the methodology described by Embrapa [34].

**Table 1.** Chemical and physical attributes of the soil used in the experiment.

| Chemical | Value | Physical | Value |
|---|---|---|---|
| pH ($CaCl_2$) | 6.20 | Sand (g $kg^{-1}$) | 444 |
| P (mg $kg^{-1}$) | 291 | Silt (g $kg^{-1}$) | 353 |
| $K^+$ (mg $kg^{-1}$) | 633 | Clay (g $kg^{-1}$) | 203 |
| $Na^+$ (mg $kg^{-1}$) | 169 | BD (g $cm^{-3}$) | 1.36 |
| $Ca^{2+}$ (mg $kg^{-1}$) | 1578 | PD (g $cm^{-3}$) | 2.59 |
| $Mg^{2+}$ (mg $kg^{-1}$) | 562 | TP ($m^3$ $m^{-3}$) | 0.47 |
| $Zn^{2+}$ (mg $kg^{-1}$) | 1.69 | FC (%) | 12.87 |
| $H^+$ + $Al^{3+}$ ($cmol_c$ $dm^{-3}$) | 2.30 | PWP (%) | 5.29 |
| OM (g $kg^{-1}$) | 6.40 | AWC (%) | 7.58 |
| V (%) | 83.0 | | |
| CEC ($cmol_c$ $dm^{-3}$) | 4.10 | - | - |

P, $K^+$, $Na^+$, and $Zn^{2+}$—Mehlich 1 extractant, $H^+$ + $Al^{3+}$—0.5 mol $L^{-1}$ calcium acetate extractant at pH 7, $Ca^{2+}$, $Mg^{2+}$—1 mol $L^{-1}$ KCl extractant, OM—organic matter, V—Base saturation, CEC—cationic exchange capacity, BD—bulk density, PD—soil particle density, TP—total porosity, FC—field capacity, PWP—permanent wilting point, AWC—available water content, pH ($CaCl_2$)—extractant $CaCl_2$ $2H_2O$ 0.01 mol $L^{-1}$.

During the experimental period, climatological data on temperature, relative humidity, and precipitation (Figure 1) were collected through the Agritempo website [35].

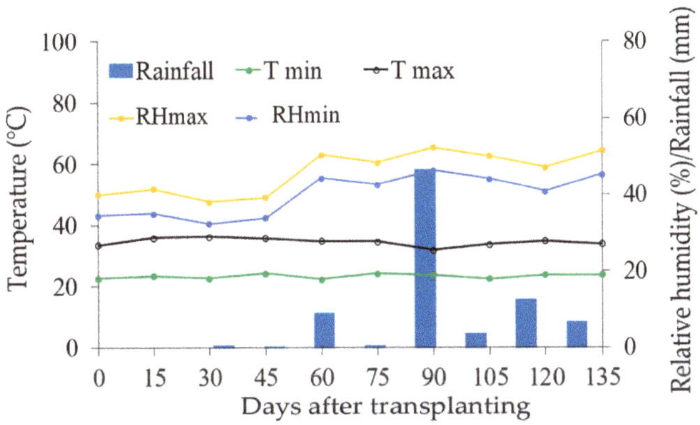


**Figure 1.** Climatological data on maximum (T max) and minimum (T min) air, maximum (RHmax) and minimum (RHmin) air relative humidity, and rainfall during the experimental period in the experimental area.

## 2.2. Treatments and Experimental Design

The experiment was carried out in a randomized block design, in a split-plot scheme, where the plots were two irrigation levels (50 and 100% of crop evapotranspiration (Etc)) and the subplots were four water deficit irrigation alleviate treatments (here named as treatments DHA) composed from nanoparticles of zinc oxide (ZnONPs), zinc sulfate ($ZnSO_4$), and plant growth-promoting bacteria (PGPB) (T1 = $ZnSO_4$ via leaves, T2 = ZnONPs via leaves, T3 = ZnONPs via leaves + PGPB via soil, T4 = $ZnSO_4$ via soil + PGPB via soil), and control treatment, with four blocks, totaling 40 experimental plots. These treatments were defined considering some premises. The supply of Zn via foliar application should be more efficient than that via soil application, and, therefore, the dose of Zn via foliar application should be many times lower than that applied to the soil. This was also considered by other authors [36–38]. Likewise, it was considered that Zn supplied as ZnONPs via foliar would be more efficient than Zn supplied by $ZnSO_4$ via soil or leaves. The biological component was to verify whether PGPB could increase the availability of Zn from the soil itself or increase the efficiency of Zn application via soil, in the form of $ZnSO_4$, since this form of Zn is susceptible to precipitation reactions, decreasing the availability of Zn for plants. In addition, the PGPB component could act as an attenuator of water deficit due to the production of beneficial substances in the root zone of green pepper.

The total area of the experiment corresponded to 22.8 m in length by 14.8 m in width (Appendix A). The plants were grown at a spacing of 1.2 m between rows and 0.5 m between plants in the row. Each subplot consisted of four 3.0 m long rows in which 20 plants were grown. A useful plot consisted of an area measuring 3.0 × 2.0 m in the central part of the plot, where 6 plants were evaluated. When installing the experiment, there was a space of 1.2 m between subplots and 2.0 m between blocks (Appendix A).

## 2.3. Seedlings' Transplanting and Soil Fertilization

Pepper seedlings (*Capsicum annuum* L.) cv. 'Kolima' were produced in expanded polystyrene trays (Isopor®), filled with coconut fiber as a substrate. The transplant was carried out when the seedlings had 4 to 6 true leaves. The seedlings were transplanted in ridges measuring 0.4 m wide and 0.30 m high. Based on the soil analysis (Table 1), planting and supplementary fertilization were carried out according to the recommendations of the Fertilization Recommendation Manual for the State of Pernambuco [39]. For planting fertilization, the formula of NPK 10-10-10 (10%N, 10% $P_2O_5$, and 10% $K_2O$) was applied at a dose of 1000 kg ha$^{-1}$ in a single dose and incorporated into the soil when lifting the planting rows. Further nitrogen and potassium fertilizations were applied via fertigation, according to the following schedule: 39.2 kg ha$^{-1}$ of urea (45% of N) and 15.8 kg ha$^{-1}$ of KCl (58% of

$K_2O$) at 15 days after transplanting (DAT), 39.2 kg ha$^{-1}$ of urea and 15.8 kg ha$^{-1}$ of KCl at 30 days after transplantation, 39.2 kg ha$^{-1}$ of urea and 15.8 kg ha$^{-1}$ of KCl at 45 DAT, and 39.2 kg ha$^{-1}$ of urea and 15.8 kg ha$^{-1}$ of KCl at 60 DAT. All fertilizers used were obtained from Heringer® Fertilizers, Rio Grande, Brazil.

### 2.4. Application of Treatments

As ZnONPs, the product nano zinc oxide (fr Sigma-Aldrich®, Darmstadt, Germany) brand was used; it had 97% purity, a particle size smaller than 100 nm, and a specific surface area of 10.8 m$^2$g$^{-1}$. ZnONPs' suspensions and $ZnSO_4$*7$H_2O$ (Vetec Fine chemistry Ltd., Duque de Caxias, Brazil)) solution (referred to in the text only as $ZnSO_4$) were prepared at concentrations of 0.125 g L$^{-1}$ (100 mg Zn L$^{-1}$) and 4.54 g L$^{-1}$ (1000 mg Zn L$^{-1}$), respectively. The treatments referring to bioinoculation (PGPB) consisted of the combination of two commercial products, one containing *Bacillus subtilis* BV-09 (Vittia Defense and Nutrition, Ltda., São Joaquim da Barra, Brazil) at a dosage of $1.0 \times 10^8$ CFU mL$^{-1}$ and the other containing *B. amyloliquefaciens* (Vittia Defense and Nutrition, Ltd., São Joaquim da Barra, Brazil) at a dosage of $1.0 \times 10^9$ CFU mL$^{-1}$. At the 15 DAT, 45 mL of each PGPB source were mixed with 10 L of water. From this mixture, 30 mL were applied to the soil, close to the stem of each plant. Each product was applied at a rate of 3 L ha$^{-1}$, totaling 6 L ha$^{-1}$.

The foliar application of the suspensions' ZnONPs or $ZnSO_4$ solution for each treatment (T1, T2, and T3) was carried out with the aid of a knapsack sprayer with a capacity of 20 L, with an average of 35 mL applied per plant. Neutral detergent was used as adjuvant in the concentration of 1 mL L$^{-1}$ in all foliar treatments. Regarding soil application, the treatments were applied close to the root zone of the plants using a beaker. For the T4 treatment, 50 mL of the $ZnSO_4$ solution was applied per plant, corresponding to a dose of 2.42 kg of $ZnSO_4$ per hectare. The mixture of bioinoculants was applied around 30 mL per plant. The treatments were applied 25 days after transplanting the seedlings, after homogenizing the plants in the field plots.

### 2.5. Irrigation Levels' Application

After transplanting, the plants were irrigated by drip, with drippers spaced 0.20 m apart and with a nominal flow of 1.6 L h$^{-1}$. After emergence and standardization of the number and size of plants in the subplots, the plants were irrigated following different levels of water depth, which ran 32 days after transplanting.

The total irrigation required (TIR) was calculated using the following expression suggested by Bernardo et al. [40].

$$\mathrm{TIR} = [(\mathrm{FC} - \mathrm{WP}) \times \mathrm{Z} \times \mathrm{BD} \times \mathrm{f})]/10$$

where:

TIR corresponds to the initial total depth of water applied in mm;
FC is the soil moisture corresponding to field capacity, %;
WP is the soil moisture corresponding to wilting point, %;
Z is the effective depth of the pepper root system (30 cm);
BD is bulk density, g cm$^{-3}$;
F is water availability factor (0.5).

The water irrigation uniformity test was carried out according to the Christiansen Uniformity Coefficient (CUC) evaluation methodology, proposed by Christiansen [41].

Control of the volume of water supplied to each irrigation level was carried out daily, at a standardized time, according to the ratio of the flow rate of the drippers to the time to reach the reference evapotranspiration proportions (Eto). As the time interval for each volume of the respective irrigation level was reached, the drip tapes corresponding to each irrigation level were successively disconnected. The irrigation depth corresponding to 100% of Etc was obtained by calculating Etc, according to the expression from Jessen [42].

$$\mathrm{Etc} = \mathrm{Kc} \times \mathrm{Eto}$$

where:

Etc is crop evapotranspiration, mm day $^{-1}$;

Eto is reference evaporation, mm day $^{-1}$;

Kc is crop coefficient (dimensionless).

The Kc values adopted for each crop depending on their phenological phases and the daily Eto values were obtained according to the FAO Penman–Monteith model [43]. During the experiment, climatological data were obtained from the automatic meteorological station in the municipality of São Gonçalo, Paraíba, as it is the closest to the experiment site, through the Agritempo platform [35].

The daily supply of irrigation depth was carried out through the irrigation time considering the characteristics of the cultivation system and the irrigation system according to the following expression suggested by Bernardo et al. [40].

$$\mathrm{It} = \frac{\mathrm{Eto} \times \mathrm{Kc} \times \mathrm{A}}{\mathrm{Ea} \times \mathrm{n} \times \mathrm{q}}$$

where:

It is irrigation time, h;

Eto is reference evaporation, mm day$^{-1}$;

Kc is crop coefficient, dimensionless;

A is area occupied by one plant, m$^2$;

Ea is application efficiency (0.90);

N is number of emitters per plant; and

Q is emitter flow rate, L h$^{-1}$.

### 2.6. Cultivation Practices and Phytosanitary Control

To avoid any interference from phytosanitary chemicals on the action of products containing PGPB, weed and pests control was carried out mechanically, manually, and/or with the use of hand hoes. During experimental period, no diseases were detected. In addition, horizontal staking was carried out using wooden stakes measuring an average of 1.60 m in length. The stakes were spread across the lines, being placed every 5 m. Ribbons were tied to the poles before the fruiting period to prevent the plants from tipping over due to the weight of the fruits.

### 2.7. Variables Evaluated

#### 2.7.1. Physicochemical Analysis

At the time of the first fruit harvest, which occurred 70 days after transplanting, physical attributes of color, ascorbic acid (vitamin C), pH, soluble solids (SS), titratable acidity (TA), and SS/TA ratio were evaluated on the six commercial fruits from useful plots.

The physical attributes of fruit color were measured using a Konica Minolta CR-400™ digital colorimeter (Konica Minolta Co., Ltd., Tokyo, Japan), using the CIELAB system, which defines a three-dimensional color space with three axes in rectangular coordinates (L*a*b*). These coordinates represent lightness (L*), tones from red to green (*a positive to −a* negative), and tones from yellow to blue (*b positive to −b* negative). Furthermore, it defines cylindrical coordinates (L*, C*, H°), and the a* and b* values were converted into Hue angle (H°), representing the intensity of the color and Chroma (C*), indicating the purity of the color, according to [44,45]. The analysis was carried out outside, with one reading per fruit for each experimental plot.

The pH was measured using a digital benchtop pH meter, with direct readings taken from the homogenized pulp, according to [46]. The determination of vitamin C was carried out using the Tillman method. The results of the analyses were expressed in mg per 100 g of a sample of fresh fruit [46]. Titratable acidity was determined following the methodology recommended by [46]. The results were expressed in g of citric acid per 100 g of a sample of fresh fruit. Soluble solids' content was determined directly in the homogenized pulp using a digital refractometer (model PR-100, Palette, Atago Co., Ltd., Tokyo, Japan). The results were expressed in °Brix [46]. The SS/TA ratio was calculated as the ratio between soluble solids and titratable acidity (SS/TA).

2.7.2. Mineral Nutrient Content

Six commercial fruits were separated from each subplot to take a sample consisting of 100 g of fresh fruit mass. These samples were dried in a forced air oven (65–70 °C) until constant weight, to later obtain the dry mass and water content. Subsequently, this material was crushed in a Wiley-type (Tecnal Scientific, Piracicaba, Brazil) mill and then sulfuric digestion, according to the methodology described in [47], was performed. Digestion was obtained in a Kjeldahl microdigester block (Tecnal Scientific, Piracicaba, Brazil), using samples of 0.2 g of dry plant material in 3 mL of concentrated sulfuric acid and 1 mL of hydrogen peroxide, with an initial temperature of 180 °C and a final temperature of 360 °C. In the sulfuric acid digestion stratum, the levels of the macronutrients calcium (Ca), potassium (K), phosphorus (P), and magnesium (Mg) and the micronutrients boron (B), manganese (Mn), iron (Fe), and zinc were determined by using the Inductively Coupled Plasma Optical Emission Spectrometry-ICP-OES (Thermo Scientific Co., Ltd., Bremen, Germany) technique. The element contents obtained in terms of dry mass content were converted to fresh mass contents, correcting the values by the water content of the fruits.

2.7.3. Overall Quality Index

A general fruit quality index (FQI), based on Funsueb [48], was generated with physical–chemical variables, physical color variables, and macro- and micronutrient content in the fruits. Initially, all values of the studied variables (X) were converted into percentages (Xi%) using the following expression:

$$\mathrm{qi} = \frac{\mathrm{Xt}}{\mathrm{Xc}} \cdot 100$$

where qi represents the value of a quality attribute X originating from treatment t and Xc represents the average value of attribute x obtained from control treatment.

FQI values were obtained, according to Funsueb [48], by using the geometric mean of the physical–chemical attributes and the mineral nutrient contents evaluated, according to the following expression:

$$\mathrm{FQI} = \left(\prod_{i=1}^{n} \mathrm{qi}\right)^{1/n}.$$

### 2.8. Statistical Analyses

The data were previously subjected to the normality test using the Shapiro–Wilk test [49]. The data obtained were subjected to analysis of variance, followed by a comparison of means using the Tukey test ($p \leq 0.05$). These statistical analyses were performed using R statistical software 4.3.0 [50].

## 3. Results

### 3.1. Physicochemical and Physical Color Parameters

Irrigation depths and water deficit alleviate treatments (here named as treatments DHA) did not influence the fruit quality parameters of pH, SS, TA, SS/TA, C*, and Hue angle (Table 2). In most of the variables evaluated, the average values obtained on the 50% Etc irrigation level were quite similar to those obtained on the 100% Etc, with variations of less than 5%. Among the DHA treatments, the T4 treatment provided an increase of 10.45% in the C* value and the T3 treatment increased the SS/TA ratio by 17.79%, both under the 100% Etc.

**Table 2.** Means ± standard error for pH values, soluble solids (SS), titratable acidity (TA), SS/TA ratio, chromaticity (C*), Hue angle (°Hue) of green peppers' fruits as a function of irrigation levels and deficit irrigation treatments' mitigation (DHA).

| Irrigation Depth (%) | Treatments DHA | pH | SS (°Brix) | TA (g per 100 g) |
|---|---|---|---|---|
| 50% Etc | Control | 5.443 ± 0.464 aA | 4.858 ± 0.175 aA | 0.109 ± 0.018 aA |
| | T1 ($ZnSO_4$ via leaves) | 5.432 ± 0.265 aA | 4.975 ± 0.325 aA | 0.109 ± 0.001 aA |
| | T2 (ZnONPs via leaves) | 5.370 ± 0.204 aA | 4.767 ± 0.639 aA | 0.130 ± 0.018 aA |
| | T3 (ZnONPs via leaves) + PGPB via soil | 5.443 ± 0.252 aA | 4.792 ± 0.382 aA | 0.119 ± 0.022 aA |
| | T4 ($ZnSO_4$ via soil) + PGPB via soil | 5.437 ± 0.388 aA | 5.289 ± 0.615 aA | 0.103 ± 0.011 aA |
| | Mean | 5.425 A | 4.936 A | 0.114 A |
| 100% Etc | Control | 5.392 ± 0.292 aA | 5.233 ± 0.183 aA | 0.119 ± 0.013 aA |
| | T1 ($ZnSO_4$ via leaves) | 5.438 ± 0.263 aA | 4.992 ± 0.185 aA | 0.114 ± 0.011 aA |
| | T2 (ZnONPs via leaves) | 5.350 ± 0.479 aA | 5.054 ± 0.042 aA | 0.109 ± 0.001 aA |
| | T3 (ZnONPs via leaves) + PGPB via soil | 5.282 ± 0.342 aA | 5.183 ± 0.745 aA | 0.114 ± 0.021 aA |
| | T4 ($ZnSO_4$ via soil) + PGPB via soil | 5.246 ± 0.138 aA | 5.042 ± 0.203 aA | 0.125 ± 0.021 aA |
| | Mean | 5.341 A | 5.101 A | 0.116 A |
| | | SS/TA | C* | °Hue |
| 50% Etc | Control | 42.792 ± 4.477 aA | 20.581 ± 0.910 aA | 50.673 ± 4.500 aA |
| | T1 ($ZnSO_4$ via leaves) | 45.820 ± 2.991 aA | 20.809 ± 1.900 aA | 53.451 ± 5.700 aA |
| | T2 (ZnONPs via leaves) | 37.420 ± 4.065 aA | 21.514 ± 0.842 aA | 50.670 ± 3.060 aA |
| | T3 (ZnONPs via leaves) + PGPB via soil | 42.878 ± 2.473 aA | 21.276 ± 2.196 aA | 50.859 ± 6.110 aA |
| | T4 ($ZnSO_4$ via soil) + PGPB via soil | 46.971 ± 5.969 aA | 21.161 ± 1.872 aA | 50.692 ± 4.560 aA |
| | Mean | 43.176 A | 21.068 A | 51.269 A |
| 100% Etc | Control | 44.246 ± 5.600 aA | 20.607 ± 2.472 aA | 50.311 ± 1.222 aA |
| | T1 ($ZnSO_4$ via leaves) | 44.029 ± 3.802 aA | 21.164 ± 1.079 aA | 50.755 ± 5.330 aA |
| | T2 (ZnONPs via leaves) | 46.549 ± 3.840 aA | 21.911 ± 2.091 aA | 50.959 ± 4.060 aA |
| | T3 (ZnONPs via leaves) + PGPB via soil | 53.822 ± 2.628 aA | 22.170 ± 0.629 aA | 49.863 ± 1.485 aA |
| | T4 ($ZnSO_4$ via soil) + PGPB via soil | 41.123 ± 6.286 aA | 23.012 ± 2.622 aA | 50.230 ± 2.891 aA |
| | Mean | 45.954 A | 21.773 A | 50.424 A |

Within each variable, means followed by the same lowercase letters do not differ for irrigation depth, and means followed by the same uppercase letters do not differ for DHA treatments by Tukey's test ($p \leq 0.05$).

The interaction of irrigation depth and DHA treatments influenced the ascorbic acid content (Figure 2a) and color lightness (L*) (Figure 2b). The 50% Etc irrigation level increased the ascorbic acid content by approximately 35.1% in the control treatment and 51.0% in the T4 treatment compared to the 100% Etc irrigation level. Under the 50% Etc irrigation level, the T3 treatment and T4 treatment promoted an 18.6% increase in the ascorbic acid content, in relation to the control, while the T1 and T2 treatments had similar performances to the control. Under the 100% Etc irrigation level, the T3 treatment increased the ascorbic acid content by 72.7% in relation to the control. At this irrigation level (100% Etc), treatments T1, T2, and T4 were inferior to treatment T3 but provided 30.5% more ascorbic acid than the control treatment. The 50% Etc irrigation level increased the L* value in treatment T1 and decreased it in treatments T3 and T4 compared to the irrigation level of 100% Etc. Under water restriction, DHA treatments did not change the L* value, while under 100% Etc, the T1 treatment decreased the value of this parameter.

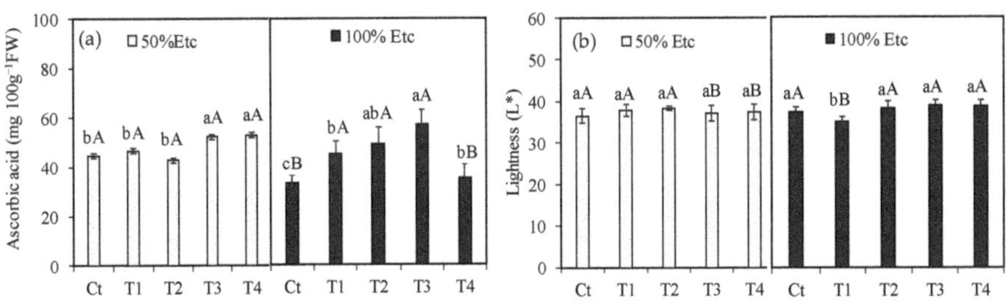


**Figure 2.** Ascorbic acid content (**a**) and luminosity (*L) of color (**b**) in green pepper fruits, as a function of irrigation depths and water deficit alleviating treatments. Control—no product application,

T1 = $ZnSO_4$ via leaves, T2 = ZnONPs via leaves, T3 = ZnONP via leaves + PGPB via soil, T4 = $ZnSO_4$ via soil + PGPB via soil. Bars with the same lowercase letters do not differ for water stress, and bars with the same uppercase letters do not differ for DHA treatments by Tukey's test ($p \leq 0.05$). Vertical bars represent the standard error.

### 3.2. Mineral Nutrient Content

The effect of DHA treatments on mineral nutrient content in green pepper fruits was dependent on irrigation depths ($p < 0.05$). The water deficit (50% Etc) reduced the levels of Ca (Figure 3a), K (Figure 3b), and P (Figure 3d) in the DHA control and T1 treatments but did not change the levels of these nutrients in the other treatments compared to the 100% Etc irrigation level. The 50% Etc irrigation level reduced the Mg (Figure 3c) and Fe (Figure 3f) contents in the treatments' control, T1, and T4 and the B contents (Figure 3e) in the T4 treatment. The water deficit reduced the levels of Mn (Figure 3g) and Zn (Figure 3h) in treatments T1 and T4. On overage, the water deficit reduced the levels of Ca by 13.2%, Mg by 8.5%, P by 8.5%, K by 8.6%, Mn by 10.5%, Fe by 12.2%, B by 12.0%, and Zn by 11.5% in the fruits.

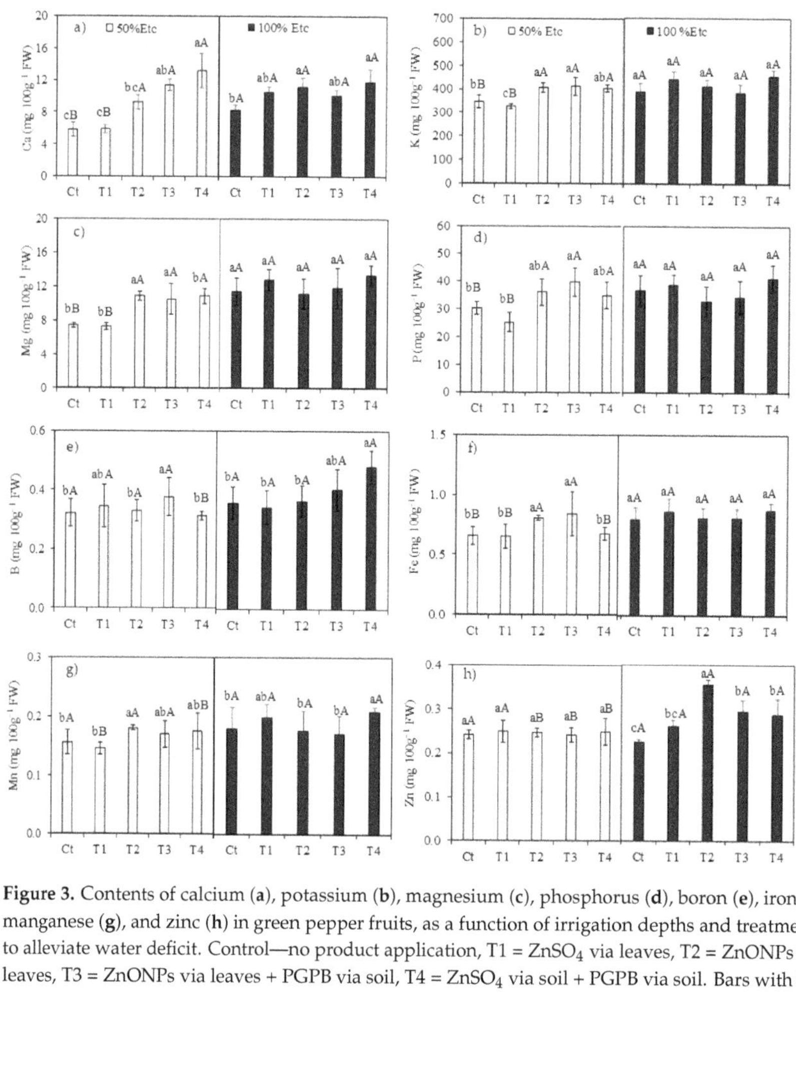

**Figure 3.** Contents of calcium (**a**), potassium (**b**), magnesium (**c**), phosphorus (**d**), boron (**e**), iron (**f**), manganese (**g**), and zinc (**h**) in green pepper fruits, as a function of irrigation depths and treatments to alleviate water deficit. Control—no product application, T1 = $ZnSO_4$ via leaves, T2 = ZnONPs via leaves, T3 = ZnONPs via leaves + PGPB via soil, T4 = $ZnSO_4$ via soil + PGPB via soil. Bars with the

same lowercase letters do not differ for water stress, and bars with the same uppercase letters do not differ for DHA treatments by Tukey's test ($p \leq 0.05$). Vertical bars represent the standard error.

Within each irrigation level, the most prominent DHA treatments were T2 and T3. Under the 50% Etc irrigation level, the T2 treatment increased the contents of K, Mg, Fe, and Mn, the T3 treatment increased the contents of Ca, K, Mg, P, B, and Fe, and the T4 treatment increased the Ca contents in green peppers. Under a 100% Etc water supply, the T2 treatment increased the Ca and Zn contents and the T4 treatment increased the Ca and Mn contents. Compared to the control treatment, the most notable increases occurred for Ca (69%) with the T4 treatment, under the level of 50% Etc, and for zinc (57%), with the T2 treatment, under 100% Etc.

There was a significant interaction between irrigation depths and DHA treatments for the general fruit quality index (FQI) (Figure 4). The water deficit decreased the FQI in the control, T1, and T4 treatments, while in the other DHA treatments, there was no difference between the irrigation levels for this variable. At the irrigation level of 50% Etc, treatments T2, T3, and T4 increased the FQI value by 8.5%, 12.3%, and 10.1%, respectively, in relation to the control treatment; however, these treatments were similar to each other. At the irrigation level of 100% Etc, treatments T1, T2, T3, and T4 increased the FQI in relation to the control treatment.

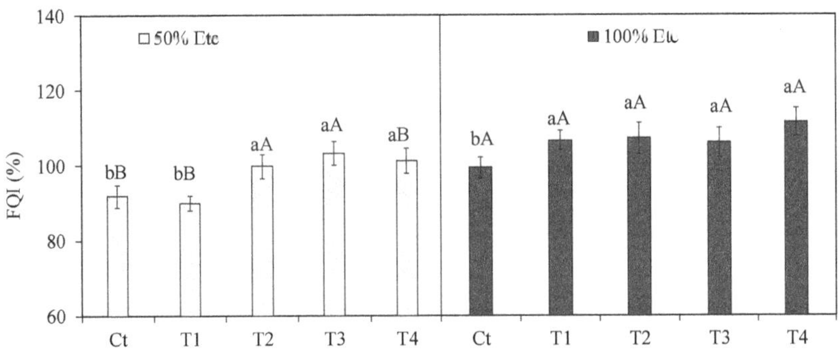


**Figure 4.** General fruit quality index (FQI) of green pepper, as a function of irrigation depths and water deficit alleviating treatments. Control—no product application, T1 = $ZnSO_4$ via leaves, T2 = ZnONPs via leaves, T3 = ZnONPs via leaves + PGPB via soil, T4 = $ZnSO_4$ via soil + PGPB via soil. Columns with the same lowercase letters do not differ for water stress, and bars with the same uppercase letters do not differ for combinations involving zinc oxide nanoparticles (ZnONPs) or bioinoculants (PGPB) by Tukey's test ($p \leq 0.05$). Vertical bars represent the standard error.

## 4. Discussion

### 4.1. Physicochemical Attributes and Color Parameters

Among the evaluated physical–chemical attributes (ascorbic acid content, pH, titratable acidity (TA), soluble solids (SS), and SS/TA ratio) and physical color parameters (L*, C*, and Hue angle), only the ascorbic acid and the L* parameter were affected by water deficit and DHA treatments. Water deficit increased the ascorbic acid in the treatments' control and T4. Water stress also increased the ascorbic acid content in mango [13], maize [51], and tomato [52]. Ascorbic acid plays crucial roles in plant metabolism, such as protecting cells against oxidative damage and acting as a cofactor for enzymes involved in the production of flavonoids and phytohormones [51,53]. The increase in ascorbic acid content under water stress, therefore, is a plant response to adjust to the stress condition, playing an important role in detoxifying reactive forms of oxygen (ROS) and thus avoiding damage to the plant's general metabolism [54–56].

Contrary to the results obtained in this present study, an increase in the levels of SS, TA, and the SS/TA ratio and the changes in the color parameters C* and Hue angle, in fruits from plants that were subjected to water deficit, was observed in other works [56]. The increase in SS content, TA (titratable acidity), and the color parameters in fruits under water deficit is generally attributed to the increase in the synthesis of metabolites and the increased activity of hexokinase enzymes due to the senescence process [57] and the loss of water by the fruit [58]. However, in eggplant, water restriction (50% Etc) decreased ascorbic acid levels, the C* parameters, and Hue angle [59]. According to these authors, this decrease was probably due to the decrease in the photosynthetic rate and, consequently, in photoassimilates and pigments. In this present study, under water restriction, the triggering processes to increase the levels of SS, TA, and the SS/TA ratio and the changes in the C* parameters and Hue angle were probably counterbalanced by the decrease in the production of photoassimilates, with the decrease in the photosynthetic rate. In this sense, an increase or decrease observed in the levels of metabolites (SS, organic acids, ascorbic acid, and carotenoids) in tomato fruits grown under water deficit (60% Etc) was dependent on the stage of fruit maturation (green or ripe) and the metabolites considered [60].

The DHA treatments affected only the ascorbic acid levels and, slightly, the fruit luminosity (L*). As already highlighted, ascorbic acid levels were the attribute most sensitive to water deficit. For this attribute, the similarity of the values obtained between the irrigation regimes (50% Etc and 100% Etc) in treatments T1 (foliar $ZnSO_4$), T2 (foliar ZnONPs), and T3 (foliar ZnONPs + PGPB) indicates that Zn applied as $ZnSO_4$ or ZnONPs possibly attenuates the negative effects of water deficit on ascorbic acid biosynthesis, which was enhanced by the inoculation of plants with BPCP. ZnONPs at a dose of 500 mg $L^{-1}$ increased the antioxidant activity in peppers in the germination phase [8] and ascorbic acid content in bell pepper fruits applied at a dose of 30 $mgL^{-1}$ via foliar [13]. The application of Zn as $ZnSO_4$ also increased SS contents in red pepper [29]. In plants, Zn performs important functions such as activating enzymes that protect against oxidative stress (such as superoxide dismutase), synthesizing tryptophan, which is a precursor of IAA (indole acetic acid), and acting on cell division and the integrity of the cell membrane [7,18,21]. In turn, the synergistic effect of PGPB with Zn may be due to the increase in the biological activity of the soil and/or production of plant growth-stimulating substances in the rhizosphere, such as organic acids, hormones, and enzymes, which can favor the photosynthetic rate and photoassimilates [25,26,61].

### 4.2. Mineral Nutrient Content

Among the minerals evaluated, green pepper exhibited higher levels of K, P, Mg, and Ca, in that order, followed by Fe, B, Zn, and Mn. Depending on the ADH treatments, the levels of these minerals were considerably reduced by the water deficit. Water deficit (80% Etc) reduced the contents of K and Ca in pepper [62] and decreased the contents of Mg, K, Cu, and Zn in tomato under 50% Etc [59] and Ca and B in strawberry under 70% Etc [63]. The process of nutrient accumulation in fruits depends on factors such as the genetic potential of the species, supply via soil or leaves, and nutrient mobility in the phloem [7]. Nutrient supply via soil depends on many factors, including soil moisture [64]. Mass flow and diffusion are the two main mechanisms that deliver mineral nutrients to plant roots [7]. The decrease in mineral nutrient content in pepper fruits was possibly due to the lower amount of these nutrients in contact with the plant's roots, since Ca, Mg, K, Fe, Mn, B, and Zn reach the roots, mainly by mass flow, while K and P reach the roots by diffusion [7,64].

Under water deficit, treatments T2 (ZnONPs via leaf + PGPB) and T3 ($ZnSO_4$ via soil+PGPB) had a greater influence on the mineral composition of pepper fruits, since they increased the levels of K, Ca, Mg, P, Fe, B, and Mn. On the other hand, when there was no water restriction, these treatments (except for Ca and Zn) did not have the same performance in relation to the control, that is, only the T4 treatment ($ZnSO_4$ via leaf) increased the Mn and B contents. The treatments T2, T3, and T4 clearly increased the Zn concentration to the irrigation level of 100% Etc, but this did not occur under water deficit. Foliar application of ZnONPs increased the contents of K, Ca, P, Mg, and Zn in pepper fruits [13]. The role of Zn in cell division and the growth of fine roots contributes to a greater absorption of water and nutrients by the pepper plants and their subsequent translocation to the fruits [65–67]. The presence of PGPB in treatments T2 and T3 enhanced the beneficial effect of zinc on the accumulation of nutrients in pepper fruits. The role of beneficial microorganisms such as *Bacillus subtilis* and *B. amyloliquefaciens* in mitigating water stress in various crops appears to be due to their direct effects on plants and indirectly on the soil [24–26]. These bacteria possibly contributed to increasing the absorption of water and nutrients by peppers, due to the increase in the photosynthetic rate with a better adaptation of metabolism and osmotic adjustment under water deficit [68,69] and/or due to the increase in the solubility of poorly soluble forms of nutrients in the soil as well as by increasing the biological activity of the soil [26].

The FQI index, as a measure of the general fruit quality, revealed, in a summarized way, the negative effect of water deficit on the quality of pepper fruits and the role of Zn as an attenuator, both in the form of ZnONPs and in the form of $ZnSO_4$, as well as PGPB. This effect was more pronounced in treatments T2 and T3, in which no significant differences were observed between the irrigation levels tested. On the other hand, the superiority of treatments T2, T3, and T4 for the FQI index revealed the stimulating effect of Zn and BPCP, even under adequate irrigation (100% Etc).

Finally, it is worth highlighting that the first hypothesis considered in this work was confirmed for the ascorbic acid content and the mineral element content. The second hypothesis can be considered confirmed based on the mineral nutrient content and the general quality index (FQI) and considering that the dose of Zn supplied via ZnONPs was 10 times lower than the dose of Zn via $ZnSO_4{*}7H_2O$, thus proving the highest efficiency of ZnONPs. Although water deficit increased the ascorbic acid content in fruits, the lack of water decreased the overall quality of fruits, especially in terms of mineral nutrients, especially when considering the role of these nutrients in human nutrition [69].

## 5. Conclusions

The principal hypothesis of this work was that water deficit would alter the nutritional quality of green pepper fruits and that the supply of zinc, associated or not with beneficial bacteria, would attenuate the possible negative effects of water scarcity on fruit quality. The study reveals significant findings regarding the effects of water deficit and zinc treatments on the nutritional quality of peppers. Water deficit increased ascorbic acid levels in green pepper fruits, without altering soluble solids' levels, color parameters (C* and Hue angle), and titratable acidity but decreased mineral nutrient levels. The application of Zn via foliar, in the form of zinc oxide nanoparticles (ZnONPs) or zinc sulfate ($ZnSO_4$), mitigated the negative effects of water deficit on the quality of pepper fruits. The ZnONPs' source was more efficient than the $ZnSO_4$ source. The water deficit alleviating effect of both zinc sources was enhanced when the plants were inoculated with *Bacillus subtilis* and *B. amyloliquefaciens*.

The findings of this present work, together with other research, are important in terms of new technological perspectives for adapting plants to water deficit, whether in a semi-arid climate or in climates that are already suffering from the effects of climate changes.

**Author Contributions:** Conceptualization, J.L.A.R.; methodology, J.L.A.R. and J.Z.L.S.; software, T.I.d.S.; validation, R.H.C.R.A., T.I.d.S. and G.L.; formal analysis, T.I.d.S.; investigation, B.L.R.M. and K.N.F.; resources, J.L.A.R.; data curation, K.N.F., W.I.d.S. and B.L.R.M.; writing—original draft preparation, J.L.A.R.; writing—review and editing, G.S.d.L.; visualization, F.B.N., F.V.d.S.S. and L.C.d.S.; supervision, F.B.N. and F.J.d.S.P.; project administration, J.L.A.R.; funding acquisition, J.L.A.R. All authors have read and agreed to the published version of the manuscript.

**Funding:** This research was funded by FAPESQ (Foundation to support research in the state of Paraíba), Fapesq/PB Edital Universal 09/2021 SEECT/Fapesq. grant number 3071/2021.

**Data Availability Statement:** Data are contained within the article.

**Acknowledgments:** We sincerely thank the Department of Soil Science at the Federal University of Lavras for laboratory support for fruit analysis.

**Conflicts of Interest:** The authors declare no conflicts of interest.

## Appendix A

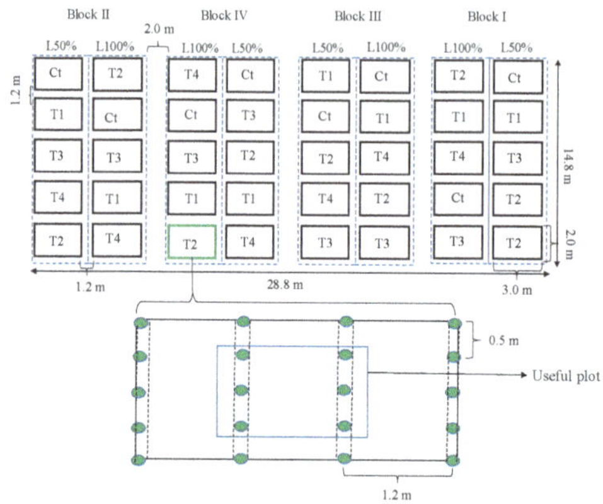


**Figure A1.** Experimental design used in the research showing the distribution of blocks, combination of irrigation levels with treatments from subplots, delimitation of experimental plots and useful plot, as well as spacing between blocks, plots, planting rows, and plants.

**Figure A2.** Partial view of the experimental area showing the installation of the drip irrigation system (**a**) and the plants in the fruit production phase (**b**).

**Figure A3.** Harvest (**a**), sanitization process and classification (**b**) of green pepper fruits before carrying out physical–chemical analyses.

## References


1. Kabir, M.Y.; Nambeesau, S.U.; Bautista, J.; Diasz-Perez, J.C. Effect of irrigation level on plant growth, physiology and fruit yield and quality in bell pepper (*Capsicum annuum* L.). *Sci. Hort.* **2021**, *281*, 109902. [CrossRef]
2. Bhardwaj, R.L.; Parashar, A.; Parewa, H.P.; Vyas, L. An alarming decline in the nutritional quality of foods: The biggest challenge for future generations' health. *Foods* **2024**, *13*, 877. [CrossRef] [PubMed]
3. Sousa, A.M.; Azevedo, B.M.; Vasconcelos, D.V.; Dantas, M.S.M.; Viana, T.V.A. Management of water deficit in the irrigated production of the green pepper. *Rev. Ciênc. Agron.* **2023**, *54*, e20228414. [CrossRef]
4. Mačkić, K.; Bajić, I.; Pejić, B.; Vlajić, S.; Adamović, B.; Popov, O.; Simić, D. Yield and water use efficiency of drip irrigation of pepper. *Water* **2023**, *15*, 2891. [CrossRef]
5. Marouelli, W.A.; Silva, L.C.W. *Irrigation in Pepper Cultivation*; Technical Circular Embrapa Hortaliças, 101, Ministério da Agricultura, Pecuária e Abastecimento: Brasília, Brazil, 2012.
6. Alordzinu, K.E.; Appiah, S.A.; AL Aasmi, A.; Darko, R.O.; Li, J.; Lan, Y.; Adjibolosoo, D.; Lian, C.; Wang, H.; Qiao, S. Evaluating the influence of deficit irrigation on fruit yield and quality indices of tomatoes grown in sandy loam and silty loam soils. *Water* **2022**, *14*, e1753. [CrossRef]
7. Marschner, H. *Mineral Nutrition of Higher Plants*, 2nd ed.; Academic Press: London, UK, 2012; p. 889.
8. García-López, J.I.; Niño-Medina, G.; Olivares-Sáenz, E.; Lira-Saldivar, R.H.; Barriga-Castro, E.D.; Vázquez-Alvarado, R.; Rodríguez-Salinas, P.A.; Zavala-García, F. Foliar application of zinc oxide nanoparticles and zinc sulfate boosts the content of bioactive compounds in habanero peppers. *Plants* **2019**, *8*, 254. [CrossRef]
9. Uresti-Porras, J.-G.; Cabrera-De-La Fuente, M.; Benavidez-Mendoza, A.; Olivares-Sáenz, E.; Cabrera, R.I.; Juárez-Maldonado, A. Effect of graft and nano ZnO on nutraceutical and mineral content in bell pepper. *Plants* **2021**, *10*, 2793. [CrossRef]
10. Semida, W.M.; Abdelkhalik, A.; Mohamed, G.F.; Abd El-Mageed, T.A.; Abd El-Mageed, S.A.; Rady, M.M.; Ali, E.F. Foliar application of zinc oxide nanoparticles promotes drought stress tolerance in eggplant (*Solanum melongena* L.). *Plants* **2021**, *10*, e421. [CrossRef]
11. Rasouli, F.; Asadi, M.; Hassanpouraghdam, M.B.; Aazami, M.A.; Ebrahimzadeh, A.; Kakaei, K.; Dokoupil, L.; Mlcek, J. Foliar application of ZnO-NPS influences chlorophyll fluorescence and antioxidants pool in *Capsicum annum* L. under salinity. *Horticulturae* **2022**, *8*, 908. [CrossRef]
12. Seleiman, M.F.; Ahmad, A.; Battaglia, M.L.; Bilal, H.M.; Alhammad, B.A.; Khan, N. Zinc oxide nanoparticles: A unique saline stress mitigator with the potential to increase future crop production. *S. Afr. J. Bot.* **2023**, *59*, 208–218. [CrossRef]
13. Durán Zuazo, V.H.; Franco, T.D.; Cárceles, R.B.; Gálvez, R.B.; Cermeño, S.P.; Cuadros, T.S.; García, T.I.F. Mango fruit quality improvements in response to water stress: Implications for adaptation under environmental constraints. *Hort. Sci.* **2021**, *48*, 1–11. [CrossRef]
14. Gholinezhad, E. Effect of drought stress and Fe nanofertilizer on seed yield, morphological traits, essential oil percentage and yield of dill (*Anethum graveolens* L.). *J. Essen. Oil Bear. Plants* **2017**, *20*, 1006–1017. [CrossRef]
15. Li, J.; Zafar, S.; Javaid, A.; Perveen, S.; Hasnain, Z.; Ihtisham, M.; Abbas, A.; Usman, M.; El-Sappah, A.H.; Abbas, M. Zinc nanoparticles (ZnNPs): High-Fidelity amelioration in turnip (*Brassica rapa* L.) production under drought stress. *Sustainability* **2023**, *15*, 6512. [CrossRef]
16. Azan, M.; Bhatti, H.N.; Khan, A.K.; Zafar, L.; Iqbal, M. Zinc oxide nano-fertilizer application (foliar and soil) effect on the growth, photosynthetic pigments and antioxidant system of maize cultivar. *Biocat. Agric. Biotechnol.* **2022**, *42*, 102343. [CrossRef]
17. Baybordi, A. *Zinc in Soils and Crop Nutrition*; Parivar Press: Tehran, Iran, 2006; p. 179.
18. Elsheery, N.I.; Helaly, M.N.; El-Hoseiny, H.M.; Alam-Eldein, S.M. Zinc oxide and silicone nanoparticles to improve the resistance mechanism and annual productivity of salt-stressed mango trees. *Agronomy* **2020**, *10*, e558. [CrossRef]
19. Ybaez, Q.E.; Sanchez, P.B.; Badayos, R.B. Synthesis and characterization of nano zinc oxide foliar fertilizer and its influence on yield and postharvest quality of tomato. *Philipp. Agric. Sci.* **2020**, *103*, 55–65.
20. Ganguly, R.; Sarkar, A.; Dasgupta, D.; Acharya, K.; Keswani, C.; Popova, V.; Minkina, T.; Maksimov, A.Y.; Chakraborty, N. Unravelling the efficient applications of zinc and selenium for mitigation of abiotic stresses in plants. *Agriculture* **2022**, *12*, e1551. [CrossRef]

21. Ahmed, N.; Zhang, B.; Chachar, Z.; Li, J.; Xiao, G.; Wang, Q.; Hayat, F.; Deng, L.; Narejo, M.; Bozdar, B.; et al. Micronutrients and their effects on horticultural crop quality, productivity and sustainability. *Sci. Hort.* **2024**, *323*, 112512. [CrossRef]
22. Pereira, S.I.A.; Abreu, D.; Moreira, H.; Veja, A.; Castro, P.M.L. Plant growth-promoting rhizobacteria (PGPR) improve the growth and nutrient use efficiency in maize (*Zea mays* L.) under water deficit conditions. *Heliyon* **2020**, *6*, 10. [CrossRef]
23. Akhtar, S.S.; Amby, D.B.; Hegelund, J.N.; Fimognari, L.; Großkinsky, D.-K.; Westergaard, J.C.; Müller, R.; Moelbak, L.; Liu, F. Roitsch, T. *Bacillus licheniformis* FMCH001 increases water use efficiency via growth stimulation in both normal and drought conditions. *Front. Plant Sci.* **2020**, *11*, 297. [CrossRef]
24. Ramakrishna, W.; Rathore, P.; Kumari, R.; Yadav, R. Brown gold of marginal soil: Plant growth-promoting bacteria to overcome plant abiotic stress for agriculture, biofuels and carbon sequestration. *Sci. Total Environ.* **2020**, *711*, 135062. [CrossRef] [PubMed]
25. Kiran, S.; Furtana, G.B.; Ellialtioglu, Ş.Ş. Physiological and biochemical assay of drought stress responses in eggplant (*Solanum melongena* L.) inoculated with commercial inoculant of *Azotobacter chroococum* and *Azotobacter vinelandii*. *Sci. Hort.* **2022**, *305*, 111394. [CrossRef]
26. Araujo, J.L.; Mesquita Alves, J.; Rocha, R.H.C.; Santos, J.Z.L.; Santos Barbosa, R.; Costa, F.M.N.; Lima, G.S.; Freitas, L.N.; Lima, A.S.; Nogueira, A.E.P.; et al. Beneficial microorganisms affect soil microbiological activity and corn yield under deficit irrigation *Agriculture* **2023**, *13*, 1169. [CrossRef]
27. Chiappero, J.; Cappellari, L.R.; Alderete, L.G.; Palermo, T.B.; Banchio, E. Plant growth promoting rhizobacteria improve the antioxidant status in *Mentha piperita* grown under drought stress leading to an enhancement of plant growth and total phenolic content. *Ind. Crops Prod.* **2019**, *139*, 111553. [CrossRef]
28. Zhang, M.; Yang, L.; Hao, R.; Bai, X.; Wang, Y.; Yu, X. Drought-tolerant plant growth-promoting rhizobacteria isolated from jujube (*Ziziphus jujuba*) and their potential to enhance drought tolerance. *Plant Soil* **2020**, *452*, 423–440. [CrossRef]
29. Ahmet, T.; Hayrettin, K.; Özmen, N. Response of red pepper (*Capsicum annuum* L.) to foliar applications of Zinc. *Comm. Soil Sci. Plant Anal.* **2021**, *52*, 1256–1263. [CrossRef]
30. Lin, Y.; Watts, D.B.; Kloepper, J.K.; Feng, Y.; Torbert, H.A. Influence of plant growth-promoting rhizobacteria on corn growth under drought stress. *Comm. Soil Sci. Plant Anal.* **2020**, *51*, 250–264. [CrossRef]
31. Kazerooni, E.A.; Maharachchikumbura, S.S.N.; Adhikari, A.; Al-Sadi, A.M.; Kang, S.M.; Kim, L.R.; Lee, I.J. Rhizospheric *Bacillus amyloliquefaciens* Protects *Capsicum annuum* cv. Geumsugangsan from multiple abiotic stresses via multifarious plant growth-promoting attributes. *Front. Plant Sci.* **2021**, *25*, 669693. [CrossRef]
32. Vinci, G.; Cozzolino, V.; Mazzei, P.; Monda, H. Effects of *Bacillus amyloliquefaciens* and different phosphorus sources on maize plants as revealed by NMR and GC-MS based metabolomics. *Plant Soil* **2018**, *429*, 437–450. [CrossRef]
33. Yu, Y.-Y.; Xu, J.-D.; Huang, T.-X.; Zhong, J.; Yu, H.; Qiu, J.-P.; Guo, J.-H. Combination of beneficial bacteria improves blueberry production and soil quality. *Food Sci. Nutr.* **2020**, *8*, 5776–5784. [CrossRef]
34. EMBRAPA. Empresa Brasileira de Pesquisa Agropecuária. Centro Nacional de Pesquisa em Solos. In *Manual of Soil Analysis Methods*, 2nd ed.; EMBRAPA: Rio de Janeiro, Brazil, 2011; 225p.
35. AGRITEMPO. Agrometeorological Monitoring System: Meteorological Stations for the State of PB. Available online: https://www.agritempo.gov.br/br/estado/PB/estacoes/ (accessed on 1 November 2023).
36. Khan, M.R.; Akram, M.S.; Moonmoon, J.F.; Tarafder, M.M.A.; Rahman, M.H.; Das, S.; Skalicka, J.; Skalicky, M.; Hossain, A. Soil and foliar zinc application techniques influence the productivity, zinc concentration, and protein content in the grains of bread wheat varieties. *Acta Agrobot.* **2023**, *76*, 1–13. [CrossRef]
37. Jalal, A.; Oliveira, C.E.S.; Fernandes, G.C.; Silva, E.C.; Costa, K.N.; Souza, J.S.; Leite, G.S.; Biagini, A.L.C.; Galindo, F.S.; Teixeira Filho, M.C.M. Integrated use of plant growth-promoting bacteria and nano-zinc foliar spray is a sustainable approach for wheat biofortification, yield, and zinc use efficiency. *Front. Plant Sci.* **2023**, *14*, 1146808. [CrossRef] [PubMed]
38. Ram, S.; Malik, V.K.; Gupta, V.; Narwal, S.; Sirohi, M.; Ankush, P.V.; Gupta, O.P.; Misra, A.K.; Singh, G. Impact of foliar application of iron and zinc fertilizers on grain iron, zinc, and protein contents in bread wheat (*Triticum aestivum* L.). *Front. Nutr.* **2024**, *11*, 1378937. [CrossRef]
39. Cavalcante, F.J.A. *Fertilization Recommendations for the State of Pernambuco (Brazil)*, 2nd ed.; Instituto Agronômico de Pernambuco—IPA: Recife, Brazil, 2008; 212p.
40. Bernardo, S.; Soares, A.A.; Mantovani, E.C. *Manual of Irrigation*, 8th ed.; Universidade Federal de Viçosa: Viçosa, Brazil, 2006; 625p.
41. Christiansen, J.E. *Irrigation by Sprinkling*; University of California: Berkeley, CA, USA, 1942.
42. Jessen, M.E. *Water Consumption by Agriculture Plants*; Kozlowski, T.T., Ed.; Water deficits and plant growth; Academic Press: New York, NY, USA; London, UK, 1968; 22p.
43. Allen, R.G.; Pereira, L.S.; Raes, D.; Smith, M. *Crop Evapotranspiration-Guidelines for Computing Crop Water Requirements*; FAO Irrigation and drainage paper 56; FAO: Rome, Italy, 1998.
44. Pinheiro, J.M.S. Post-Harvest Technology for Preserving Bananas of the Cultivar Tropical. Master's Degree, Universidade Estadual de Montes Claros, Janaúba, Brazil, 2009; 59p.
45. IAL—Instituto Adolfo Lutz. Normas analíticas do Instituto Adolfo Lutz. In *Physicochemical Methods for Food Analysis*, 4th ed.; IAL—Instituto Adolfo Lutz: São Paulo, Brazil, 2008; 1020p.
46. AOAC—Association of Official Analytical Chemists. *Official Methods of Analysis of the Association of Official Analytical Chemists*; AOAC: Gaithersburg, Maryland, 2012.

47. Tedesco, M.J.; Milan, P.A.; Ernani, P.R. *Analysis of Soil, Plants and Other Materials*, 2nd ed.; UFRGS. Boletim Técnico, 5; Universidade Federal do Rio Grande do Sul, Departamento de Solos: Porto Alegre, Brazil, 1995; 118p.
48. Funsueb, S.; Thanavanich, C.; Theanjumpol, P.; Kittiwachana, S. Development of new fruit quality indices through aggregation of fruit quality parameters and their predictions using near-infrared spectroscopy. *Postharvest Biol. Technol.* **2023**, *204*, 112438. [CrossRef]
49. Shapiro, S.S.; Wilk, M.B. An Analysis of Variance Test for Normality (Complete Samples). *Biom. Trust.* **1965**, *52*, 591–609. [CrossRef]
50. R Core Team. *R: A Language and Environment for Statistical Computing*; R Core Team: Viena, Austria, 2023.
51. Zhang, H.; Xiang, Y.; He, N.; Liu, X.; Liu, H.; Fang, L.; Zhang, F.; Sun, X.; Zhang, D.; Li, X.; et al. Enhanced vitamin C production mediated by an ABA-Induced PTP-like nucleotidase improves plant drought tolerance in *Arabidopsis* and maize. *Mol. Plant.* **2020**, *13*, 760–776. [CrossRef]
52. Medyouni, I.; Zouaoui, R.; Rubio, E.; Serino, S.; Ahmed, H.B.; Bertin, N. Effects of water deficit on leaves and fruit quality during the development period in tomato plant. *Food. Sci. Nutr.* **2021**, *9*, 1949–1960. [CrossRef]
53. Agbemafle, R.; Owusu-Sekyere, J.D.; Bart-Plange, A. Effect of deficit irrigation and storage on the nutritional composition of tomato (*Lycopersicon esculentum* Mill. cv. Pectomech). *Croatian J. Food Technol. Biotechnol. Nutr.* **2015**, *10*, 59–65.
54. Al-Selwey, W.A.; Alsadon, A.A.; Al-Doss, A.A.; Solieman, T.H.; Dewir, Y.F.; Ibrahim, A.A. Effect of deficit irrigation on total yield, fruit physical characteristics, and nutritional value of four drought tolerant tomato (*Solanum lycopersicum* L.) genotypes. *J. Agric. Sci. Technol.* **2021**, *23*, 1105–1118.
55. Nurzy'nska-Wierdak, R.; Buczkowska, H.; Sałata, A. Do AMF and Irrigation Regimes Affect Sweet Pepper Fruit Quality under Open Field Conditions? *Agronomy* **2021**, *11*, 2349. [CrossRef]
56. Toumi, I.; Zarrouk, O.; Ghrab, M.; Nagaz, K. Improving peach fruit quality traits using deficit irrigation strategies in southern Tunisia arid area. *Plants* **2022**, *11*, 1656. [CrossRef] [PubMed]
57. Dai, N.; Schaffer, A.; Petreikov, M.; Shahak, Y.; Giller, Y.; Ratner, K. Overexpression of Arabidopsis hexokinase in tomato plants inhibits growth, reduces photosynthesis, and induces rapid senescence. *Plant Cell* **1999**, *11*, 1253–1266. [CrossRef] [PubMed]
58. Mira-García, A.B.; Conejero, W.; Vera, J.; Ruiz-Sánchez, M.C. Effect of water stress and shading on lime yield and quality. *Plants* **2023**, *12*, 503. [CrossRef] [PubMed]
59. Martins, B.L.R.; Araujo, R.H.C.R.; Rocha, J.L.A.; Silva, T.I.; Alves, K.A.; Almeida, E.S.; Ferreira, K.N. Zinc oxide nanoparticles and bioinoculants on the postharvest quality of eggplant subjected to water deficit. *Rev. Bras. Eng. Agric. Amb.* **2024**, *28*, e279019. [CrossRef]
60. Silva, T.I.; Sousa, S.V.; Cruz, R.R.P.; Costa, F.B.; Silva, L.A.; Sales, J.N.B.; Brito, M.E.B.; Ribeiro, W.S.R.; Soares, L.A.A.; Soares-Filho, W.S. Fruit quality of 'Tahiti' lime grafted onto different rootstocks grown under salt stress. *Arid Land Res. Man.* **2024**, *38*, 447–463. [CrossRef]
61. Lu, H.; Qi, X.; Rahman, S.; Qiao, D.; Li, P.; Han, Y.; Zhao, Z. Rice physiological response with *Bacillus subtilis* and saccharomyces cerevisiae inoculation into soil under reclaimed water–fresh water combined irrigation. *Water* **2021**, *13*, e773. [CrossRef]
62. Albuquerque, F.S.; Silva, E.F.F.; Bezerra Neto, E.; Souza, A.E.R.; Santos, A.N. Mineral nutrients in pepper fertigated under irrigation slides and doses of potassium. *Hort. Bras.* **2012**, *30*, 681–687. [CrossRef]
63. Perin, E.C.; Messoas, R.S.; Gallo, V.; Bdrdwsko, J.M.; Sduza, E.R.; Avola, L.D.; Bamberg, A.L.; Rdmbaldo, C.V. Mineral content and antioxidant compounds in strawberry fruit submitted to drough stress. *Food Sci. Technol.* **2019**, *39*, 245–254. [CrossRef]
64. Silva, E.C.; Nogueira, R.J.; Silva, M.A.; Albuquerque, M.B. Drought stress and plant nutrition. Plant stress. *Glob. Sci. Books* **2011**, *5*, 32–41. [CrossRef]
65. Farooq, A.; Javad, S.; Jabeen, K.; Shah, A.A.; Ahmad, A.; Shah, A.N.; Alyemeni, M.N.; Walid, F.A.M.; Abbas, A. Effect of calcium oxide, zinc oxide nanoparticles and their combined treatments on growth and yield attributes of *Solanum lycopersicum* L. *J. King Saud Univ.-Sci.* **2023**, *35*, e102647. [CrossRef]
66. Dimkpa, C.O.; Singh, U.; Bindraban, P.S.; Elmer, W.H.; Gardea-Torresdey, J.L.; White, J.C. Zinc oxide nanoparticles alleviate drought-induced alterations in sorghum performance, nutrient acquisition, and grain fortification. *Sci. Total Environ.* **2019**, *688*, 926–934. [CrossRef] [PubMed]
67. Mahouachi, J.; Marrero-Díaz, E. Plant growth and fruit nutrient changes in *Carica papaya* L. genotypes subjected to regulated deficit irrigation. *Life* **2022**, *12*, 1831. [CrossRef] [PubMed]
68. Cui, J.; Shao, G.; Lu, J.; Keabetswe, L.; Hoogenboom, G. Yield, quality and drought sensitivity of tomato to water deficit during different growth stages. *Sci. Agric.* **2020**, *77*, e20180390. [CrossRef]
69. Maret, W.; Sandstead, H.H. Zinc requirements and the risks and benefits of zinc supplementation. *J. Trace Elem. Med. Biol.* **2006**, *20*, 3–18. [CrossRef]

*horticulturae*

MDPI

*Article*

# Optimizing Lemon Balm (*Melissa Officinalis* L.) Cultivation: Effects of Different Manures on Plant Growth and Essential Oil Production During Consecutive Harvests

**Sina Fallah [1,*], Filippo Maggi [2,*], Askar Ghanbari-Odivi [1] and Maryam Rostaei [1]**

1 Department of Agronomy, Faculty of Agriculture, Shahrekord University, Shahrekord 115, Iran; askarghanbariodivi@gmail.com (A.G.-O.); maryamrosta318@yahoo.com (M.R.)
2 Chemistry Interdisciplinary Project (ChIP) Research Center, School of Pharmacy, University of Camerino, 62032 Camerino, Italy
* Correspondence: fallah-s@sku.ac.ir or falah1357@yahoo.com (S.F.); filippo.maggi@unicam.it (F.M.)

**Abstract:** This study examined the impact of organic manures from different sources (poultry, sheep, and cattle) on lemon balm (*Melissa officinalis* L., Lamiaceae) during different harvests. Manure application increased the photosynthetic pigments levels (chlorophyll-a, 9–41%; chlorophyll-b, 24–60%), biomass (41–60%), and essential oil yield (60–71%). Sheep manure treatment exhibited the highest antioxidant capacity among all the manures tested. Through GC-MS and GC-FID analysis, 10 chemical constituents were identified in the essential oil, accounting together for 91–95% of the total composition. The primary chemical component was geranial (39–46%), followed by neral (28–35%), (*E*)-caryophyllene (4.7–11%), geranyl acetate (2.7–5.9%), and caryophyllene oxide (1.7–4.8%). The utilization of livestock manures significantly improved the quality of the essential oil in terms of neral and geranial percentages compared to the control. Notably, during mid-August and early October, there was a substantial rise in these valuable compounds. However, a decrease in geranyl acetate and oxygenated monoterpenes resulted in a decline of the antioxidant capacity to 3%. Consequently, it is recommended to utilize essential oils from the second and third harvests for industrial purposes. Overall, the use of livestock manures, especially sheep manure, as a nutrient source for lemon balm cultivation, proves to be a viable approach for producing high-quality essential oils.

**Keywords:** essential oil composition; biomass; photosynthesis; antioxidant capacity; lemon balm

## 1. Introduction

Medicinal plant growers heavily rely on a variety of fertilizers to meet the demands of a rapidly growing global population [1]. Throughout history, manure has served as a crucial source of plant nutrition. Its ability to enhance and sustain soil productivity has been acknowledged since ancient times. The presence of fertilizer materials, humus, and other organic components makes manure a valuable tool for soil improvement [2]. Animal manure, including those deriving from poultry, sheep, and cattle, not only provide essential macro and micronutrients but also enhances soil properties, creating an optimal growth environment [3–5]. Consequently, this results in enhanced plant productivity while avoiding any negative impacts on human health and the environment. Whereas, inadequate handling of animal manure could potentially endanger human health and the environment.

In relation to medicinal plants, the application of organic manures becomes even more crucial as it not only enhances the growth and yield of plants but also improves the overall quality of final products like essential oils [1,6,7]. Medicinal plants hold increasing importance in various industries worldwide, especially those with higher levels of secondary metabolites [8]. The quality of medicinal plants has garnered significant attention in research circles, with a particular focus on the effects of organic manures, which have been found to have a positive association with medicinal plant quality [6,9]. Various studies have demonstrated a boost in the yield of medicinal plants when organic fertilizers are applied [10–12]. However, it is highly recommended to use organic manures to enhance the quality and productivity of medicinal plants.

*Melissa officinalis* L., mostly known as lemon balm, is a fragrant herb belonging to the mint family (Lamiaceae). Throughout history, lemon balm has been utilized as a natural remedy to aid sleep, alleviate stress and anxiety, enhance digestion and appetite, and combat viral and other infections. In addition to its medicinal uses, lemon balm is frequently employed as a flavoring agent in food and beverages as well as in the production of health products and cosmetics [13]. Scientific research has revealed that lemon balm possesses sedative properties, acts as a muscle relaxant (antispasmodic), aids in digestion, relieves gas (carminative), exhibits antioxidant effects, and possesses antimicrobial activity [14]. The therapeutic effects of lemon balm can be ascribed to the existence of various chemical compounds, including flavonoids, tannins, phenolics like rosmarinic acid and caffeic acid, terpenes such as ursolic acid and oleanolic acid, and essential oils [15]. *M. officinalis* exhibits higher antioxidant activity and levels of total phenolic content compared to other Lamiaceae species, including *Salvia officinalis* L., *Lavandula angustifolia* L., and *Origanum vulgare* L. [16].

In their investigation, Sodré et al. [15] found that citral and citronellal are the primary constituents of the lemon balm essential oil (LEO) in organic and inorganic treatments. They observed that, as the amount of cattle manure (CA) increased, the citronellal content in LEO obtained from dried leaves decreased. On the other hand, the content of geraniol showed a progressive increase with higher doses of cattle manure compared to the control treatment. According to de Assis et al. [17], the application of 900 kg ha$^{-1}$ of cattle manure on lemon balm led to a remarkable improvement in plant height (17%), root biomass (237%), and shoot biomass (128%) in comparison to the control. Within natural antioxidants, citral plays a pivotal role. It effectively hampers the oxidation of linoleic acid and acts as a safeguard for IEC-6 cells, shielding them from the oxidative stress induced by aspirin [18]. The cultivation of plants rich in citral has become increasingly prevalent in recent times and citral, as a monoterpene, is capturing the attention of people globally [19].

In Iran, lemon balm is a perennial plant that is planted in the Spring and it has the capacity to produce more than one harvest annually. However, there is limited information available on the effects of different types of organic manures on the productivity and quality of lemon balm, especially during consecutive harvests. We hypothesize that (i) the use of manures can augment nutrient availability for lemon balm; (ii) the absorption of nutritional elements, along with the production of essential oil and its various compounds, fluctuates during different harvesting periods; additionally, (iii) an enhancement in the quality of essential oil is a beneficial sign for the preservation of plant productivity in agroecosystems. Based on these hypotheses, this study aims to provide the initial findings regarding the reaction of lemon balm to different manure sources across three harvests. The objective was to evaluate the impact of organic manures on biomass, nutrient content, yield of LEO, chemical compositions, and antioxidant activity in consecutive harvests.

## 2. Materials and Methods

### 2.1. Description of Study Site

This field experiment took place in Koohrang (50°16′ E; 32°29′ N), Chaharmahal, and Bakhtiari Province, Iran, in 2019, from April to October. The site is situated at an altitude of 2234 m above sea level. The study site exhibits a cold semiarid climate, known for its moderate winters and comparatively humid and hot summers. Table A1 shows the average temperature and monthly total precipitation for each month of the lemon balm's growing season.

### 2.2. Field Experiment and Cultivation Management

This research was carried out as a factorial experiment utilizing a randomized complete block design (RCBD) with 3 replications (Table 1). The initial factor consisted of four nutrient sources: poultry manure (PO) (3200 kg $ha^{-1}$), sheep manure (SH) (5400 kg $ha^{-1}$), cattle manure (CA) (6200 kg $ha^{-1}$), and control (CO) (without manure). The second factor was harvest, specifically HAR1, HAR2, and HAR3 (June 21, August 14, and October 6, 2019, respectively). To estimate the quantity of animal manure, the provision of 100 kg N $ha^{-1}$ (equivalent to chemical fertilizer) throughout the lemon balm's growing season was considered. It was postulated that approximately 50% of the total organic nitrogen present in the livestock manures would be released through mineralization, as suggested by Fallah et al. [6] and Alizadeh et al. [20]. Matching the nitrogen release with the needs of the plant contributes to its superior efficiency compared to chemical fertilizer [20].

**Table 1.** Animal manures and harvest treatments tested on lemon balm in 2019.

| Manure Treatment | Harvest Treatment | | |
|---|---|---|---|
| | HAR1 | HAR2 | HAR3 |
| Poultry (PO) | June 21 | August 14 | October 6 |
| Sheep (SH) | June 21 | August 14 | October 6 |
| Cattle (CA) | June 21 | August 14 | October 6 |
| Control (CO) | June 21 | August 14 | October 6 |

On 23 April 2018, lemon balm transplants were planted into the experimental plots. The transplants selected for this research were derived from seeds sourced from the local genotype of lemon balm, which were nurtured in a greenhouse environment for a period of 6 weeks. By the end of the nursery stage, the transplants reached a height of 10 cm. On 24 April 2019, organic fertilizers were applied in close proximity to the root of the plants. Drip tape irrigation was used immediately after fertilization and manual weed control was performed multiple times throughout the growth season. Herbicides, fungicides, pesticides, and micronutrients were not applied in any of the treatments. The size of every plot was carefully determined to be 9 $m^2$, with measurements of 3 m by 3 m. The spacing between and within rows was consistently maintained at 50 cm. The plots were arranged with a distance of 1 m between them while the replications were spaced 2 m apart. Soil samples were taken randomly at a depth of 0–30 cm to assess the soil properties before seedling transplantation. Table 2 displays the characteristics of the soil and animal manure utilized. The soil in the area possessed a clay texture with a pH of 7.75, an electrical conductivity of 760 μS $cm^{-1}$, and an organic carbon content of up to 7.6 g $kg^{-1}$. There were no restrictions on the salinity levels of the animal manure employed. Poultry manure contained twice the amount of nitrogen compared to cattle and sheep manures, while sheep manure had a higher potassium content than cattle and poultry manures.

**Table 2.** Some chemical properties of the soil and livestock manures used in the experiment.

| Parameter | Unit | Soil | PO | SH | CA |
|---|---|---|---|---|---|
| pH | | 7.8 | 6.7 | 7.9 | 7.9 |
| EC | μS $cm^{-1}$ | 760 | 4750 | 4380 | 1980 |
| OC | g $kg^{-1}$ | 7.6 | 312 | 175 | 195 |
| Nitrogen | % | 0.08 | 4.5 | 2.6 | 2.3 |
| Phosphorus | % | $8 \times 10^{-4}$ | 1.7 | 0.59 | 0.56 |
| Potassium | % | $328 \times 10^{-4}$ | 9.8 | 12.5 | 6.2 |
| Iron | mg $kg^{-1}$ | 3.6 | 1475 | 3812 | 1718 |
| Zinc | mg $kg^{-1}$ | 0.70 | 425 | 120 | 206 |
| Copper | mg $kg^{-1}$ | 0.90 | 117 | 28.1 | 51.7 |
| Manganese | mg $kg^{-1}$ | 8.1 | 493 | 331 | 220 |

EC: electrical conductivity; OC: organic carbon; PO: poultry manure; SH: sheep manure; CA: cattle manure.

## 2.3. Measurements

### 2.3.1. Chlorophyll

Simultaneously with each harvest, the quantification of the photosynthetic pigments was conducted using the method described by Rajput and Patil [21]. In summary, 1 g of three pairs of young leaves developed was crushed and blended in 10 mL of 80% acetone. The resulting mixture was then centrifuged at a speed of 5000 revolutions per minute for a duration of 5 min. The liquid portion, known as the supernatant, was carefully collected for further analysis. To determine the concentration of chlorophyll-a (Cha) and chlorophyll-b (Chb), the absorbance of the supernatant was measured at two specific wavelengths, namely 645 nm and 663 nm using a UV-Vis spectrophotometer (AE-UV 1606). These measurements were then used in the following equations to determine the levels of photosynthetic pigments.

$$\mathrm{Cha}(mg/g\ FW) = 12.7 \times \mathrm{OD}_{663} - 2.69 \times \mathrm{OD}_{645} \times \frac{V}{1000} \times \mathrm{W} \quad (1)$$

$$\mathrm{Chb}(mg/g\ FW) = 22.9 \times \mathrm{OD}_{645} - 4.68 \times \mathrm{OD}_{663} \times \frac{V}{1000} \times \mathrm{W} \quad (2)$$

The variables OD, V, and W represent the absorbance values obtained at particular wavelengths, the total volume of chlorophyll extracted using 80% acetone, and the fresh weight of the extracted tissue, respectively.

### 2.3.2. Macro- and Micronutrient Concentration

To determine the nutrient concentration at different harvests, samples (lemon aerial parts) were collected by randomly selecting plants (simultaneously with each harvest) from each plot to ensure a representative sample. The samples were then washed with distilled water to eliminate any surface contaminants, air-dried, and finely ground to a uniform powder. The ground samples were digested using the Kjeldahl method [22] to determine the N concentration. According to Ostadi et al. [23] and Jackson [24], the potassium and phosphorus concentration was determined utilizing the flame photometer (Fater Electronic Model 620G, Fater Electronic, Iran) and spectrophotometric method using a UV-Vis spectrophotometer (Model AE-UV 1606, A & E Laboratory Instruments Co., China). The concentrations of copper, manganese, zinc, and iron were determined using an atomic absorption spectrometer following the method outlined by Asadi et al. [25].

### 2.3.3. Air-Dried Biomass and Essential Oil (LEO) Isolation

The shoots (stem, leaf, and flower) of lemon balm plants were harvested at the initial flowering stage on June 21 (HAR1), August 14 (HAR2), and October 6 (HAR3). Harvesting was carried out manually by using a sickle to cut all the plants in each plot from a height of 5 cm above the soil surface. The plants were dried in a shaded environment post-cutting to ensure the moisture content reached a stabilized level of 8%. Subsequently, the aboveground air-dried herbal product was quantified in kilograms per hectare.

The LEO was isolated using the hydro-distillation method. A total of 50 g of crushed shoot samples were combined with 500 mL of distilled water and it then underwent hydro-distillation for 3.5 h utilizing a Clevenger apparatus following the recommendations of the British Pharmacopoeia [26]. The LEO that was gathered underwent dehydration with anhydrous sodium sulfate and was subsequently placed in opaque vials at a temperature of 4 °C for the purpose of chemical composition analysis. The calculation of the LEO yield involved the multiplication of air-dried shoot biomass (kg $ha^{-1}$) with the LEO content (g $kg^{-1}$) and expressed as kg $ha^{-1}$.

### 2.3.4. Qualitative and Semiquantitative Analysis of LEO Samples

Utilizing a Thermo-UFM gas chromatography system, gas chromatography (GC) analysis was carried out with a flame ionization detector (FID) and a Ph-5 column

(10 m × 0.10 mm i.d., film thickness 0.25 μm). The LEO samples from each treatment were introduced into the GC/FID system. Helium served as the carrier gas, flowing at a rate of 0.5 mL $min^{-1}$. The oven temperature was varied from 60 to 285 °C while both the injector and FID detector were held at a constant temperature of 280 °C.

Gas chromatography-mass spectrometry analysis was carried out employing an Agilent 7890A/5975C GC-MS system with a DB-5 fused silica column. The LEO samples from different treatments were injected into the GC/FID system for quantification of the compounds while one of the samples was injected into GC/MS for identification of the compositions. The oven temperature and other parameters were set according to the established protocol. Identification of the chemical constituents in LEO samples involved matching their mass spectra with those stored in a computer library or obtained from authentic components. Subsequently, retention indices (RI) were determined by comparing the retention times of *n*-alkanes ($C_8$–$C_{24}$) injected under identical conditions. The confirmation of chemical compositions was achieved through the use of relative retention indices and references from reputable literature sources [27–29]. The area normalization method was applied to ascertain the percentage content of the compounds present in the essential oil, without any corrections made [30].

#### 2.3.5. Antioxidant Capacity

The assessment of antioxidant capacity was carried out by combining 20 μL of LEO from each treatment with 100 μL of 0.5 mM 2,2-diphenyl-1-picrylhydrazil (DPPH solution) in methanol. The resulting mixture was then adjusted to a final volume of 200 μL through microdilution. Afterward, the mixture was vigorously shaken and placed in a dark environment at room temperature for 15 min. The absorbance of the solutions, including a blank (without sample) and various concentrations, was measured using an Awareness Technology, Inc. (Palm City, FL, USA), Elisa reader at a wavelength of 517 nm. The inhibition percentage for the samples was calculated according to the formula proposed by Xiao et al. [31].

$$\text{Inhibition} = \left(\frac{AB - AA}{AB}\right) \times 100 \tag{3}$$

AB and AA were used to represent the absorbance values of the DPPH radical in the control and presence of the LEO sample, respectively. The resulting inhibition percentages were then plotted against the concentration of the samples. To determine the $IC_{50}$ value, which signifies the concentration at which 50% inhibition occurs, linear regression analysis was utilized, as described by Rostaei et al. [32]. The antioxidant capacity of the samples was evaluated using the DDPH assay following the methodology outlined by Fallah et al. [6].

### *2.4. Statistical Analysis*

The data were organized and subjected to analysis of variance (ANOVA) as an RCBD with the manure type (Mt) and harvest (H) as fixed effects. For the analysis, the PROC GLM procedure of SAS version 9.1 (SAS Institute Inc., Cary, NC, USA) was used. The means were compared and significant differences were determined using the LSD test at a significance level of $p < 0.05$. Various parameters, including biomass, photosynthetic pigments, macro-micronutrients, antioxidant capacity, LEO content, and yield, were measured over three consecutive harvests with 3 repetitions in each treatment. The essential oil compounds (%) during the first and second harvests were expressed as the mean ± SD.

## 3. Results

### *3.1. Effect on Photosynthesis Pigments*

Chlorophyll-a (Cha) and Chlorophyll-b (Chb) were affected by manure type and harvest (Table 3). In the HAR1 and HAR3, there were no significant differences observed in Cha levels between plants treated with livestock manures and the control ($p > 0.05$). However, in the HAR2, the Cha content in plants treated with poultry manure was higher compared to the CO treatment, measuring 6.6 and 4.5 μg $mL^{-1}$, respectively ($p < 0.05$;

Table 4). The Cha levels in plants treated with sheep manure did not show a significant variance from the CO treatment ($p > 0.05$; Table 4).

**Table 3.** Analysis of variance for photosynthesis pigments, nutrients concentration, biomass, essential oil, and antioxidant activity of different harvests of lemon balm cultivated in soil amended with livestock manures.

| Source of Variance | Cha | Chb | N | P | K | Fe | Zn | Cu | Mn | Biomass | LEO Content | LEO Yield | AC |
|---|---|---|---|---|---|---|---|---|---|---|---|---|---|
| Block | NS | NS | NS | NS | NS | NS | NS | NS | NS | NS | NS | NS | NS |
| Manure type (Mt) | ** | * | NS | * | ** | NS | *** | ** | * | *** | NS | *** | ** |
| Harvest (H) | *** | ** | NS | NS | *** | ** | *** | *** | NS | *** | *** | *** | * |
| Mt × H | NS | NS | ** | ** | * | NS | * | ** | NS | NS | ** | NS | * |

NS, *, ** and ***, nonsignificant, significant $p < 0.05$, $p < 0.01$, and $p < 0.001$, respectively. Cha: chlorophyll-a; Chb: chlorophyll-b. N: nitrogen; P: phosphorus; K: potassium, Fe: iron; Zn: zinc; Cu: copper; Mn: manganese; LEO: lemon balm's essential oils; AC: antioxidant activity.

**Table 4.** Mean comparison for photosynthesis, iron concentration, manganese concentration, biomass, and essential oil yield of different harvests of lemon balm cultivated in soil amended with livestock manures.

| Treatments | Cha ($\mu g\ mL^{-1}$) | Chb ($\mu g\ mL^{-1}$) | Iron ($mg\ kg^{-1}$) | Manganese ($mg\ kg^{-1}$) | Biomass ($kg\ ha^{-1}$) | LEO Yield ($kg\ ha^{-1}$) |
|---|---|---|---|---|---|---|
| Harvest 1 | | | | | | |
| PO | 5.72 ab | 3.95 a | 417 c | 27.0 ab | 1448 a | 3.22 cd |
| SH | 4.68 b–d | 2.96 a–c | 345 c | 24.6 ab | 1228 ab | 3.84 b–d |
| CA | 5.41 a–c | 2.99 a–c | 278 c | 22.6 b | 1282 ab | 2.88 c–e |
| CO | 4.29 b–f | 2.40 b–d | 327 c | 21.6 b | 895 cd | 1.54 f |
| Harvest 2 | | | | | | |
| PO | 6.61 a | 3.15 ab | 399 c | 28.0 ab | 1152 bc | 5.43 a |
| SH | 4.87 b–d | 2.56 b–d | 1129 a | 30.9 a | 1088 bc | 3.89 bc |
| CA | 5.35 a–c | 2.71 bc | 945 ab | 31.6 a | 1164 b | 4.97 ab |
| CO | 4.48 b–e | 2.04 b–d | 492 bc | 24.8 ab | 771 d | 3.32 cd |
| Harvest 3 | | | | | | |
| PO | 4.07 c–f | 2.34 b–d | 440 c | 26.8 ab | 702 d | 2.64 d–f |
| SH | 3.15 ef | 1.82 cd | 443 ab | 26.5 ab | 729 d | 3.73 cd |
| CA | 3.55 d–f | 1.88 cd | 480 bc | 23.5 b | 678 d | 2.95 c–f |
| CO | 2.90 f | 1.47 d | 388 c | 27.0 ab | 386 e | 1.89 ef |

Means with a similar letter are not significantly different according to LSD test ($p \leq 0.05$). PO: poultry manure; SH: sheep manure; CA: cattle manure; CO: control (without manure); Cha: chlorophyll-a; Chb: chlorophyll-b.

In the HAR1, the Chb content in plants treated with poultry manure was found to be similar to those treated with sheep and cattle manure ($p > 0.05$) but significantly different from the unfertilized plants ($p < 0.05$; Table 4). Conversely, in the HAR2 and HAR3, there were no significant differences observed in Cha levels between plants treated with livestock manures and the CO treatment ($p > 0.05$; Table 4).

### 3.2. *Effect on Nutrient Concentration*

As shown in Table 3, the effect of manure treatments on the concentration of potassium, phosphorus, zinc, manganese, and copper in lemon balm leaves was statistically significant. The amount of potassium, iron, zinc, copper, and manganese was affected by harvest. The concentration of nitrogen ($p < 0.01$), phosphorus ($p < 0.01$), potassium ($p < 0.05$), copper ($p < 0.01$), and zinc ($p < 0.05$) was influenced by manure type and harvest interaction.

At each harvest, the concentration of manganese in plants treated with manures did not differ significantly (Table 4). In the HAR1, the iron concentration of plants treated with manure was found to be similar to the unfertilized plants ($p > 0.05$; Table 4). Conversely, in the HAR2 and HAR3, the iron concentration in plants treated with sheep manure (1129 and 443 mg kg$^{-1}$, respectively) was significantly higher than those treated with poultry manure and the control treatment ($p < 0.05$). However, there was no significant difference in iron concentration between the cattle manure treatment and the control treatment ($p > 0.05$; Table 4).

Figures 1 and 2 show the changes in the concentration of potassium, phosphorus, nitrogen, zinc, and copper in different treatments during consecutive harvests. In the HAR1, the levels of Cu, Zn, K, P, and N in the leaves of the manure-treated plants showed no significant variation compared to those in the CO treatment, except for the concentration of P and Zn in the CA and CO treatments, respectively ($p > 0.05$; Figures 1 and 2). In the HAR2, the nitrogen concentration in the SH and CA treatments and the phosphorus concentration in the CA treatment decreased compared to the PO treatment ($p < 0.05$; Figure 1) but they did not have significant differences with the CO treatment ($p > 0.05$). In the HAR2, the leaf potassium concentration in manure treatments was similar to the control ($p > 0.05$; Figure 1). In the HAR2, the concentration of zinc in the PO and SH treatments was significantly lower than in the CO treatment. The concentration of copper in the SH treatment was significantly lower than in the PO and CA treatments ($p < 0.05$; Figure 2).

In the HAR3, the highest concentrations of nitrogen, phosphorus, and zinc belonged to the SH treatment ($p < 0.05$) and there was no significant change between the PO and CA treatments compared with the CO ($p > 0.05$). In this harvest, the potassium concentration of manure treatments was considerably higher than the CO treatment ($p < 0.05$; Figures 1 and 2). However, the copper concentration of the CA treatment was higher compared to the PO and CO treatments ($p < 0.05$; Figure 2).

### 3.3. *Effect on Biomass*

The biomass of the lemon balm's aerial parts was altered by manure types ($p < 0.001$) and harvest ($p < 0.001$). The biomass in the manure treatments was in the range of 1015–1100 kg ha$^{-1}$ ($p > 0.05$). The difference between the manure treatments and the CO treatment was 48–61% ($p < 0.05$; Table 3). Across all harvests, the amount of biomass from manure treatments exhibited a higher level compared to the CO treatment ($p < 0.05$; Table 4). There was an observable decline in biomass from the initial to the final harvest ($p < 0.05$). The variation in biomass levels for plants treated with manure in the HAR1, HAR2, and HAR3 ranged from 1282–1448, 1088–1164, and 678–729 kg ha$^{-1}$, respectively ($p > 0.05$; Table 4).

### 3.4. *Effect on Lemon Balm's Essential Oils*

As presented in Table 3, the LEO content was influenced by harvest ($p < 0.001$) and manure type × harvest interaction ($p < 0.01$), while the LEO yield was altered by manure type ($p < 0.001$) and harvest ($p < 0.001$). In the HAR2 and HAR3, the LEO content significantly increased compared to the HAR1 ($p < 0.05$; Figure 3a). In the HAR1, the LEO content of the plants treated with PO and CA treatments did not show a significant difference compared with the CO treatment ($p > 0.05$) but in plants treated with SH it was 83% higher than the CO treatment ($p < 0.05$). In the HAR2, the LEO content in manure treatments was not significantly different from that of the CO treatment ($p > 0.05$). In the HAR3, the LEO content in the CA and SH treatments was similar to the CO treatment ($p > 0.05$) but in the PO treatment, the LEO content was 24% lower than that of the CO treatment ($p < 0.05$; Figure 3a).

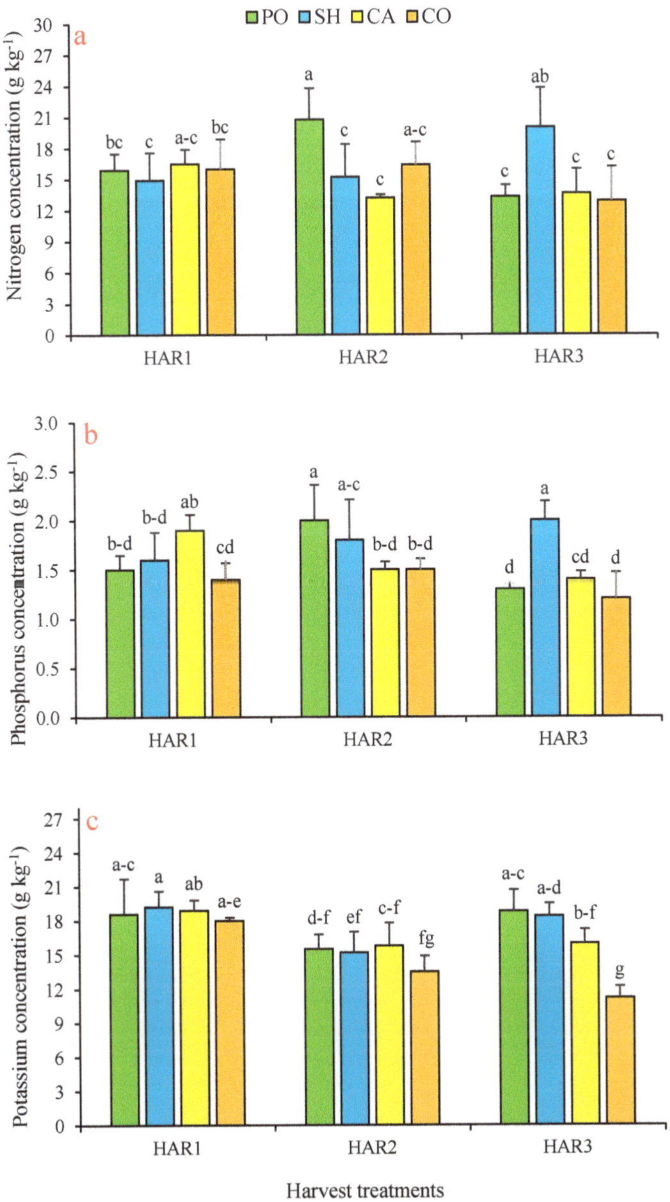


**Figure 1.** Concentration of nitrogen (**a**), phosphorus (**b**), and potassium (**c**) in different harvests of lemon balm cultivated in soil amended with livestock manures. Means with a similar letter are not significantly different according to LSD test ($p \leq 0.05$). Bars represent standard deviation. PO: poultry manure; SH: sheep manure; CA: cattle manure; CO: control (without manure); HAR1: first harvest; HAR2: second harvest; HAR3: third harvest.

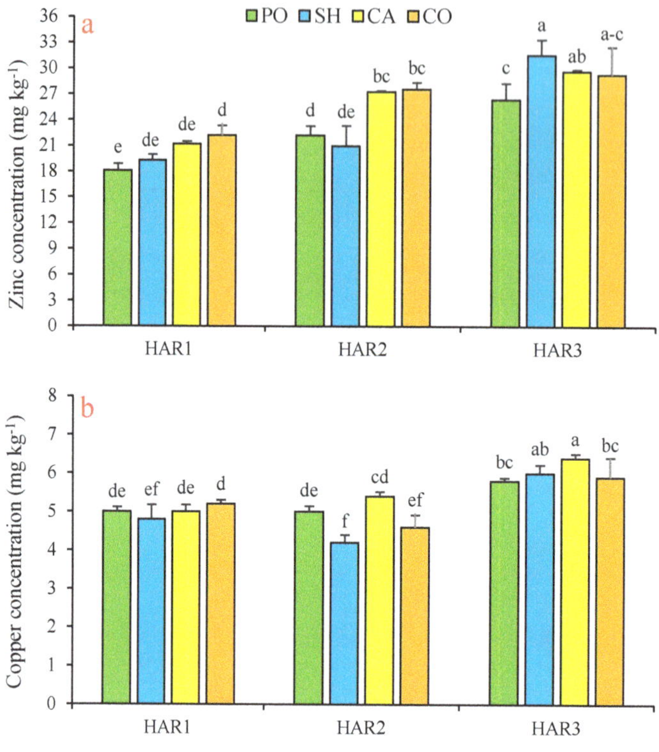


**Figure 2.** Concentration of zinc (**a**) and copper (**b**) in different harvests of lemon balm cultivated in soil amended with livestock manures. Means with a similar letter are not significantly different according to LSD test ($p \leq 0.05$). Bars represent standard deviation. PO: poultry manure; SH: sheep manure; CA: cattle manure; CO: control (without manure); HAR1: first harvest; HAR2: second harvest; HAR3: third harvest.

The LEO yield was changed by the manure type and harvest ($p < 0.001$; Table 3). In the HAR1 and HAR3, no significant difference among manure treatments was observed for the LEO yield ($p > 0.05$). Nevertheless, the LEO yield in the manure treatments showed a substantial increase of 87–109% and 44–97% compared to the CO treatment ($p < 0.05$; Table 4). In the HAR2, the greatest LEO yield was recorded in poultry manure at 5.43 kg ha$^{-1}$, which was statistically similar to the EO yield from cow manure at 4.97 kg ha$^{-1}$ (Table 4). Poultry manure exhibited EO yield values that were 40 and 64% higher than those of sheep manure and the control treatment, respectively ($p < 0.05$; Table 4).

Based on the GC-MS and GC-FID analyses, 10 chemical compounds were identified in LEO with a total percentage ranging from 91 to 95% (Table 5, Figure A1). Table 5 shows the effects of manure type (Mt) and harvest (H) and their interaction effects for different chemical compositions. Compounds such as 1-octan-3-ol, 6-methyl-5-hepten-2-one, and nerol were found to be unaffected by the type of manure and harvest (Table 5). The type of fertilizer had a significant effect on the levels of linalool, citronellal, neral, geranial, geranyl acetate, (*E*)-caryophyllene, and caryophyllene oxide (Table 5). The main chemical component observed was geranial, accounting for 39–46% of the total composition. This was followed by neral (28–35%), (*E*)-caryophyllene (4.7–11%), geranyl acetate (2.7–5.9%), and caryophyllene oxide (1.7–4.8%) (Table 6; Figures 4 and 5). The lemon balm treated with livestock manures exhibited a decrease in the amount of (*E*)-caryophyllene compared to CO of HAR1 (Table 6). Additionally, the plants treated with PO and SH showed lower levels of geranyl acetate and citronellal, respectively, compared to the HAR1 control treatment (Table 6).

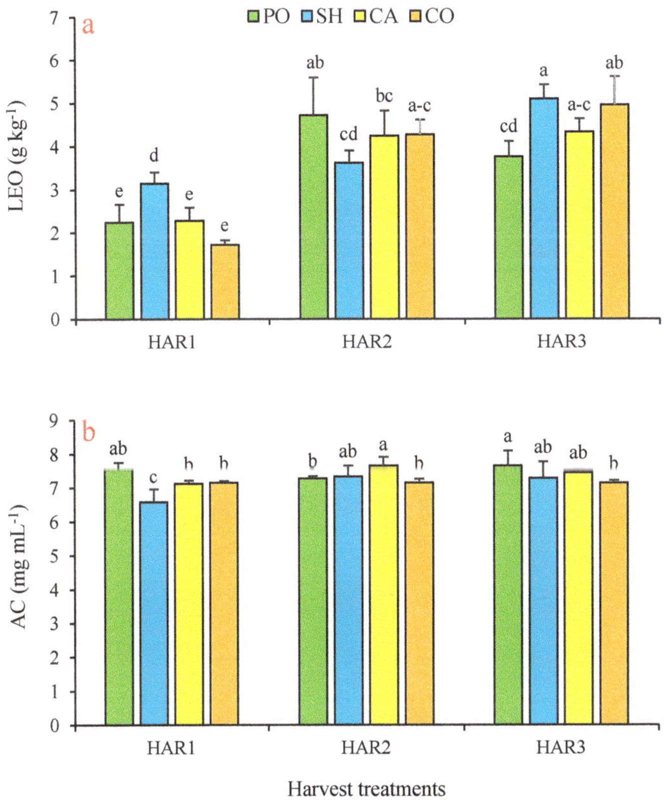


**Figure 3.** Essential oil content (LEO) (**a**) and antioxidant activity (AC) (**b**) in different harvests of lemon balm cultivated in soil amended with livestock manures. Means with a similar letter are not significantly different according to LSD test ($p \leq 0.05$). Bars represent standard deviation. PO: poultry manure; SH: sheep manure; CA: cattle manure; CO: control (without manure); HAR1: first harvest; HAR2: second harvest; HAR3: third harvest.

**Table 5.** Analysis of variance for the chemical compositions of different harvests of lemon balm cultivated in soil amended with livestock manures.

| Source of Variation | Retention Index (RI) | | | | | | | | | |
|---|---|---|---|---|---|---|---|---|---|---|
| | 975 | 987 | 1100 | 1155 | 1230 | 1240 | 1265 | 1383 | 1415 | 1580 |
| Block | NS | NS | NS | NS | NS | NS | NS | NS | NS | NS |
| Manure type (Mt) | NS | NS | *** | ** | NS | *** | * | ** | ** | *** |
| Harvest (H) | NS | NS | *** | NS | NS | *** | *** | *** | *** | *** |
| Mt × H | NS | NS | *** | NS | NS | ** | * | NS | NS | ** |

RI: Kovats index on the DB-5 column. NS, *, **, and ***, nonsignificant, significant $p < 0.05$, $p < 0.01$, and $p < 0.001$, respectively. 975: 1-octan-3-ol; 987: 6-methyl-5-hepten-2-one; 1100: linalool; 1155: citronellal; 1230: nerol; 1240: neral; 1265: geranial; 1383: geranyl acetate; 1415: (*E*)-caryophyllene; 1580: caryophyllene oxide.

**Table 6.** Mean comparison for citronellal, geranyl acetate, and (*E*)-caryophyllene of different harvests of lemon balm cultivated in soil amended with livestock manures.

| Treatments | | Citronellal (%) | Geranyl Acetate (%) | (*E*)-Caryophyllene (%) |
|---|---|---|---|---|
| Harvest 1 | | | | |
| | PO | 1.22 a–d | 3.82 cd | 7.45 b |
| | SH | 0.92 d | 5.90 a | 7.56 b |
| | CA | 1.23 a–d | 4.34 bc | 7.22 bc |
| | CO | 1.38 ab | 5.29 ab | 10.7 a |
| Harvest 2 | | | | |
| | PO | 1.09 b–d | 2.74 e | 4.72 d |
| | SH | 1.04 cd | 3.59 c–e | 6.15 b–d |
| | CA | 1.50 a | 3.12 de | 5.72 cd |
| | CO | 1.30 a–c | 3.29 c–e | 6.41 bc |

Means with a similar letter are not significantly different according to LSD test ($p \leq 0.05$). PO: poultry manure; SH: sheep manure; CA: cattle manure; CO: control.

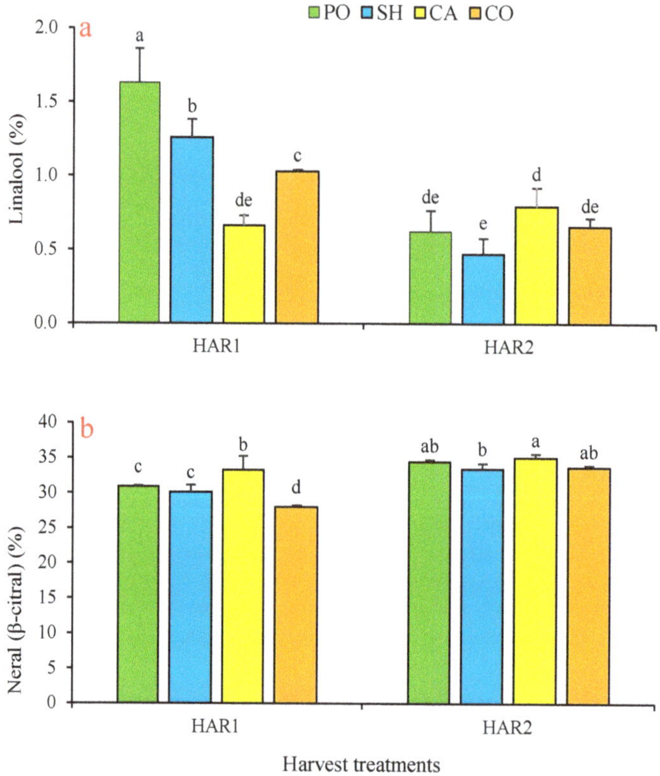


**Figure 4.** Linalool (**a**) and neral (**b**) in different harvests of lemon balm cultivated in soil amended with livestock manures. Means with a similar letter are not significantly different according to LSD test ($p \leq 0.05$). Bars represent standard deviation. PO: poultry manure; SH: sheep manure; CA: cattle manure; CO: control (without manure); HAR1: first harvest; HAR2: second harvest.

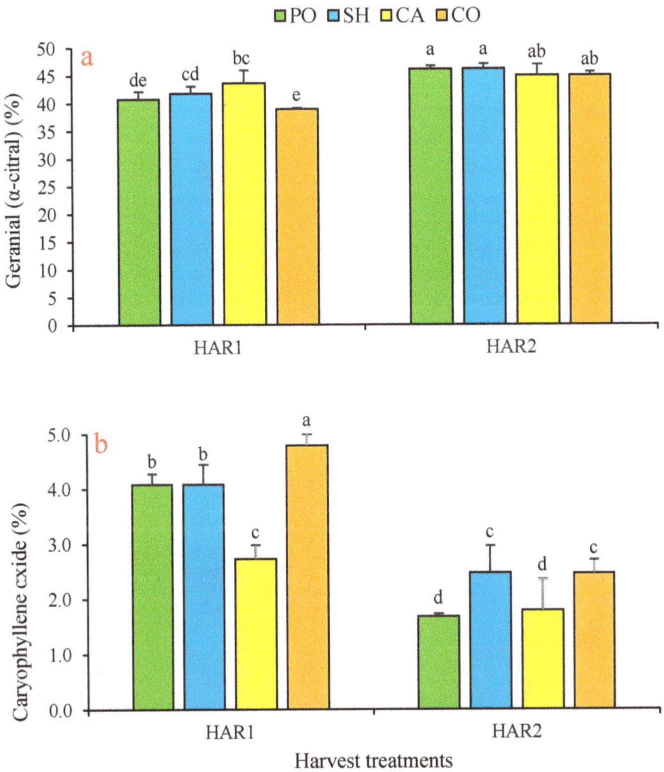


**Figure 5.** Geranial (**a**) and caryophyllene oxide (**b**) in different harvests of lemon balm cultivated in soil amended with livestock manures. Means with a similar letter are not significantly different according to LSD test ($p \leq 0.05$). Bars represent standard deviation. PO: poultry manure; SH: sheep manure; CA: cattle manure; CO: control (without manure); HAR1: first harvest; HAR2: second harvest.

The interaction effect of Mt × H was found to be significant for linalool ($p < 0.001$), neral ($p < 0.01$), geranial ($p < 0.05$), and caryophyllene oxide ($p < 0.01$) (Table 5). In the HAR1, the plants treated with PO and SH exhibited higher levels of linalool, neral, and geranial compared to the control, although no significant difference was observed in the HAR2 (Figures 4 and 5a). Moreover, in the HAR1, the plants treated with CA had lower levels of linalool and caryophyllene oxide but higher levels of geranial and neral compared to the control. In the HAR2 plants (Figures 4 and 5), the levels of linalool, neral, and caryophyllene oxide in the plants treated with CA were similar to the control (Figures 4 and 5b).

### 3.5. Effect on Antioxidant Capacity

The effect of Mt, H, and Mt × H on the LEO's antioxidant capacity (AC) was significant (Table 3). In the HAR1, the AC of the SH treatment was 8% less than that of the CO treatment ($p < 0.05$). But the AC of the PO and CA treatments was similar to that of the CO treatment ($p > 0.05$; Figure 3b). In the HAR2, the AC of the CA treatment exhibited a higher value compared to the CO treatments ($p < 0.05$). Conversely, the PO and SH treatments did not show any significant difference when compared with the CO treatment ($p > 0.05$; Figure 3b). In the HAR3, the AC of the PO treatment was significantly higher than the CO treatment; other manures did not show a significant contrast when compared with the CO treatment (Figure 3b).

## 4. Discussion

### 4.1. Photosynthesis Pigments

The increase in chlorophyll content in the PO treatment and the subsequent decrease in chlorophyll levels during the third harvest can be attributed to the role of nitrogen in chlorophyll and protein molecules. Nitrogen is a crucial component for the formation of chloroplasts and the accumulation of chlorophyll within them [33]. According to Padilla et al. [34], there was a strong and positive relationship between chlorophyll a + b content and all chlorophyll measurements in sweet pepper with curvilinear relationships. Furthermore, there was a slight inclination toward stronger correlations with chlorophyll-a content compared to chlorophyll-b content for all chlorophyll measurements. In the PO treatment, the smaller particle size led to faster mineralization, resulting in higher nitrogen levels compared to other animal manure [32,35]. The continuous uptake of nitrogen across multiple harvests resulted in a decline in nitrogen concentrations within the rhizosphere, subsequently causing a reduction in chlorophyll content [36].

### 4.2. Nutrient Concentration

The elevated levels of nitrogen and phosphorus in lemon balm, supplied with PO and SH at HAR2, and HAR3, respectively, can be attributed to the availability of higher mineralized nitrogen during an extended period of plant growth. This was corroborated by the reports of Alizadeh et al. [20] and Baghdadi et al. [37], who also observed an improvement in soil conditions for nitrogenase activity in these specific treatments. In separate investigations conducted by Javanmard et al. [27] and Asadi et al. [25] on peppermint (*Mentha × piperita* L.), the application of vermicompost resulted in higher nitrogen levels compared to the control plants. Similarly, Bajeli et al. [38] found that the application of PO and CA treatments increased the nitrogen content in the leaves of mint (*Mentha arvensis* L.) compared to the CO treatment. Anwar et al. [39] described that the utilization of vermicompost also enhanced the nitrogen concentration in French basil (*Ocimum basilicum* L.) when compared to the control plants.

The phosphorus and nitrogen concentration patterns observed in the three harvests suggest that the peak mineralization of the PO and SH treatments aligns with the second and third harvests, respectively. However, in the case of the CA treatment, it is likely that after the mineralization of rapidly decomposing substances, the mineralization process of lignin materials has decreased. As a result, the release of phosphorus and nitrogen is restricted [26].

Zhu et al. [40] discovered that cattle manure (CA) had higher initial polysaccharide (cellulose and hemicellulose) concentrations compared to sheep manure across all sites. The order of initial cellulose concentrations was found to be cattle manure > sheep manure, while the opposite order was observed for initial crude protein concentrations: sheep manure > cattle manure. However, there were no significant changes in initial lignin concentrations among the different types of manure and the four sites. According to the findings of Markewich et al. [41], fresh manure exhibits a significant $NH_4$+ concentration resulting from the high decomposition rate of urea derived from urine. These processes include immobilization by the microbial decomposer population, volatilization as $NH_3$, nitrification followed by leaching, or further denitrification resulting in the emission of $N_2$, or NOx [42].

The potassium requirement of the plant has frequently been met through manure treatments, resulting in no significant alteration in potassium concentration. Nevertheless, the reduced potassium concentration observed in lemon balm harvested in the HAR3 of CO treatment can be attributed to an augmentation in plant biomass during the two preceding harvests, leading to a dilution of potassium content in the plants. Asadi et al. [25] found that the potassium concentration in peppermint harvested from HAR1 was higher than that in peppermint harvested from HAR2.

Iron and manganese levels in the HAR2 were found to be higher compared to the first and third harvests. This could be ascribed to the existence of substances like several secondary metabolites bearing an azetidine ring in the rhizosphere of some Lamiaceae species, which are more abundant in a larger volume of soil and modify the soil nutrient availability. Consequently, the plant absorbed a greater quantity of these two elements [43–45].

In the HAR3, the prevailing environmental conditions could have led to a decrease in the quantity and rate of recent carbon allocation during the belowground growth. Consequently, there has been an elevation in the allocation of this carbon toward respiratory use or sugar and starch reserves in roots, aiming to enhance winter hardiness [46,47]. The high levels of zinc and copper in plant tissue of the HAR3 in comparison to iron and manganese suggest that the expanded rhizosphere enhances the presence of these two elements (Zn and Cu). Additionally, the successive harvests resulted in a rise in their concentration in the plant, particularly in the CA and CO treatment, which exhibited lower biomass [32].

### 4.3. Lemon Balm Biomass

The utilization of manure led to significant improvements in various soil properties and biological activities [48]. These improvements included an increase in water-stable aggregation by 29%, soil organic carbon by 18%, available nitrogen by 16%, available phosphorus by 66%, and available potassium by 19% as well as the activities of urease by 25%, sucrase by 18%, and catalase by 16%. Additionally, the abundances of fungi, actinomyces, and bacteria were found to increase by 28, 60, and 38%, respectively. These findings suggest that the application of manure enhances the nutrient reservoir and decomposition capability of the soil, which in turn positively impacts crop yield. Abd-ElKader et al. [49] also demonstrated that the utilization of compost sharply raised peppermint biomass in three harvests in two consecutive seasons when compared to the control plants. Furthermore, another investigation revealed that the maximum biomass of *Thymus vulgaris* L. was achieved in thyme that received treatments of quail and cattle manure [50].

Typically, the decline in biomass over consecutive harvests can be ascribed to the plant's aging along with variations in temperature and lighting duration [23]. Likewise, several studies have indicated that peppermint harvested in the HAR2 exhibited lower biomass than in the HAR1 [23,51].

### 4.4. Lemon Balm's Essential Oils

The rise in LEO content in the HAR2 and HAR3, particularly in the CA and CO treatments, may be attributed to the limited nutrient availability (CO treatment), slow manure decomposition rate (CA treatment), and increased nutrient uptake by previous harvests (PO and SH treatments). Additionally, the leaves in the HAR3 were immature and the temperature was decreasing. These factors combined have resulted in the allocation of photosynthetic assimilates toward LEO production to enhance tolerance to unfavorable environmental conditions. Tripathi and Hazarika [52] found that the essential oil concentration of immature leaves (4.5%) of *Pogostemon cablin* (Blanco) Benth. was higher compared to vegetative and fully mature leaves (4.0, and 3.8%, respectively). This higher concentration in immature leaves can be attributed to the biogenetic activity of these leaves. Young leaves exhibit greater biogenetic activity compared to mature leaves, which means they have a higher rate of essential oil synthesis and accumulation [53].

In a separate study, Manjunatha et al. [54] observed a similar oil percentage for harvests conducted 6, 9, and 12 months after planting *P. cablin*. However, there was a significant decline in the essential oil yield for these harvests, with a more pronounced decrease noted from the initial harvest to the subsequent one. The LEO yield is dependent on both the biomass and the LEO content. As there is no significant variation in LEO content among manure treatments, it can be inferred that changes in biomass are the primary factor influencing the LEO yield. Consequently, the application of manure treatments, which provide essential nutrients and other advantageous effects [48], has resulted in a substantial increase in LEO yield in plants.

In terms of the plant's age (harvests), it is evident that the impact of the LEO content in determining the LEO yield is more significant when compared to the biomass. Thus, the LEO content not only offsets the reduction in biomass but also leads to a substantial increase in the LEO yield during the second harvest. This outcome highlights the importance of synthesizing primary and secondary metabolites under unfavorable conditions, as the temperature and daylight duration decrease in the second and third harvests along with a decline in nutrient availability.

Employing sequential harvesting can be an effective method for generating a superior end product that possesses enhanced bioactive properties [55]. Citral is a blend of geranial and neral that occurs naturally in various essential oils derived from citrus fruits, herbs, or spices [56]. The antimicrobial properties of citral have been proven effective against bacteria and fungi under various circumstances [57]. Given its widespread use across multiple industries, citral has gained approval from regulatory authorities in Europe and the US, earning the designation of being generally recognized as a safe (GRAS) substance [56]. Furthermore, it is a compound that finds extensive application in the pharmaceutical and cosmetic sectors [58]. Limonene, citral, and linalool are frequently found as nonphenolic terpenoid constituents in essential oils. These components have been associated with antioxidant properties, although there is some debate surrounding their effectiveness. Among these compounds, citral exhibits the highest reactivity due to the presence of the aldehyde group [59].

The presence of nutrients can impact the chemical compositions of the LEO. Phosphorus and nitrogen are crucial in improving plant photosynthesis and the synthesis of primary and secondary metabolites in plants, thereby supporting terpenoid production [60,61]. Terpenoid biosynthesis takes place via the mevalonic pathway and methylerythritol phosphate pathways, which do not conflict directly with nitrogen resource availability [61]. Additionally, the terpenoid precursors, such as isopentenyl diphosphate, farnesyl diphosphate, dimethylallyl pyrophosphate, and geranyl diphosphate, contain phosphorus as a structural component. Therefore, an increase in phosphorus availability could potentially lead to a rise in these pathways and subsequent terpene production [61,62]. Micronutrients in plants act as cofactors for important enzymes involved in secondary metabolism, thereby positively influencing the synthesis and accumulation of plant secondary metabolites [63]. Furthermore, potassium application has been found to enhance the content and quality of essential oils extracted from *Mentha spicata* L. according to Chrysargyris et al. [64]. Therefore, it is important for lemon balm plants to uptake sufficient nutrients to meet their growth and terpenoid synthesis requirements [65].

### 4.5. Antioxidant Capacity (AC)

The chemical components of LEO may contribute to plant antioxidant activity, particularly the main compounds [66]. It should be emphasized that this property is not solely dependent on the high concentration of a single compound [67]. This is because both the minor and major components can work together synergistically in the LEO, enhancing the overall antioxidant activity of plant essential oil [68]. Sheep and poultry manure gave a greater percentage in geranyl acetate and linalool compared to the control treatment. As a result, it exhibited the strongest antioxidant capacity. Moreover, it is evident that there is a direct correlation between the level of essential oil and antioxidant capacity. Hence, sheep and poultry manures demonstrated the highest antioxidant capacity in the HAR1 and HAR2, respectively. The enhanced antioxidant activity of this manure could be attributed to its ability to supply a significant amount of geranyl acetate and oxygenated monoterpenes, as suggested by Deba et al. [69], Fallah et al. [70], and Tohidi et al. [71]. Ozliman et al. [72] demonstrated that the utilization of various doses of farmyard manure not only enhances the EO content and components but also serves as a means to augment the antioxidant activity of *Anethum graveolens* L.

## 5. Conclusions

Our research findings indicate that lemon balm can thrive in cold semiarid climates and can be harvested three times a year. The application of animal manure resulted in a substantial increase in biomass (41–60%) and essential oil yield (60–71%) compared to the control. The use of livestock manure, particularly sheep and/or poultry manure, demonstrated a notable improvement in the essential oil's quality, as evidenced by the increase in its main components and enhanced antioxidant capacity. Notably, during mid-August and early October, there was a significant boost in the essential oil yield and citral levels. However, the decrease in geranyl acetate and oxygenated monoterpenes led to a decline in antioxidant capacity. Consequently, it is recommended to utilize essential oils from the second and third harvests for industrial purposes. Overall, employing animal manures, especially sheep manure, as a nutrient source for lemon balm cultivation proves to be a viable approach for producing high-quality essential oils and brings about the sustainability of the agricultural ecosystem.

**Author Contributions:** S.F.: project administration, data analysis, conceptualization, writing—reviewing and editing, supervision. F.M.: conceptualization, writing—reviewing and editing. A.G.-O.: investigation, methodology, data curation, writing—original draft preparation. M.R.: writing—original draft preparation. All authors have read and agreed to the published version of the manuscript.

**Funding:** This study was funded, in part, by the Shahrekord University (grant #96GRN1M731 to SF).

**Data Availability Statement:** The raw data supporting the conclusions of this article will be made available by the authors on request.

**Acknowledgments:** The authors would like to thank the Shahrekord University for providing financial support.

**Conflicts of Interest:** The authors declare no conflicts of interest.

## Appendix A

**Table A1.** Monthly average temperature and total monthly precipitation in 2019 and long-term (2016 –2019).

| | Average Temperature (°C) | | Average Minimum Temperature (°C) | | Average Maximum Temperature (°C) | | Precipitation (mm) | |
|---|---|---|---|---|---|---|---|---|
| **Month** | **2016–2019** | **2019** | **2016–2019** | **2019** | **2016–2019** | **2019** | **2016–2019** | **2019** |
| January | −0.06 | 0.1 | −6.6 | −5.5 | 5.5 | 5.7 | 262.6 | 615.2 |
| February | 0.2 | −1.1 | −6.3 | −7.4 | 6.8 | 5.2 | 143.2 | 204.7 |
| March | 4.4 | 0.3 | −1.7 | −6.5 | 10.5 | 7.1 | 281.1 | 561.4 |
| April | 8.7 | 6.2 | 2.9 | 1.7 | 14.4 | 10.8 | 181.1 | 278.87 |
| May | 14.3 | 14.1 | 7.2 | 7.1 | 21.4 | 21.1 | 76.3 | 17.1 |
| June | 19.2 | 19.8 | 10.2 | 11.0 | 28.1 | 28.6 | 1.5 | 0.2 |
| July | 22.7 | 22.8 | 14.0 | 14.4 | 31.4 | 31.3 | 0.0 | 0.0 |
| August | 22.1 | 21.9 | 13.1 | 13.1 | 31.0 | 30.8 | 0.0 | 0.0 |
| September | 18.6 | 18.4 | 9.6 | 9.1 | 27.5 | 27.8 | 0.3 | 0.0 |
| October | 12.8 | 13.0 | 5.3 | 6.2 | 20.2 | 19.9 | 26.5 | 58.5 |
| November | 5.2 | 2.5 | −1.6 | −5.8 | 12.1 | 10.7 | 123.5 | 53.5 |
| December | 1.7 | −2.3 | −4.7 | −9.2 | 8.0 | 4.6 | 235.8 | 229.8 |

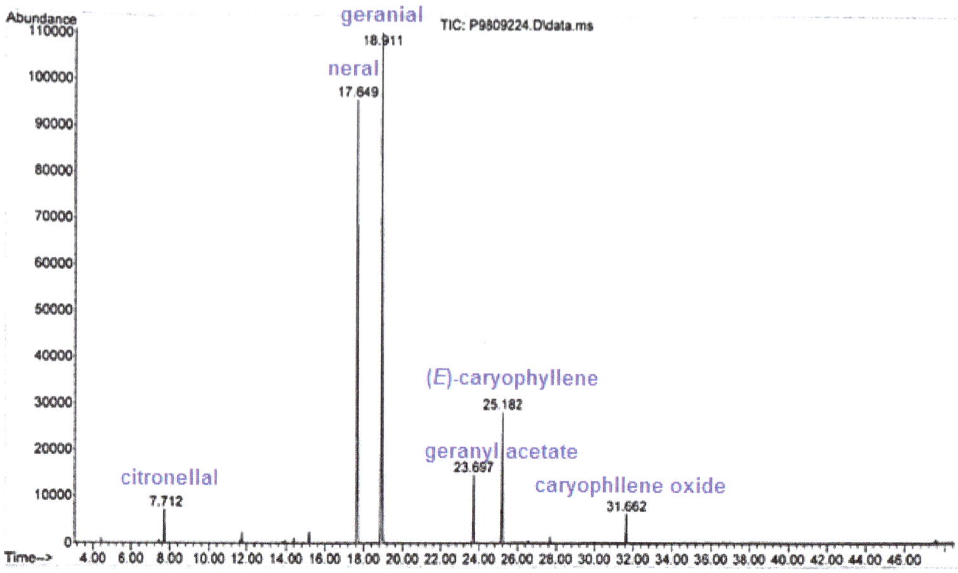


**Figure A1.** GC-MS profile of essential oil from aerial parts of *Melissa officinalis* treated with livestock manure.

## References


1. Rostaei, M.; Fallah, S.; Lorigooini, Z.; Surki, A.A. Crop productivity and chemical compositions of black cumin essential oil in sole crop and intercropped with soybean under contrasting fertilization. *Ind. Crop. Prod.* **2018**, *125*, 622–629. [CrossRef]
2. Hatfield, J.L.; Stewart, B.A. *Animal Waste Utilization: Effective Use of Manure as a Soil Resource*; CRC Press: Boca Raton, FL, USA, 2019; p. 328.
3. Aziz, T.; Ullah, S.; Sattar, A.; Nasim, M.; Farooq, M.; Khan, M.M. Nutrient availability and maize (*Zea mays*) growth in soil amended with organic manures. *Int. J. Agric. Biol.* **2010**, *12*, 621–624.
4. Pandey, V.; Patel, A.; Patra, D.D. Integrated nutrient regimes ameliorate crop productivity, nutritive value, antioxidant activity and volatiles in basil (*Ocimum basilicum* L.). *Ind. Crop. Prod.* **2017**, *87*, 124–131. [CrossRef]
5. Almasi, F. Organic fertilizer effects on morphological and biochemical traits and yield in coriander (*Coriandrum sativum* L.) as an industrial and medicinal plant. *Agrotech. Ind. Crop.* **2021**, *1*, 19–23. [CrossRef]
6. Fallah, S.; Ghanbari-Odivi, A.; Rostaei, M.; Maggi, F.; Shahbazi, E. Improvement of production and quality of essential oils in multi-cut peppermint (*Mentha × piperita* L.) through eco-friendly fertilizers in the semi-arid highlands. *Ind. Crop. Prod.* **2024**, *216*, 118801. [CrossRef]
7. Ghanbari-Odivi, A.; Fallah, S.; Carrubba, A. Optimizing hyssop (*Hyssopus officinalis* L.) cultivation: Effects of eifferent manures on plant growth and essential oil yield. *Horticulturae* **2024**, *10*, 894. [CrossRef]
8. Nasiri, Y. Crop productivity and chemical compositions of dragonhead (*Dracocephalum moldavica* L.) essential oil under different cropping patterns and fertilization. *Ind. Crop. Prod.* **2021**, *171*, 113920. [CrossRef]
9. Habibi Sharafabad, Z.; Abdipour, M.; Hosseinifarahi, M.; Kelidari, A.; Rashidi, L. Integrated humic acid and vermicomposting changes essential oil quantity, and quality in field-grown *Lavandula angustifolia* L. intercropped with *Brassica nigra* L. *Ind. Crop. Prod.* **2022**, *178*, 114635. [CrossRef]
10. Sadat Darakeh, S.A.S.; Weisany, W.; Diyanat, M.; Ebrahimi, R. Bio-organic fertilizers induce biochemical changes and affect seed oil fatty acids composition in black cumin (*Nigella sativa* Linn). *Ind. Crop. Prod.* **2021**, *164*, 113383. [CrossRef]
11. Serri, F.; Souri, M.K.; Rezapanah, M. Growth, biochemical quality and antioxidant capacity of coriander leaves under organic and inorganic fertilization programs. *Chem. Biol. Technol. Agric.* **2021**, *8*, 1–8. [CrossRef]
12. Sajwan, M.Y.; Mishra, D.; Negi, M.S.; Bisht, P.S. Effect of organic manures on antioxidant activity and essential oil composition of *artemisia annua* cv. CIM arogya. *Int. J. Curr. Microbiol. App. Sci.* **2020**, *9*, 3922–3930. [CrossRef]
13. Khan, S. Lemon Balm. Available online: https://www.medicinenet.com/lemon_balm/article.htm (accessed on 15 September 2024).
14. Miraj, S.; Rafieian-Kopaei, M.; Kiani, S. *Melissa officinalis* L: A review study with an antioxidant prospective. *J. Evid. Based Complement. Altern. Med.* **2017**, *22*, 385–394. [CrossRef] [PubMed]
15. Sodre, A.; Luz, J.; Haber, L.; Marques, M.; Rodrigues, C.; Blank, A.F. Organic and mineral fertilization and chemical composition of lemon balm (*Melissa officinalis* L.) essential oil. *Rev. Bras. Farmacogn.* **2012**, *22*, 40–44. [CrossRef]

16. Yaldiz, G.; Kasko Arici, Y.; Yilmaz, G. Phytochemical analysis, antioxidant and antibacterial activities of four Lamiaceae species cultivated in barnyard manure. *J. Agric. Sci.* **2017**, *23*, 95–108.
17. de Assis, R.M.A.; Carneiro, J.J.; Medeiros, A.P.R.; de Carvalho, A.A.; da Cunha Honorato, A.; Carneiro, M.A.C.; Bertolucci, S.K.V.; Pinto, J.E.B.P. Arbuscular mycorrhizal fungi and organic manure enhance growth and accumulation of citral, total phenols, and flavonoids in *Melissa officinalis* L. *Ind. Crop. Prod.* **2020**, *158*, 112981. [CrossRef]
18. Bouzenna, H.; Hfaiedh, N.; Giroux-Metges, M.A.; Elfeki, A.; Talarmin, H. Biological properties of citral and its potential protective effects against cytotoxicity caused by aspirin in the IEC-6 cells. *Biomed. Pharmacother.* **2017**, *87*, 653–660. [CrossRef]
19. Ling, Q.; Zhang, B.; Wang, Y.; Xiao, Z.; Hou, J.; Xiao, C.; Liu, Y.; Jin, Z. Chemical composition and antioxidant activity of the essential oils of citral-rich chemotype *Cinnamomum camphora* and *Cinnamomum bodinieri*. *Molecules* **2022**, *27*, 7356. [CrossRef]
20. Alizadeh, P.; Fallah, S.; Raiesi, F. Potential N mineralization and availability to irrigated maize in a calcareous soil amended with organic manures and urea under field conditions. *Int. J. Plant Prod.* **2012**, *6*, 493–512. [CrossRef]
21. Rajput, R.D.; Patil, R.P. The comparative study on spectrophotometric analysis of chlorophyll and carotenoids pigments from non-leguminous fodder crops. *Int. J. Innov. Res. Sci. Eng. Technol.* **2017**, *7*, 140–148.
22. Bremner, J.M. Nitrogen-total. In *Methods of Soil Analysis*; Sparks, D.L., Ed.; part 3. SSSA book ser. 5; SSSA and ASA: Madison, WI, USA, 1996; pp. 1085–1121.
23. Ostadi, A.; Javanmard, A.; Machiani, M.A.; Morshedloo, M.R.; Nouraein, M.; Rasouli, F.; Maggi, F. Effect of different fertilizer sources and harvesting time on the growth characteristics, nutrient uptakes, essential oil productivity and composition of *Mentha × piperita* L. *Ind. Crop. Prod.* **2020**, *148*, 112290. [CrossRef]
24. Jackson, M.L. *Soil Chemical Analysis*; Prentice-Hall Inc.: Englewood Cliffs, NJ, USA, 1958; pp. 178–182.
25. Asadi, M.; Nasiri, Y.; Maggi, F.; Rasouli, F.; Morshedloo, M.R. Biomass yield and essential oil chemical composition of *Mentha x piperita* as affected by amino acids and different fertilizer resources. *J. Soil Sci. Plant Nut.* **2023**, *23*, 668–682. [CrossRef]
26. British pharmacopoeia. *British Pharmacopoeia*; HMSO: London, UK, 1988; Volume 2, pp. 137–138.
27. Javanmard, A.; Amani Machiani, M.; Haghaninia, M.; Pistelli, L.; Najar, B. Effects of green manures (in the form of monoculture and intercropping), biofertilizer and organic manure on the productivity and phytochemical properties of peppermint (*Mentha piperita* L.). *Plants* **2022**, *11*, 2941. [CrossRef] [PubMed]
28. Keshavarz-Mirzamohammadi, H.; Tohidi-Moghadam, H.R.; Hosseini, S.J. Is there any relationship between agronomic traits, soil properties and essential oil profile of peppermint (*Mentha piperita* L.) treated by fertiliser treatments and irrigation regimes? *Annal. Appl. Bio.* **2021**, *179*, 331–344. [CrossRef]
29. Adams, R.P. *Identification of Essential Oil Components by Gas Chromatography/Mass Spectrometry*, 4th ed.; Allured Publishing Corporation: Carol Stream, IL, USA, 2007.
30. Ranđelović, M.; Dimitrijević, M.; Otašević, S.; Stanojević, L.; Išljamović, M.; Ignjatović, A.; Arsić-Arsenijević, V.; Stojanović-Radić, Z. Antifungal activity and type of interaction of *Melissa officinalis* essential oil with antimycotics against biofilms of multidrug-resistant candida isolates from vulvovaginal mucosa. *J. Fungi.* **2023**, *9*, 1080. [CrossRef]
31. Xiao, F.; Xu, T.; Lu, B.; Liu, R. Guidelines for antioxidant assays for food components. *Food Front.* **2020**, *1*, 60–69. [CrossRef]
32. Rostaei, M.; Fallah, S.; Lorigooini, Z.; Surki, A.A. The effect of organic manure and chemical fertilizer on essential oil, chemical compositions and antioxidant activity of dill (*Anethum graveolens*) in sole and intercropped with soybean (*Glycine max*). *J. Clean. Prod.* **2018**, *199*, 18–26. [CrossRef]
33. Bojović, B.; Marković, A. Correlation between nitrogen and chlorophyll content in wheat (*Triticum aestivum* L.). *Kragujevac J. Sci.* **2009**, *31*, 69–74.
34. Padilla, F.M.; de Souza, R.; Peña-Fleitas, M.T.; Gallardo, M.; Giménez, C.; Thompson, R.B. Different responses of various chlorophyll meters to increasing nitrogen supply in sweet pepper. *Front. Plant Sci.* **2018**, *9*, 1752. [CrossRef]
35. Zandvakili, O.R.; Barker, A.V.; Hashemi, M.; Etemadi, F. Biomass and nutrient concentration of lettuce grown with organic fertilizers. *J. Plant Nutr.* **2019**, *42*, 444–457. [CrossRef]
36. Fallah, S.; Mouguee, S.; Rostaei, M.; Adavi, Z.; Lorigooini, Z.; Shahbazi, E. Productivity and essential oil quality of Dracocephalum kotschyi under organic and chemical fertilization conditions. *J. Clean. Prod.* **2020**, *255*, 120189. [CrossRef]
37. Baghdadi, A.; Halim, R.A.; Ghasemzadeh, A.; Fauzi Ramlan, M.; Sakimin, S.Z. Impact of organic and inorganic fertilizers on the yield and quality of silage corn intercropped with soybean. *Peer J.* **2018**, *6*, 5280. [CrossRef] [PubMed]
38. Bajeli, J.; Tripathi, S.; Kumar, A.; Tripathi, A.; Upadhyay, R.K. Organic manures a convincing source for quality production of Japanese mint (*Mentha arvensis* L.). *Ind. Crop. Prod.* **2016**, *83*, 603–606. [CrossRef]
39. Anwar, M.; Patra, D.D.; Chand, S.; Alpesh, K.; Naqvi, A.A.; Khanuja, S.P.S. Effect of organic manures and inorganic fertilizer on growth, herb and oil yield, nutrient accumulation, and oil quality of French basil. Commun. *Soil Sci. Plant Anal.* **2005**, *36*, 1737–1746. [CrossRef]
40. Zhu, Y.; Merbold, L.; Leitner, S.; David, E.; Pelster, D.E.; Abwanda Okoma, S.; Ngetich, F.; Onyango, A.A.; Pellikka, P.; Butterbach-Bahl, K. The effects of climate on decomposition of cattle, sheep and goat manure in Kenyan tropical pastures. *Plant Soil.* **2020**, *451*, 325–343. [CrossRef]
41. Markewich, H.A.; Pell, A.N.; Mbugua, D.M.; Cherney, D.J.R.; van Es, H.M.; Lehmann, J.; Robertson, J.B. Effects of storage methods on chemical composition of manure and manure decomposition in soil in small-scale Kenyan systems. *Agric. Ecosyst. Environ.* **2010**, *139*, 134–141. [CrossRef]

42. Zhu, Y.; Merbold, L.; Leitner, S.; Xia, L.; Pelster, D.E.; Diaz-Pines, E.; Abwanda, S.; Mutuo, P.M.; Butterbach-Bahl, K. Influence of soil properties on $N_2O$ and $CO_2$ emissions from excreta deposited on tropical pastures in Kenya. *Soil Biol. Biochem.* **2020**, *140*, 107636. [CrossRef]
43. Singh, G.S. Advances in synthesis and chemistry of azetidines. In *Advances in Heterocyclic Chemistry*; Scriven, E.F.V., Ramsden, A.C., Eds.; Elsevier: Amsterdam, The Netherlands, 2020; Volume 130.
44. Ehlers, B.K.; Berg, M.P.; Staudt, M.; Holmstrup, M.; Glasius, M.; Ellers, J.; Tomiolo, S.; Madsen, R.B.; Slotsbo, S.; Penuelas, J. Plant secondary compounds in soil and their role in belowground species interactions. *Trends Ecol. Evol.* **2020**, *35*, 716–730. [CrossRef]
45. Al-Amier, H.; Toaima, V.; Mansour, B.M.M.; El Hela, A.A.; Sastry, K.S.; Craker, L. Use of mutations to improve essential oil yield and constituency in Pennyroyal. *J. Herbs Spices Med. Plants.* **2006**, *11*, 91–101. [CrossRef]
46. Hasibeder, R.; Fuchslueger, L.; Richter, A.; Bahn, M. Summer drought alters carbon allocation to roots and root respiration in mountain grassland. *New Physiologist.* **2015**, *205*, 1117–1127. [CrossRef]
47. Barthel, M.; Cieraad1, E.; Zakharova, A.; Hunt, J.E. Sudden cold temperature delays plant carbon transport and shifts allocation from growth to respiratory demand. *Biogeosciences* **2014**, *11*, 1425–1433. [CrossRef]
48. Du, Y.; Cui, B.; Zhang, Q.; Wang, Z.; Sun, J.; Niu, W. Effects of manure fertilizer on crop yield and soil properties in China: A meta-analysis. *Catena* **2020**, *193*, 104617. [CrossRef]
49. Abd-Elkader, E.H.; Ali, A.F.; Tawfik, O.H. Growth and essential oil of peppermint (*Mentha piperita* L.) plants as influenced *by compost* and some biostimulants. *Arch. Agri. Sci. J.* **2022**, *5*, 53–76. [CrossRef]
50. Honorato, A.D.C.; de Assis, R.M.A.; Maciel, J.F.A.; Nohara, G.A.; de Carvalho, A.A.; Pinto, J.E.B.P.; Bertolucci, S.K.V. Fertilization with different manure sources and doses provides quantitative-qualitative gains in the production of *Thymus vulgaris* L. *S. Afr. J. Bot.* **2024**, *164*, 345–355. [CrossRef]
51. Amani Machiani, M.; Javanmard, A.; Morshedloo, M.R.; Maggi, F. Evaluation of yield, essential oil content and compositions of peppermint (*Mentha piperita* L.) intercropped with faba bean (*Vicia faba* L.). *J. Clean. Prod.* **2018**, *171*, 529–537. [CrossRef]
52. Tripathi, Y.C.; Hazarika, P. Impact of harvesting cycle, maturity stage, drying and storage on essential oil content of patchouli leaves grown in Northeast region of India. *J. Essent. Oil-Bear Plants* **2014**, *17*, 1389–1396. [CrossRef]
53. Eiasu, B.K.; Dyafta, V.; Araya, H.T. Effect of Leaf Age on Essential Oil Yield and Composition in Rose-scented Geranium. *HortScience* **2022**, *57*, 1524–1528. [CrossRef]
54. Manjunatha, R.; Farooqi, A.A.; Vasundhara, M.; Srinivasappa, K.N. Effect of biofertilizers on growth, yield and essential oil content in patchouli (*Pogostemon cablin* Pellet.). *Indian Perfum.* **2002**, *46*, 97–104.
55. Petropoulos, S.; Fernandes, Â.; Karkanis, A.; Ntatsi, G.; Barros, L.; Ferreira, I.C.F.R. Successive harvesting affects yield, chemical composition and antioxidant activity of *Cichorium spinosum* L. *Food Chem.* **2017**, *237*, 83–90. [CrossRef]
56. Shi, C.; Zhao, X.; Liu, Z.; Meng, R.; Chen, X.; Guo, N. Antimicrobial, antioxidant, and antitumor activity of epsilon-poly-L-lysine and citral, alone or in combination. *Food Nutr. Res.* **2016**, *60*, 31891. [CrossRef]
57. Clara, M.B.G.; María, C.P.P.; Anders, L.; Antonio, M.; Dolores, R. Impact assessment of carvacrol and citral effect on *Escherichia coli* K12 and *Listeria innocula* growth. *Food Control* **2013**, *33*, 536–544. [CrossRef]
58. Hirai, M.; Ota, Y.; Ito, M. Diversity in principal constituents of plants with a lemony scent and the predominance of citral. *J. Nat. Med.* **2022**, *76*, 254–258. [CrossRef] [PubMed]
59. Baschieri, A.; Daci Ajvazi, M.; Folifack Tonfack, J.L.; Valgimigli, L.; Amorati, R. Explaining the antioxidant activity of some common non-phenolic components of essential oils. *Food Chem.* **2017**, *232*, 656–663. [CrossRef]
60. Saloner, A.; Bernstein, N. Nitrogen supply affects cannabinoid and terpenoid profile in medical cannabis (*Cannabis sativa* L.). *Ind. Crop. Prod.* **2021**, *167*, 113516. [CrossRef]
61. Bustamante, M.Á.; Michelozzi, M.; Barra Caracciolo, A.; Grenni, P.; Verbokkem, J.; Geerdink, P.; Safi, C.; Nogues, I. Effects of soil fertilization on terpenoids and other carbon-based secondary metabolites in *Rosmarinus officinalis* plants: A comparative study. *Plants* **2020**, *9*, 830. [CrossRef]
62. Cockson, P.; Schroeder-Moreno, M.; Veazie, P.; Barajas, G.; Logan, D.; Davis, M.; Whipker, B.E. Impact of phosphorus on *Cannabis sativa* reproduction, cannabinoids, and terpenes. *Appl. Sci.* **2020**, *10*, 7875. [CrossRef]
63. Bhat, B.A.; Islam, S.T.; Ali, A.; Sheikh, B.A.; Tariq, L.; Ul Isalam, S.; Ul Hassan Dar, T. Role of micronutrients in secondary metabolism of plants. In *Plant Micronutrients*; Aftab, T., Hakeem, K.R., Eds.; Springer: Cham, Switzerland, 2020. [CrossRef]
64. Chrysargyris, A.; Xylia, P.; Botsaris, G.; Tzortzakis, N. Antioxidant and antibacterial activities, mineral and essential oil composition of spearmint (*Mentha spicata* L.) affected by the potassium levels. *Ind. Crop. Prod.* **2017**, *103*, 202–212. [CrossRef]
65. Blanch, J.S.; Peñuelas, J.; Sardans, J.; Llusià, J. Drought, warming and soil fertilization effects on leaf volatile terpene concentrations in *Pinus halepensis* and *Quercus ilex*. *Acta Physiol. Plant.* **2009**, *31*, 207–218. [CrossRef]
66. Derwich, E.; Benziane, Z.; Boukir, A. Antibacterial activity and chemical composition of the leaf essential oil of *Mentha rotundifolia* from Morocco. *Electronic J. Environ. Agric. Food Chem.* **2011**, *9*, 19–28.
67. Benabdallah, A.; Boumendjel, M.; Aissi, O.; Rahmoune, C.; Boussaid, M.; Messaoud, C. Chemical composition, antioxidant activity and acetylcholinesterase inhibitory of wild *Mentha* species from northeastern Algeria. *S. Afr. J. Bot.* **2018**, *116*, 131–139. [CrossRef]
68. Wang, S.Y.; Chen, C.T.; Sciarappa, W.; Wang, C.Y.; Camp, M.J. Fruit quality, antioxidant capacity and flavonoid content of organically and conventionally grown blueberries. *J. Agric. Food Chem.* **2008**, *56*, 5788–5794. [CrossRef]

69. Deba, F.; Xuan, T.D.; Yasuda, M.; Tawata, S. Chemical composition and antioxidant, antibacterial and antifungal activities of the essential oils from *Bidens pilosa* linn. Var. Radiata. *Food Control* **2008**, *19*, 346–352. [CrossRef]
70. Fallah, S.; Rostaei, M.; Lorigooini, Z.; Surki, A.A. Chemical compositions of essential oil and antioxidant activity of dragonhead (*Dracocephalum moldavica*) in sole crop and dragonhead-soybean (*Glycine max*) intercropping system under organic manure and chemical fertilizers. *Ind. Crop. Prod.* **2018**, *115*, 158–165. [CrossRef]
71. Tohidi, B.; Rahimmalek, M.; Arzani, A. Essential oil composition, total phenolic, flavonoid contents, and antioxidant activity of Thymus species collected from different regions of Iran. *Food Chem.* **2017**, *220*, 153–161. [CrossRef] [PubMed]
72. Ozliman, S.; Yaldiz, G.; Camlica, M.; Ozsoy, N. Chemical components of essential oils and biological activities of the aqueous extract of *Anethum graveolens* L. grown under inorganic and organic conditions. *Chem. Biol. Technol. Agric.* **2021**, *8*, 20. [CrossRef]

horticulturae

MDPI

*Article*

# Effects of Irrigation Methods on Growth and Water Productivity in Bell Pepper Cultivation in Northern South Korea

**Hwichan Yang [1,†], Yunhyeong Bae [1,†], Youngho Kim [1], Soonjae Hyeon [1], Minseong Choi [2], Seongho Yang [2,3], Daehyun Kim [4] and Dongcheol Jang [1,2,*]**

[1] Department of Interdisciplinary Program in Smart Agriculture, Kangwon National University, Chuncheon 24341, Republic of Korea; jack955@kangwon.ac.kr (H.Y.); xhaahxj@kangwon.ac.kr (Y.B.); rladydgh99@kangwon.ac.kr (Y.K.); tnseo175@kangwon.ac.kr (S.H.)
[2] Department of Horticulture, Kangwon National University, Chuncheon 24341, Republic of Korea; peter6569@kangwon.ac.kr (M.C.); grower8812@iocrops.com (S.Y.)
[3] Agricultural Company ioCrops, Inc., Seoul 06694, Republic of Korea
[4] Department of Biosystems Engineering, Kangwon National University, Chuncheon 24341, Republic of Korea; daekim@kangwon.ac.kr
* Correspondence: jdc@kangwon.ac.kr; Tel.: +82-03-250-6426
† These authors contributed equally to this work.

**Abstract:** Although the bell pepper (*Capsicum annuum* L.) is sensitive to water stress, little information is available on proper irrigation management methods for bell pepper cultivation in the northern South Korean climate. We compared the effects of different irrigation methods on crop growth and water productivity in two bell pepper varieties (Maldonado and Nagano) at different irrigation durations (ending 3 h before sunset and ending 4 h before sunset) and irrigation quantities (placing two, three, and four drippers capable of irrigating at 2 $L \cdot h^{-1}$) over approximately 280 days by performing in-depth analysis of various growth indicators. The plant height of Maldonado increased as the irrigation amount increased in all irrigation treatments of T1 and T2. In Nagano, there was no significant difference in plant height between D3 and D4. Overall, the irrigation treatments produced a higher difference in yield in Maldonado plants than in Nagano plants. WP tended to increase inversely to reduction in irrigation quantity in Groups 1 (May–June) and 4 (November–December) of both varieties, and response to irrigation stop time varied among the varieties. We inferred that the optimal irrigation method for bell pepper cultivation in northern South Korea is to supply irrigation at the D3 level and adjust the irrigation end time according to the variety and crop strength.

**Keywords:** *Capsicum annuum* L.; greenhouse; gross revenue; water management; yield

## 1. Introduction

Bell pepper (*Capsicum annuum* L.) is a fruit vegetable crop from the eggplant family. It is an economically important fruit vegetable and is perceived as a high-quality food by consumers owing to its high water content and various components such as vitamins, minerals, and antioxidants [1]. In South Korea, the scale of bell pepper cultivation has increased significantly from 1.1 hectares (ha) in 1995 to 726 ha in 2022 with 82,042 tons of bell pepper produced that year [2]. Bell pepper is cultivated during the summer or winter growing season in South Korean depending on the latitude and altitude. The cold climate of the northern region is conducive to crop cultivation in summer, whereas the warm climate of the southern region is conducive to bell pepper cultivation in winter [3]. Bell pepper cultivation involves a relatively shorter introduction period in northern South Korea than in the southern region. Furthermore, the quality and quantity of bell peppers is relatively low because cultivation techniques have not been properly established [4]. Additionally, recent extreme weather events have resulted in a sharp decline in summer harvests, causing a sharp rise in prices during certain periods [5]. Therefore, summer cultivation in northern South Korea, which is performed under a relatively cool climate and contributes a large proportion of the summer harvest, is becoming increasingly essential to stabilize the production and supply of bell peppers in the summer.

Bell pepper is a fruit vegetable crop that exhibits both vegetative and reproductive growth. The balance between the two is important for maximizing yield [6]. Methods for controlling vegetative and reproductive growth include environmental control, crop management (number of leaves or fruit set, etc.), and irrigation control. Among these, irrigation control is a major factor that controls the supply of moisture and nutrients to the crops, which in turn contribute to crop vigor and fruit quantity and quality [6,7]. However, bell pepper is a plant that is very vulnerable to water stress [8], and care should be taken when controlling irrigation. Bell pepper is sensitive to both low and high water stress [9,10]. Bell pepper is highly susceptible to water deficits during flowering and fruit set but is also susceptible to excessive moisture and the subsequent reduction in slab aeration [8]. Lack of moisture reduces fruit size and increases sunburn and blossom end rot (BER), which reduces yields owing to the increase in proportion of non-commodity fruits [10,11]. Excess moisture affects the slab temperature and breathability, which inhibits root growth. This makes the plants susceptible to pests, which adversely affect crop growth [12]. To prevent water stress-related problems, plants must be properly irrigated; however, plant water consumption varies depending on climatic factors and stage of plant growth [13]. However, since South Korea currently uses European irrigation methods with very different climate conditions, appropriate irrigation management methods are needed to achieve optimal growth and high yields [10].

Irrigation water scarcity is being increasingly reported globally. Agriculture accounts for >80% of total water use. Hence, it is the most affected activity and is facing a major crisis [14]. Agriculture is a water-intensive industry, and shortage of available water due to industrialization and climate change needs to be addressed to achieve sustainable agriculture [15]. Additionally, global food demand is increasing owing to continued population growth, which requires efficient crop production [16]. Therefore, improving crop water productivity (WP) is critical for addressing the rising food demand and declining yields due to climate change and irrigation scarcity [17]. Moreover, improving WP is key to the sustainable development of the facility fruit vegetable cultivation industry [18]. However, determining the optimal timing and range of irrigation to achieve adequate yield while reducing irrigation quantity requires careful evaluation of growth and yield under different irrigation conditions [19].

South Korea has four distinct seasons; therefore, the light intensity is very different for each season. In summer, it is hot and humid due to the rainy season called the "Changma" affected by the East Asian monsoon system [1]. Currently, South Korea grows bell peppers using European irrigation management methods with very different climates [6]. If the European irrigation management method is used in South Korea's unusual climate, irrigation management cannot be carried out efficiently, leading to wastewater or reduced yields. In addition, the quality and quantity of bell peppers are relatively low because the cultivation technology has not been properly established due to the relatively shorter introduction period in northern South Korea than in southern regions [4]. Related studies to solve these problems include analyzing the effect of the summer greenhouse environment on bell pepper productivity; it has been said that the reduction in solar radiation should be minimized in July when the temperature is high [4]. The night temperature should be kept low. Bell pepper varieties suitable for northern South Korea were selected by comparing the cultivation of domestic and foreign varieties [3]. The authors in [1] reported that cool-white inter-lighting supplemental lighting in bell pepper summer cultivation yields high-quality fruit. Many studies have been conducted in the northern environment of South Korea, but there is a lack of research on how to manage irrigation while growing for a long time. So, in the field, we are supplementing irrigation management methods based on the accumulated experience of growers and advisors. So, sustainable agriculture should be realized by improving WP without wasting irrigation by finding irrigation management methods suitable for the climate in northern South Korea. This study attempted to explore the appropriate irrigation management method by analyzing the effect of irrigation end time and irrigation amount on growth and yield when growing bell pepper in northern South Korea using two bell pepper varieties with different growth forms.

## 2. Materials and Methods

### *2.1. Experimental Site and Design*

This study was conducted from February to December 2018 in a multi-span plastic greenhouse (latitude, 38°14′55″; longitude, 128°12′15″; elevation, 309 m) located in Inje, Gangwon state, South Korea (Figure 1A). The following bell pepper cultivars were used: red varieties 'Maldonado' (*C. annuum* L, Enza Zaden Global, Enkhuizen, The Netherlands) and 'Nagano' (*C. annuum* L, Rijk Zwaan, De Lier, The Netherlands) (Table S1). The slab was made of rock wool (W 120 × L 12 × H 7.5 cm) (Grodan Co., Roermond, The Netherlands). In total, 240 plants were seeded on 15 December 2017 on rockwool seed plates (W 410 × L 600 × H 50 mm; Grodan Co., Roermond, The Netherlands) and germinated for 7 days at 26–27 °C and 80–90% humidity under dark conditions. On day 13 after germination, the plants were potted in Rockwool cubes (W 10 × L 10 × H 10 cm; Expert Grodan Co., Roermond, The Netherlands) and grown in a greenhouse for 60 days. The bell peppers were transplanted on 14 February 2018, at a stem density of 6.8 stems·m$^{-2}$. To observe variations over time, we divided the months into Group 1: May–June; Group 2: July–August; Group 3: September–October; and Group 4: November–December.

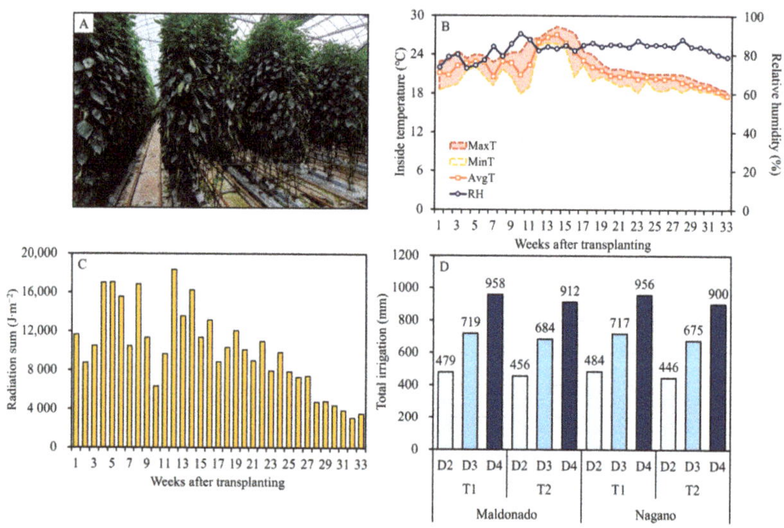


**Figure 1.** Growing bell pepper in Gangwon state, South Korea. (**A**) The bell pepper plants were grown on a rockwool slab with applied hydroponics system. The image was captured on 15 October 2018. (**B**) Weekly mean inside temperature. The red line indicates maximum temperature; yellow line indicates minimum temperature; orange marker indicates 24 h average temperature, and blue marker indicates relative humidity. (**C**) Average weekly radiation sum in the greenhouse. Data were collected from 1 to 33 weeks after transplantation (WAT). (**D**) Total irrigation was based on the difference between irrigation stop time and quantity.

### *2.2. Greenhouse Environment*

The 24 h average inside temperature of the greenhouse was maintained at a temperature of 17–28 °C using an environmental control system (CONNEXT 906 Version, Priva B.V, Den Haag, The Netherlands). The inside carbon dioxide concentration was maintained at 400–700 ppm using liquid carbon dioxide (Sundo Chemical, Seosan, Republic of Korea).

External weather stations (Priva B.V, Den Haag, The Netherlands) and measuring boxes (Priva B.V, Den Haag, The Netherlands) were used to collect environmental data such as cumulative light intensity and internal humidity (Figure 1B,C). Irrigation was monitored daily for the duration of the study by measuring the number of cups in each treatment zone (Figure 1D). To determine the slab environment in each treatment according to differences in irrigation, a moisture content meter (Grodan Co., Roermond, The Netherlands) was used to measure the moisture content, electrical conductivity (EC), and temperature.

### 2.3. Irrigation Treatments

Irrigation was automatically controlled using cumulative insolation to achieve an irrigation amount of 2.3 cc·J$^{-1}$. From the time of transplanting until the end of the growing season, EC 2.0–2.8 dS·m$^{-1}$ was maintained using Dutch PBG non-circulating culture water. Different slab environments were created based on the different irrigation stop times by setting the irrigation to Time 1 (T1): ends 3 h before sunset or T2: ends 4 h before sunset (Figure 2). To provide each slab with a different quantity of irrigation, we set up two, three, and four drippers capable of irrigating at 2 L·h$^{-1}$. These were denoted as Dripper 2 (D2), D3, and D4, respectively.

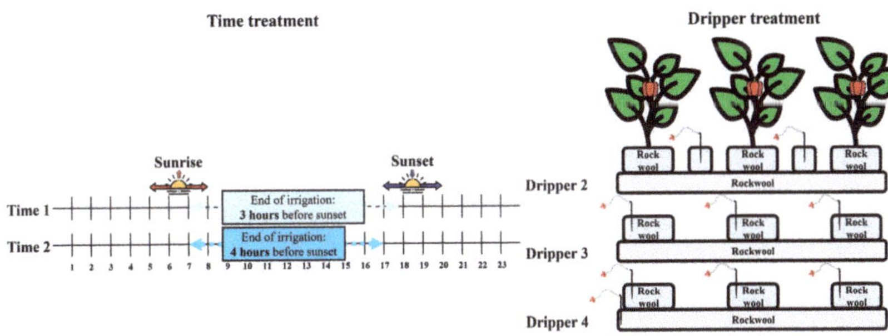


**Figure 2.** Treatment method according to irrigation stop time and quantity. Two groups were formed to observe the effect of irrigation stop time—T1: ending 3 h before sunset and T2: ending 4 h before sunset, a difference of approximately 1 h. Three groups were created to observe the effect of different irrigation quantities by installing two, three, or four drippers on the slabs for the respective group.

### 2.4. Growth and Harvested Fruit

Plant height, plant growth height, flower position, stem diameter, and leaf width were measured. Similarly, the number of leaves, leaf blades, nodes, and fruit set were counted. All measurements were made once a week for 28 weeks. In total 108 stems were examined, which includes 9 stems·treatment$^{-1}$. Additionally, the average fruit weight and number of harvested fruits were investigated once a week for 35 weeks. Thus, 1985 fruit samples were examined at an average rate of 56.7 fruits·week$^{-1}$. Plant height was defined as the length from the slab surface to the growth point, and plant growth height was defined as the length measured from the previous week's growth-point height to the height observed on the current week. Flower position was measured as the length from the growth point to the fully opened flower. Leaf length and width were measured as the size of the leaf produced at the node that determined the flowering position. Stem diameter was measured as the height of the growth point a week prior to flowering. Fruit set was defined as fruit with a length of $\geq$1 cm. Harvest was defined as normally harvested fruit with at least 90% coloration.

Leaf area index (LAI) (1) was calculated as follows [20]:

$$LAI = Leaflength \times (\text{cm})\ Leafwidth\ (\text{cm}) \times 0.6 \times SD \times \{L_1 + (L_2 \times 0.7)\} \quad (1)$$

where SD indicates stem density (ea·m$^{-2}$), $L_1$ indicates number of the leaves on the main stem, and $L_2$ indicates number of leaves of the lateral stem.

Yield (2) was calculated as follows:

$$Yield\left(\mathrm{kg{\cdot}m^{-2}}\right) = Averageharvestfruit\ (ea) \times Averagefruitfreshweight\ (\mathrm{g}) \times SD \times 0.001 \quad (2)$$

ET was calculated as shown in Equation in (3). However, a simplified formula was used by omitting the parameters such as precipitation, groundwater recharge, surface runoff, and deep drainage because this study was conducted in a greenhouse facility that was isolated from the outside [21].

$$ET\ (\mathrm{mm}) = I\ (\mathrm{mm}) \pm \Delta W\ (\mathrm{mm}) - D(\mathrm{mm}) \quad (3)$$

where I indicates total irrigation water (mm), ΔW indicates change in the water content (mm), and D indicates drainage below root zone (mm).

WP (4) was calculated as follows [22]:

$$WP\ (\mathrm{kg{\cdot}m^{-3}}) = (Yield\ (\mathrm{kg{\cdot}m^{-2}})/ET\ (\mathrm{mm})) \times 100 \quad (4)$$

Gross revenue (5) was calculated as mentioned [23]. As variations in bell pepper prices in Korea greatly depend on time, the price was organized by group and multiplied by yield (Table S2).

$$Grossrevenue\ (\mathrm{won{\cdot}m^{-2}}) = Averagepriceover3years\ (\mathrm{won{\cdot}kg^{-1}}) \times Yield\ (\mathrm{kg{\cdot}m^{-2}}) \quad (5)$$

*2.5. Statistical Analysis*

Experiments were conducted using a randomized complete block design with 9 samples per treatment, replicated three times with three plants each. The area was divided into several plots to reduce variation between plant locations. These plots were further subdivided to create as many plots as there were treatments, and each plot was randomly assigned a treatment. Three replicates were performed with three plants. Statistical analysis was performed using SPSS (IBM SPSS Statistics 26, IBM Corporation, New York, NY, USA). Any significant difference between treatments was tested at $p < 0.05$ using Duncan's multiple range test. SRplot (https://www.bioinformatics.com.cn/srplot (accessed on 9 November 2024)) was used to generate heat maps that visualized the data for quick and easy understanding of the overall impact of irrigation conditions on the survey metrics [24].

## 3. Results

*3.1. Slab Environment and ET Changes*

Slab temperature, slab EC, ΔW, and ET were compared for the two bell pepper varieties to confirm the changes induced in the slab environment and ET by different irrigation stop times and quantities (Figure 3). The temperature of the Maldonado slab was higher for T1D2 treatment than for the other irrigation treatments (Figure 3A). The EC in the slab was lower for T1D4 and T2D4 treatments, which provided higher irrigation quantities throughout the growth period. ET was lower for T2 than for T1 treatment and tended to increase as irrigation quantity increased for both T1 and T2 treatments. With respect to slab EC, both T1 and T2 treatments showed a trend of increasing EC as irrigation quantity decreased at the early stage of growth (Figure 3B). The effect of ΔW did not show any difference based on irrigation treatment; however, after 15 WAT, ΔW tended to increase for T2 than for T1 treatment. Additionally, ET tended to increase as irrigation quantity increased, and ET was higher for T1 than for T2 treatment.

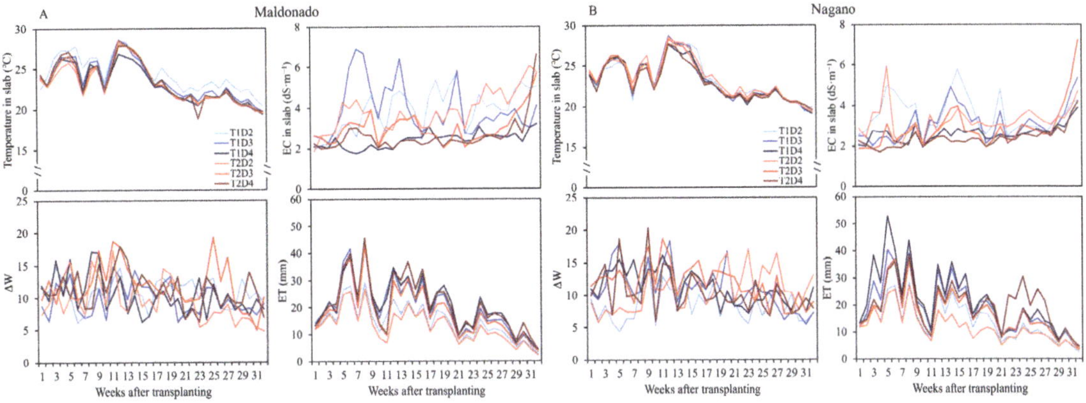


**Figure 3.** Effects of irrigation stop time and quantity on root zone environment and evapotranspiration (ET) of bell pepper grown in a greenhouse. The parameters were measured for 32 weeks from the time of transplanting to the end of cultivation. The treatments based on the difference in irrigation stop times were: T1 (ending 3 h before sunset) and T2 (ending 4 h before sunset). The treatments based on difference in irrigation were: D2 (two drippers in a Slab), D3 (three drippers in a Slab), and D4 (four drippers in a Slab). (**A**) Slab temperature, slab EC, ΔW, and ET for Maldonado. (**B**) Slab temperature, slab EC, ΔW, and ET for Nagano.

### *3.2. Plant Height and LAI*

The effect of irrigation stop time and quantity on plant height and LAI were determined in the Maldonado and Nagano varieties of bell peppers (Figure 4). In all treatment groups (T1 and T2), grass length increased in proportion to the irrigation quantity for Maldonado (Figure 4A). Plant height was highest for T1D4 treatment at 11 WAT and was highest among all irrigation treatments at 259 cm at 28 WAT. T1D2 produced the shortest plant height at 28 WAT, which was 41.8 cm shorter (approximately 16%) than the highest length of 217.2 cm observed for T1D4 treatment. LAI was widest for T1D4 treatment from 9 WAT onwards; it was significantly high at 28 WAT, which was 5.01-times higher than those observed for the other irrigation treatments (Figure 4B). T1D2 and T2D2 treatments produced the smallest LAI of 3.78, which was 1.23 (approximately 33%) smaller than that for T1D4 treatment, which exhibited the widest LAI. D2 and D3 treatments did not differ in LAI according to irrigation stop time, but LAI for D4 treatment was 0.6 (approximately 12%) larger for T1 than that for T2 treatment. Plant height for Nagano showed similar trends across various irrigation treatments until 17 WAT, after which plant heights began to vary between treatments (Figure 4C). Plant heights for T1D4, T2D3, and T2D4 treatments were higher than those of other irrigation treatments at 29 WAT. Similarly to that observed for Maldonado plants, D2 treatment produced shorter plant heights. However, the difference between T1 and T2 treatments was 34.3 cm (approximately 21%), which is a large difference in response to irrigation stop time. Furthermore, irrigation stop time exerted a higher impact than that of irrigation quantity on LAI in Nagano plants for all treatments, with the exception of D2 treatment (Figure 4D). LAI was similar between T1D4 and T1D3 treatments (approximately 4.2) and between T2D3 and T2D4 treatments (approximately 3.8). LAI was significantly different between D3 and D4 treatments at 0.26 (approximately 6%) for T1 and 0.38 (approximately 9%) for T2 treatment. In contrast, the LAI for D2 treatment did not show a significant difference with respect to the irrigation stop time.

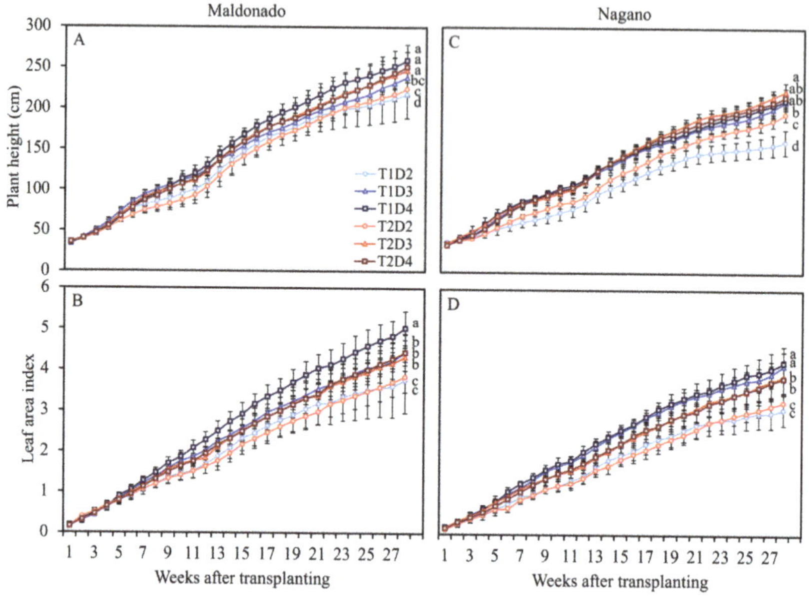


**Figure 4.** The effect of irrigation stop time and quantity on plant height and leaf area index (LAI) of bell pepper observed through weekly measurements. Plant height was calculated as the length from the slab surface to growing point. LAI was calculated as leaf length (cm) × leaf width (cm) × 0.6 × stem density (stem·m$^{-2}$) × {number of leaves on the main stem + (number of leaves on the lateral stems × 0.7)}. Error bars indicate the standard deviation of sample means. (**A**) Plant height for Maldonado. (**B**) LAI for Maldonado. (**C**) Plant height for Nagano. (**D**) LAI for Nagano. Different lowercase letters indicate significant differences according to irrigation methods as determined using Duncan's multiple range test ($p < 0.05$).

### 3.3. Correlation of Flower Position with Plant Growth Height and Stem Diameter

Flower position was analyzed in correlation with plant growth height and stem diameter and compared between groups to determine the growth behavior at different irrigation stop times and quantities (Figure 5). In Maldonado bell pepper, T1D2 and T2D2 treatments, which provided less irrigation, tended to show flowers distributed at lower positions compared with that observed for other irrigation treatments (Figure 5A). The slope of the correlation for flower position and plant growth height was steepest at 1.16 in T2D3, whereas a moderate slope of 0.59 was observed for T1D4 treatment. The correlation between flower position and stem diameter was relatively high ($R^2$ = 0.38–0.64) compared with that with plant growth height (Figure 5B). T1D2 treatment predominantly produced short flower positions and stem diameters, whereas D3 treatment groups predominantly showed high flower positions and stem diameters. Analysis of flower position and plant growth height distribution in Nagano showed that flower positions occurred predominantly in the 2–4 cm range and plant growth height was 4–8 cm·week$^{-1}$. Furthermore, D2 treatment plants showed lower distribution than those of the other treatments (Figure 5C). The slope of the correlation for flower position and plant growth height was steepest at 1.09 for T2D3 treatment, whereas a moderate slope of 0.53 was observed for T1D2 treatment. Similarly to that observed for Maldonado, Nagano plants showed higher correlation for flower position with stem diameter ($R^2$ = 0.33–0.66) than with plant growth height ($R^2$ = 0.10–0.39) (Figure 5D). Distribution of flower position and stem diameter was generally lower for T1D2 than for T2D4 treatment.

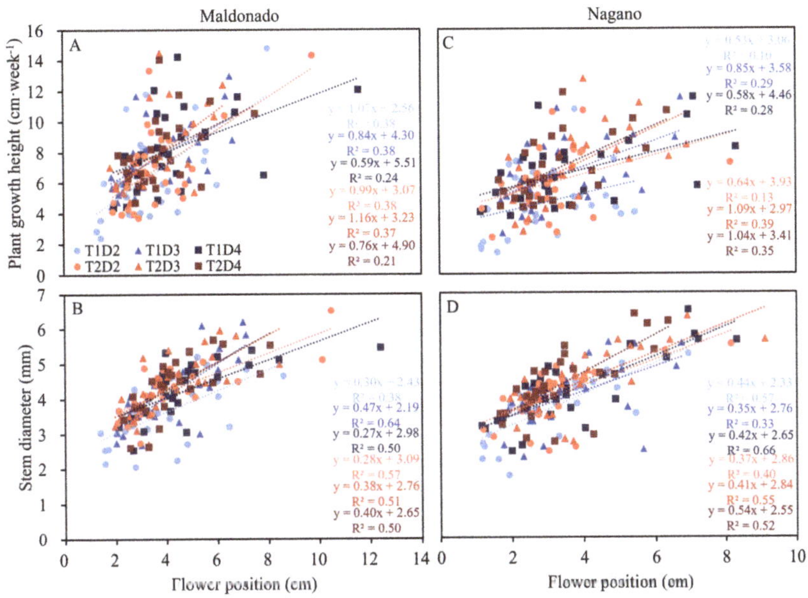


**Figure 5.** Correlation analyses of plant growth height with flower position in Maldonado (**A**), stem diameter with flower position in Maldonado (**B**), plant growth height with flower position in Nagano (**C**), and stem diameter with flower position in Nagano (**D**). The values at each point are the average plant growth height, stem diameter, and flower position by week from 1 WAT to 28 WAT and reflect the impact of irrigation stop time and quantity.

### *3.4. Harvested Fruit*

The effects of irrigation stop time and quantity on yield, WP, and gross revenue were analyzed and compared between the two varieties to check fruit productivity (Figure 6). T1D4 treatment, which provided the highest quantity of irrigation, produced the highest yield over time in Group 4 at 25.9 kg·m$^{-2}$, although it did not produce the highest yield in Group 1 (Figure 6A). Among both T1 and T2 treatments, D2 treatment produced the lowest yield because the amount of irrigation was low. For both T1 and T2 treatments at the early growth stage in Group 1, WP tended to decrease as irrigation quantity increased (Figure 6B). However, WP did not vary significantly between Groups 2 and 3 in terms of irrigation stop time or quantity. For T2D2 treatment, WP was significantly high in all groups, with the highest cumulative WP being 1604.6 kg·m$^{-3}$. In Group 1, T2D3 treatment produced the highest gross revenue; however, in the remaining groups, T1D4 treatment produced the highest gross revenue because irrigation quantity was higher (Figure 6C). Total gross revenue, which is the sum of the values from all groups, was higher for D3 and D4 treatments than for T1 and T2 treatments. In all groups except Group 1, yield was lowest for both T1D2 and T2D2 treatments, whereas it was highest for T1D3 and T1D4 treatments in Group 4, which represented the total yield (Figure 6D). The difference between the lowest yield of 53.3 kg·m$^{-2}$ for T1D2 treatment and highest yield of 66.4 kg·m$^{-2}$ for T1D3 treatment was 13.1 kg·m$^{-2}$ (about 25%). In Group 1, the D2 treatments for T1 and T2 exhibited the highest WP of 420.9 and 430.6 kg·m$^{-3}$, respectively (Figure 6E). However, no significant differences were observed in Groups 2 and 3 for any of the treatments. For Group 4, WP decreased as irrigation quantity increasing at T2, with T2D2 treatment producing the highest WP at 645.3 kg·m$^{-3}$ compared with 454.6 kg·m$^{-3}$ (about 70%) for T2D4 treatment. In Groups 2 and 3, the highest gross revenue was observed for T1D4 treatment, whereas the lowest was observed for T1D2 and T2D2 treatments (Figure 6F). Total gross revenue was highest at approximately 74,980 won·m$^{-3}$ for T1D3 and T1D4 treatments and lowest at approximately 57,176 won·m$^{-3}$ for T1D2 and T2D2 treatments; the difference was 17,804 won·m$^{-3}$ (approximately 24%).

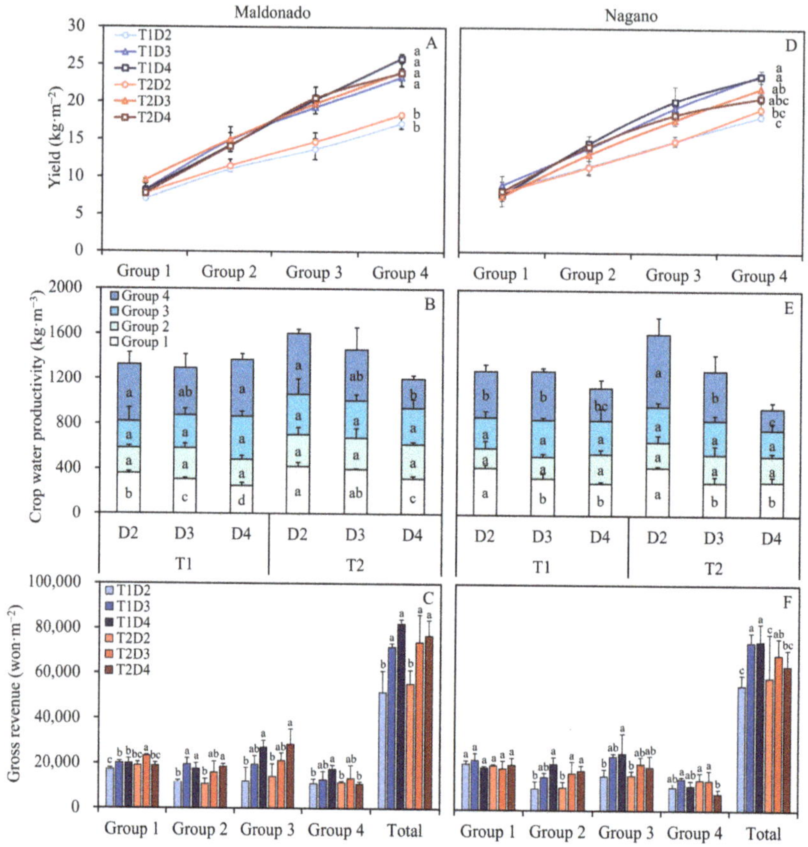

**Figure 6.** Yield, WP, and gross revenue across the growing season. Group 1: May–June; Group 2: July–August; Group 3: September–October; and Group 4: November–December. Yield was calculated as average harvest fruit (ea) × average fruit fresh weight (g) × stem density (ea·m$^{-2}$) × 0.001. Crop water productivity was calculated as yield (kg·m$^{-2}$) / ET (mm) × 100. Gross revenue was calculated as average price over 3 years (won·kg$^{-1}$) × yield (kg·m$^{-2}$). Error bars indicate the standard deviation of sample means. (**A**) Yield for Maldonado. (**B**) Crop water productivity for Maldonado. (**C**) Gross revenue for Maldonado. (**D**) Yield for Nagano. (**E**) Crop water productivity for Nagano. (**F**) Gross revenue for Nagano. Different lowercase letters indicate significant differences according to irrigation methods as determined using Duncan's multiple range test ($p < 0.05$).

### 3.5. Heatmap

Heatmaps were used to evaluate the responses of bell pepper varieties to different irrigation stop times, irrigation quantity-induced growth variation, harvest metrics, slab conditions, and ET (Figure 7). Maldonado plant height, flower position, stem diameter, and LAI increased as irrigation quantity increased, and the differences were more distinct for T1 than for T2 treatment (Figure 7A). Harvest-related metrics (number of harvested fruits, fruit fresh weight, yield, and gross revenue) were positively impacted by increase in irrigation quantity for both T1 and T2 treatments. Slab EC was similar for both T1 and T2 treatments, and it decreased as irrigation quantity increased. In contrast, slab temperature did not significantly differ with irrigation quantity for T2 treatment, whereas it decreased as irrigation quantity increased for T1 treatment. T2 treatment exerted a higher impact than that of T1 on ET. A trend similar to that observed for Maldonado was found in Nagano; however, the variation due to the difference in irrigation quantity was not as significant as that observed for Maldonado (Figure 7B). Irrigation quantity during T2 treatment exerted a significantly higher impact than that during T1 on plant growth height, plant height, flower position, and stem diameter. In contrast, leaf length, leaf width, and LAI metrics related to leaf growth were significantly impacted by irrigation quantity during T1. Similar outcomes were observed for harvest-related metrics, which were positively affected by increase in irrigation quantity during T1 treatment than during T2. Unlike in Maldonado plants, slab EC and slab temperature decreased for Nagano plants as irrigation quantity increased both in T1 and T2 treatments.

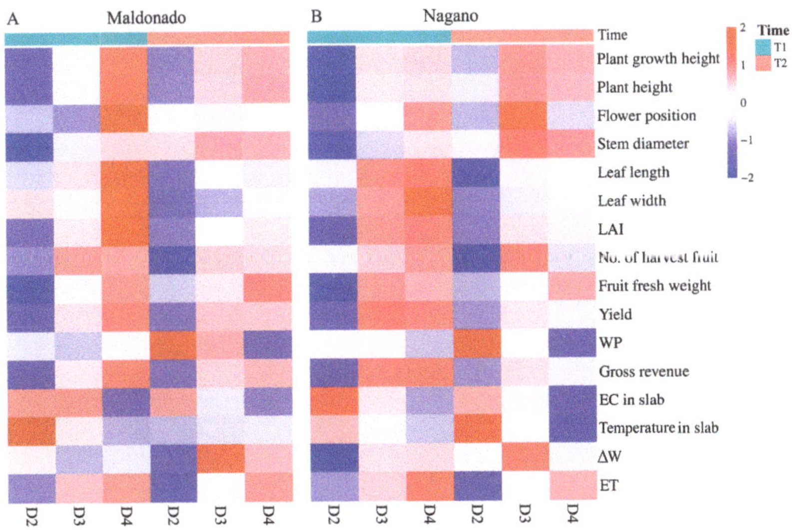


**Figure 7.** Heatmap of growth and yield indices, environment in slab, and ET analysis according to irrigation stop time and quantity in Maldonado (**A**) and Nagano (**B**) plants. Each column represents a different combination of irrigation stop time and quantity, and each row represents the measured parameter. Red cells indicate an increase in value; blue cells indicate a decrease in value.

## 4. Discussion

The global rise in the cultivation and production of fruit vegetables in greenhouses under controlled environments has enabled year-round production of high-quality crops and increased the efficient use of water and fertilizer, thereby making sustainable agriculture possible [25]. Increasing food demand and water shortage have highlighted the growing importance of efficient water use for maximizing yields with less water [15,16]. South Korea uses the same irrigation management system as that used in Europe, which has a very different climate; hence, the amount of irrigation may possibly be excessive or insufficient depending on the weather [6]. Therefore, in this study, we explored an appropriate irrigation management method suitable for northern South Korean climate by comparing the growth and yield responses according to irrigation conditions in two bell pepper genotypes.

Plant height and plant growth height are indicators of plant vitality [26]. Additionally, plant height is an indicator of crop conditions [27]. The height of Maldonado plants increased in proportion to the irrigation quantity for all T1 and T2 treatments (Figure 4A). This was confirmed using a heat map (Figure 7A). For T1D2 treatment, plant height was the shortest at 28 WAT, which was 41.8 cm (approximately 16%) shorter than the highest plant growth height observed for T1D4 treatment, which was 217.2 cm. For the Nagano variety, plant height did not differ between D3 and D4 treatments until 17 WAT, and no significant difference was observed between D3 and D4 treatments even at 28 WAT, which was the end of the cultivation period (Figure 4C). Similar to that observed for Maldonado plants, D2 treatment showed smaller plant height than that observed for D3 and D4 treatments; however, the difference between T1 and T2 treatment was 34.3 cm (approximately 21%), implying that irrigation stop time significantly impacted D2 treatment. Alshami et al. [28] have reported that plant height tended to decrease when irrigation quantity reduces from 100% to 20% during tomato cultivation. In contrast, Souza et al. [29] have stated that plant height did not vary depending on irrigation level in bell peppers. Although irrigation quantity induced changes in plant height in the Maldonado genotype, the same could not be confirmed in Nagano for D3 and D4 treatments. This may be attributed to the growth characteristics of bell pepper varieties (Table S1), although differences in cultivation environments may have exerted an effect.

In plants, leaves are the most important organs that absorb light energy and produce the assimilates necessary for growth [30]. LAI affects various processes such as photosynthesis, canopy interception of solar radiation, and ET [31,32]. During bell pepper cultivation, LAI value decreased with earlier irrigation stop time and lesser irrigation quantity (Figure 4). In Maldonado, the LAI was lowest at 3.78 for D2 treatment, which was approximately 33% lower than that observed for T1D4 treatment (Figure 4B). Moreover, irrigation stop time exerted a higher impact than irrigation quantity on LAI of Nagano with an approximate difference of 6% at T1 and 9% at T2 between D3 and D4 treatments (Figure 4D). The heatmap of Nagano LAI showed similar responses for D3 and D4 treatments for both T1 and T2 treatment groups (Figure 7B). Atilgan et al. [30] have reported that leaf area and LAI decreased as irrigation quantity decreased from 399 to 279 mm during tomato cultivation. Water stress reduces the leaf area in plants, which negatively affects their growth and fruit yield [33]. In the current study, LAI decreased as irrigation quantity decreased for both varieties, and the yield was also negatively affected (Figure 6).

Bell peppers are sensitive to the balance between vegetative and reproductive growth; hence, maintaining a balance between both growth types is essential to maximize yield [6]. In the field, indicators such as plant growth height, stem diameter, and flower position are used to assess crop balance [34]. Farmers have determined that smaller flower position, plant growth height, and stem diameter indicate the focus of the plant on reproductive growth. The correlation between flower position, plant growth height, and stem diameter indicated low distribution in the D2 treatment groups in both Maldonado and Nagano plants (Figure 5). Additionally, the correlation between flower position and stem diameter was stronger than that between flower position and plant growth height in both Maldonado and Nagano varieties. Furthermore, plants conserve water as irrigation quantity decreases by reducing their height, stem diameter, and leaf area [35]. In contrast, increase in irrigation level from 33 to 133% during bell pepper cultivation, tended to increase plant height and stem diameter [10]. Additionally, Alshami et al. [28] have reported that plant height and stem diameter decreased as the quantity of irrigation decreased from 100% to 20%. Moreover, plants subjected to water stress show a proportional increase in reproductive growth [36]. Therefore, the low irrigation volume of the D2 treatment may have exerted a higher impact on growth pattern in terms of plant growth height, stem diameter, and flower position of bell peppers in D2 treatment than in D3 and D4 treatments. This may explain the difference in the balance between vegetative and reproductive growth between treatments.

Irrigation is one of the most important inputs in cultivation for improving yields, and the irrigation schedule (amount and timing) is an important factor for obtaining high yields per unit area [37]. Yield and harvest timing depend on water stress levels, which greatly affect the economic benefits for growers [21]. This study explored irrigation management methods that can increase yield and gross revenue by considering WP. Maldonado yield was the highest after T1D4 treatment. It did not differ significantly from the D3 treatment plots but differed significantly from the D2 treatment plots (Figure 6A). The overall trend was similar in Nagano, but the yield difference according to irrigation treatment was not large compared with that observed for Maldonado variety (Figure 6D). Yield was greatly affected by irrigation quantity, and yield increased as irrigation quantity increased [38]. However, the occurrence of water stress occurs because of insufficient irrigation decreased photosynthetic capacity by inducing stomatal closure, leaf damage, and growth reduction, leading to a decrease in yield [39]. Additionally, water shortage increases the risk of an increase in the proportion of non-commercial fruits because water shortage increases the production of small fruits and occurrence of BER [40]. A previous study reported that reducing irrigation from 100% to 80% during bell pepper cultivation decreased the yield [41], which is similar to the result of this study. Reducing the irrigation quantity (D2 treatment) when cultivating bell peppers in the northern South Korea led to a significant decrease in yield; hence, the non-use of D2 treatment is deemed safer in the time periods other than that of Group 2, which is the "Changma" period with little sunlight and heavy rainfall.

Farmers irrigate using large quantities of water to maximize the yields based on their cultivation experience, crop condition checks, and ET comparisons [42]. In the present study, increasing the irrigation quantity from D2 to D3 increased the yield significantly; however, increasing the irrigation quantity from D3 to D4 only produced a similar or slightly increased yield. Thus, the water quantity used for D3 treatment has been determined to be the appropriate irrigation quantity required to increase the yield, whereas the quantity of the D4 treatment has been confirmed to significantly reduce the water use efficiency. Furthermore, both Maldonado and Nagano varieties showed the tendency of increasing WP as irrigation quantity decreased in Groups 1 and 4 (Figure 6B,E). Additionally, the heatmap showed that various harvest indices of both varieties responded to an increase in irrigation quantity (Figure 7). In the case of tomatoes, WP was significantly higher when irrigation quantity was reduced compared with that at the general irrigation level [15]. Shammout et al. [38] have reported that WP increased further when bell pepper plants were irrigated with 80% ET. Plant water consumption varies depending on climatic factors and plant growth stage [37]. In this study, WP did not change according to irrigation treatment in Groups 2 and 3. These groups correspond to the "Changma" period in South Korea when sunlight is scarce. This may be attributed to the fact that the greenhouse where the experiment was conducted was irrigated using ISR; hence, the difference in irrigation quantity may not have been high in Groups 2 and 3; thus, no difference was observed in WP.

Bell pepper market price in South Korea fluctuates significantly throughout the year because bell pepper production seasons in the north and south differ (Table S2). The gross revenue of Maldonado was not significantly different between the D3 and D4 treatments; however, T1D4 treatment showed the highest gross revenue (Figure 6C). Nagano plants showed the highest gross revenue for T1D3 and T1D4 treatments (Figure 6F). Therefore, gross revenue was higher in both varieties for D3 and D4 treatments because of higher yields. In contrast, the high gross revenue in Groups 3 and 4, which had high bell pepper prices, may be attributed to T1 treatment, which produced high yields (Table S1). In summary, WP was highest in the T2D2 treatment group, which was supplied an extremely small quantity of irrigation. Nevertheless, its yield was the lowest, which resulted in negative results for gross revenue. Therefore, the irrigation method of T2D2 is not appropriate. D4 treatment group, which was supplied with the largest irrigation quantity, showed the highest gross revenue profitability. However, with the exception of the Maldonado T1D4 treatment group, the D4 treatment groups among Nagano and Maldonado plants showed poor WP for all treatments. Therefore, the D4 irrigation method is not appropriate. Finally, D3 treatment produced a lower WP than that of D2 treatment, but the corresponding gross revenue was higher and yield was stable throughout the entire growing period. Specifically, the 'Maldonado' variety showed the best WP and gross revenue for T2D3 treatment, in which irrigation ended early, and the 'Nagano' variety showed the best WP and gross revenue for T1D3 treatment, in which irrigation ended late. These differences may be attributed to the growth characteristics of each variety [28,43]. In conclusion, we believe that supplying irrigation at the D3 level and adjusting the irrigation stop time according to the variety and vigor of the crop is the optimal irrigation method for northern South Korean bell pepper cultivation.

In this study, we investigated the response of two bell pepper varieties to different irrigation methods in terms of growth and yield in northern South Korea. However, we could not identify distinct indicators that can indicate water stress according to irrigation quantity. This study shows that additional investigation, such as analysis of chlorophyll fluorescence to determine plant stress and measurement of photosynthetic rate to identify changes in photosynthesis, is needed to obtain detailed information on water stress.

## 5. Conclusions

Different irrigation end times and amounts affected the growth and yield of the two bell pepper genotypes during the summer cultivation in northern South Korea. For Maldonado, plant height increased with irrigation amount in all T1 and T2. Nagano showed no significant plant height differences between the D3 and D4. Both cultivars experienced a decrease in LAI with reduced irrigation end time and amount, which was more pronounced in Nagano. Maldonado yield was similar between the D3 and D4 but differed at D2. Nagano yield followed a similar trend, with less variation between treatments than that of Maldonado. WP increased with reduced irrigation amount in Group 1 and 4 but remained unchanged in Group 2 and 3 due to climatic factors. At D2, growth and yield significantly decreased due to water stress, whereas D4 resulted in only slightly higher growth and yield than D3, indicating a decrease in water-use efficiency. Thus, in northern South Korea, optimal irrigation can be achieved by providing D3 level irrigation and adjusting the irrigation end time based on the crop variety and vigor.

**Supplementary Materials:** The following supporting information can be downloaded at: https://www.mdpi.com/article/10.3390/horticulturae10121353/s1, Table S1: Differences in characteristics according to red bell pepper varieties.; Table S2: Monthly wholesale prices of red bell peppers in South Korea from May to December. Group 1 is May–June, Group 2 is July–August, Group 3 is September–October, and Group 4 is November–December.

**Author Contributions:** Conceptualization, H.Y., Y.B. and D.J.; methodology, H.Y. and Y.K.; software, S.Y. and D.K.; validation, D.J. and D.K.; formal analysis, Y.B. and D.J.; investigation, H.Y., Y.B. and S.H.; resources, S.H. and M.C.; data curation, Y.B. and M.C.; writing—original draft preparation, H.Y. and Y.B.; writing—review and editing, D.J.; visualization, H.Y., Y.B. and S.H.; supervision, S.Y. and D.J.; project administration, H.Y., Y.B. and D.J.; funding acquisition, D.J. and D.K. All authors have read and agreed to the published version of the manuscript.

**Funding:** This work was supported by the Korea Institute of Planning and Evaluation for Technology in Food, Agriculture and Forestry (IPET) through the Smart Farm Innovation Technology Development Program funded by Ministry of Agriculture, Food and Rural Affairs (MAFRA) (grant number 421040-04).

**Institutional Review Board Statement:** Not applicable.

**Informed Consent Statement:** Not applicable.

**Data Availability Statement:** The raw data supporting the conclusions of this article will be made available by the authors on request.

**Conflicts of Interest:** The authors declare no conflict of interest.

## References

1. Kwon, Y.B.; Lee, J.H.; Roh, Y.H.; Choi, I.-L.; Kim, Y.; Kim, J.; Kang, H.-M. Effect of Supplemental Inter-Lighting on Paprika Cultivated in an Unheated Greenhouse in Summer Using Various Light-Emitting Diodes. *Plants* **2023**, *12*, 1684. [CrossRef] [PubMed]
2. Korean Statistical Information Service (KOSIS), Fruit Vegetables. Available online: https://kosis.kr/statHtml/statHtml.do?orgId=114&tblId=DT_114018_005&vw_cd=MT_ZTITLE&list_id=F1_AAA_010&seqNo=&lang_mode=ko&laguage=kor&obj_var_id=&itm_id=&conn_path=MT_ZTITLE (accessed on 31 October 2024).

3. Kim, Y.-H.; Yang, H.-C.; Yang, S.-H.; Kim, D.H.; Jang, D.-C. Comparison of Growth and fruiting characteristics between cultivars for the selection of appropriate domestic cultivars when cultivating paprika in high-temperature periods. *Hortic. Sci. Technol.* **2023**, *41*, 674–685. [CrossRef]
4. Kim, M.; Kim, K.H.; An, J.H.; Choi, K.-Y.; Yeo, K.-H.; Bae, J.H.; Lee, Y.-B. Effects of greenhouse environmental factors on the productivity of summer-cultivated sweet pepper and influential factors. *Hortic. Sci. Technol.* **2023**, *41*, 202–213. [CrossRef]
5. Kim, M.; Lee, Y.-B.; Choi, K.-Y. Effects of Greenhouse Temperatures and Solar Irradiance Levels on the Monthly Production of Summer-Cultivated Paprika. *Hortic. Sci. Technol.* **2024**, *42*, 339–349. [CrossRef]
6. Yeo, K.-H.; Choi, G.-L.; Lee, J.-H.; Park, K.-S.; Choi, K.-Y. Development of nutrient solution compositions for paprika cultivation in a closed coir substrate hydroponic system in Republic of Korea's winter cropping season. *Horticulturae* **2023**, *9*, 412. [CrossRef]
7. Schröder, F.-G.; Lieth, J.H. Irrigation control in hydroponics. *Hydroponic Prod. Veg. Ornam.* **2002**, *52*, 263–298.
8. Mačkić, K.; Bajić, I.; Pejić, B.; Vlajić, S.; Adamović, B.; Popov, O.; Simić, D. Yield and Water Use Efficiency of Drip Irrigation of Pepper. *Water* **2023**, *15*, 2891. [CrossRef]
9. Bruce, R.R. *Irrigation of Crops in the Southeastern United States: Principles and Practice*; Agricultural Research, Southern Region, Science and Education Administration, US Department of Agriculture: New Orleans, LA, USA, 1980; Volume 9.
10. Kabir, M.Y.; Nambeesan, S.U.; Bautista, J.; Díaz-Pérez, J.C. Effect of irrigation level on plant growth, physiology and fruit yield and quality in bell pepper (*Capsicum annuum* L.). *Sci. Hortic.* **2021**, *281*, 109902. [CrossRef]
11. Gisbert-Mullor, R.; Martín-García, R.; Bažon Zidarić, I.; Pascual-Seva, N.; Pascual, B.; Padilla, Y.G.; Calatayud, Á.; López-Galarza, S. A water stress–tolerant pepper rootstock improves the behavior of pepper plants under deficit irrigation through root biomass distribution and physiological adaptation. *Horticulturae* **2023**, *9*, 362. [CrossRef]
12. Zhang, H.; Wang, Y.; Yu, S.; Zhou, C.; Li, F.; Chen, X.; Liu, L.; Wang, Y. Plant Photosynthesis and Dry Matter Accumulation Response of Sweet Pepper to Water–Nitrogen Coupling in Cold and Arid Environment. *Water* **2023**, *15*, 2134. [CrossRef]
13. Brendel, O. The relationship between plant growth and water consumption: A history from the classical four elements to modern stable isotopes. *Ann. Sci.* **2021**, *78*, 47. [CrossRef]
14. Tzanakakis, V.A.; Paranychianakis, N.V.; Angelakis, A.N. Water supply and water scarcity. *Water* **2020**, *12*, 2347. [CrossRef]
15. Mukherjee, S.; Dash, P.K.; Das, D.; Das, S. Growth, yield and water productivity of tomato as influenced by deficit irrigation water management. *Environ. Process.* **2023**, *10*, 10. [CrossRef]
16. Liao, R.; Zhang, S.; Zhang, X.; Wang, M.; Wu, H.; Zhangzhong, L. Development of smart irrigation systems based on real-time soil moisture data in a greenhouse: Proof of concept. *Agric. Water Manag.* **2021**, *245*, 106632. [CrossRef]
17. Yang, F.; Wu, P.; Zhang, L.; Wei, Y.; Tong, X.; Wang, Z. Effects of subsurface irrigation types on root distribution, leaf photosynthetic characteristics, and yield of greenhouse tomato. *Sci. Hortic.* **2024**, *328*, 112883. [CrossRef]
18. Wu, Y.; Yan, S.; Fan, J.; Zhang, F.; Zhao, W.; Zheng, J.; Guo, J.; Xiang, Y.; Wu, L. Combined effects of irrigation level and fertilization practice on yield, economic benefit and water-nitrogen use efficiency of drip-irrigated greenhouse tomato. *Agric. Water Manag.* **2022**, *262*, 107401. [CrossRef]
19. Parkash, V.; Singh, S. A review on potential plant-based water stress indicators for vegetable crops. *Sustainability* **2020**, *12*, 3945. [CrossRef]
20. Jang, D.-C.; Yang, S.-H.; Na, W.-H.; Jeong, M.-H.; Yang, H.-C.; Kim, Y.-H.; Hong, E.-H.; Choi, K.-Y.; Kim, I.-S.; Vu, N.-T. Optimum management of leaf removal and lateral branch fruit set in winter-planted cultivation of paprika. *Horticulturae* **2021**, *7*, 348. [CrossRef]
21. Wu, Y.; Yan, S.; Fan, J.; Zhang, F.; Xiang, Y.; Zheng, J.; Guo, J. Responses of growth, fruit yield, quality and water productivity of greenhouse tomato to deficit drip irrigation. *Sci. Hortic.* **2021**, *275*, 109710. [CrossRef]
22. Li, H.; Liu, H.; Gong, X.; Li, S.; Pang, J.; Chen, Z.; Sun, J. Optimizing irrigation and nitrogen management strategy to trade off yield, crop water productivity, nitrogen use efficiency and fruit quality of greenhouse grown tomato. *Agric. Water Manag.* **2021**, *245*, 106570. [CrossRef]
23. Abdelkhalik, A.; Pascual, B.; Nájera, I.; Domene, M.; Baixauli, C.; Pascual-Seva, N. Effects of deficit irrigation on the yield and irrigation water use efficiency of drip-irrigated sweet pepper (*Capsicum annuum* L.) under Mediterranean conditions. *Irrig. Sci.* **2020**, *38*, 89–104. [CrossRef]
24. Tang, D.; Chen, M.; Huang, X.; Zhang, G.; Zeng, L.; Zhang, G.; Wu, S.; Wang, Y. SRplot: A free online platform for data visualization and graphing. *PLoS ONE* **2023**, *18*, e0294236. [CrossRef]
25. Rasouli, F.; Babla, M.; Li, L.; Liang, W.; Yong, M.-T.; Ahmed, T.; Tissue, D.; Huda, S.; Chen, Z.-H. Physiological and yield performance is partially linked to water use efficiency of eggplant genotypes in a high-tech glasshouse. *Horticulturae* **2022**, *9*, 19. [CrossRef]
26. Aminifard, M.; Aroiee, H.; Nemati, H.; Azizi, M.; Khayyat, M. Effect of nitrogen fertilizer on vegetative and reproductive growth of pepper plants under field conditions. *J. Plant Nutr.* **2012**, *35*, 235–242. [CrossRef]
27. Jang, D.-C.; Choi, K.-Y.; Heo, J.-Y.; Kim, I.-S. Comparison of growth and fruit setting characteristics for selecting the optimum winter-planted paprika cultivars. *Hortic. Sci. Technol.* **2016**, *34*, 424–432. [CrossRef]
28. Alshami, A.K.; El-Shafei, A.; Al-Omran, A.M.; Alghamdi, A.G.; Louki, I.; Alkhasha, A. Responses of Tomato Crop and Water Productivity to Deficit Irrigation Strategies and Salinity Stress in Greenhouse. *Agronomy* **2023**, *13*, 3016. [CrossRef]
29. Souza, Á.H.C.d.; Rezende, R.; Lorenzoni, M.Z.; Santos, F.A.S.; Oliveira, J.M.d. Response of bell pepper to water replacement levels and irrigation times. *Pesqui. Agropecuária Trop.* **2019**, *49*, e53662. [CrossRef]

30. Atilgan, A.; Rolbiecki, R.; Saltuk, B.; Jagosz, B.; Arslan, F.; Erdal, I.; Aktas, H. Deficit Irrigation Stabilizes Fruit Yield and Alters Leaf Macro and Micronutrient Concentration in Tomato Cultivation in Greenhouses: A Case Study in Turkey. *Agronomy* **2022**, *12*, 2950. [CrossRef]
31. Bonan, G.B. Importance of leaf area index and forest type when estimating photosynthesis in boreal forests. *Remote Sens. Environ.* **1993**, *43*, 303–314. [CrossRef]
32. Jose, S.; Gillespie, A.R. Leaf area-productivity relationships among mixed-species hardwood forest communities of the central hardwood region. *For. Sci.* **1997**, *43*, 56–64. [CrossRef]
33. Kaman, H.; Kurunç, A.; Demır, H.; Tezcan, A.; Sayıcı, A.; Can, M.; Gökçen, U. *Effects of Different Irrigation Water Salinity on Leaf Area and Dry Mater of Tomato*; Gaziosmanpaşa Üniversitesi Matbaası: Tokat, Turkey, 2017.
34. Jang, D.-C.; Choi, K.-Y.; Lee, S.M.; Heo, J.-Y.; Kim, I.-S. The effect of transplant age on growth and fruit yield in winter-planted paprika cultivation. *Hortic. Sci. Technol.* **2018**, *36*, 470–477. [CrossRef]
35. Alghamdi, A.G.; Alshami, A.K.; El-Shafei, A.; Al-Omran, A.M.; Alkhasha, A.; Aly, A.A.; Alharbi, A.R. Evaluating Tomato Performance: A Novel Approach of Combining Full and Deficit Irrigation with Saline Water. *Agronomy* **2024**, *14*, 559. [CrossRef]
36. López-Marín, J.; Gálvez, A.; del Amor, F.M.; Albacete, A.; Fernández, J.A.; Egea-Gilabert, C.; Pérez-Alfocea, F. Selecting vegetative/generative/dwarfing rootstocks for improving fruit yield and quality in water stressed sweet peppers. *Sci. Hortic.* **2017**, *214*, 9–17. [CrossRef]
37. Kırnak, H.; Gökalp, Z.; Demir, H.; Kodal, S.; Yıldırım, E. Paprika pepper yield and quality as affected by different irrigation levels. *J. Agric. Sci.* **2016**, *22*, 77–88.
38. Shammout, M.W.; Qtaishat, T.; Rawabdeh, H.; Shatanawi, M. Improving water use efficiency under deficit irrigation in the Jordan Valley. *Sustainability* **2018**, *10*, 4317. [CrossRef]
39. Ferreira, C.S.; Soares, P.R.; Guilherme, R.; Vitali, G.; Boulet, A.; Harrison, M.T.; Malamiri, H.; Duarte, A.C.; Kalantari, Z.; Ferreira, A.J. Sustainable Water Management in Horticulture: Problems, Premises, and Promises. *Horticulturae.* **2024**, *10*, 951. [CrossRef]
40. Fernández, M.; Gallardo, M.; Bonachela, S.; Orgaz, F.; Thompson, R.; Fereres, E. Water use and production of a greenhouse pepper crop under optimum and limited water supply. *J. Hortic. Sci. Biotechnol.* **2005**, *80*, 87–96. [CrossRef]
41. Obadi, A.; AlHarbi, A.; Abdel-Razzak, H.; Al-Omran, A. Biochar and compost as soil amendments: Effect on sweet pepper (*Capsicum annuum* L.) growth under partial root zone drying irrigation. *Arab. J. Geosci.* **2020**, *13*, 508. [CrossRef]
42. Bierer, A.M. Development of an open-source soil water potential management system for horticultural applications, "Open_Irr". *HardwareX* **2023**, *15*, e00458. [CrossRef]
43. Xiao, X.; Liu, X.; Jin, N.; Wu, Y.; Tang, Z.; Khan, K.S.; Lyu, J.; Yu, J. Two Genotypes of Tomato Cultivated in Gobi Agriculture System Show a Varying Response to Deficit Drip Irrigation under Semi-Arid Conditions. *Agronomy* **2024**, *14*, 2133. [CrossRef]

*horticulturae*

MDPI

*Article*

# Effects of Light Intensity and Irrigation Method on Growth, Quality, and Anthocyanin Content of Red Oak Lettuce (*Lactuca sativa* var. *cripspa* L.) Cultivated in a Plant Factory with Artificial Lighting

**Thanit Ruangsangaram [1,2], Pariyanuj Chulaka [2], Kriengkrai Mosaleeyanon [3], Panita Chutimanukul [3], Michiko Takagaki [1] and Na Lu [4,*]**

1 Graduate School of Horticulture, Chiba University, 648 Matsudo, Matsudo 271-8510, Chiba, Japan; rsar.thanit@gmail.com (T.R.); mtgaki@faculty.chiba-u.jp (M.T.)
2 Department of Horticulture, Faculty of Agriculture, Kasetsart University, Bangkok 10900, Thailand; pariyanuj.c@ku.th
3 National Center for Genetic Engineering and Biotechnology, National Science and Technology Development Agency, Klong Nueng, Klong Luang, Pathum Thani 12120, Thailand; krieng@biotec.or.th (K.M.); panita.chu@biotec.or.th (P.C.)
4 Center for Environment, Health and Field Sciences, Chiba University, 6-2-1 Kashiwanoha, Kashiwa 277-0882, Chiba, Japan
* Correspondence: na.lu@chiba-u.jp; Tel.: +81-471-378-103

**Abstract:** Cultivating red oak lettuce in plant factories often encounters challenges in achieving the desired red leaf coloration. To make the leaves a pleasant red color, anthocyanins are key substances that need to be induced. This study aimed to evaluate the effects of increasing light intensity and irrigation methods on the growth and leaf color of red oak lettuce in a controlled environment. Two light intensities (300 and 400 μmol $m^{-2}$ $s^{-1}$) with white LEDs and two irrigation methods (circulating vs. non-circulating irrigation) were applied seven days before harvesting. The results indicated that plants grown with circulating irrigation exhibited significantly higher fresh and dry weights than those grown under non-circulating conditions, regardless of light intensity. When non-circulating irrigation was applied, shoot fresh weight decreased by approximately 22% on the harvesting day compared to the circulating treatments. Under the 400 μmol $m^{-2}$ $s^{-1}$ light intensity with non-circulating irrigation (400N-C), plants displayed the lowest lightness (L*) at 40.7, increased redness (a*) to −7.4, and reduced yellowness (b*) to 11.0. These changes in coloration were optimized by day 5 after treatment. Additionally, spectral indices, including normalized difference vegetation index and photochemical reflectance index, varied significantly among treatments. The 400N-C treatment also resulted in the highest anthocyanin content and antioxidant activity in red oak lettuce. These findings suggest that combining high light intensity with non-circulating irrigation before harvest can improve both the coloration and quality of red oak lettuce in plant factories with artificial lighting.

**Keywords:** hydroponic; hypoxia stress; pre-harvest; red coloration; PFAL; light conditions

## 1. Introduction

A plant factory with artificial lighting (PFAL) is a crop production system isolated from external environmental conditions [1]. PFALs offer the advantage of precise control over various environmental factors, regardless of weather or seasonal changes. Environmental conditions such as light, carbon dioxide levels, temperature, humidity, water, and nutrients are regulated using advanced technology, enabling the production of high-quality crops by optimizing their growth requirements [2].

Lettuce (*Lactuca sativa* L.) is widely regarded as one of the most suitable crops for production in PFAL due to its adaptability to indoor growing conditions. Its short height makes it ideal for multi-tier cultivation systems, and it grows rapidly even under low light intensity and high density [3,4]. Additionally, lettuce can be produced year-round and sold fresh at premium prices. It is also commonly used as a model plant in controlled environment research [5–7]. Lettuce is a valuable source of vitamins A and K, fiber, and minerals, with its nutritional quality varying based on the cultivar, growing conditions, and season [8,9]. Green leaf lettuce is known for high concentrations of calcium, magnesium, ascorbic acid, lutein, and beta-carotene. In contrast, red leaf lettuce contains high levels of antioxidant compounds such as phenolic acids, flavonoids, and anthocyanins [10]. However, red lettuce grown in PFAL is particularly sensitive to environmental factors, especially light, which can influence its coloration [11,12].

Light is a critical factor affecting the growth, development, physiology, and chemical composition of lettuce [6,13]. In PFAL, artificial light is the primary energy source, accounting for approximately 25% of total production costs due to electricity consumption [14]. To reduce energy costs, many commercial lettuce growers using PFALs apply a relatively low photosynthetic photon flux density (PPFD), typically not exceeding 300 $\mu$mol m$^{-2}$ s$^{-1}$ [15]. Many studies indicated that low light intensities result in low yields compared to high light levels [6,16,17]. Besides yield, light intensity also influences the coloration of red leaf lettuce [18]. Under low light conditions, red leaf lettuce often fails to fully complete its red coloration due to reduced anthocyanin accumulation, which can affect consumer acceptance of the produce.

To enhance the red coloration, promoting the production of anthocyanins plays a crucial role [19]. Anthocyanins are natural pigments belonging to the flavonoid family and help plants tolerate different stresses, such as excessive light exposure [20]. Anthocyanin accumulation is often triggered by reactive oxygen species (ROS) in response to various abiotic stresses, including UV damage, extreme temperatures, salinity, and drought [21–23]. ROS act as signaling molecules that activate stress tolerance mechanisms and antioxidant activities by scavenging free radicals, though the plant's sensitivity to stress varies based on the severity, frequency, and duration of the stress, impacting growth and development by reducing chlorophyll content and photosynthetic efficiency [24]. As noted, ROS can be triggered by many abiotic stresses [25], leading to enhanced anthocyanin accumulation and consequently, deeper red coloration in red lettuce. To minimize the risk of plant deterioration caused by ROS, many researchers employed short-term treatments just before harvest to improve plant quality using stress conditions [23,26–28].

By manipulating light intensity and stopping circulating irrigation, stress conditions can be generated to enhance anthocyanin accumulation in plants. When non-circulating irrigation is applied, plants experience an oxygen deficiency, leading to hypoxia stress [29,30]. This condition will gradually trigger self-defense mechanisms to counteract the effects of ROS. To explore this, we designed an experiment applying short-term high light intensity and non-circulating irrigation before harvesting in order to enhance anthocyanin accumulation. The main goal of this study is to examine how light intensity and irrigation method affect the growth, quality, and anthocyanin content of red oak lettuce grown in PFAL. The findings from this research will provide alternative solutions for enhancing the red coloration and overall quality of red oak lettuce, benefiting both fundamental research and commercial production in the future.

## 2. Materials and Methods

### 2.1. Plant Materials and Seedling Preparation

Seeds of red oak lettuce (*Lactuca sativa* var. cripa L. 'Mondai RZ'), obtained from Rijk Zwaan Co. Ltd., De Lier, The Netherlands, were sown in a seeding mat (polyurethane sponge foam) consisting of 300 cuboids (2.3 × 2.3 × 2.8 cm). The seeding mat contains 90–95% moisture and was kept under dark conditions for 48 h in a controlled environment room (a transplant preparation room of a PFAL). After germination, the seedlings were exposed to white LED lamps (spectral distribution is shown in Figure S1) (AGRI-OPTECH Co., Ltd., Tainan, Taiwan) at a light intensity of 150 μmol $m^{-2}$ $s^{-1}$ with a 14 h $d^{-1}$ photoperiod. The air temperature, relative humidity (RH%), and carbon dioxide ($CO_2$) concentration in the room were 25 ± 1.5 °C, 70 ± 5%, and 400 ± 10 μmol $mol^{-1}$, respectively. At 10 days after sowing, seedlings with 1–2 fully expanded leaves were transplanted into a deep flow technique (DFT) hydroponic system in a PFAL. The planting density was 22.2 plants $m^{-2}$ (12 seedlings on a 0.6 × 0.9 $m^2$ panel).

### 2.2. Growth Conditions and Treatment Design

After transplanting, the seedlings were cultivated for 24 days with the Enshi nutrient solution [31] under circulating conditions. The electrical conductivity (EC) of the nutrient solution was maintained at 1.2 ± 0.5 dS $m^{-1}$, with a pH of 6.0 ± 0.5, and the cultivation room was enriched with $CO_2$ at 1000 ± 10 μmol $mol^{-1}$ throughout the cultivation. Lighting conditions were set at an intensity of 300 μmol $m^{-2}$ $s^{-1}$ using white LEDs, the same model as described in Section 2.1 with a photoperiod of 16 h per day.

On the 25th day after transplanting, half the amount of the red oak lettuces was exposed to a light intensity of 400 μmol $m^{-2}$ $s^{-1}$, while the remaining half were kept growing under 300 μmol $m^{-2}$ $s^{-1}$. These light intensities were combined with two irrigation methods (circulating and non-circulating) for seven days until 32 days after transplanting (DAT). The four treatments were designated as 300C, 300N-C, 400C, and 400N-C (Figure 1), and the 300C served as the control in this experiment. During the seven-day treatment prior to harvest, the new nutrient solution supply (fertilizers) was stopped from being fed to keep the same amount of nutrients in solution between circulating and non-circulating treatments. The electrical conductivity (EC), pH, and dissolved oxygen (DO) levels of the nutrient solution were monitored regularly.

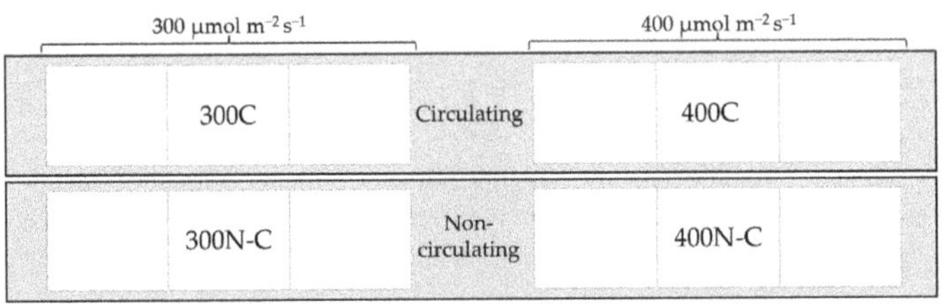


**Figure 1.** Representative illustration of light intensity and irrigation treatments: two light intensities in the same row and two irrigation methods across rows. The 300C, 300N-C, 400C, and 400N-C 400 represent the light intensity at 300 μmol $m^{-2}$ $s^{-1}$ with circulating irrigation, 300 μmol $m^{-2}$ $s^{-1}$ with non-circulating irrigation, 400 μmol $m^{-2}$ $s^{-1}$ with circulating irrigation, and 400 μmol $m^{-2}$ $s^{-1}$ with non-circulating irrigation, respectively.

### 2.3. Measurements

#### 2.3.1. Color Attributes, Gas Exchange, and Growth

The determination of color characteristics was measured by using the colorimeter (CR-300 Konica Minolta, Bangkok, Thailand) that operates within the CIELAB color space diagram. This color system represents the quantitative relationships of colors across three axes: the L* value indicates lightness, while the a* and b* are the chromaticity coordinates representing red (+a), green (−a), yellow (+b), and blue (−b).

A portable photosynthesis system (LI-6800, LICOR Inc., Lincoln, NE, USA) was used to measure gas exchange parameters, including photosynthetic rate, stomatal conductance, transpiration rate, $CO_2$, and internal $CO_2$ concentration. Additionally, chlorophyll fluorescence parameters, such as Fv/Fm, ETR, and NPQ, were assessed using the same equipment. All measurements were performed on fully expanded leaves, specifically from the 9th to the 12th outermost leaves.

The Poly Pen RP 410 UVIS, a handheld spectroradiometer (Photon Systems Instruments, Brno, Czech Republic), was employed to measure spectral responses within the range of 380–790 nm. Measurements were conducted on three fully expanded leaves of red oak lettuce, with each leaf measured three times per plant. Reflectance indices derived from these measurements were analyzed using Spectra Pen software version 1.1.0.21.

Normalized Difference Vegetation Index (NDVI) [32]

- = (RNIR − RRED)/(RNIR + RRED)

Photochemical Reflectance Index (PRI) [33]

- = (R531 − R570)/(R531 + R570)

Anthocyanin Reflectance Indices (ARI) [34]

- ARI1 = 1/R550 − 1/R700
- ARI2 = R790 × (1/R550 − 1/R700)

Plants were harvested 32 days after transplanting. The total shoot fresh weight of red oak was recorded, with five plants per replication and a total of 4 replications. Moreover, dry plant material was dried by a freeze dryer machine (SP VirTis Genesis Pilot Lyophilizer, Warminster, PA, USA) and kept at −60 °C for biochemical analysis.

#### 2.3.2. Biochemical Determination

##### Chlorophyll a, b, and Total Carotenoids Contents

The contents of chlorophyll a, chlorophyll b, and total carotenoids were determined using a spectrophotometric method with modifications based on the literature [26]. First, 0.1 g of powdered sample was mixed with 5 mL of 80% acetone. The mixture was vortexed and then incubated in the dark at room temperature for 3 h. After incubation, the samples were centrifuged at 5000× *g* for 10 min and filtered through Whatman No. 1 filter paper. The supernatants were analyzed using a spectrophotometer (Multiskan Sky, Thermo Scientific, MA, USA). The maximum absorbance values were recorded at 663 nm for chlorophyll a, 645 nm for chlorophyll b, and 470 nm for total carotenoids. The pigment concentrations were calculated using the formulas provided in [26] as follows:

Chlorophyll a (mg $g^{-1}$ DW) = $(12.21A_{663} - 2.81A_{645}) \times (v/w)$
Chlorophyll b (mg $g^{-1}$ DW) = $(20.13A_{645} - 5.03A_{663}) \times (v/w)$
Total chlorophyll (mg $g^{-1}$ DW) = Chlorophyll a + Chlorophyll b
Carotenoids (mg $g^{-1}$ DW) = $[1000A_{470} - 3.27\text{Chl.a} - 104\text{Chl.b}/229] \times (v/w)$

##### Anthocyanin Content

The anthocyanin content was measured according to established methods [35], with minor modifications. The anthocyanins were extracted by incubating 250 mg of leaf powder in 1500 μL of extraction solution (1% HCl in methanol) overnight at 4 °C in darkness. Following this incubation, 1000 μL of water and 1000 μL of chloroform were added, and the mixture was centrifuged at 10,000 rpm for 5 min to remove the suspended substances. The supernatant was used to measure the anthocyanin content, which was quantified using a spectrophotometer and calculated with the equation ($A_{530} - 0.33 \times A_{657}$) provided in [36]. The anthocyanin concentration is reported as the value derived from the equation per gram of dry weight.

2,2-Diphenyl-1-Picrylhydrazyl (DPPH)

The modified method described in [35] was used to estimate the free radical scavenging activity of red oak lettuce. A DPPH solution (60 µM, 1.95 mL) was added to 0.05 mL of the extract. The reaction mixture was stirred using a vortex mixer and incubated at 25 °C for 15 min. An 80% ethanol solution was used as the blank. A control was prepared using the solution without any test samples. Trolox (0.5 mM) served as a reference antioxidant. The absorbance of the solution was measured at 515 nm. Antioxidant activity was expressed as the inhibition percentage of DPPH absorbance, calculated as follows:

$$(A_{control} - A515)/A_{control}) \times 100.$$

Malondialdehyde (MDA)

Lipid peroxidation in the leaves of red oak lettuce was measured using the thiobarbituric acid (TBA) test to determine malondialdehyde (MDA), an end product of lipid peroxidation, following a modified method [37]. First, 30 mg of dried leaf powder was mixed with 5 mL of a 0.1% (*w*/*v*) trichloroacetic acid (TCA) solution. The mixture was then centrifuged at 14,000 rpm for 15 min. Next, 0.5 mL of the clear supernatant was added to 1 mL of 0.5% (*w*/*v*) TBA in 20% TCA. The mixture was incubated in boiling water (95 °C) for 30 min, after which the reaction was stopped by placing the reaction tubes in an ice bath. The samples were then centrifuged again at 14,000 rpm for 5 min, and the absorbance of the supernatant was measured at 532 nm and 600 nm using a spectrophotometer. MDA content was calculated using an extinction coefficient of 155 $mM^{-1}$ $cm^{-1}$, as described in [38].

### 2.4. Statistical Analysis

The experiment was conducted using a completely randomized design (CRD) with four replications, corresponding to the four tiers in the cultivation system. Five plants per treatment were sampled for overall growth and non-destructive measurements in each replication, while three plants per treatment were selected for gas exchange analysis and biochemical assessments. Data were analyzed using the Statistical Analysis System (SAS) software, and mean differences between treatments were evaluated using an independent sample *t*-test. Statistical significance was determined at *p*-values of $\leq$0.05 and $\leq$0.01.

## 3. Results

### 3.1. Plant Growth and Biomass

Increasing light intensity and stopping circulating nutrient solution for seven days before harvest affected the growth of red oak lettuce (Figure 2). Increasing light intensity did not affect shoot fresh or dry weights, regardless of irrigation condition; however, stopping circulating nutrient solution significantly reduced fresh weight and dry weight of the lettuce, regardless of light intensity (Figure 3A). Specifically, under the light intensity of 300 µmol $m^{-2}$ $s^{-1}$ and 400 µmol $m^{-2}$ $s^{-1}$, stopping circulating the nutrient solution decreased the yield of lettuce (shoot fresh weight) by 21% and 23%, respectively.

However, red oak lettuce grown under 400 μmol m$^{-2}$ s$^{-1}$ exhibited a significantly higher root fresh weight compared to those grown under 300 μmol m$^{-2}$ s$^{-1}$. In the 400C treatment, red oak lettuce exhibited the highest root fresh weight (26 g) and dry weight (0.76 g).

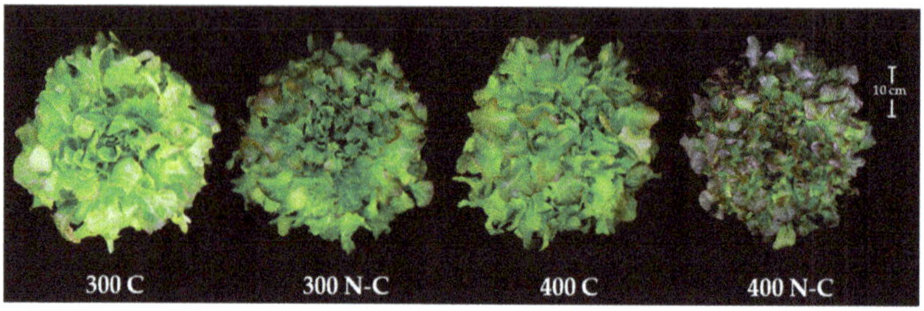


**Figure 2.** The top view of red oak lettuce grown under different light intensities (300 and 400 μmol m$^{-2}$ s$^{-1}$) and irrigation methods (circulating and non-circulating) for 7 days before harvesting. The numbers before the letters represent light intensities in the PPFD unit. C and N-C after numbers mean circulating and non-circulating irrigations. The 300C was served as the control.

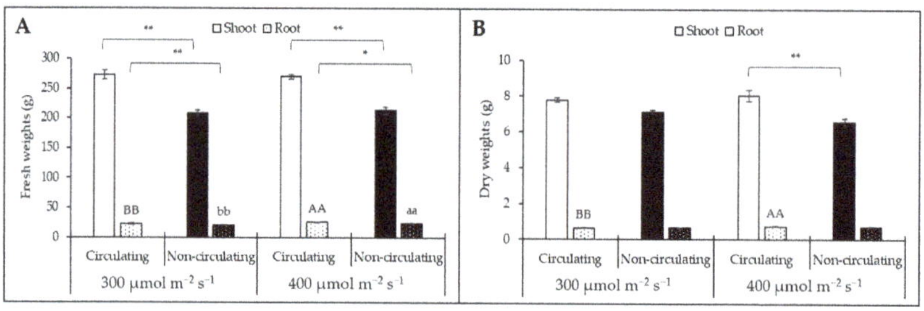


**Figure 3.** Fresh weights (**A**) and dry weights (**B**) of red oak lettuce when treated under 300 and 400 μmol m$^{-2}$ s$^{-1}$ with circulating and non-circulating irrigation for 7 days before harvesting. The vertical bars are presented as mean value ± SE; n = 20, according to the independent samples *t*-test. The asterisks indicate a significant difference between the circulation methods in each lighting of 300 and 400 μmol m$^{-2}$ s$^{-1}$ by * ($p \leq 0.05$) and ** ($p \leq 0.01$). Uppercase and lowercase letters indicate a significant difference between the lighting under circulating and non-circulating irrigations, respectively. When $p \leq 0.01$ is represented by double letters. Moreover, 300C served as the control in this experiment.

### 3.2. Gas Exchange Rate, Chlorophyll Fluorescence, and Spectral Indices

As shown in Figure 4, key gas exchange parameters, including net $CO_2$ assimilation rate (A), transpiration rate (E), internal $CO_2$ concentration ($C_i$), and stomatal conductance ($g_{sw}$), fluctuated over the seven days after the treatments. Statistical analysis revealed that E, $C_i$, and $g_{sw}$ were unaffected by light intensity or irrigation methods from the beginning until the harvesting day in this experiment. However, on the 7th day, A showed a significant difference between the 300C and 400C treatments (11.1 and 12.6 μmol m$^{-2}$ s$^{-1}$), while the treatments with N-C showed a slightly greater trend than those with C. The trend in A across treatments demonstrated that 400N-C had the highest value (13.38 μmol m$^{-2}$ s$^{-1}$), followed by 400C, 300N-C, and 300C, respectively.

The variable maximal fluorescence ratio (Fv/Fm) exhibited a slight increase over the 7-day period following treatment, with no significant differences observed between the treatments. Figure 5 (B and C) illustrates the electron transport rate (ETR) and nonphotochemical quenching (NPQ), showing two significant differences between treatments influenced by light intensity. Furthermore, treatments with higher light intensity led to a significant increase in the mean values of ETR and NPQ in red oak lettuce from day 3 to day 7 after the treatments began.

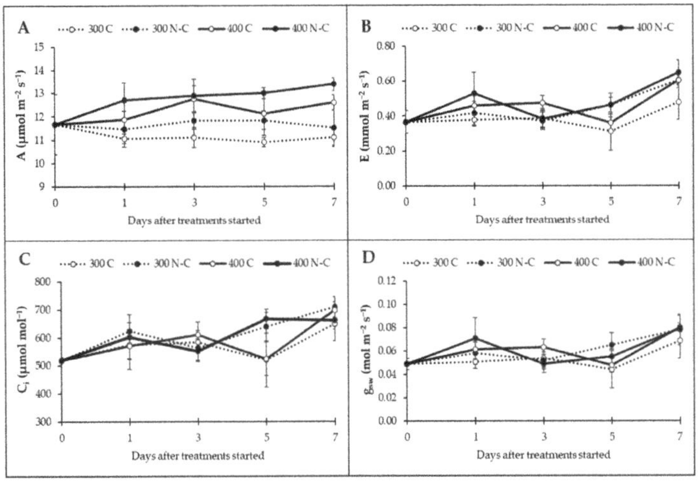


**Figure 4.** Leaf gas exchange parameters of red oak lettuce when treated under 300 and 400 μmol $m^{-2}$ $s^{-1}$ of light intensities with circulating and non-circulating irrigation for 7 days before harvesting. Those were (**A**) the net $CO_2$ assimilation rate, A; (**B**) the transpiration rate, E; (**C**) the internal $CO_2$ concentration, $C_i$; and (**D**) the stomatal conductance, $g_{sw}$. The points are presented as mean value ± SE; n = 12. The numbers 300 and 400 before the letters represent light intensities in the PPFD unit. C and N-C after numbers mean circulating and non-circulating irrigations. The 300C was served as the control.

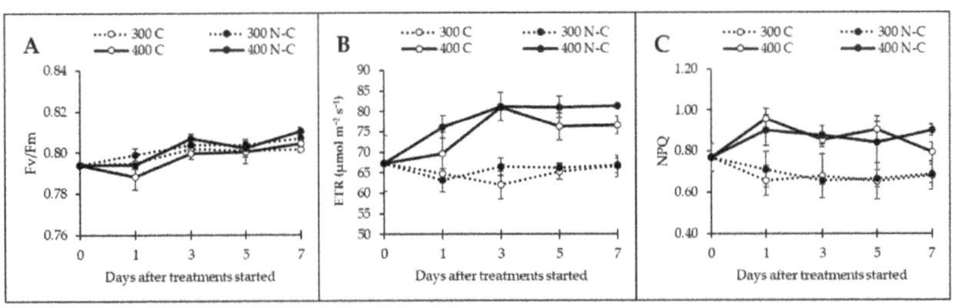


**Figure 5.** Chlorophyll fluorescence parameters of red oak lettuce when treated under 300 and 400 μmol $m^{-2}$ $s^{-1}$ of light intensities with circulating and non-circulating irrigation for 7 days before harvesting. Those were (**A**) the variable/maximal fluorescence ratio, Fv/Fm; (**B**) the electron transport rate, ETR; and (**C**) the non-photochemical quenching, NPQ. The points are presented as mean value ± SE; n = 12. The numbers 300 and 400 before the letters represent light intensities in the PPFD unit. C and N-C after numbers mean circulating and non-circulating irrigations.

The reflectance spectrum indices for each treatment are shown in Figure 6. The highest NDVI value of red oak lettuce was recorded in the 400N-C treatment (0.7043), while the lowest was in the 300C (0.6737) on the last day. Over the 7-day treatment period, NDVI showed slight fluctuations. Significant differences in the photochemical reflectance index (PRI) emerged between days 5 and 7, with 400C fluctuating, 300C remaining stable, and 300N-C declining. The anthocyanin reflectance index (ARI1) showed the highest values in the 400N-C treatment on days 5 and 7 (3.02 and 2.19, respectively), opposite to PRI trends, where 400N-C had the lowest values. By day 7, the PRI of 400N-C increased slightly, while ARI decreased.

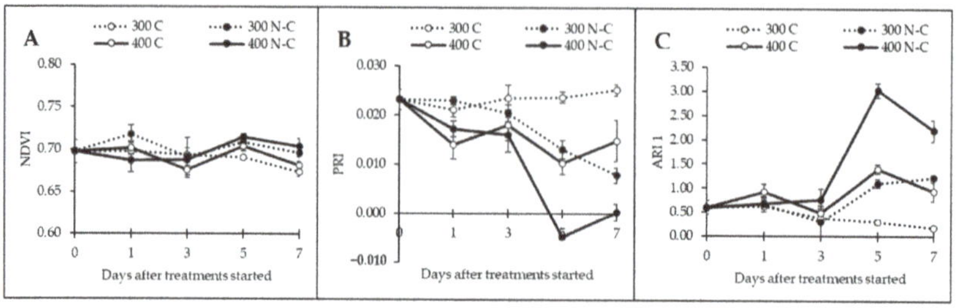


**Figure 6.** Spectral indices of red oak lettuce when treated under 300 and 400 µmol $m^{-2}$ $s^{-1}$ of light intensities with circulating and non-circulating irrigation for 7 days before harvesting. Those were (**A**) the normalized difference vegetation index, NDVI; (**B**) the photochemical reflectance index, PRI; and (**C**) the anthocyanin reflectance index, ARI1. The points are presented as mean value ± SE; n = 20. The numbers 300 and 400 before the letters represent light intensities in the PPFD unit. C and N-C after numbers mean circulating and non-circulating irrigations. The 300C was served as the control.

*3.3. Leaf Color and Biochemicals*

Photographs of red oak lettuce, separated into three layers for each treatment, were taken on the final day and are presented in Figure 7 for comparison.

**Figure 7.** Red oak lettuce leaves when grown under different light intensities (300 and 400 µmol $m^{-2}$ $s^{-1}$) and irrigation methods (circulating and non-circulating) for 7 days before harvesting. The leaves were divided into three parts, from lower to upper. The numbers before the letters represent light intensities in the PPFD unit. C and N-C after numbers mean circulating and non-circulating irrigations. The 300C was served as the control.

When comparing treatments, the average L* value for the 400N-C treatment exhibited a trend toward a darker color (Figure 8A), while the 300C treatment maintained a consistent average value over the 7-day period. Additionally, the a* and b* values began to diverge significantly from day 3 and deepest on day 5. By day 5, the color differences in red oak lettuce became visually apparent. The lowest significant lightness value was recorded for the 400N-C treatment at 40.7, indicating a 13.95% decrease compared to the 300C treatment (47.3) on day 7. Furthermore, the 400N-C treatment exhibited a significantly redder (+a*) and bluer (−b*) hue (−9.8 and 13.3, respectively) compared to the 400C, 300N-C, and 300C treatments.

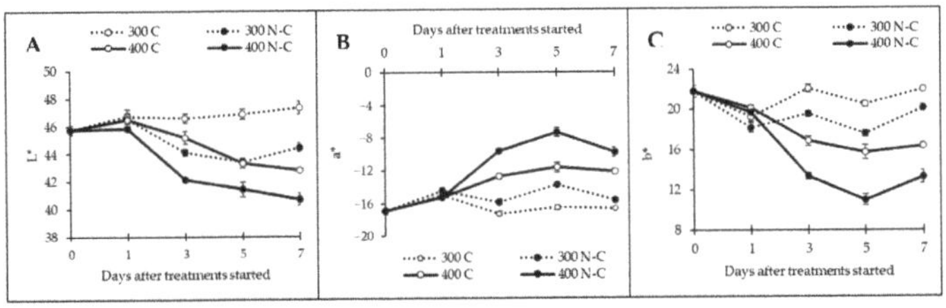


**Figure 8.** Leaf color value of red oak lettuce when treated under 300 and 400 μmol m$^{-2}$ s$^{-1}$ of light intensities with circulating and non-circulating irrigation for 7 days before harvesting. Those were (**A**) lightness; L*, (**B**) a*, and (**C**) b*. The points are presented as mean value ± SE (n = 20). The numbers 300 and 400 before the letters represent light intensities in the PPFD unit. C and N-C after numbers mean circulating and non-circulating irrigations. The 300C was served as the control.

The anthocyanin content in red oak lettuce was the highest under the 400N-C treatment (4.31), exhibiting a 24% increase compared to the control (3.26) (Figure 9A). An increasing trend in anthocyanin content was observed as light intensity increased, especially when grown under high light intensity with non-circulating irrigation.

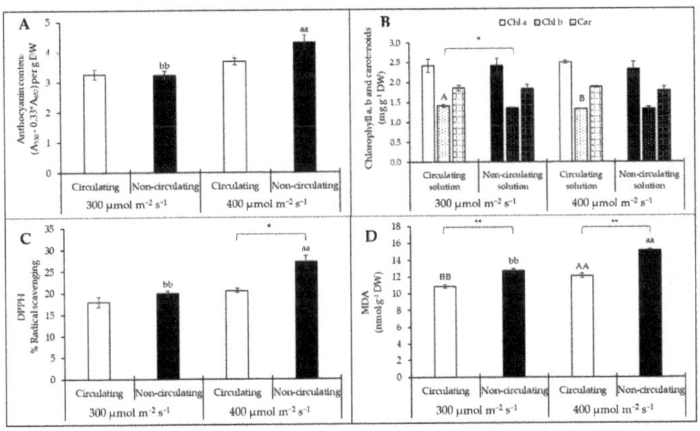


**Figure 9.** (**A**) Anthocyanin content, (**B**) photosynthetic pigments, (**C**) % radical scavenging, and (**D**) malondialdehyde of red oak lettuce when treated under 300 and 400 μmol m$^{-2}$ s$^{-1}$ with circulating and non-circulating irrigation for 7 days before harvesting. The vertical bars are presented as mean value ± SE (n = 12) according to independent samples *t*-test. The asterisks indicate a significant difference between the circulation methods in each lighting of 300 and 400 μmol m$^{-2}$ s$^{-1}$ by * ($p \leq 0.05$) and ** ($p \leq 0.01$). Uppercase and lowercase letters indicate a significant difference between the lighting under circulating and non-circulating irrigations. When $p \leq 0.05$ and $p \leq 0.01$ are represented by single and double letters, respectively. The 300C was served as the control.

An analysis of the photosynthetic pigments in red oak lettuce across all treatments revealed a slight decrease in chlorophyll b content under the 300N-C treatment compared to the control (1.35 mg g$^{-1}$ and 1.41 mg g$^{-1}$, respectively). Moreover, the 400C treatment resulted in a significant reduction in chlorophyll b content relative to the control, as illustrated in Figure 9B.

Furthermore, the percentage of radical scavenging activity (DPPH) was significantly highest in the 400N-C treatment (27.15%), surpassing that of the other treatments (Figure 9C). This was significantly observed when comparing irrigation methods under high light intensity, as well as between light intensities of 300 and 400 μmol m$^{-2}$ s$^{-1}$ under non-circulating irrigation methods.

Furthermore, the lipid peroxidation (MDA) level was significantly elevated in the 400N-C treatment, reaching 15.15 nmol $g^{-1}$ DW, compared to the 400C and 300N-C treatments (12.09 and 12.70 nmol $g^{-1}$ DW, respectively) (Figure 9D). These results indicate that increased light intensity, combined with non-circulating irrigation, led to a significant increase in MDA content in red oak lettuce.

### 3.4. Interactive Heat Map Analysis

The results presented in Figure 10 display a heat map generated through cluster analysis, which includes three color space values, four gas exchange parameters, three chlorophyll fluorescent parameters, and five reflectance indices of red oak plants. These were measured under two different light intensities and two irrigation methods at 1, 3, 5, and 7 days after the treatment methods. The first cluster showed light intensity at 300 μmol $m^{-2}$ $s^{-1}$. The second and third clusters contained measures for 400 μmol $m^{-2}$ $s^{-1}$. Notably, the red oak plants grown under 300 μmol $m^{-2}$ $s^{-1}$ of light intensity and non-circulating irrigation at 7 days were grouped in the third cluster. Conversely, the fourth cluster contained only red oak lettuce plants exposed to 400 μmol $m^{-2}$ $s^{-1}$ of light intensity with non-circulating irrigation, specifically on days 5 and 7. The correlation plot in Figure 11 illustrates the relationships among 15 key parameters measured in red oak lettuce under different light intensities and irrigation methods at 1, 3, 5, and 7 days after treatment began. The strongest positive correlations are observed between the following variables: ARI1 and ARI2 (r = 0.9941, $p < 0.001$), PRI and G (r = 0.9366, $p < 0.001$), A and ETR (r = 0.9358, $p < 0.001$), and L* and G (r = 0.9002, $p < 0.001$). Conversely, the most significant negative correlations occur between the following variables: a* and b* (r = −0.9924, $p < 0.001$), PRI and ARI2 (r = −0.9642, $p < 0.001$), PRI and ARI1 (r = −0.9585, $p < 0.001$), G and ARI2 (r = −0.94101, $p < 0.001$), and G and ARI1 (r = −0.9126, $p < 0.001$).

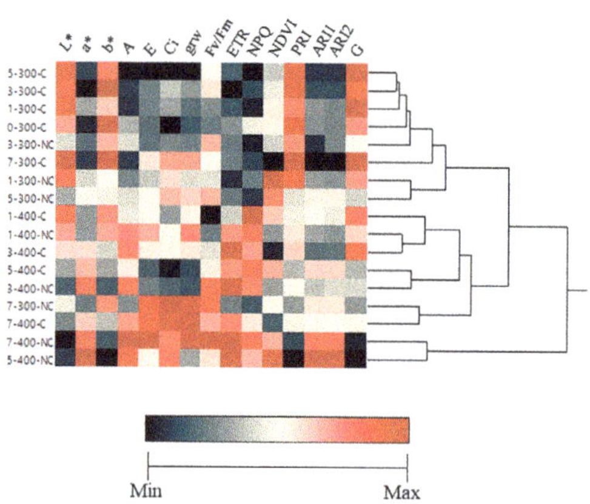


**Figure 10.** Heat map produced by cluster analysis, lightness (L*), red/green (a*), blue/yellow (b*), photosynthesis rate (A), transpiration rate (E), internal $CO_2$ concentration ($C_i$), stomatal conductance ($g_{sw}$), PSII maximum efficiency (Fv′/Fm′), non-photochemical quenching (NPQ), normalized difference vegetation index (NDVI), photochemical reflectance index (PRI), anthocyanin reflectance index 1 (ARI1), anthocyanin reflectance index 2 (ARI2), and greenness index (G) under two different light intensities (300 and 400 μmol $m^{-2}$ $s^{-1}$) and two irrigation methods (circulating; C and non-circulating; N-C) treatments of red oak lettuce. The 300C was served as the control. The hierarchical clustering analysis uses color gradients to represent the magnitude of relationships, with Max (high value) in red and Min (low value) in dark blue.

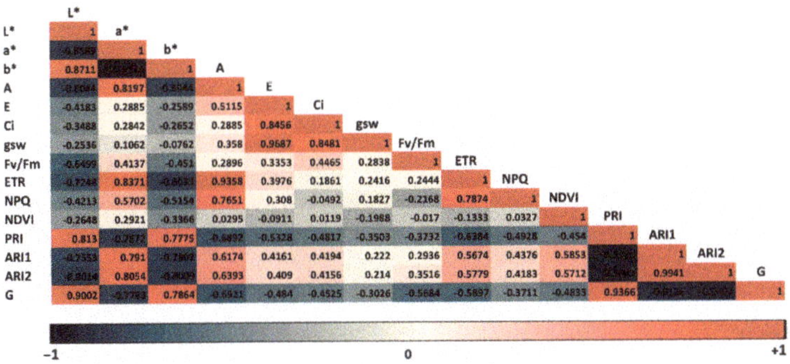


**Figure 11.** A correlational plot from 15 parameters recorded for red oak: lightness (L*), red/green (a*), blue/yellow (b*), photosynthesis rate (A), transpiration rate (E), internal $CO_2$ concentration ($C_i$), stomatal conductance ($g_{sw}$), PSII maximum efficiency (Fv′/Fm′), non-photochemical quenching (NPQ), normalized difference vegetation index (NDVI), photochemical reflectance index (PRI), anthocyanin reflectance index 1 (ARI1), anthocyanin reflectance index 2 (ARI2), and greenness index (G) under two different light intensities (300 and 400 μmol $m^{-2}$ $s^{-1}$) and two irrigation methods (circulating; C and non-circulating; N-C) treatments of red oak lettuce. The 300C was served as the control.

Among photosynthetic traits, A displayed a strong positive correlation with a* (r = 0.8197). In contrast, A showed a strong negative correlation with L* (r = −0.8044, $p < 0.001$) and a strong positive correlation with b* (r = 0.8044, $p < 0.001$). The stomatal conductance ($g_{sw}$) was positively correlated with $C_i$ (r = 0.8481, $p < 0.001$). Similarly, E was also positively correlated with $C_i$ (r = 0.8456, $p < 0.001$). ETR displayed a positive correlation with a* (r = 0.8371, $p < 0.001$) but had a negative correlation with b* (r = −0.8033, $p < 0.001$).

The heat map and correlation analysis provide a comprehensive overview of the relationships and clustering among these parameters, offering valuable insights into the physiological responses of red oak lettuce to different light intensities and irrigation methods over time.

Dissolved oxygen (DO) and electrical conductivity (EC) in the nutrient solution were continuously monitored from the first day of treatment (Figure 12). After the nutrient supply was halted and treatments began, DO levels decreased gradually from 7.36 mg $L^{-1}$ to 4.82 mg $L^{-1}$ within one day, eventually dropping to 3.83 mg $L^{-1}$ by day 7. In contrast, regular irrigation consistently maintained DO levels above 7 mg $L^{-1}$.

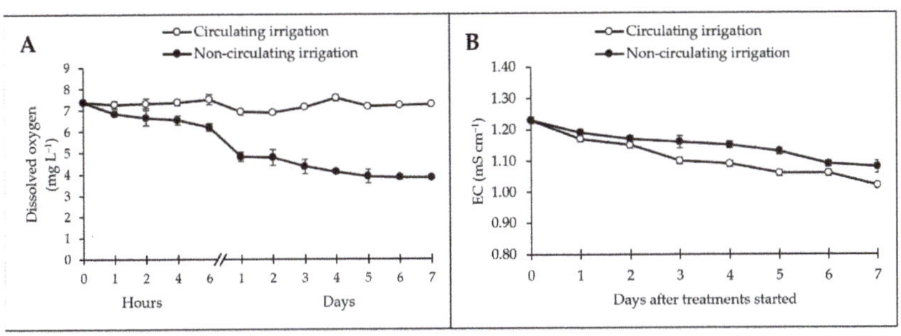


**Figure 12.** Changes in (**A**) dissolved oxygen concentration (DO) and (**B**) electrical conductivity (EC) in nutrient solution between circulating and non-circulating irrigation after treatment.

EC also showed a slight decline after nutrient feeding was stopped for both irrigation methods to maintain equilibrium between the circulating (C) and non-circulating treatments. However, circulating irrigation exhibited a greater reduction in EC than non-circulating conditions.

## 4. Discussion

### 4.1. Growth and Leaf Coloration

To enhance the red coloration of red oak lettuce cultivated under artificial lighting, increased light intensity and stopping circulating irrigation were employed to promote anthocyanin accumulation. In this study, elevating the light intensity to 400 μmol $m^{-2}$ $s^{-1}$ for seven days before harvest could enhance redder coloration, although there was no increment in shoot weight. Stopping circulating irrigation reduced yield but improved the coloration significantly. This effect was particularly pronounced when the plants were grown under the treatment of high light intensity combined with stopping circulation.

Previous studies have shown that root oxygen deficiency leads to hypoxia, often resulting in yield loss [39,40], but it could significantly induce higher anthocyanin content [41]. As shown in Figure 12, non-circulating irrigation resulted in a 35% reduction in dissolved oxygen (DO) within the first day. After seven days of treatment, DO levels dropped below 5 mg $L^{-1}$, which is generally considered insufficient to support optimal plant growth [42]. However, plant adaptability to stress conditions, such as hypoxia, is highly variable and depends on multiple factors, including the plant species, the specific cultivar or variety, and the physiological and biochemical survival strategies utilized. Tang et al. [43] reported that hypoxic conditions inhibited the growth and biomass of lettuce due to a reduction in the activities of catalase and total superoxide dismutase. These strategies may involve morphological adaptations, metabolic adjustments, or the activation of stress-responsive pathways to mitigate the adverse effects of hypoxic environments [44].

Increasing light intensity and non-circulating irrigation (400N-C) most effectively enhanced the red coloration in red oak lettuce compared to other treatments (Figure 2). Zhang et al. reported that high light intensity positively influences red coloration by increasing anthocyanin accumulation [19]. In our study, red oak lettuce displayed enhanced redness with low lightness (L*) and shifted toward red (+a) and blue (-b) under 400 μmol $m^{-2}$ $s^{-1}$. At the same time, red coloration of lettuce also had more improvement when high light intensity combined with non-circulating irrigation was applied. The red coloration of red oak lettuce in the 400N-C treatment extended further into the entire leaves, whereas only increasing light intensity (400C), the red coloration was concentrated primarily around the leaf margins (Figure 7).

Thus, raising light intensity with non-circulating irrigation (400N-C) had a more significant impact on red coloration than only non-circulating irrigation was solely applied, as seen when comparing the 400C and 300N-C treatments (Figure 8). Higher light intensity may trigger self-defense mechanisms in lettuce due to excess energy, as suggested by Fu et al. [17], whereas hypoxic conditions trigger increased anthocyanin accumulation (Figure 9). However, to achieve optimal red coloration, shortening the time of using the non-circulating method for 5 days may be better than 7 days (Figure 8). In addition, shortening the time of stopping circulating irrigation may also alleviate yield loss.

### 4.2. Gas Exchange and Chlorophyll Fluorescence

The photosynthetic capacity of red oak lettuce increased under high light intensity and non-circulating irrigation (Figure 4A), whereas the lowest photosynthetic rate was observed in the control treatment. Many studies reported that a high photosynthetic rate in plants can be increased when light intensity is raised as long as it does not reach the saturation point [45,46]. To fully understand the results of why non-circulating irrigation made lettuce have a higher photosynthetic capacity than growing under circulating irrigation, leaf area indexes should have been measured in order to calculate or assess the overall growth response. According to the report from Barickman et al. [47], low oxygen concentration by using waterlogging could gradually reduce the photosynthetic rate in cucumbers over a period of nine days. However, He et al. [48] reported that hypoxia during a short time (3 days) before harvest did not reduce photosynthesis or plant biomass in lettuce. In this study, lettuce grown with non-circulating irrigation achieved higher photosynthesis rates than the circulating irrigation, but the yield and biomass did not increase. Furthermore, there were no significant differences in E, Ci, and $g_{sw}$ values between treatments over the seven days.

To evaluate the red oak lettuce stress responses when increasing light intensity and stopping circulating irrigation, we found that the maximum quantum yield of PSII (Fv/Fm) differed significantly only on day 7, without consistent trends among treatments. From day 1 to day 7 of treatment, photosynthetic electron transport rates (ETR) and non-photochemical chlorophyll fluorescence quenching (NPQ) showed significant differences across light treatments, independent of the irrigation method. ETR and NPQ increased with light intensity [49], indicating that treatments with 400 µmol $m^{-2}$ $s^{-1}$ supported higher photosynthetic activity compared to 300 µmol $m^{-2}$ $s^{-1}$. NPQ in this study is one of the key parameters to understand the photoprotective mechanism that dissipates excess light energy in PSII as heat, which helps prevent oxidative damage [50]. According to Osmond's report [51], plants under high light generally exhibit higher NPQ values than those in low light. This explains why red oak lettuce exposed to high light intensity exhibited significantly higher NPQ levels.

### 4.3. Spectral Reflectance

To assess the health of red oak lettuce, we used spectral reflectance indices as non-destructive methods. This approach was inspired by the work of Lin et al. [52], who applied NDVI and PRI to examine waterlogging tolerance in cucurbit species. Our study found similar trends in NDVI value with no significant differences under non-circulating irrigation treatments compared to circulating irrigation. Notably, NDVI values tended to be higher in the 400N-C and 300N-C treatments than in the 400C and 300C treatments. This suggests that red oak lettuce did not exhibit signs of poor health when non-circulating was applied as a hypoxia condition. In addition, PRI value can be used to reflect short-term reversible changes in xanthophyll pigments in response to plant stress [53]. Our results showed a continuous decrease in the PRI starting from day 3 in non-circulating treatments, particularly in the 400N-C; that means xanthophyll pigments (e.g., violaxanthin) are being converted to photoprotective forms (e.g., zeaxanthin). This conversion is part of the plant's stress response to dissipate excess light energy as heat [54].

### 4.4. Leaf Pigments, Anthocyanin, Antioxidant Capacity, and Lipid Peroxidation

In this study, the leaf pigments, anthocyanin, antioxidant capacity, and lipid peroxidation were examined for the biochemical changes and qualities of red oak lettuce. The results indicated that increasing light intensity and stopping circulating irrigation did not significantly differ in the levels of chlorophyll and carotenoids but increased anthocyanin levels and enhanced red coloration in red oak lettuce. In this context, anthocyanins serve as antioxidants, protecting plant cells from reactive oxygen species (ROS) [20]. Similarly, Sakamoto and Suzuki [55] found that root stress conditions using low root-zone temperatures (10 °C) increased anthocyanin levels and improved radical scavenging in red lettuce because of oxidative stress. In addition, our results are similar to those of Rajapakse et al. [41], who reported that hypoxia increased anthocyanin levels, total phenolic compounds, carbohydrate concentration, and free radical scavenging capacity in lettuce.

In our results, the 400N-C treatment showed the highest levels of malondialdehyde (MDA), a key marker of lipid oxidation associated with cell membrane damage [56]. This suggests that anthocyanin, DPPH, and MDA directly responded to oxidative stress. Thus, our findings indicate that only non-circulating irrigation did not significantly increase anthocyanin and antioxidant capacity if the light intensity did not increase (Figure 9A,B). However, the highest occurred when increasing light intensity combined with the non-circulating irrigation.

## 5. Conclusions

This study examined the effects of light intensities and irrigation methods on the growth, quality, and coloration of red oak lettuce growing in a plant factory. Increasing light intensity significantly enhanced red coloration but did not significantly affect yield. Non-circulating irrigation reduced plant growth while significantly improving red coloration. The most pronounced red coloration was observed when high light intensity was combined with non-circulating irrigation before harvest. This study further demonstrated that anthocyanin content was maximized under the combined application of increasing light intensity and non-circulating irrigation, while neither factor alone produced a significant enhancement. This result was associated with increased radical scavenging activity and elevated MDA levels, triggered by oxidative stress. We suggest that integrating high light intensity with non-circulating irrigation presents an alternative and practical strategy for plant factory producers to enhance red oak lettuce coloration and anthocyanin accumulation under irrigation-controlled conditions without additional operational costs.

**Supplementary Materials:** The following supporting information can be downloaded at: https://www.mdpi.com/article/10.3390/horticulturae11010075/s1, Figure S1: The spectral distribution of LED used in the experiment (peak wavelength = 451 nm).

**Author Contributions:** Conceptualization, T.R. and K.M.; methodology, P.C. (Pariyanuj Chulaka); software, P.C. (Panita Chutimanukul); validation, M.T. and N.L.; formal analysis, T.R.; investigation, P.C. (Pariyanuj Chulaka); resources, K.M.; data curation, P.C. (Pariyanuj Chulaka) and P.C. (Panita Chutimanukul); writing—original draft preparation, T.R.; writing—review and editing, M.T. and N.L.; visualization, T.R.; supervision, N.L.; project administration, T.R.; funding acquisition, K.M. All authors have read and agreed to the published version of the manuscript.

**Funding:** This research was funded by the National Center for Genetic Engineering and Biotechnology (BIOTEC), Thailand, grant number P1951788.

**Data Availability Statement:** The original contributions presented in this study are included in the article. Further inquiries can be directed to the corresponding author.

**Acknowledgments:** The authors sincerely thank the National Center for Genetic Engineering and Biotechnology for their support in graduate studies and research on Plant Factories with Artificial Lighting (PFALs), as well as for providing access to the facility used in this study. Deep gratitude is extended to Praderm Wanichananan and Supattana Janta for their kind assistance with laboratory instruments. Special appreciation goes to Suchalee Sueachuen and Jakkaphattara Kas-Osot for their invaluable contributions to materials and data collection. Lastly, we also earnestly thank "The Capacity Building of KU Students on Internationalization Program: KUCSI" for their support on the collaborative research between Kasetsart University and Chiba University.

Conflicts of Interest: The authors declare no conflicts of interest.

## References


1. Kozai, T.; Niu, G. Chapter 2—Role of the plant factory with artificial lighting (PFAL) in urban areas. In *Plant Factory*, 2nd ed.; Kozai, T., Niu, G., Takagaki, M., Eds.; Academic Press: Cambridge, MA, USA, 2020; pp. 7–34. [CrossRef]
2. Rizkiana, A.; Prima Nugroho, A.; Salma, N.; Afif, S.; Masithoh, R.; Sutiarso, L.; Okayasu, T. Plant growth prediction model for lettuce (Lactuca sativa.) in plant factories using artificial neural network. *IOP Conf. Ser. Earth Environ. Sci.* **2021**, *733*, 012027. [CrossRef]
3. Azad, M.O.; Kjaer, K.H.; Adnan, M.; Naznin, M.T.; Lim, J.D.; Sung, I.J.; Park, C.H.; Lim, Y.S. The Evaluation of Growth Performance, Photosynthetic Capacity, and Primary and Secondary Metabolite Content of Leaf Lettuce Grown under Limited Irradiation of Blue and Red LED Light in an Urban Plant Factory. *Agriculture* **2020**, *10*, 28. [CrossRef]
4. Sugimura, N.; Thinh, N.Q.; Kohama, S.; Fukui, Y.; Iwamura, K. A Study on Demand Forecasting of Wholesale Markets of Lettuces for Production Planning in Plant Factories. *Int. J. Autom. Technol.* **2022**, *16*, 218–229. [CrossRef]
5. Li, Q.; Kubota, C. Effects of supplemental light quality on growth and phytochemicals of baby leaf lettuce. *Environ. Exp. Bot.* **2009**, *67*, 59–64. [CrossRef]
6. Kang, J.H.; KrishnaKumar, S.; Atulba, S.L.S.; Jeong, B.R.; Hwang, S.J. Light intensity and photoperiod influence the growth and development of hydroponically grown leaf lettuce in a closed-type plant factory system. *Hortic. Environ. Biotechnol.* **2013**, *54*, 501–509. [CrossRef]
7. Moon, T.; Choi, W.-J.; Jang, S.-H.; Choi, D.-S.; Oh, M.-M. Growth Analysis of Plant Factory-Grown Lettuce by Deep Neural Networks Based on Automated Feature Extraction. *Horticulturae* **2022**, *8*, 1124. [CrossRef]
8. Souza, A.S.N.d.; Schmidt, H.d.O.; Pagno, C.; Rodrigues, E.; Silva, M.A.S.d.; Flôres, S.H.; Rios, A.d.O. Influence of cultivar and season on carotenoids and phenolic compounds from red lettuce influence of cultivar and season on lettuce. *Food Res. Int.* **2022**, *155*, 111110. [CrossRef]
9. Mampholo, B.; Maboko, M.; Soundy, P.; Sivakumar, D. Phytochemicals and Overall Quality of Leafy Lettuce (*Lactuca sativa* L.) Varieties Grown in Closed Hydroponic System: LETTUCE PHYTOCHEMICAL QUALITY. *J. Food Qual.* **2016**, *39*, 805–815. [CrossRef]
10. Kim, D.-E.; Shang, X.; Assefa, A.D.; Keum, Y.-S.; Saini, R.K. Metabolite profiling of green, green/red, and red lettuce cultivars: Variation in health beneficial compounds and antioxidant potential. *Food Res. Int.* **2018**, *105*, 361–370. [CrossRef]
11. Shioshita, R.; Enoka, J.; Aiona, D.; Young, C.; Sakai, W.; Wall, M. Coloration and growth of red lettuce grown under UV-radiation transmitting and non-transmitting covers. *Acta Hortic.* **2006**, *761*. [CrossRef]
12. Owen, W.G.; Lopez, R.G. End-of-production Supplemental Lighting with Red and Blue Light-emitting Diodes (LEDs) Influences Red Pigmentation of Four Lettuce Varieties. *HortScience Horts* **2015**, *50*, 676–684. [CrossRef]
13. Cheng, D.M.; Pogrebnyak, N.; Kuhn, P.; Krueger, C.G.; Johnson, W.D.; Raskin, I. Development and phytochemical characterization of high polyphenol red lettuce with anti-diabetic properties. *PLoS ONE* **2014**, *9*, e91571. [CrossRef] [PubMed]
14. Kozai, T. Resource use efficiency of closed plant production system with artificial light: Concept, estimation and application to plant factory. *Proc. Jpn. Acad. Ser. B Phys. Biol. Sci.* **2013**, *89*, 447–461. [CrossRef] [PubMed]
15. Xin, Z.; He, D.; Niu, G.; Yan, Z.; Song, J. Effects of environment lighting on the growth, photosynthesis, and quality of hydroponic lettuce in a plant factory. *Int. J. Agric. Biol. Eng.* **2018**, *11*, 33–40. [CrossRef]
16. Zhou, J.; Li, P.; Wang, J. Effects of Light Intensity and Temperature on the Photosynthesis Characteristics and Yield of Lettuce. *Horticulturae* **2022**, *8*, 178. [CrossRef]
17. Fu, W.; Li, P.; Wu, Y. Effects of different light intensities on chlorophyll fluorescence characteristics and yield in lettuce. *Sci. Hortic.* **2012**, *135*, 45–51. [CrossRef]
18. Modarelli, G.C.; Paradiso, R.; Arena, C.; De Pascale, S.; Van Labeke, M.-C. High Light Intensity from Blue-Red LEDs Enhance Photosynthetic Performance, Plant Growth, and Optical Properties of Red Lettuce in Controlled Environment. *Horticulturae* **2022**, *8*, 114. [CrossRef]
19. Zhang, Y.; Xu, S.; Cheng, Y.; Peng, Z.; Han, J. Transcriptome profiling of anthocyanin-related genes reveals effects of light intensity on anthocyanin biosynthesis in red leaf lettuce. *PeerJ* **2018**, *6*, e4607. [CrossRef]
20. Gould, K.S. Nature's Swiss Army Knife: The Diverse Protective Roles of Anthocyanins in Leaves. *J. Biomed. Biotechnol.* **2004**, *2004*, 314–320. [CrossRef]
21. Winkel-Shirley, B. Biosynthesis of flavonoids and effects of stress. *Curr. Opin. Plant Biol.* **2002**, *5*, 218–223. [CrossRef]
22. Kovinich, N.; Kayanja, G.; Chanoca, A.; Otegui, M.S.; Grotewold, E. Abiotic stresses induce different localizations of anthocyanins in Arabidopsis. *Plant Signal. Behav.* **2015**, *10*, e1027850. [CrossRef] [PubMed]

23. Skowron, E.; Trojak, M.; Pacak, I. Effects of UV-B and UV-C Spectrum Supplementation on the Antioxidant Properties and Photosynthetic Activity of Lettuce Cultivars. *Int. J. Mol. Sci.* **2024**, *25*, 9298. [CrossRef] [PubMed]
24. Naing, A.H.; Kim, C.K. Abiotic stress-induced anthocyanins in plants: Their role in tolerance to abiotic stresses. *Physiol. Plant.* **2021**, *172*, 1711–1723. [CrossRef]
25. Nadarajah, K.K. ROS Homeostasis in Abiotic Stress Tolerance in Plants. *Int. J. Mol. Sci.* **2020**, *21*, 5208. [CrossRef]
26. Wittayathanarattana, T.; Wanichananan, P.; Supaibulwatana, K.; Goto, E. Enhancement of bioactive compounds in baby leaf Amaranthus tricolor L. using short-term application of UV-B irradiation. *Plant Physiol. Biochem.* **2022**, *182*, 202–215. [CrossRef]
27. Zhang, L.; Zhang, Q.; Contreras, V.H.E.; Huang, T.; Jiang, H.; Song, B.; Duan, Z.; Li, Y.; Yang, X.; Song, H.; et al. Short-term high-light intensity and low temperature improve the quality and flavor of lettuce grown in plant factory. *J. Sci. Food Agric.* **2024**, *104*, 9046–9055. [CrossRef]
28. Zeljković, S.Ć.; Štefelová, N.; Hron, K.; Doležalová, I.; Tarkowski, P. Preharvest Abiotic Stress Affects the Nutritional Value of Lettuce. *Agronomy* **2023**, *13*, 398. [CrossRef]
29. Loreti, E.; Striker, G. Plant Responses to Hypoxia: Signaling and Adaptation. *Plants* **2020**, *9*, 1704. [CrossRef]
30. Yemelyanov, V.; Puzanskiy, R.; Shishova, M. Plant Life with and without Oxygen: A Metabolomics Approach. *Int. J. Mol. Sci.* **2023**, *24*, 16222. [CrossRef]
31. Asao, T.; Kitazawa, H.; Washizu, K.; Ban, T.; Pramanik, H. Effect of different nutrient levels on anthocyanin and nitrate-N contents in turnip grown in hydroponics. *J. Appl. Hortic.* **2005**, *7*, 87–89. [CrossRef]
32. Rouse, J.W.; Haas, R.H.; Schell, J.A.; Deering, D.W. Monitoring vegetation systems in the great plains with ERTS. In *Third Earth Resources Technology Satellite-1 Symposium: Section A-B. Technical presentations*; NASA: Washington, DC, USA, 1974.
33. Gamon, J.A.; Peñuelas, J.; Field, C.B. A narrow-waveband spectral index that tracks diurnal changes in photosynthetic efficiency. *Remote Sens. Environ.* **1992**, *41*, 35–44. [CrossRef]
34. Gitelson, A.A.; Merzlyak, M.N.; Chivkunova, O.B. Optical properties and nondestructive estimation of anthocyanin content in plant leaves. *Photochem. Photobiol.* **2001**, *74*, 38–45. [CrossRef] [PubMed]
35. Chutimanukul, P.; Wanichananan, P.; Janta, S.; Toojinda, T.; Darwell, C.; Mosaleeyanon, K. The influence of different light spectra on physiological responses, antioxidant capacity and chemical compositions in two holy basil cultivars. *Sci. Rep.* **2022**, *12*, 588. [CrossRef] [PubMed]
36. Liang, J.; He, J. Protective role of anthocyanins in plants under low nitrogen stress. *Biochem. Biophys. Res. Commun.* **2018**, *498*, 946–953. [CrossRef]
37. Velikova, V.; Yordanov, I.; Edreva, A. Oxidative stress and some antioxidant systems in acid rain-treated bean plants: Protective role of exogenous polyamines. *Plant Sci.* **2000**, *151*, 59–66. [CrossRef]
38. Heath, R.L.; Packer, L. Photoperoxidation in isolated chloroplasts. I. Kinetics and stoichiometry of fatty acid peroxidation. *Arch. Biochem. Biophys.* **1968**, *125*, 189–198. [CrossRef]
39. Paes de Melo, B.; Carpinetti, P.d.A.; Fraga, O.T.; Rodrigues-Silva, P.L.; Fioresi, V.S.; de Camargos, L.F.; Ferreira, M.F.d.S. Abiotic Stresses in Plants and Their Markers: A Practice View of Plant Stress Responses and Programmed Cell Death Mechanisms. *Plants* **2022**, *11*, 1100. [CrossRef]
40. Armstrong, W.; Hodgson, J.G.; Booth, R.E.; Gaitens, P.V.; Grime, J.P.; Band, S.R.; Mowforth, M.A.; Crawford, R.M.M.; Bossard, C.C.; Hillier, S.H.; et al. Methods of Comparative Study. In *Methods in Comparative Plant Ecology*; Hendry, G.A.F., Grime, J.P., Eds.; Springer: Dordrecht, The Netherlands, 1993.
41. Rajapakse, N.C.; He, C.; Cisneros-Zevallos, L.; Davies, F.T. Hypobaria and hypoxia affects growth and phytochemical contents of lettuce. *Sci. Hortic.* **2009**, *122*, 171–178. [CrossRef]
42. Becker, K. Understanding Dissolved Oxygen. Available online: https://www.growertalks.com/Article/?srch=1&articleID=22058&highlight=understanding+dissolved+oxygen (accessed on 29 November 2024).
43. Tang, Y.; Gao, F.; Guo, S.; Li, F. The morphology, physiology and nutritional quality of lettuce grown under hypobaria and hypoxia. *Acta Astronaut.* **2015**, *112*, 29–36. [CrossRef]
44. Nagai, K.; Hattori, Y.; Ashikari, M. Stunt or elongate? Two opposite strategies by which rice adapts to floods. *J. Plant Res.* **2010**, *123*, 303–309. [CrossRef]
45. Cammarisano, L.; Körner, O. Response of Cyanic and Acyanic Lettuce Cultivars to an Increased Proportion of Blue Light. *Biology* **2022**, *11*, 959. [CrossRef] [PubMed]
46. Zhou, J.; Wang, J.; Hang, T.; Li, P. Photosynthetic characteristics and growth performance of lettuce (Lactuca sativa L.) under different light/dark cycles in mini plant factories. *Photosynthetica* **2020**, *58*, 740–747. [CrossRef]
47. Barickman, T.C.; Simpson, C.R.; Sams, C.E. Waterlogging Causes Early Modification in the Physiological Performance, Carotenoids, Chlorophylls, Proline, and Soluble Sugars of Cucumber Plants. *Plants* **2019**, *8*, 160. [CrossRef]

48. He, C.; Jacobo-Velázquez, D.; Cisneros-Zevallos, L.; Davies, F. Hypobaria and hypoxia affects phytochemical production, gas exchange, and growth of lettuce. *Photosynthetica* **2013**, *51*, 465–473. [CrossRef]
49. Kitao, M.; Yasuda, Y.; Kodani, E.; Harayama, H.; Awaya, Y.; Komatsu, M.; Yazaki, K.; Tobita, H.; Agathokleous, E. Integration of electron flow partitioning improves estimation of photosynthetic rate under various environmental conditions based on chlorophyll fluorescence. *Remote Sens. Environ.* **2021**, *254*, 112273. [CrossRef]
50. Amir, E.; Shaked, R.; Eshel, G.; Barak, S.; Rachmilevitch, S. Low induction of non-photochemical quenching and high photochemical efficiency in the annual desert plant Anastatica hierochuntica. *Physiol. Plant.* **2013**, *151*, 544–558. [CrossRef]
51. Osmond, C.B.; Ramus, J.; Levavasseur, G.; Franklin, L.A.; Henley, W.J. Fluorescence quenching during photosynthesis and photoinhibition of Ulva rotundata blid. *Planta* **1993**, *190*, 97–106. [CrossRef]
52. Lin, H.H.; Lin, K.H.; Huang, M.Y.; Su, Y.R. Use of Non-Destructive Measurements to Identify Cucurbit Species (Cucurbita maxima and Cucurbita moschata) Tolerant to Waterlogged Conditions. *Plants* **2020**, *9*, 1226. [CrossRef]
53. Peñuelas, J.; Garbulsky, M.F.; Filella, I. Photochemical reflectance index (PRI) and remote sensing of plant $CO_2$ uptake. *New Phytol.* **2011**, *191*, 596–599. [CrossRef]
54. Wong, C.Y.S.; Gamon, J.A. The photochemical reflectance index provides an optical indicator of spring photosynthetic activation in evergreen conifers. *New Phytol.* **2015**, *206*, 196–208. [CrossRef]
55. Sakamoto, M.; Suzuki, T. Effect of Root-Zone Temperature on Growth and Quality of Hydroponically Grown Red Leaf Lettuce (Lactuca sativa L. cv. Red Wave). *Am. J. Plant Sci.* **2015**, *06*, 2350–2360. [CrossRef]
56. Sharma, P.; Jha, A.B.; Dubey, R.S.; Pessarakli, M. Reactive Oxygen Species, Oxidative Damage, and Antioxidative Defense Mechanism in Plants under Stressful Conditions. *J. Bot.* **2012**, *2012*, 217037. [CrossRef]

*Article*

# Compost and Vermicompost as Substrates Enriched with *Trichoderma asperellum* for the Production of Basic Potato Seed in the Venezuelan Andes

**Haydee Peña [1], Fernando Diánez [2], Beatriz Ramírez [1], José Sulbarán [1], Karen Arias [1], Victoria Huertas [2] and Mila Santos [2,*]**

[1] Laboratory for the Recovery of Compostable Waste, Universidad Nacional Experimental del Táchira, San Cristóbal 5001, Venezuela; hbpena@gmail.com (H.P.); bramirez@unet.edu.ve (B.R.); jsulbaran@unet.edu.ve (J.S.); karias@unet.edu.ve (K.A.)
[2] Departamento de Agronomía, Escuela Superior de Ingeniería, Universidad de Almería, 04120 Almería, Spain; fdianez@ual.es (F.D.); vhn616@ual.es (V.H.)
* Correspondence: msantos@ual.es

**Abstract:** The agricultural sector faces significant pressure to meet the growing global demand for food while managing the planet's limited natural resources. Addressing this challenge requires the strategic use of abundant materials and accessible biotechnologies that farmers can effectively implement. This study evaluated the effects of various substrate mixtures, including combinations of coconut fiber with compost or vermicompost (50:50 *v/v*). It also assessed the impact of heat treatment and inoculation with the antagonist *Trichoderma asperellum* on the production of basic potato seeds grown in 70 L plastic baskets. Statistical analysis revealed that the vermicompost-based mixture outperformed others, demonstrating superior biometric variables for potato plants. The treatment with sterilized vermicompost led to a 41% increase in seed weight, underscoring the beneficial effects of vermicompost. Correlation analysis indicated a positive relationship between the phosphorus content of the mixture and a negative relationship with the E4/E6 ratio—a parameter recognized as a reliable and easily measurable indicator of substrate quality, along with pH and electrical conductivity.

**Keywords:** vermicompost; plant growth-promoting fungi; *Solanum tuberosum*; substrate; organic matter; bio-waste; composting

## 1. Introduction

The increasing global population demands a steady supply of high-quality food. Agronomists, researchers, and agricultural producers have a responsibility to meet this demand while minimizing the environmental impact of their practices. Traditional agriculture, with its heavy reliance on chemical inputs and limited control over environmental variables, often fails to provide sustainable strategies to combat plant diseases and generate sufficient quantities of quality food [1,2]. However, emerging technological developments present a range of possibilities that are beginning to be implemented in Venezuela. These technologies must be validated to generate the necessary knowledge that can positively influence rural development [1]. One of the key challenges facing the agricultural sector is the need for sustainable agronomic practices that can increase crop productivity while addressing the impacts of climate change. Sustainable agronomic practices, such as selecting suitable crop varieties resistant to biotic and abiotic stresses, implementing sustainable intensification strategies, and adopting resource-conserving agriculture, can play a crucial role in improving food security [1].

The production of high-quality potato seed is a critical aspect of sustainable agriculture, as it ensures the availability of disease-free and genetically superior planting material. Traditional methods of potato seed production can be resource-intensive, often relying on synthetic fertilizers and pesticides. In the technological context described, the production of potato seed developed in Mucuchíes, Venezuela, executed by PROINPA [3], is framed within the Venezuelan National Seed Plan. The objective is to contribute to the supply of quality seed demand at the national level, articulating inter-institutional actions of the public sector and cooperatives that cover the selection of the best germplasm according to national needs, to the commercialization of certified seed among producers [4].

This initiative in Mucuchíes, Venezuela, is part of a broader effort to address the lack of high-quality planting material for farmers' potato varieties, produced locally and at low cost, which is a major constraint limiting the expansion of potato production in the region. The practices implemented in Mucuchíes should be promoted among Andean producers to increase the productivity and sustainability of their potato–grain–pasture systems. Preparing suitable substrates is of utmost importance for the acclimatization of in vitro grown plantlets and the production of pre-basic and basic potato seeds [5]. The production of compost with sufficient quality attributes is extremely valuable to ensure the seed production chain, while also leveraging the beneficial nutrients and biological principles contained within [5,6]. Compost not only nourishes plants but also protects them from pathogens, as these are emergent properties of the agroecosystem that arise from its beneficial interactions [7]. Alongside the use of compost and vermicompost-based substrates, the possibility of enriching the culture medium with plant growth-promoting microorganisms presents additional advantages in terms of nutrition and antagonism against pathogens, such as *Trichoderma* [8–10]. Enriching the beneficial microbiome associated with plants is a promising approach to developing sustainable crop production, as microbes residing in the soil system have immense potential to increase crop growth, nutrient acquisition, stress tolerance, soil fertility, and disease resistance [11]. Rhizospheric soil, the region influenced by plant roots, is typically richer in nutrients compared to bulk soil due to the accumulation of various organic compounds released by roots.

The use of vermicompost in organic farming has been well-documented, with improvements in soil's physical, chemical, and biological properties that enhance crop yield. While vermicompost has been shown to significantly improve plant growth, its application at high concentrations may impede growth due to high levels of soluble salts. Therefore, vermicomposts should be applied at moderate concentrations to obtain maximum plant yield. Additionally, studies have demonstrated the positive effects of soil drenching with vermicompost leachate on the growth of greenhouse-grown tomato seedlings, alleviating deficiencies of phosphorus and potassium.

To address this research gap, the present study evaluated the characteristics of compost and vermicompost and their influence on the growth and yield of potato seedlings in a protected cultivation environment. The integration of these eco-friendly inputs into potato seed production systems could potentially reduce the reliance on synthetic inputs while maintaining or even enhancing crop performance [12–15].

## 2. Materials and Methods

### 2.1. Characterization of Compost-Vermicompost Blends with Coconut Fiber for Plant Growth Media

The study was conducted in a greenhouse located at the La Pradera Academic Unit of the Universidad Nacional Experimental del Táchira in the Jáuregui municipality of Táchira state, Venezuela. The site is situated at coordinates 8°11′21.5″ N/71°57′54.4″ W, at an elevation of 1750 m above sea level, with an average temperature of 18 °C and a relative humidity of 73%. According to the Holdridge life zone classification, the area is characterized as a premontane wet forest [16]. In a greenhouse with a maximum average temperature of 32.7 °C and a minimum of 15.5 °C with humidity levels ranging from a minimum of 36.07% to a maximum of 90.21%. The laboratory analyses were carried out at the Compostable Waste Valorization Laboratory at UNET. As the first step, 50:50 blends of compost and vermicompost with coconut fiber were prepared. Subsequent to the preparation of compost and vermicompost mixtures, moist heat treatment was applied to half of each of the blends [17–20]. This was completed by transferring 40 L portions into clean nylon bags, which were then placed in a 150 L metal container with a perforated base to create a double-bottom setup. Water was added to the container to a height of 0.25 m, and the level was maintained as needed, similar to a laboratory water bath. The treatment was continued for two hours after the onset of boiling and was repeated twice. The study involved planting five mini-tubers of the Iniafrit potato variety in plastic baskets with a capacity of 70 L, filled with the described substrate mixtures. The initial volume of the mixture was 20 L, with two additional inputs of 10 L each at 30 and 60 days after sowing to simulate hilling. The cultivated potato of the Iniafrit variety was released by the CIP in 2006 and introduced in Venezuela by the Venezuelan Institute of Agricultural Investigations INIA in 2006; by its acronym in Spanish, for industrial processing because it has elongated tubers with characteristics for cane-type processing, tolerant to *P. infestans* [21]. At 60 days after sowing, the number of leaves and the height of the main stem from the base to the tip were measured for each plant [20]. It was measured: thickness of the stems (G-Steam), number of leaves (N° Leaves), dry weight of the roots (radical dry weight), number of tubers per basket (N° tuber box), and weight of the tubers per basket (weight box) were measured. A wettable powder formulation of the rice-based fungus *T. asperellum* [21], containing $1.3 \times 10^9$ spores per gram, was used. Suspensions with $10^7$ spores mL were prepared from this product and applied as a spray to the point of runoff at the time of sowing in potato seed treatments [22] and plants. Two complementary treatments were performed at 15 and 30 days after planting with 5 mL/plant of the *Trichoderma* suspension. The *T. asperellum* strain was obtained from the Microbiological Control Laboratory of the National Experimental University of Táchira (UNET), Venezuela of soils cultivated with potatoes in the municipality of José María Vargas in the state of Táchira.

The experiment was conducted using a completely randomized factorial design with three treatments and two levels each: (1) two organic materials, Cp (agro-industrial compost) and Cv (vermicompost), both mixed with coconut fiber in a 50:50 *v/v* ratio; (2) two levels of physical treatment: sterilized and non-sterilized; and (3) two levels of biological treatment: with and without *T. asperellum*.

This research was conceived as a continuation of the evaluation of the quality parameters of compost from three compostable mixtures, prepared from agro-industrial waste, biowaste, and agricultural waste [22,23]. The Laboratory for the Recovery of Compostable Waste of the UNET was used to carry out the analyses described unless otherwise indicated. The proportions of waste used to obtain compostable mixtures were calculated according to Richards for their optimization in terms of C/N and moisture content [24]. Therefore, the compost used in the test was wastewater sludge compost from the food industry (Cod. LER 02 01 03) (European Parliament 2014) [25], designated Cp; and vermicompost comprising coffee pulp (Cod. LER 02 01 03), designated Cv. The pulp is a byproduct of the wet processing of the coffee fruit and is composed of the mesocarp of the cherry or berry, which is located just below the pericarp or skin [26]. This product was purchased from the National Institute of Agricultural Research INIA located in the Junín municipality of the state of Táchira, Venezuela.

## 2.2. Physical, Chemical, and Biological Analysis of Substrate Mixtures

### 2.2.1. Determination of Organic Carbon in Compost Samples

The determination of organic carbon (OC) in compost was carried out following the procedure described by Sadzawka et al. [27]. This method assumes that the organic matter in the compost contains 56% carbon and uses a conversion factor of 1.8 to calculate the organic carbon content from the organic matter content. Specifically, the organic carbon content was calculated as OC (%): OM/1.8, where OM represents the concentration of organic matter in the sample and was calculated as OM (%) $[(a - b)/a] \times 100$ where **a** is mass (g) of the dry sample at 70 ± 5 °C before calcination and **b** is mass (g) of the calcined sample at 550 °C. To measure the organic matter content, 10 g of the sample, dried at 36 °C and passed through a 16 mm sieve, was placed in a crucible and dried at 70 ± 5 °C for 24 h. The sample was then slowly heated in a muffle furnace up to 550 °C and held at that temperature for 2 h. After this, the temperature was slowly decreased to 200 °C. Finally, the samples were placed in a desiccator until room temperature.

### 2.2.2. Nitrogen Content (%)

For N determination, the procedure described by [28] was followed. A 0.1 g sample was weighed into a digestion tube. Then, 2 mL of the digestion mixture containing 2.5 g of copper sulfate diluted in 2.5 mL of deionized distilled water was added; subsequently, 0.15 g of selenium was added to the previous preparation. In 100 mL of concentrated sulfuric acid, 2.5 g of potassium sulfate was dissolved, the selenium solution was added, and then 1.0 g of mercury oxide was added. The sample was placed in the digestion apparatus at 350–380 °C for 90 min or until the mixture became transparent. While still warm, distilled water was slowly added to the tubes. To perform the distillation phase, 15 mL of 40% NaOH was slowly added to the flask. The flask was placed in the distillation system for 10 min, and at the opposite end of the condensation system, a flask containing 25 mL of 2% boric acid solution (with methyl red indicator (0.1%) and bromocresol green (0.1%) in 100 mL of 95% ethanol) was placed. The captured ammonium was titrated with 0.1 N $H_2SO_4$. *N* (%) was calculated as $[(V \times N \times 0.014)/W] \times 100$, where V is sulfuric acid consumption in mL; N is normality of sulphuric acid; milliequivalent of nitrogen; and W: sample weight (g).

### 2.2.3. Phosphorous Content (%)

The determination of P followed the procedure described by Sadzawka et al. [27]. Approximately 2 g of fresh sample dried at 70 °C for 24 h was subjected to calcination at 500 °C. The sample was weighed after drying at 70 °C, then placed in a muffle furnace where the temperature was gradually increased to 500 °C and maintained for 2 h. The sample was then allowed to cool slowly and transferred to a desiccator. The ash was analyzed for its P and Ca content [28]. Once the ash from the samples was obtained, 40 mL of HCl 1:3 with a few drops of concentrated $HNO_3$ was added. The mixture was heated on a hot plate in crucibles until boiling point was reached. After filtration, the extracts were transferred to 250 mL volumetric flasks and topped up with distilled water once cooled. For P determination, 10 mL aliquots were transferred to 50 mL volumetric flasks, followed by the addition of 5 mL of molybdenum vanadate solution [29]. After 10 min, the absorbance of the samples and corresponding standards was read at 400 nm using a 6405 UV-vis brand Jenway (USA).

### 2.2.4. Calcium Content (%)

Ca (%) was measured by complexometry. In this procedure, the calcium present in the sample is forced to form a calcium oxalate precipitate by adding a saturated solution of ammonium oxalate to the ash solution of the evaluated material. This precipitate is thoroughly washed with ammonium hydroxide to remove excess ammonium oxalate. Through the action of sulfuric acid, calcium oxalate forms oxalic acid and calcium sulfate. The oxalic acid is determined using a standardized $KMnO_4$ solution [28].

#### 2.2.5. Proportion of Humic and Fulvic Substances: E4/E6 Ratio

A 1 g sample of compost or vermicompost was weighed into 250 mL polyethylene flasks, and 50 mL of 0.5 M NaOH solution was added. The mixture was shaken for 2 h using a horizontal shaker at 60 cycles per min and left to rest for 12 h. The sample was then transferred to 50 mL centrifuge tubes and centrifuged at 3000 rpm for 25 min, followed by filtration. The absorbance of the supernatant was determined at 465 nm (E4) and 665 nm (E6) using a 6405 UV-vis brand Jenway (USA). The E4/E6 ratio was calculated using these values [30].

#### 2.2.6. pH and Electrical Conductivity

To measure the pH, the sample sieved at 16 mm humidity was mixed and stirred with water at a 1:5 ratio. Then, the pH and electrical conductivity of the same extract were measured with a Hanna Instrument brand pH meter. The mass of the sample <16 mm and wet (A) equivalent to 40 g of dry sample at 70 ± 5 °C and the volume of water (B) were calculated to achieve a ratio of 1:5 according to $A = (40/ST) \times 100$ $B = 200 - (A - 40)$, where A = mass, in g, of a sample <16 mm and wet, ST = total solids, in % for a wet sample, and B = volume, in mL. For water, A = mass, in g, of a sample <16 mm and wet [27].

#### 2.2.7. Physical Properties

Total porosity, aeration porosity, moisture retention capacity, bulk density, and particle density were determined using the method of Pire and Pereira [31]. The material was kept in the laboratory at an average temperature of 24 °C during the study to measure the physical properties of the mixtures, including total porosity, TPS (%) = (Va + (WW − DW)/Pa)/Vc × 100; aeration porosity (AP (%) = Va/Vc × 100%); apparent density (Bd ($Mg{\cdot}m^{-3}$) = DW/Vc); particle density (Pd ($Mg{\cdot}m^{-3}$) = Bd/(1 − PT/100)); and water retention capacity H_W(%) = (WW − DW)/Vc × 100, where Va is the volume drained ($cm^3$), WW is the wet weight of the sample (g), DW is the dry weight of the sample (g), Pa is the specific weight of water (1 g $cm^{-3}$), and Vc is the volume of the tube or cylinder ($cm^3$). For the determination of physical properties, porometers made of PVC tubes were prepared. Each had a diameter of 7.5 cm, 15 cm, and a foot composed of the same material, which was sealed as a base and perforated on the side to allow water circulation. A ring was added as an extension of the cylinder to prevent sample loss due to expansion when it became wet. The samples were used to fill the porometers to their maximum capacity. Settling was promoted by dropping the porometers twice from a height of 7.5 cm and then again filling them to the upper edge. The cylinders with the samples and the upper rings were placed in a container with a water level just below the upper edge. This process forced moisture into the sample via the bottom holes, allowing the air to escape freely through the upper face. The samples were left in the water until the next day to standardize the saturation process. After 24 h, they were extracted from the container with water and leveled with a spatula. The samples were fitted with a cloth attached to a rubber band on top and completely immersed for 30 min. Rubber plugs were then placed in the bottom holes before removal from the water. Once removed, each porometer with its sample was placed on paths, the trays were removed to measure the volume of water drained in ten minutes, the wet sample was extracted from the porometers, and its dry weight was determined at 105 °C.

2.2.8. Microbial Populations

A suspension was prepared with 1 g of each sample of air-dried substrate mixed in 10 mL of a sterile 0.89% NaCl solution, shaking at 200 rpm for 30 min [32]. The resulting suspension constituted the 1/10 dilution, from which serial dilutions were performed up to $1 \times 10^{-6}$. Total fungal counts were carried out using PDA (potato dextrose agar) medium with streptomycin sulfate (100 $\mu g \cdot mL^{-1}$) and chloramphenicol (30 $\mu g \cdot mL^{-1}$). Total bacterial counts were performed using nutrient broth at 1/10 of the standard concentration [32]. For the evaluation of *Trichoderma*, a specific medium was prepared with the following composition: 40 g of PDA, 0.15 g of Bengal Rose, 0.05 g of Captan, and 0.001 g of Benomyl. After sterilization, 0.25 g of chloramphenicol, 0.10 g of streptomycin, and 1 mL of lactic acid per liter were added. In the case of Azotobacter, the culture medium included 20 g mannitol, 0.2 g $K_2HPO_4$, 0.2 g $MgSO_4 \cdot 7H_2O$, 0.2 g NaCl, 0.1 g $K_2SO_4$, 5 g $CaCO_3$, and 15 g agar. The pH was adjusted to 7.4 ± 0.2 [33]. The results aim to determine whether agro-industrial waste compost and/or coffee pulp vermicompost, with or without *Trichoderma*, can be used as substrates for producing basic seed potatoes of the Iniafrit variety under the studied conditions.

*2.3. Data Analysis*

The experimental design was factorial with three factors, two levels each, and three replicates: (1) Two levels of organic materials: Cp (agro-industrial compost) and Cv (vermicompost), both mixed with coconut fiber in a 50:50 *v/v* ratio; (2) Two levels of physical treatment: heat treated and non-heat treated; and (3) Two levels of biological treatment: with and without *T. asperellum*. The objective was to measure the effect of compost type in each mixing ratio and its interaction with the antagonist *T. asperellum*. Descriptive analyses were performed, followed by tests to verify the assumptions of normality, homogeneity of variances using the Kolmogorov–Smirnov and Levene tests, and error independence through graphical analysis. To identify the variables with the greatest influence on the measured biological response, principal component analysis (PCA) was applied, followed by analysis of variance, correlation analysis, and regression analysis using the statistical software InfoStat v. 2022 [34].

## 3. Results

*3.1. Evaluation of Biometric Variables Measured on Plants and Substrate Characteristics Affected by Treatments and Their Interactions*

All the results obtained were subjected to a principal component analysis (PCA) (Figure 1) with the intention of determining the best associations between the variables, as well as with the treatments that have shown the best performance, measured as the number of tubers per basket. First, it was verified how much variability was represented in the first components after the analysis. It was verified that the cumulative variability of the first two components, Cp1 (56.0%) and Cp2 (20.5%), exceeded 70%, reaching 76.5%. It was also confirmed that the correlation of the variables included in the analysis with the original variables was greater than 70%. According to the PCA biplot (Figure 1), treatments with vermicompost, both sterilized and non-sterilized, are concentrated in quadrants 1 and 3, close to the biometric variables and specifically to the number of mini tubers. This variable was selected for its importance in seed potato production, being close to the variables P and OM, revealing the importance of good P supply and organic matter content in the biometric variables measured. On the opposite side, in quadrant 4, the E4/E6 proportion is located, showing the importance of the evolution of the humification of organic matter in compost for its performance as a substrate and its benefit to yield.

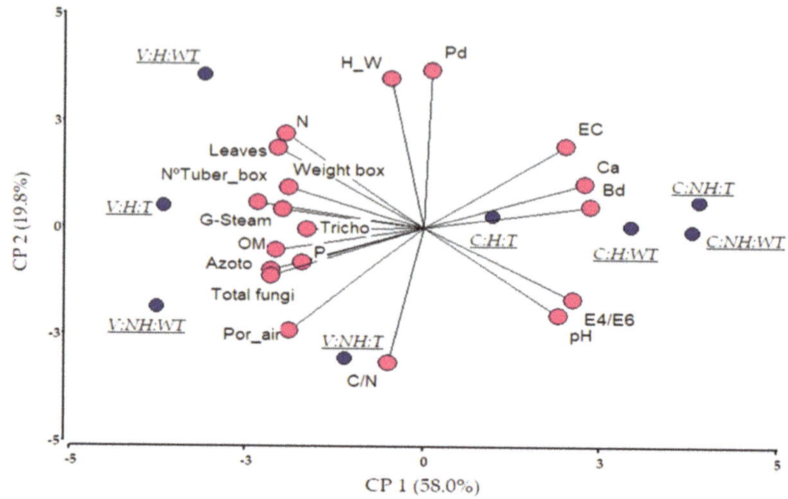


**Figure 1.** Biplot of the PCA applied to the set of variables measured on the plants, substrates, and mini tubers produced in the trial. Where C: compost; V: vermicompost; NH: non-heat; H: heat treatment; T: with *T. asperelllum*; WT: without *T. asperelllum*.

The best treatments were the substrates composed of sterilized vermicompost, both with and without *T. asperellum*, followed by treatments with vermicompost, but not sterilized. It was also observed that the other biometric variables measured were notably associated with the number of mini tubers produced per basket. Likewise, the variables N (%) and P (%) content, total fungal count, *Trichoderma*, and *Azotobacter* revealed a positive association with yield, but a negative one with E4/E6, pH, Ca, and EC. The latter was associated with the worst biological expression. Given the need to better identify the most favorable treatments, a Multivariate Analysis of Variance (MANOVA) (Table 1) was conducted for the variables number and weight of mini tubers, and the number of stems per basket, due to their great agronomic importance. The results show the significance of the sources of variation as well as the compost variable in combination with the antagonist.

**Table 1.** Result of the Multivariate Analysis of Variance (Wilks, df num = 2, df den = 15) for the variables number and weight of mini tubers per basket (with five plants and 40 L of substrate) in the trial for the evaluation of substrate type (compost and vermicompost), heat treatment, and inoculation with *T. asperellum* for the production of potato seed tubers of the INIAFrit variety.

| Variable | Statistic | F | *p*-Value |
|---|---|---|---|
| Substrate | 0.03 | 237.16 | <0.0001 |
| Heat treatment | 0.08 | 89.83 | <0.0001 |
| *T. asperellum* | 0.010 | 68.79 | <0.0001 |
| Substrate*Heat treatment | 0.65 | 4.04 | 0.0395 |
| Substrate**T. asperellum* | 0.37 | 12.52 | 0.0006 |
| Heat treatment**T. asperellum* | 0.06 | 117.42 | <0.0001 |
| Heat treatment*Substrate**T. asperellum* | 0.08 | 82.82 | <0.0001 |

In the global analysis, the effect of the type of compost used to prepare the substrate stands out, with respect to sterilization and inoculation with *T. asperellum*, measured as the number and weight of mini tubers per basket, all at $p < 0.001$. Similarly, the analysis detected an interaction between the type of compost and inoculation, as well as between sterilization and inoculation. Sterilization had a significant impact on the yield in both composts. In the expression of the antagonist, double significance was found with the sterilization of the mixture and triple significance among the evaluated variables. To determine which treatment configuration resulted in the best yield, a Bonferroni-type mean comparison test was applied, with the results summarized in the corresponding table (Table 2). The tests for the individual factors provided significant information according to the Ho telling test with Bonferroni correction ($\alpha$ = 0.05) regarding the effect of the type of compost on the expression of the variable number of mini tubers. A 151.70% increase ($p < 0.05$) in the number of mini tubers was produced in vermicompost compared to compost. As for the heat treatment factor (humid heat treatment), a 67% effect ($p < 0.05$) was quantified in the overall measurement of the number of mini tubers. Treatments such as VHT and VHWT show the highest values for the number and weight of mini tubers, positioning themselves as the best resulting options.

**Table 2.** Hotelling test with Bonferroni-corrected $\alpha$ = 0.05 and 16 df for the variables weight number of mini tubers and weight produced per basket with 5 plants from the treatments in the evaluation of compost and vermicompost as substrates for potato seed production in a protected environment.

| Treatment | N°Tuber Box | Weight Box (g) |
|---|---|---|
| VHWT | 51.67 ± 0.58 *a* | 345.2 ± 4.44 *c* |
| VHT | 45 ± 3 *ab* | 434.3 ± 33.2 *b* |
| CHT | 43.67 ± 4.51 *abc* | 176.89 ± 3.92 *d* |
| VNHT | 37 ± 4 *bcd* | 82.27 ± 24.84 *ef* |
| VNHWT | 34.33 ± 3.79 *cd* | 540.52 ± 36.89 *a* |
| CHWT | 26.67 ± 3.51 *de* | 193.6 ± 17.06 *d* |
| CNHWT | 23 ± 4 *e* | 159.1 ± 42.53 *de* |
| CNHT | 5.67 ± 2.52 *f* | 27.28 ± 8.78 *f* |

Data presented are means ± SE, n = 3. Means within the same column in each studied factor followed by the different letters indicate significant differences at $p \leq 0.05$ according to Bonferroni test. C: compost; V: vermicompost; NH: Non-heat; H: heat treatment; T: with *T. asperelllum;* WT: without *T. asperelllum.*

The variables associated with the best yields are displayed in the Spearman correlation analysis in Table 3. There, the positive relationship of the measured biometric variables can be seen, including the number of leaves, number of stems, number of tubers per basket, and total tuber weight (per basket with five plants).

**Table 3.** Spearman correlations for the variables measured in the trial.

| | Leaves | Steams | N°-Tuber | Weight Box | Total Fungi | Azoto-Bacter | Bd | pH | EC | N | P | E4/E6 | OM | $Ca^{++}$ |
|---|---|---|---|---|---|---|---|---|---|---|---|---|---|---|
| Leaves | 1 | | | | | | | | | | | | | |
| Steams | 0.61 | 1 | | | | | | | | | | | | |
| No tuber | 0.7 | 0.64 | 1 | | | | | | | | | | | |
| Weight box | 0.64 | 0.63 | 0.55 | 1 | | | | | | | | | | |
| Total fungi | 0.59 | 0.72 | 0.61 | 0.71 | 1 | | | | | | | | | |
| Azotobac | 0.53 | 0.35 | 0.54 | 0.7 | 0.69 | 1 | | | | | | | | |
| Bd | −0.51 | −0.74 | −0.52 | −0.66 | −0.71 | −0.51 | 1 | | | | | | | |
| pH | −0.62 | −0.66 | −0.66 | −0.59 | −0.35 | −0.21 | 0.69 | 1 | | | | | | |
| EC | −0.33 | −0.52 | −0.42 | −0.5 | −0.66 | −0.64 | 0.84 | 0.39 | 1 | | | | | |
| N | 0.63 | 0.55 | 0.4 | 0.56 | 0.34 | 0.16 | −0.59 | −0.72 | −0.24 | 1 | | | | |
| P | 0.39 | 0.39 | 0.57 | 0.47 | 0.49 | 0.63 | −0.63 | −0.42 | −0.68 | 0.21 | 1 | | | |
| E4/E6 | −0.62 | −0.7 | −0.62 | −0.68 | −0.45 | −0.32 | 0.76 | 0.9 | 0.46 | −0.85 | −0.47 | 1 | | |
| OM | 0.36 | 0.41 | 0.41 | 0.56 | 0.47 | 0.39 | −0.8 | −0.68 | −0.66 | 0.43 | 0.67 | −0.66 | 1 | |
| $Ca^{++}$ | −0.46 | −0.77 | −0.54 | −0.54 | −0.64 | −0.47 | 0.78 | 0.51 | 0.68 | −0.34 | −0.53 | 0.5 | −0.4 | 1 |

The results showed a positive correlation of the four measured biometric variables with the total fungal count, *Azotobacter*, the nitrogen content of the mixture, phosphorus, and organic matter. In contrast, yield was negatively affected, according to the analyzed correlations, by the bulk density of the mixture, electrical conductivity, E4/E6 ratio, pH, and Ca content of the substrate. Some of these relationships were assessed through regression analysis (Table 3) to better understand the effect of the variables on yield. Likewise, it is observed that as the pH value increases, the yield of mini tubers decreases, just as it decreases with the increase in electrical conductivity, the E4/E6 proportion, and the concentration of Ca in the mixture used in the treatments. On the other hand, the higher the P and N content of the mixtures, the greater the number of mini tubers. The correlations with the highest values, both positive and negative, were selected for the application of linear regression analysis: pH, EC, E4/E6, total fungi, *Azotobacter*, N, and P.

### 3.2. Evaluation of Growth Indices in Potato Plants

The information derived from the calculation of physiological growth indices helps to more clearly explain the mechanisms activated in the plant through the mediation of the physics, chemistry, and biology of the substrates. Figure 2A shows an expression of the efficiency of plants that grew in vermicompost-based substrates, treated or not with heat, and in the absence of *T. asperellum*.

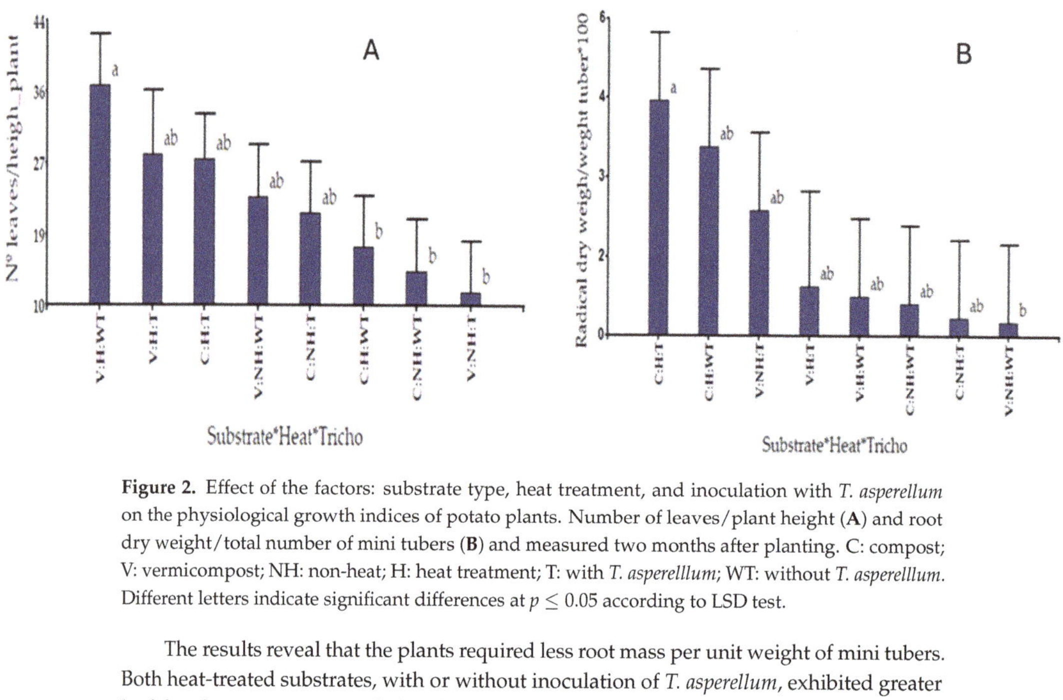


**Figure 2.** Effect of the factors: substrate type, heat treatment, and inoculation with *T. asperellum* on the physiological growth indices of potato plants. Number of leaves/plant height (**A**) and root dry weight/total number of mini tubers (**B**) and measured two months after planting. C: compost; V: vermicompost; NH: non-heat; H: heat treatment; T: with *T. asperelllum*; WT: without *T. asperelllum*. Different letters indicate significant differences at $p \leq 0.05$ according to LSD test.

The results reveal that the plants required less root mass per unit weight of mini tubers. Both heat-treated substrates, with or without inoculation of *T. asperellum*, exhibited greater leaf development per unit of plant height, indicating increased foliage (Figure 2B and 3).

### 3.3. Effect of Substrate Characteristics on the Yield of Seed Potatoes of the INIAfrit Variety

Two important variables that have been shown to influence mini tuber yield are the ratio of E4/E6 (Figure 4A), which is an indicator of substrate humification, and the phosphorus (P) (Figure 4B) content of the growing medium.

**Figure 3.** Differential aspects of potato plants developed under the treatments applied in the trial. C: compost; V: vermicompost; NH: non-heat; H: heat treatment; T: with *T. asperelllum*; WT: without *T. asperelllum*.

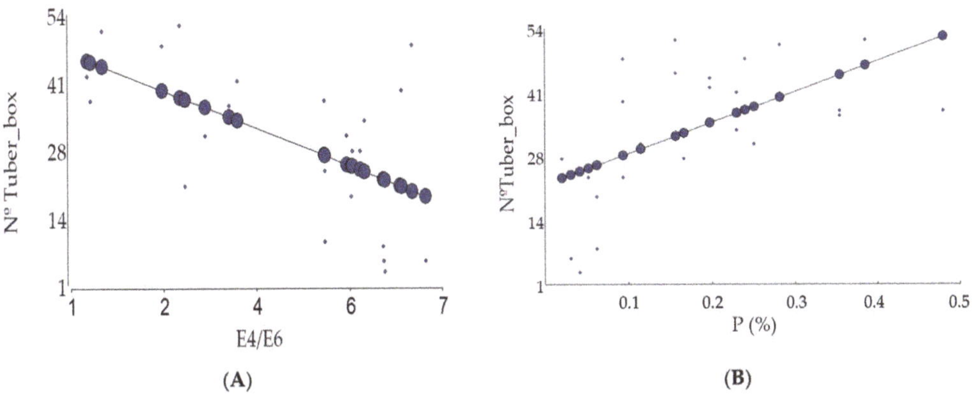


**Figure 4.** Adjustment of the linear regression analysis for the variable number of mini tubers per box against the variables (**A**) E4/E6 proportion and (**B**) P concentration in the substrate as regressor variables.

## 4. Discussion

The study of plant biometric variables, substrate characteristics, and their interactions is crucial for understanding the complex relationships between plants and their growing environment. This research paper aims to investigate the effects of various treatments on plant biometric variables and substrate characteristics, as well as the interactions between these factors. The composition of compost can vary significantly depending on the materials from which it is derived, leading to diverse effects on plant growth and disease suppression [35,36]. Compost can directly provide plants with essential nutrients, while also indirectly influencing the soil microbiome in ways that promote plant health and protect against pathogens [37–39]. Several studies have highlighted the benefits of compost application to soil. Compost can help reduce the financial costs associated with chemical fertilizers and mitigate the environmental impacts of their production and use [38]. Furthermore, compost-amended soils have shown significant improvements in overall soil health compared to untreated soils [38]. The physical, chemical, and biological properties of the growth medium are crucial factors that shape the complex web of interactions decisive for the development and agronomic performance of crops [40–43]. Linear Discriminant Analysis (LDA) was employed to identify the minimal set of variables capable of effectively discriminating between groups and establishing quality ranges, as illustrated in Figure 5. The analysis suggests that biological and chemical variables were the most influential in the observed response, as they could not be disregarded, unlike other factors such as apparent density or real density.

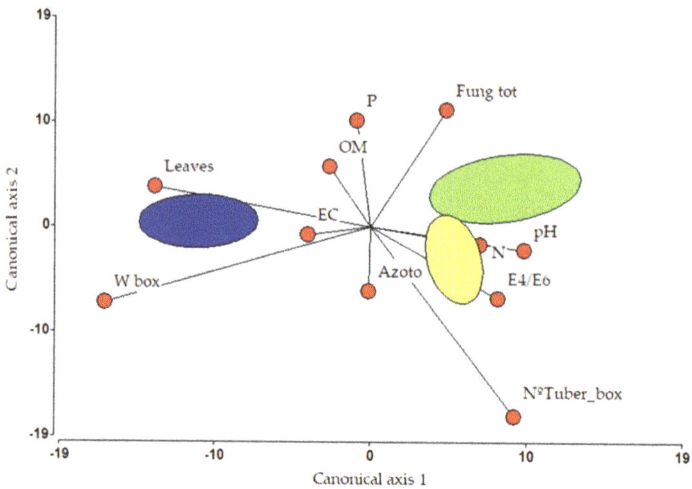


**Figure 5.** Groups for three qualities of substrates obtained according to an LDA applied to the variables measured on Iniafrit potato plants. Blue = a; green = b; yellow = c.

The use of bioorganic fertilizers has emerged as a promising approach to promote sustainable agriculture and address the growing concerns associated with the excessive use of chemical fertilizers [44]. These bioorganic fertilizers, often combined with compost, have been found to enhance plant growth and yield by improving the composition and functionality of the rhizosphere microbiome. The rhizosphere, the zone immediately surrounding plant roots, harbors a diverse community of microorganisms that play a crucial role in nutrient cycling, plant development, and disease resistance [45]. By modulating the rhizosphere microbiome, bioorganic fertilizers can improve the availability and uptake of essential nutrients, leading to increased plant growth and productivity [46].

Several studies have demonstrated the benefits of bioorganic fertilizers in enhancing plant performance. It has been observed that the application of bioorganic fertilizers alters the soil's chemical properties, thereby selectively favoring specific microbial taxa and functions within the rhizosphere microbiome. This targeted manipulation of the rhizosphere microbiome composition and activity can lead to improved nutrient availability, increased biological nitrogen fixation, and enhanced phosphorus solubilization, all of which contribute to the stimulation of plant growth and development [45]. Furthermore, the presence of plant growth-promoting microorganisms, such as *Trichoderma*, in combination with compost, has been found to be particularly promising in reducing the reliance on chemical fertilizers while maintaining high yields and protecting soil environments from agricultural contamination [45,46].

As suggested by various authors, *Trichoderma* strains can be defined as symbiotic opportunistic organisms of plants, capable of colonizing plant roots through mechanisms similar to those of mycorrhizal fungi to stimulate plant growth [47,48]. Previous research has demonstrated that *Trichoderma* species can rapidly colonize niches left vacant by other organisms, and their TasHyd1 gene plays a crucial role in the colonization of plant roots [49,50]. Additionally, *T. harzianum* can efficiently colonize tomato roots and soil under field conditions, independent of the inoculation formulation. The presented experiment aimed to assess the impact of applying a spore suspension to an organic substrate on the growth-promoting potential of the microorganism within the root system of potato plants. However, the observed response did not align with the expected outcomes in terms of agronomic performance under the prevailing conditions of the trial.

*Trichoderma* spp. can also act as biocontrol agents and induce disease resistance in symbiotic plants. The introduction of *Trichoderma* has been found to correlate positively with the abundance of microorganisms and the level of available nutrients, similar to the findings in this study where an organic substrate and potato were used instead of tomato [50]. The results of our study suggest that *Trichoderma* can stimulate plant growth primarily through the production of hormone-like substances and/or the supply of phosphorus, which promotes root system development and overall plant growth.

The characteristics of compost and/or substrate play a crucial role in mediating the development and performance of the plant grown in it [51]. For instance, in the case of our study, the vermicomposted coffee pulp residue was a better substrate than the composted wastewater sludge from the food processing industry. Similarly, the treatment applied to control or limit pathogens in the mixtures can either promote or hinder the desired growth-promoting effect, as observed in the ranges presented in Table 4 [52]. The characteristics of compost and substrate have a significant influence on the development and yield of plants [53]. The composition and quality of compost can vary depending on the raw materials used and the composting process, which can impact its effects on plant growth [54]. For instance, in the case of our study, the vermicomposted coffee pulp residue was a better substrate than the composted wastewater sludge from the food processing industry. Similarly, the treatment applied to control or limit pathogens in the mixtures can either promote or hinder the desired growth-promoting effect, as observed in the ranges presented in Table 4.

**Table 4.** Ranges generated in Linear Discriminant Analysis (LDA) classifications (Group a, b, and c) applied to biometric variables.

| Variable | Group a | | Group b | | Group c | |
|---|---|---|---|---|---|---|
| | V:H:T *<br>V:H:WT<br>V:NH:WT | | C:H:T<br>C:H:WT<br>C:NH:WT | | C:NH:T<br>V:NH:T | |
| | Min | Max | Min | Max | Min | Max |
| N°Tuber_box | 34.33 | 51.67 | 29.67 | 33.67 | 5.67 | 37 |
| N° Leaves/plant | 37.33 | 47.33 | 17 | 28.33 | 14.67 | 20 |
| Weight mini tubers box | 345.2 | 540.52 | 149.55 | 196.39 | 27.28 | 82.27 |
| E4/E6 | 1.79 | 4.07 | 5.78 | 6.89 | 5.06 | 6.25 |
| EC | 1.41 | 2.59 | 3.32 | 3.72 | 1.62 | 3.9 |
| OM | 74.14 | 78.6 | 28.46 | 39.35 | 43.07 | 74.55 |
| N | 2.53 | 3.76 | 1.16 | 2.36 | 1.79 | 2.07 |
| P | 0.17 | 0.29 | 0.05 | 0.09 | 0.02 | 0.28 |
| pH | 3.90 | 6.70 | 6.70 | 6.90 | 5.80 | 6.90 |

* C: compost; V: vermicompost; NH: Non-heat; H: heat treatment; T: with *T. asperelllum*; WT: without *T. asperelllum*.

The positive effect of sterilizing composted and vermicomposted materials is explained by the elimination of nutrient competition, including pathogenic competition, by the native microbiota [55]. Other authors have reported that heat treatment, even microfiltration, impairs the biological effect of compost [53]. In our case, we evaluated the growth-promoting effect, which persisted in sterilized treatments and even surpassed that of non-sterilized ones. The E4/E6 proportion, which indicates the degree of humification of the composted material, showed an inverse linear relationship with yield. This is understood as the ratio between lower molecular weight chain compounds detected at 465 nm absorbance and more humified, condensed compounds with higher molecular weight measured at 665 nm [56]. An E4/E6 proportion of less than 5 is typically associated with humic acids, while a ratio between 6.6 and 8.5 indicates the presence of fulvic acids [57–59]. The study found that the lower values of the E4/E6 proportion were obtained in the treatments classified as quality "a", while the higher values were observed in the quality "c" group, which had the lowest agronomic yield. How does this affect growth? The relationship between the E4/E6 index and plant growth can be interpreted as follows: Low E4/E6: substrate rich in humified organic matter, providing better growth conditions due to higher nutrient availability and improved soil structure. High E4/E6: soils with organic matter in a less advanced stage of humification. A low E4/E6 value indicates that compost is highly mature and stabilized. In this study, we observed that the promotion of plant growth was attributed to the use of compost, whether treated with *T. asperellum* or not, and always with heat treatment. The growth-promoting effect can be attributed to the combined benefits of the addition of nitrogen (N) and phosphorus (P), primarily provided in a stable matrix with a higher content of humic acids. The data presented in the analysis indicates that the total nitrogen levels ranged from 1.16 to 3.76, which corresponds to a ratio of 10:6:8 (C:H:N) to 37.6 kg per ton (V:H:N). Similarly, the total phosphorus levels were observed to be in the range of 0.46 to 6.64 kg/ton $P_2O_5$. These findings suggest that the treatments with the highest levels of these essential nutrients (a and b) exhibited the best response in terms of crop quality and yield. The importance of phosphorus in crop production is further supported by the literature. Phosphorus is the second most critical plant nutrient after nitrogen, as it directly affects root proliferation, nitrogen fixation, and overall plant growth and development [59]. Moreover, the efficient utilization of atmospheric nitrogen and the symbiotic process are enhanced by adequate phosphorus levels, reducing the dependence on nitrogen-based fertilizers and promoting sustainable farming practices [59]. The present study investigated the nutrient content of substrates prepared with vermicompost as a soil amendment. Vermicompost has been shown to increase the uptake of essential nutrients in greenhouse crops [60] and improve soil fertility through its physical, chemical, and biological properties [6]. In particular, the incorporation of vermicompost has been reported to increase the availability of nitrogen, phosphorus, and other micronutrients in the growing medium [61,62]. The results of this study indicated that both nitrogen and phosphorus levels in substrates amended with vermicompost are above the regulatory requirements established in Spain and Germany [6,20,63–65]. The nutrient concentrations were not only sufficient to meet the needs of the target crops but were present in markedly higher quantities in the vermicompost treatments. The nutrient concentrations were not only sufficient to meet the needs of the target crops but were present in markedly higher quantities in the vermicompost treatments.

## 5. Conclusions

Ultimately, compost and vermicompost can be used as substrates for potato seed production, with vermicompost being the most effective. Vermicompost, especially when heat-treated and enriched with beneficial microorganisms such as *T. asperellum*, was found to be the most effective in increasing both the quantity and weight of mini tubers per plant. This approach ensures higher seed yield and quality, thereby improving crop productivity and profitability. The nutrient content of the compost played a key role, particularly adequate levels of nitrogen (N) and phosphorus (P), along with good organic matter content. An important advantage is the application of heat treatment to the substrate: heat treatment demonstrated a significant benefit to productivity as optimal humification level (E4/E6 ratio) improves the plant's efficiency in tuber production. Based on these findings, it is recommended that producers choose high-quality substrates, preferably with treated vermicompost and adequate levels of key nutrients. Additionally, monitoring factors such as pH, density, humification, and nutrient content—particularly N and P—of the substrate can significantly improve harvest outcomes. Implementing these practices can help increase both the quantity and quality of potato seeds, making the crop more competitive in terms of yield and sustainability.

**Author Contributions:** Conceptualization, H.P.; methodology, H.P., B.R. and J.S.; software, H.P.; validation, M.S., H.P. and F.D.; formal analysis, H.P. and B.R.; investigation, H.P. and B.R.; resources, H.P. and K.A.; data curation, H.P. and M.S.; writing—original draft preparation, H.P.; writing—review and editing, H.P., M.S. and V.H.; visualization, H.P., F.D. and M.S.; supervision and administration, H.P., F.D. and M.S. All authors have read and agreed to the published version of the manuscript.

**Funding:** This research received no external funding.

**Data Availability Statement:** No new data were created or analyzed in this study.

**Conflicts of Interest:** The authors declare no conflicts of interest.

## References

1. Amanullah; Khalid, S. Agronomy—Climate Change & Food Security. In *Agronomy-Food Security-Climate Change and the Sustainable Development Goals*; IntechOpen: London, UK, 2020.
2. Ghasemzadeh, A. Global Issues of Food Production. *Agrotechnol* **2012**, *1*, 1–2. [CrossRef]
3. Escalamiento Industrial, Red Socialista de Innovacion Productiva de Papa del Municipio Rangel, Estado Merida. Available online: https://proinpameridavenezuela.blogspot.com/2014/02/escalamiento-industrial-red-socialista.html (accessed on 23 August 2024).
4. Barrera Mosquera, V.H.; Delgado, J.A.; Alwang, J.R.; Escudero López, L.O.; Cartagena Ayala, Y.E.; Domínguez Andrade, J.M.; D'Adamo, R. Conservation Agriculture Increases Yields and Economic Returns of Potato, Forage, and Grain Systems of the Andes. *Agron. J.* **2019**, *111*, 2747–2753. [CrossRef]
5. Hatfield, J.L.; Walthall, C.L. Soil Biological Fertility: Foundation for the Next Revolution in Agriculture? *Commun. Soil Sci. Plant Anal.* **2015**, *46*, 753–762. [CrossRef]
6. Lim, S.L.; Wu, T.Y.; Lim, P.N.; Shak, K.P.Y. The Use of Vermicompost in Organic Farming: Overview, Effects on Soil and Economics. *J. Sci. Food Agric.* **2015**, *95*, 1143–1156. [CrossRef] [PubMed]
7. Neher, D.A.; Harris, J.M.; Horner, C.E.; Scarborough, M.J.; Badireddy, A.R.; Faulkner, J.W.; White, A.C.; Darby, H.M.; Farley, J.C.; Bishop-von Wettberg, E.J. Resilient Soils for Resilient Farms: An Integrative Approach to Assess, Promote, and Value Soil Health for Small- and Medium-Size Farms. *Phytobiomes J.* **2022**, *6*, 201–206. [CrossRef]
8. Zayed, G.; Abdel-Motaal, H. Bio-Active Composts from Rice Straw Enriched with Rock Phosphate and Their Effect on the Phosphorous Nutrition and Microbial Community in Rhizosphere of Cowpea. *Bioresour. Technol.* **2005**, *96*, 929–935. [CrossRef] [PubMed]
9. Roy, R.; Singh, K. Effect of Integrated Nutrient Management on Sequential Productivity, Economics and Nutrient Uptake of Rice (Oryza Sativa)—Potato (Solanum Tuberosum) Cropping Sequence. *Indian J. Agric. Sci.* **2014**, *84*, 1096–1101. [CrossRef]
10. Ala, A.; Musa, Y.; Kadola, L. Effect of Compost Enrichment with Trichoderma Sp on the Growth of Arabica Coffee Seedlings. *IOP Conf. Ser. Earth Environ. Sci.* **2020**, *575*, 12145. [CrossRef]
11. Sudheer, S.; Bai, R.G.; Usmani, Z.; Sharma, M. Insights on Engineered Microbes in Sustainable Agriculture: Biotechnological Developments and Future Prospects. *Curr. Genom.* **2020**, *21*, 321–333. [CrossRef]
12. Singh, Y.P.; Arora, S.; Mishra, V.K.; Singh, A.K. Synergizing Microbial Enriched Municipal Solid Waste Compost and Mineral Gypsum for Optimizing Rice-Wheat Productivity in Sodic Soils. *Sustainability* **2022**, *14*, 7809. [CrossRef]
13. Arthur, G.D.; Aremu, A.O.; Kulkarni, M.G.; Staden, J.V. Vermicompost Leachate Alleviates Deficiency of Phosphorus and Potassium in Tomato Seedlings. *HortScience* **2012**, *47*, 1304–1307. [CrossRef]
14. Qasim, M.; Ju, J.; Zhao, H.; Bhatti, S.M.; Saleem, G.; Memon, S.P.; Ali, S.; Younas, M.U.; Rajput, N.; Jamali, Z.H. Morphological and Physiological Response of Tomato to Sole and Combined Application of Vermicompost and Chemical Fertilizers. *Agronomy* **2023**, *13*, 1508. [CrossRef]

15. Singh, D.P.; Prabha, R.; Renu, S.; Sahu, P.K.; Singh, V. Agrowaste Bioconversion and Microbial Fortification Have Prospects for Soil Health, Crop Productivity, and Eco-Enterprising. *Int. J. Recycl. Org. Waste Agric.* **2019**, *8*, 457–472. [CrossRef]
16. Holdridge, L. *Ecología Basada En Zonas de Vida*; Publicaciones y documentos; IICA: San José, Costa Rica, 1978; ISBN 92-9039-131-6.
17. Devi, J.; Deb, U.; Barman, S.; Das, S.; Sundar Bhattacharya, S.; Fai Tsang, Y.; Lee, J.-H.; Kim, K.-H. Appraisal of Lignocellusoic Biomass Degrading Potential of Three Earthworm Species Using Vermireactor Mediated with Spent Mushroom Substrate: Compost Quality, Crystallinity, and Microbial Community Structural Analysis. *Sci. Total Env.* **2020**, *716*, 135215. [CrossRef] [PubMed]
18. Zhou, X.; Yang, J.; Xu, S.; Wang, J.; Zhou, Q.; Li, Y.; Tong, X. Rapid In-Situ Composting of Household Food Waste. *Process Saf. Environ. Prot.* **2020**, *141*, 259–266. [CrossRef]
19. Pane, C.; Spaccini, R.; Piccolo, A.; Celano, G.; Zaccardelli, M. Disease Suppressiveness of Agricultural Greenwaste Composts as Related to Chemical and Bio-Based Properties Shaped by Different on-Farm Composting Methods. *Biol. Control* **2019**, *137*, 104026. [CrossRef]
20. Arancon, N.; Edwards, C.; Atiyeh, R.; Metzger, J. Effects of Vermicomposts Produced from Food Waste on the Growth and Yields of Greenhouse Peppers. *Bioresour. Technol.* **2004**, *93*, 139–144. [CrossRef]
21. Diánez, F.; Santos, M.; Carretero, F.; Marín, F. Trichoderma Saturnisporum, a New Biological Control Agent. *J. Sci. Food Agric.* **2016**, *96*, 1934–1944. [CrossRef]
22. Peña, H.; Mendoza, H.; Dianez, F.; Santos, M. Parameter Selection for the Evaluation of Compost Quality. *Agronomy* **2020**, *10*, 1567. [CrossRef]
23. Peña, H.; Santos, M.; Ramírez, B.; Sulbarán, J.; Arias, K.; Huertas, V.; Diánez, F. Essential Quality Attributes of Culture Media Used as Substrates in the Sustainable Production of Pre-Basic Potato Seeds. *Sustainability* **2024**, *16*, 8552. [CrossRef]
24. Richards, T. *Compost Mixture Calculation Spreadsheet*; Cornell Waste Management Institute: Ithaca, NY, USA, 2014; Volume 18, p. 2014.
25. *2014/955/UE*; Decisión de La Comisión, de 18 de Diciembre de 2014, Por La Que Se Modifica La Decisión 2000/532/CE, Sobre La Lista de Residuos, de Conformidad Con La Directiva 2008/98/CE Del Parlamento Europeo y Del Consejo Texto Pertinente a Efectos Del EEE. Publications Office of the EU: Luxembourg, 2014; p. Diario Oficial, L 370. pp. 44–86.
26. Martínez-Alemán, S.R.; Hernández-Castillo, F.D.; Aguilar-González, C.N.; Rodríguez-Herrera, R. Extractos de Pulpa de Café: Una Revisión Sobre Antioxidantes Polifenólicos y Su Actividad Antimicrobiana. *Investig. Y Cienc. De La Univ. Autónoma De Aguascalientes* **2019**, 73–79. [CrossRef]
27. Sadzawka, A.; Carrasco, M.; Grez, R.; Mora, M. *Metodos de Analisis de Compost*; INIA, Ed.; Serie Actas INIA; Instituto de Investigaciones Agropecuarias: Santiago de Chile, Chile, 2005.
28. De García, M. *Guía Para El Análisis Bromatológico De Muestras De Forrajes*; Universidad de Panamà: Panama City, Panama, 2015; p. 55.
29. Murphy, J.; Riley, J. A Modified Single Solution Method for the Determination of Phosphate in Natural Waters. *Anal. Chim. Acta* **1962**, *27*, 31–36. [CrossRef]
30. Zbytniewski, R.; Buszewski, B. Characterization of Natural Organic Matter (NOM) Derived from Sewage Sludge Compost. Part 2: Multivariate Techniques in the Study of Compost Maturation. *Bioresour. Technol.* **2005**, *96*, 479–484. [CrossRef] [PubMed]
31. Pire, R.; Pereira, A. Propiedades Físicas de Composnentes de Sustratos de Uso Comun En al Horticultura Del Estado Lara, Venezuela. Propuesta Metodológica. *Bioagro* **2003**, *15*, 55–64.
32. Reyes, I.; Valery, A.; Valduz, Z. Phosphate-Solubilizing Microorganisms Isolated from Rhizospheric and Bulk Soils of Colonizer Plants at an Abandoned Rock Phosphate Mine Phosphate-Solubilizing Microorganisms Isolated from Rhizospheric and Bulk Soils of Colonizer Plants at an Abandoned Roc. *Plant Soil* **2006**, *287*, 69–75. [CrossRef]
33. Gerhardt, P.; Murray, R.; Wood, W.; Krieg, N. *Methods for General and Molecular Bacteriology | Wageningen University and Research Library Catalog*; American Society for Microbiology: Washington, DC, USA, 1994; ISBN 1-55581-048-9.
34. Di Rienzo, J.; Balzarini, M.; Casanoves, F.; Gonzalez, L.; Tablada, E. *InfoStat*; Universidad Nacional de Córdoba: Córdoba, Argentina, 2021; p. 336.
35. D'Hose, T.; Cougnon, M.; Vliegher, A.; Willekens, K.; Bockstaele, E.; Reheul, D. Farm Compost Application: Effects on Crop Performance. *Compos. Sci. Util* **2013**, *20*, 49–56. [CrossRef]
36. van der Gaag, D.J.; van Noort, F.R.; Stapel-Cuijpers, L.H.M.; de Kreij, C.; Termorshuizen, A.J.; van Rijn, E.; Zmora-Nahum, S.; Chen, Y. The Use of Green Waste Compost in Peat-Based Potting Mixtures: Fertilization and Suppressiveness against Soilborne Diseases. *Sci. Hortic.* **2007**, *114*, 289–297. [CrossRef]
37. Mehta, C.M.; Palni, U.; Franke-Whittle, I.H.; Sharma, A.K. Compost: Its Role, Mechanism and Impact on Reducing Soil-Borne Plant Diseases. *Waste Manag.* **2014**, *34*, 607–622. [CrossRef]
38. Jain, M.S.; Paul, S.; Kalamdhad, A.S. Kinetics and Physics during Composting of Various Organic Wastes: Statistical Approach to Interpret Compost Application Feasibility. *J. Clean. Prod.* **2020**, *255*, 120324. [CrossRef]

39. Bahramisharif, A.; Rose, L.E. Efficacy of Biological Agents and Compost on Growth and Resistance of Tomatoes to Late Blight. *Planta* **2019**, *249*, 799–813. [CrossRef]
40. Leifert, C.; Pryce, S.; Lumsden, P.J.; Waites, W.M. Effect of Medium Acidity on Growth and Rooting of Different Plant Species Growing in Vitro. *Plant Cell Tiss. Organ. Cult* **1992**, *30*, 171–179. [CrossRef]
41. Haney, R.L.; Haney, E.B.; Smith, D.R.; Harmel, R.D.; White, M.J. The Soil Health Tool—Theory and Initial Broad-Scale Application. *Appl. Soil Ecol.* **2018**, *125*, 162–168. [CrossRef]
42. de Moraes, M.T.; Debiasi, H.; Franchini, J.C.; Bonetti, J.d.A.; Levien, R.; Schnepf, A.; Leitner, D. Mechanical and Hydric Stress Effects on Maize Root System Development at Different Soil Compaction Levels. *Front. Plant Sci.* **2019**, *10*, 1358. [CrossRef] [PubMed]
43. Kaur, R.; Goyal, D. Soil Application of Fly Ash Based Biofertilizers for Increased Crop Production. *Vegetos- Int. J. Plant. Res.* **2014**, *27*, 291. [CrossRef]
44. Kaza, S.; Yao, L.; Bhada-Tata, P.; Woerden Van, F. *What a Waste 2.0. A Global Snapshot of Solid Waste Management to 2050*; World Bank Publications: Herndon, VA, USA, 2018; ISBN 978-1-4648-1329-0.
45. Wang, Z.; Yang, T.; Mei, X.; Wang, N.; Li, X.; Yang, Q.; Dong, C.; Jiang, G.; Lin, J.; Xu, Y.; et al. Bio-Organic Fertilizer Promotes Pear Yield by Shaping the Rhizosphere Microbiome Composition and Functions. *Microbiol. Spectr.* **2022**, *10*, e0357222. [CrossRef]
46. Bhardwaj, D.; Ansari, M.W.; Sahoo, R.K.; Tuteja, N. Biofertilizers Function as Key Player in Sustainable Agriculture by Improving Soil Fertility, Plant Tolerance and Crop Productivity. *Microb. Cell. Fact* **2014**, *13*, 66. [CrossRef]
47. Liu, B.; Glenn, D.; Buckley, K. *Trichoderma* communities in soils from organic, sustainable, and conventional farms, and their relation with Southern blight of tomato. *Soil Biol. Biochem.* **2008**, *40*, 1124–1136. [CrossRef]
48. Harman, G.E.; Howell, C.R.; Viterbo, A.; Chet, I.; Lorito, M.; Pathology, P. *Trichoderma* Species—Oportunistic, Avirulent. *Plant Symbionts.* **2004**, *2*, 43–56. [CrossRef]
49. Schuster, A.; Schmoll, M. Biology and Biotechnology of Trichoderma. *Appl. Microbiol. Biotechnol.* **2010**, *87*, 787–799. [CrossRef]
50. Vinci, G.; Cozzolino, V.; Mazzei, P.; Monda, H.; Spaccini, R.; Piccolo, A. An Alternative to Mineral Phosphorus Fertilizers: The Combined Effects of Trichoderma Harzianum and Compost on Zea Mays, as Revealed by 1H NMR and GC-MS Metabolomics. *PLoS ONE* **2018**, *13*, e0209664. [CrossRef]
51. Priadi, D.; Mulyaningsih, E.S. Effects of Compost Type and Rootstock Length on Fruit and Vegetable Seedlings Growth in the Nursery. *Biosaintifika J. Biol. Biol. Educ.* **2016**, *8*, 301–307. [CrossRef]
52. Lenhart, J.H. Compost as a Soil Amendment for Water Quality Treatment Facilities. In *Low Impact Development*; American Society of Civil Engineers: Wilmington, NC, USA, 2008; pp. 212–219.
53. Chaudhuri, P.S.; Paul, T.K.; Dey, A.; Datta, M.; Dey, S.K. Effects of Rubber Leaf Litter Vermicompost on Earthworm Population and Yield of Pineapple (Ananas Comosus) in West Tripura, India. *Int. J. Recycl. Org. Waste Agric* **2016**, *5*, 93–103. [CrossRef]
54. Miyamoto, H.; Shigeta, K.; Suda, W.; Ichihashi, Y.; Nihei, N.; Matsuura, M.; Tsuboi, A.; Tominaga, N.; Aono, M.; Sato, M.; et al. An Agroecological Structure Model of Compost—Soil—Plant Interactions for Sustainable Organic Farming. *ISME Commun.* **2023**, *3*, 1–12. [CrossRef] [PubMed]
55. Pugliese, M.; Liu, B.P.; Gullino, M.L.; Garibaldi, A. Microbial Enrichment of Compost with Biological Control Agents to Enhance Suppressiveness to Four Soil-Borne Diseases in Greenhouse. *J. Plant Dis. Prot.* **2011**, *118*, 45–50. [CrossRef]
56. Campitelli, P.; Ceppi, S. Effects of Composting Technologies on the Chemical and Physicochemical Properties of Humic Acids. *Geoderma* **2008**, *144*, 325–333. [CrossRef]
57. Bedolla-Rivera, H.I.; Conde-Barajas, E.; Galván-Díaz, S.L.; Gámez-Vázquez, F.P.; Álvarez-Bernal, D.; Xochilt Negrete-Rodríguez, M. de la L. Compost Quality Indexes (CQIs) of Biosolids Using Physicochemical, Biological and Ecophysiological Indicators: C and N Mineralization Dynamics. *Agronomy* **2022**, *12*, 2290. [CrossRef]
58. Bernal, M.P.; Paredes, C.; Sánchez-Monedero, M.A.; Cegarra, J. Maturity and Stability Parameters of Composts Prepared with a Wide Range of Organic Wastes. *Bioresour. Technol.* **1998**, *63*, 91–99. [CrossRef]
59. Watham, L.; Athokpam, H.S.; Chongtham, N.; Devi, K.N.; Singh, N.B.; Singh, N.G.; Sharma, P.T.; Heisnam, P. Phosphorus Status in the Soils of Imphal West District, Manipur (India). *Int. J. Curr. Microbiol. App. Sci.* **2018**, *7*, 3871–3877. [CrossRef]
60. Yang, F.; Li, G.X.; Yang, Q.Y.; Luo, W.H. Chemosphere Effect of Bulking Agents on Maturity and Gaseous Emissions during Kitchen Waste Composting. *Chemosphere* **2013**, *93*, 1393–1399. [CrossRef]
61. Liu, X.; Geng, B.; Zhu, C.; Li, L.; Francis, F. An Improved Vermicomposting System Provides More Efficient Wastewater Use of Dairy Farms Using Eisenia Fetida. *Agronomy* **2021**, *11*, 833. [CrossRef]
62. Wang, X.-X.; Zhao, F.; Zhang, G.; Zhang, Y.; Yang, L. Vermicompost Improves Tomato Yield and Quality and the Biochemical Properties of Soils with Different Tomato Planting History in a Greenhouse Study. *Front. Plant Sci.* **2017**, *8*, 1978. [CrossRef] [PubMed]
63. Bandera, A. Inorganic Fertilizers (Ground and Foliar Application) Andorganic Fertilizer: Their Effects on the Growthand Yield of Pechay (Brassica Napus L. Subsp. Chinensis Var. Black Behi). *Int. J. Agric. Sci.* **2020**, *6*, 38–55. [CrossRef]

64. Arancon, N.Q.; Edwards, C.A.; Babenko, A.; Cannon, J.; Galvis, P.; Metzger, J.D. Influences of Vermicomposts, Produced by Earthworms and Microorganisms from Cattle Manure, Food Waste and Paper Waste, on the Germination, Growth and Flowering of Petunias in the Greenhouse. *Appl. Soil. Ecol.* **2008**, *39*, 91–99. [CrossRef]
65. He, Z.; Yang, X.; Kahn, B.; Stoffella, P.; Calvert, D. Plant Nutrition Benefits of Phosphorus, Potassium, Calcium, Magnesium, and Micronutrients from Compost Utilization. In *Compost Utilization in Horticultural Cropping Systems*; CRC Press: Boca Raton, FL, USA, 2001; pp. 307–317, ISBN 9780367397593.

horticulturae

Article

# Evaluation of Growth and Production Parameters of Raspberries and Blackberries Cultivated in Romania

Augustina Pruteanu [1], Gabriel Alexandru Constantin [2,*], Nicoleta Alexandra Vanghele [1,*] and Valentin Vlăduț [1]

[1] Testing Department, National Institute of Research—Development for Machines and Installations Designed for Agriculture and Food Industry—INMA Bucharest, 013811 Bucharest, Romania; pruteanuaugustina@gmail.com (A.P.); vladut@inma.ro (V.V.)

[2] Department of Biotechnical Systems, Faculty of Biotechnical Systems Engineering, National University of Science and Technology Politehnica Bucharest, 006042 Bucharest, Romania

* Correspondence: gabriel.constantin@upb.ro (G.A.C.); vanghelelrc@gmail.com (N.A.V.)

**Abstract:** There are a multitude of raspberry and blackberry varieties, and each of them develops differently depending on environmental factors and cultivation technology, so much research is needed to see which variety has the best yield in a desired area. This paper studied the growth under natural soil and specific climate conditions in the Bucharest–Ilfov region of Romania of a raspberry plantation and a blackberry plantation, both in their first year of vegetation. The studied interval, the period of June to October 2024, was established from the beginning of the ripening of the first fruits to the late ripening of the fruits. The study analyzed the correlations between the vegetative and productive parameters of the raspberry variety "*Delniwa*" and the blackberry variety "*Thornfree*" on productivity per plant. During the study period, good shoot formation was observed, with an average height of 1400 mm for raspberries and 3474 mm for blackberries ($r = 0.99$ to raspberries and $r = 0.98$ to blackberries); a good development of the average stem diameter of 8.54 mm for raspberries and 12.78 mm for blackberries ($r = 0.96$ la zmeur si $r = 0.89$ la mur), of the number of ripe fruits harvested ($r = 0.68$ to raspberries and $r = 0.58$ to blackberries), all of which are correlated with increased productivity of 820 g/plant for raspberries and 2050 g/plant for blackberries. The experimental data were statistically analyzed using linearized, polynomial and hyperbolic models to identify the relationships between the studied variables and to highlight growth variations and fruit production in raspberries and blackberries throughout the season. In the first year of vegetation, both crops recorded constant growth but with different rhythms: raspberries showed constant fruit production, with a peak in June–July and a slight resumption in September, while blackberries had high production at the beginning of summer, followed by a significant decrease in August. The results obtained support farmers who grow raspberries and blackberries under similar natural pedo-climatic conditions, contributing to crop planning and production optimization.

**Keywords:** raspberries; blackberries; vegetative and productive indices; soil and climate conditions; Romania

## 1. Introduction

Growing demand for fresh berries has led to the rapid development of advanced production methods for raspberries and blackberries, ensuring healthy products, rich in antioxidants, vitamins and minerals [1] and phenolic compounds [2], throughout the year. In addition, raspberry and blackberry plants can offer a wide range of bioactive substances

 https://doi.org/10.3390/horticulturae11040369

that can be obtained from different vegetative organs, such as young shoots, leaf buds or leaves, which are available throughout the growing season [3,4].

Raspberries, *Rubus idaeus*, and blackberries, *Rubus fruticosus*, both species described by Linnaeus, are plants of the genus Rubus which have a perennial root system that produces biennial stems, called "canes", which usually grow vegetative in the first year, and are called "primocanes"; in the second year, after dormancy, they are called "floricanes", and produce flowers and fruit [5].

Mexico is the main producer of fresh blackberries, supplying the United States and Europe, while Russia is the largest producer of raspberries, with over 100,000 tonnes annually, followed by Serbia and Poland, each with over 80,000 tonnes [5]. In Romania, in 2023, the area under raspberries was 190 hectares, with a production of 430 tonnes, and for blackberries, 1200 hectares, with a production of 3840 tonnes [6].

In countries such as Norway, production is limited by low temperatures but in recent decades, Poland, the Netherlands, the United Kingdom and Belgium have invested in growing raspberries and blackberries in tunnels and plastic greenhouses to extend the season and meet market demand [7,8]. At the same time, research indicates the possibility of cultivating blackberries in shaded systems, integrated into productive forestry, to obtain commercially viable crops [9].

The expansion of berry cultivation areas around the world has led farmers to overuse nitrogen fertilizers in a bid to increase yields. However, proper fertilizer use is essential for healthy crops [10,11] and an appropriate dosage depends on the level of organic matter in the soil and the plant's nutrient reserves [12].

The external natural factors that influence the processes of initiation, growth and development of shoots of raspberry varieties and floral buds of blackberry varieties are temperature and photoperiod [1,13,14], light, humidity and carbon dioxide [5].

Climate change has brought and will bring warmer weather, more variable climate and different precipitation, thus, significant challenges for the cultivation of berries, especially in temperate regions. Plants will have to adapt to much more variable weather conditions, and changes in temperature and precipitation will influence both plant health and crop quality [15].

There are many studies [16–22] regarding the vegetative and productive parameters of berries under certain growing conditions, in different climate and soil areas, to obtain optimal yields to satisfy the demands of consumers around the world.

Recent studies have analyzed various aspects of raspberry and blackberry growth and productivity under different climatic and planting conditions. In Moldova, the study [17] evaluated the growth phenophases and drought resistance of 29 raspberry varieties. In Bulgaria, the studies [16] and [18] examined the vegetative and productive characteristics of the raspberry varieties "*Magdalena*" and "*Shopska Alena*", highlighting the influence of planting distance (0.3 m and 0.5 m) on the number (35.5), height (1.83 m, 1.38 m, respectively) and thickness of shoots (6.31–7.52 mm, respectively, 8.41 mm), as well as yield (1.69 kg/1 $m^2$, for the variety "*Shopska Alena*"). In the "*Magdalena*" variety, the highest number of shoots was recorded for a planting distance of 0.3 m and the highest shoot height was for a planting distance of 0.5 m. Also, a correlation was observed between the height, thickness of the shoots and yield in the second and third years after planting.

In Kosovo, the study [19] evaluated the fruit quality of five raspberry cultivars, identifying the cultivar "*Delniwa*" as having the largest fruit diameter (24.79 mm, and the cultivar "*Mapema*" with the highest values of fruit length (29.84 mm) and fruit weight (6.10 g). In another study [22], the impact of raspberry cultivars on growth parameters and biomass characteristics was analyzed, with the cultivar "*Polana*" having the highest number of lateral shoots and "*Polonez*" having the highest shoot length and thickness.

For blackberries, the study [20] identified the genotype "*Chester Thornless*" as the most suitable for commercial cultivation in the Mediterranean region of Turkey, due to its superior fruit yield and quality. In Serbia, the study [21] demonstrated that seven blackberry cultivars ("*Dirksen Thornless, Thornfree, Čačanska Bestrna, Black Satin, Loch Ness, Chester Thornless, Navaho*") have adapted well to acidic soil and difficult climatic conditions, and are recommended for commercial cultivation.

Previous studies have focused on analyzing the vegetative and productive parameters of raspberries and blackberries starting from the second or third year of production without correlating the influence of environmental factors on the development of plants in the first year and without applying mathematical modeling for yield forecasting. The current research is innovative, analyzing for the first time the development of raspberries "*Delniwa*" and blackberries "*Thornfree*" from the first year of vegetation, under the climatic and pedological conditions of Bucharest. It also uses advanced statistical models for the forecast of staggered production, thus contributing to the optimization of production and the forecast of yield per plant depending on the vegetative parameters.

In this context, the present work shows the potential of Romania for the development of raspberry and blackberry crops under natural climate and soil conditions. The experiment was carried out using raspberry and blackberry seedlings, respectively, in the first year of vegetation, which developed well and produced fruits from the first year of planting. In this study, an in-depth analysis of the vegetative and productive parameters (shoot heights, stem diameters, number of fruits and their sizes) was carried out in raspberry and blackberry plants in the first year of vegetation. The aim of our study was to investigate how the stem, shoots and fruits of raspberry and blackberry plants develop under natural environmental conditions (air temperature, precipitation, atmospheric humidity, solar radiation, evapotranspiration, soil temperature and humidity, volume content of ions). Another objective was to evaluate the growth and development indices of raspberry and blackberry plants under the influence of climate on the yield and quality of fruits. Also, statistical modeling was performed for the staggered production of a raspberry and blackberry plant, depending on the vegetative and productive indices, which resulted in a polynomial and a hyperbolic formula, suitable for both scientific research and agricultural management. Therefore, this study provides some new clues for studying the correlation of raspberry and blackberry growth indices with plant productivity in the future. The main objectives of the study were:

- determining the impact of climatic conditions on the development and productivity of "*Delniwa*" raspberry and "*Thornfree*" blackberry plants in the first year of vegetation;
- analysis of the correlation between vegetative parameters and fruit yield;
- development of statistical formulas: linear, polynomial and hyperbolic, to predict staggered production per plant depending on vegetative parameters and environmental conditions and to compare the performance of the two crops;
- establishing recommendations for optimizing yield from the first year of vegetation in the climatic conditions of Romania (Bucharest–Ilfov area).

The research hypothesis for the study of raspberry and blackberry crops under natural climatic and soil conditions in the Bucharest–Ilfov area consists in the adaptability of the plants to the specific conditions of the capital, given the warmer and more urbanized climate, the yield and quality of production obtained on a sandy–loamy soil. This hypothesis could contribute to a deeper understanding of how urban, climatic and pedological conditions in the studied area influence the development and productivity of these crops and could provide recommendations for local producers or for the development of sustainable urban agriculture strategies.

## 2. Materials and Methods

### *2.1. Research Area and Growing Conditions*

The research was carried out in open field conditions (agro-climatic), in the Bucharest–Ilfov Region, location Figure 1, southern Romania (latitude 44°28′56″ N; longitude 26°04′41″ E) and north-eastern Europe.

**Figure 1.** The location of the experimental study sites (Sursa: earth.google.com [23]).

The experimental plantation has a humid continental climate with very hot summers, with temperature variations between a minimum of −30 °C (in January and February 1942) and a maximum of +42 °C (in July and August 2000), with an average annual temperature in the period from 2019 to 2022 of +13 °C and total precipitation of 623.8 mm. The experimental area is a lowland with terraces, characterized by the presence of numerous terraces along the rivers that drain it, an area made up of exclusively quaternary deposits represented by loess and loessoid deposits, with a pH of 6.5–7.2 and an average humus content of 3.7–4.3% [24].

### *2.2. The Varieties Used in the Experiments*

The plantations were established in the spring of 2024, the raspberry plantation on an area of 1000 $m^2$ and the blackberry plantation on an area of 500 $m^2$, the distance between the rows being 3.3 m and the distance between the plants being 0.75 m. The study included two varieties, one of raspberry "*Delniwa*" and one of blackberry "*Thornfree*". After planting the seedlings, the trellis support system, essential for ensuring real growing conditions, and drip irrigation system were established for both plantations. The irrigation system was prepared as a precaution in case of extreme drought but no irrigation was done during the experiment.

The cuttings of the raspberry variety "*Delniwa*", used to establish the raspberry crop, are part of the primocane variety, in which the appearance of the stems, flowering and fruiting took place in the same vegetation season. The cycle of shoot growth and fruiting takes place in a single year, on the first-year wood, fruiting from June 15 to October 15. It is a hybrid created by crossing the "*Polka, Polana*" and "*Himbotop*" varieties, a premium variety, with fruits resistant to transport and handling; they do not darken after harvesting but retain their natural color 2 to 3 days after harvesting, pink [22]. Usually, the fruits are large, with large drupes with small seeds inside each one welded together, red in color, rounded conical in shape, intensely aromatic, sweet–sour in taste, weighing between 4.3 and 6.1 g [25].

The cuttings of the blackberry variety "*Thornfree*", used to establish the blackberry crop, are part of a vigorous and productive variety, which develops very well even in arid areas with harsh winters, and usually do not need abundant irrigation. This blackberry variety, originating in Europe, is perennial, multiplies itself, giving new shoots from the root every year, which produce fruit, has no thorns, bears fruit from July 15 to August 25 and develops a root up to 60 to 70 cm deep; the shoots can reach a height of up to 3 to 4 m, needing a support system. The fruits are medium to large in size (12 g), have an elongated conical shape, black epidermis with a very shiny purple tint, slightly acidic taste, are very aromatic and juicy [25,26].

In the images in Figures 2 and 3, the two crops and their fruits can be observed during their monitoring.

**Figure 2.** Raspberry variety culture "*Delniwa*".

**Figure 3.** Blackberry variety culture "*Thornfree*".

The requirements of raspberries and blackberries for ecological factors are average for light; the average optimum atmospheric temperature is 16 to 17 °C, the soil temperature is approximately 16 °C; the optimum atmospheric humidity is 70 to 80% during flowering and 65 to 70% during the rest of the vegetation period. They grow well on sandy–loamy, fertile, drained, aerated soil, with pH 5.6 to 6.5, rich in nitrogen and proteins [26].

The experiment monitored the growth and fruiting of the raspberry variety *"Delniwa"* and the blackberry variety *"Thornfree"* under the influence of climatic factors in the experimental site of the crops developed within the National Institute of Agricultural Machinery in the Bucharest–Ilfov area, Romania. Thus, monthly measurements were made with a tape measure on the height of the shoots, with a caliper on the diameter of the stem at 10 cm above the ground, and the fruits harvested throughout the ripening period were weighed with a Kern PCB 6000-0 balance (precision 1 g) (Kern&Sohn GmbH, Ziegelei, Balingen, Germany). To establish the average values with the heights, thicknesses and weights shown in the Section 3, measurements were made on five plants for each parameter.

The raspberry varieties *"Delniwa"* and *"Thornfree"* were selected for this study based on the following criteria: their adaptability to local conditions (Poland is the country of origin for the *"Delniwa"* raspberry variety, and the United States for the *"Thornfree"* blackberry variety), both having good tolerance to heat and cold, resistance to common forest berry diseases, high yield and large, sweet and aromatic, juicy fruits, as well as low maintenance requirements. Also, both varieties adapt well to different soil types, including slightly acidic or neutral ones, which are common in the Bucharest region, well-drained and fertile, making them suitable for this area. Both varieties have flowering and harvesting periods that are well suited to the climatic conditions of Bucharest–Ilfov. *"Delniwa"* produces early fruits, while *"Thornfree"*, also known for its lack of thorns, produces fruits later, being able to ensure a prolonged harvest throughout the season. In addition, raspberries and blackberries are highly valued on the local and international market; therefore, selecting these varieties can ensure a stable income for growers, due to high demand and competitive prices. In conclusion, these varieties offer an optimal combination of fruit quality and economic efficiency, which makes them ideal for cultivation in this region.

### *2.3. Soil Analysis*

Five soil samples were randomly taken from the field at a depth of 30 cm before the establishment of the two crops. After sampling, the soil was ground, air-dried and filtered through a 2.2 mm sieve before analysis. The analysis methods and the values obtained for general soil parameters are presented in Table 1.

**Table 1.** General soil parameters.

| Parameter | Analysis Method | Value Obtained | References |
|---|---|---|---|
| pH | Potentiometric method in aqueous suspension | 6.6 | [27,28] |
| Bulk density | Gravimetric method | 1.31 $g/cm^3$ | [29] |
| Porosity | Bulk density and soil solid particle density method | 53% | [28,30] |
| Organic matter | Walkley-Black method | 3.08% | [31,32] |
| Soil texture | Sedimentation method | Sandy loamy | [33] |

These values suggest that the analyzed soil presents suitable conditions for the cultivation of raspberries and blackberries, with a good balance between moderate acidity, water and nutrient retention capacity and a physical structure favorable to root development. A sandy–loamy texture ensures good drainage and proper aeration of the roots and allows efficient absorption of water and nutrients.

The analysis of heavy metals and minerals in the soil was carried out using standardized procedures in the field of soil analysis, using an atomic absorption spectrometer. The results are shown in Table 2.

**Table 2.** Soil analysis before the establishment of the two crops.

| Heavy Metals and Minerals | Content, mg/kg | Macro and Microelements | Content, mg/kg |
|---|---|---|---|
| Lead | 17.00 | Aqueous extract chlorides | 88.6 |
| Nickel | 18.20 | Sulfates (S-$SO_4^{2-}$) | 72.0 |
| Copper | 18.62 | Soluble calcium | 269.0 |
| Cadmium | 0.33 | Soluble magnesium | 31.0 |
| Manganese | 493.3 | Soluble sodium | 11.3 |
| Zinc | 79.8 | Exchangeable calcium | 4310.0 |
| **Nitrogen and Nitrogen Compounds** | **Content, mg/kg** | Exchangeable magnesium | 400.6 |
| Ammonium (N-$NH_4^+$) | 5.77 | Exchangeable sodium | 161.7 |
| Nitrite (N-$NO_2^-$) | 0.73 | Exchangeable potassium | 331.6 |
| Nitrate (N-$NO_3^-$) | 3.42 | | |

Analysis of the values obtained for heavy metals and minerals in the soil indicates that the soil has a balanced profile in terms of heavy metals and micronutrients. The low levels of lead and cadmium ensure a safe environment, and the values for nickel, copper, manganese and zinc are within normal limits for soils intended for raspberry and blackberry crops.

The analysis of macro and micronutrients shows that the soil has a good content of calcium and potassium, which is beneficial for raspberries and blackberries. Magnesium could be slightly limiting compared to calcium and the low level of sodium is positive; and sulfates are within a safe range, nitrogen is low, so a supplement with compost, manure or organic fertilizers could be beneficial. In conclusion, the soil is suitable for the development of these crops, providing both food security and the nutrients necessary for optimal growth.

Meteorological parameters, humidity, temperature and volumetric ion content in the soil were measured by the meteorological station located in the field, next to the two crops.

### 2.4. Monitoring Growing Conditions

During the growing period of the two crops, meteorological parameters were recorded by a weather station located 50 m from the two research crops and are presented in Table 3.

**Table 3.** Meteorological parameters in the experimental area, period June–October 2024.

| Months | Average Air Temperature, °C | Total Precipitation, mm | Relative Air Humidity, % | Solar Radiation, W/m$^2$ | Evapotranspiration, mm |
|---|---|---|---|---|---|
| June | 25.24 | 87.6 | 55.95 | 265 | 121.1 |
| July | 26.86 | 61.6 | 53.26 | 262 | 151.3 |
| August | 26.16 | 18.2 | 49.07 | 216 | 123.4 |
| September | 19.87 | 68.4 | 63.4 | 135 | 70.3 |
| October | 12.43 | 7.6 | 76.33 | 101 | 36.1 |

The month of October was marked by the lowest average temperatures during the growing season of the two crops, which only reached the value of 12.43 °C, compared to the highest values of over 26 °C, in the months of July and August. Especially in summer, air temperature ranged between +25.24 °C and +26.86 °C. The phases of phenological development of raspberry plants depend on the variety and other factors, but especially on the climatic conditions of the year [17,34].

According to Table 1, the analysis of meteorological data shows that the minimum monthly precipitation accumulated in October was 7.6 mm, and the largest amount was collected in June in the amount of 87.6 mm, although precipitation during the vegetation period was evenly distributed except for August in the amount of 18.2 mm, this month being a critical period for the growth and development of raspberry and blackberry plants.

Air humidity in August was 49%, which is at the lower limit for raspberry and blackberry plants. Lower relative humidity can favor dehydration of the plants and growth can be slower, especially in warmer periods. In contrast, the maximum value of 76% is an optimal humidity, especially during fruiting periods. This helps maintain plant health and prevent premature drying of the fruits [26].

Raspberries and blackberries need good sun exposure to produce quality fruit. Plants should be exposed to sunlight for at least 6 h per day to ensure adequate photosynthesis. Red raspberry fruit is prone to a disorder that has been identified as a form of sun injury. The condition, most commonly seen in areas with high temperatures and solar radiation, is characterized by drupes that enlarge but fail to turn red [35].

Solar radiation of 265 $W/m^2$ in June is sufficient to support plant growth, but not ideal for maximum production. Plants could benefit from more sun exposure or a sunnier location to reach their maximum growth potential [9].

Also, a balance between temperature, humidity and light is essential to ensure a good harvest [36,37]. In the biennial raspberry, flower initiation occurs when temperatures are lower and/or photoperiods are shorter than in the case of primocane fruiting varieties, which are closely related to solar radiation, day length and temperature [14].

Raspberries and blackberries require a significant amount of water to support their growth and produce quality fruit. Their evapotranspiration will depend on climatic conditions and water management. During periods of active growth (spring and summer), when solar radiation is high, in the study it was over 200 $W/m^2$ and temperatures are higher; in the study it was over 25 °C, and evapotranspiration will be more intense, with over 120 mm recorded. The water requirement for these plants varies, but in general, to support them during the growing season, frequent watering is recommended, especially during hot or dry periods [38,39].

Air temperature and soil moisture are important in the development and implementation of phenological phases in raspberries and blackberries [17,34]. The values of the monitored soil parameters are presented in Table 4.

**Table 4.** Physico–chemical parameters of the soil during the experimental study.

| Months | Average Soil Temperature, °C | Soil Moisture, % | Volume Content of Ions |
|---|---|---|---|
| June | 20.6 | 48.76 | 1382 |
| July | 23.7 | 27.9 | 1361 |
| August | 24.7 | 26.79 | 1327 |
| September | 20.7 | 26.32 | 1375 |
| October | 14.6 | 21.74 | 1236 |

It is known [26] that maintaining a balance between temperature and humidity is essential for healthy development of raspberry and blackberry plants. Constant monitoring of these conditions can help prevent heat stress or irrigation problems that can affect the harvest. If the soil temperature was within optimal parameters for raspberry and blackberry growth (14.6–24.7), a soil humidity between 21% and 48% can be considered suboptimal for optimal raspberry and blackberry growth and development. Ideally, for both plants, soil humidity should be between 60 and 80% of field capacity to support healthy growth and good fruit production.

According to the FAO [6], if your soil is sandy loam and contains 3.08% organic matter, we can estimate field capacity based on the standard values for this type of soil. Typically, sandy loam soils have a field capacity of 15 to 25%, and the organic matter contributes slightly to water retention.

Volumetric ion content refers to the total concentration of ions (cations and anions) in a volume of soil or soil solution. This is an important measure of soil fertility, nutrient availability and salinity, for monitoring to adjust fertilization accordingly [40].

Higher values of ion volume in June (1382) suggest a relatively fertile soil, but this fertility gradually decreases throughout the season as plants consume nutrients. If ion concentrations become too high, there is a risk that the soil will become saline, which can affect water uptake by plants. Lower ion values in August and October suggest that salinity is not a major problem, but it is important to monitor. A low value in October (1236) could signal a nutrient deficiency or a need for additional fertilizer application, especially for crops that have consumed a lot of soil resources (raspberries and blackberries).

### 2.5. Statistical Analysis

The parameters investigated in the study were the average height of the shoots, the average diameter of the stems, the average weight of the fruits and the number of fruits, and measurements were made on five plants for each parameter.

The correlation analyses of the growth parameters were performed using Microsoft Excel 2010 software. To study the correlation between variables, plant characteristics such as height, stem diameter, fruit weight and number of fruits were used, then the correlation coefficient between these variables was calculated [41,42].

Also, logarithmic linearization calculations were performed using Microsoft Excel 2010 software. The linearized model presented in this paper allows the identification and understanding of the seasonal behaviors of growth parameters (height, diameter) and production (weight, number of fruits). The model allows the comparison of the performance of the two crops (raspberries and blackberries) and decision-making based on experimental data.

The calculations of the coefficients in the interpolation polynomials were performed using the interpolation function of the MathCAD 15 program. In the analysis performed, we applied polynomial interpolation to estimate the staggered production of the first year of vegetation of a raspberry or blackberry plant, respectively, depending on the following vegetative and productive parameters: average plant height, average stem diameter and number of ripe fruits harvested. Polynomial (second, third or fourth degree) and hyperbolic models are more flexible and can better describe complex relationships, such as rapid increases followed by sudden decreases or curvilinear trends; therefore, they can better capture the real behavior of crops, identify critical points of growth and production (for example, the month with maximum production or the period of sudden decrease), providing precise analyses and optimized planning.

## 3. Results

### 3.1. Experimental Data Obtained and Correlations Between Them

Depending on the temperatures during this period, the ripening of raspberry fruits begins in the first decade of June and continues until October, and the ripening of blackberry fruits begins in the second decade of June and ends at the end of August, both crops with some deviations depending on the temperatures established during this period.

Table 5 presents the vegetative and productive parameters monitored and measured in raspberry and blackberry plants.

**Table 5.** Vegetative and productive parameters of the raspberry variety *"Delniwa"* and the blackberry variety *"Thornfree"*.

| Months | Raspberries | | | | Blackberries | | | |
|---|---|---|---|---|---|---|---|---|
| | Average Plants Height, mm | Average Stem Diameter, mm | Weight of Fruits, g | Number of Fruits, Pieces | Average Plants Height, mm | Average Stem Diameter, mm | Weight of Fruits, g | Number of Fruits, Pieces |
| June | 1315 | 7.11 | 283.5 | 77 | 2685 | 11.27 | 743.0 | 128 |
| July | 1367 | 8.36 | 248.1 | 98 | 3243 | 11.81 | 1252.3 | 327 |
| August | 1390 | 8.51 | 53.2 | 32 | 3642 | 12.88 | 55.2 | 18 |
| September | 1447 | 9.18 | 161.3 | 115 | 3897 | 13.74 | 0.0 | 0 |
| October | 1482 | 9.56 | 73.7 | 67 | 3905 | 14.22 | 0.0 | 0 |
| TOTAL | - | - | 819.8 | 389 | - | - | 2050.5 | 473 |
| MEDIA | 1400 | 8.54 | 163.96 | 77.80 | 3474 | 12.78 | 683.50 | 94.60 |
| STDEV | 65.85 | 0.94 | 102.21 | 31.65 | 516.91 | 1.25 | 600.76 | 156.62 |
| COVAR | 4.70 | 10.99 | 62.34 | 40.68 | 14.88 | 9.76 | 87.90 | 99.34 |

The results regarding the vegetative indicator average number of ripe fruits report that there are significant variations in the values over the five-month period, which ranged from 32 to 115 for raspberries and 18 to 327 for blackberries (Table 5). The average for the five-month period of the first year of cultivation of the two plants was 389 pieces for raspberries and 473 pieces for blackberries, which shows that the two varieties of berries were very productive from the first year of fruiting. The average height of the shoots at the end of the vegetation period for the two plants studied was increasing for both raspberries (1400 mm) and blackberries (3474 mm).

Regarding average stem thickness at the end of the vegetation period, it was 8.54 mm for raspberries and 12.78 mm for blackberries, significantly higher between June and July for raspberries (the increase being 1.25 mm between the two months), with the smallest increase in stem diameter for raspberries being recorded between July and August (0.15 mm). For blackberries, a significant increase in stem diameter was recorded between July and August (1.07 mm) and the smallest between September and October (0.48 mm). Therefore, the values recorded for thickness required the application of a support system for the two crops from the first year of vegetation to ensure plant growth. According to [18], large, thick shoots produce more fruit-bearing branches.

The productivity of the plants was at its maximum in June and July for raspberries (≈250 g/month/plant), the first harvest; and the second harvest was lower, ≈70–160 g/plant. In the case of blackberries, only in the first three months of the five were fruits harvested; maximum productivity was reached in July and was 1252 g/plant.

By applying an ANOVA test, it was observed that there were significant differences between the measurement months for all analyzed parameters (plant height, stem diameter, fruit weight and number of fruits), with a statistical significance level of $p < 0.05$. Then, to identify between which months there were significant differences for each parameter separately and for each crop, the post hoc Tukey test was applied, which identified significant differences between the measurement months for all parameters, both for raspberries and blackberries. In raspberries, plant height and fruit weight varied significantly between June and October, and the number of fruits was higher in July and September than in August. In blackberries, plant height and fruit weight showed significant differences between July and subsequent months (August, September), and the number of fruits decreased significantly after July. Therefore, these differences reflect seasonal variations in plant growth and fruit production, with maxima in June and July and significant decreases in August and September.

Next, a *t*-test was applied to compare the means of the two crops, raspberries and blackberries, for each parameter and for each month and to determine whether the differences between them are statistically significant. The results showed that blackberries are significantly taller and have thicker stems than raspberries in all months. In terms of fruit production, blackberries have heavier and more numerous fruits in June and July, but the differences disappear in August, September and October, when production decreases significantly for both crops.

Applying the *t*-test, it was found that the differences between raspberries and blackberries were statistically significant for plant height and stem diameter in all months ($p < 0.05$). In the case of fruit weight and number of fruits, the differences were significant in June and July ($p < 0.05$) but not in August, September or October ($p > 0.05$)

Table 6 presents the correlations between the parameters investigated in this paper.

**Table 6.** Correlation of dependencies between vegetative and productive indicators in raspberries and blackberries.

| Indicators | Raspberries | | | | Blackberries | | | |
|---|---|---|---|---|---|---|---|---|
| | Plants Height | Stem Diameter | Weight of Fruits | Number of Fruits | Plants Height | Stem Diameter | Weight of Fruits | Number of Fruits |
| Plants height | 1 | | | | 1 | | | |
| Stem diameter | 0.99 | 1 | | | 0.98 | 1 | | |
| Weight of fruits | 0.98 | 0.96 | 1 | | 0.99 | 0.89 | 1 | |
| Number of fruits | 0.99 | 0.96 | 0.68 | 1 | 0.91 | 0.93 | 0.58 | 1 |

In raspberries, a very high correlation dependence was reported from correlation analyses, between height indicators and shoot thickness but also between height and the number of harvested raspberry fruits (0.99) (Table 6), a fact also supported in the work [18].

In blackberries, a very high correlation dependence was reported from correlation analyses, between height indicators and fruit weight (0.99). Small correlations exist between the number of harvested fruits and fruit weights of both studied berry varieties.

Therefore, a high to very high correlation dependence between shoot height and thickness, shoot height and ripe fruit weight, and shoot height and ripe fruit number was reported throughout the growing season. There is a very strong relationship between plant height and stem diameter ($r = 0.99$ for raspberry and $r = 0.98$ for blackberry), suggesting that taller plants usually have thicker stems. Fruit weight and number of fruits have a moderate correlation ($r = 0.68$ for raspberry and $r = 0.58$ for blackberry), indicating that a higher number of fruits does not necessarily imply a higher weight per fruit. All variables are intercorrelated, showing that plant size influences fruit production and the selection of plants with desired characteristics in agriculture. There are significant relationships between these plant characteristics throughout the growing season, and they may be particularly useful for researchers and farmers studying plant behavior under specific environmental conditions.

These correlations could suggest that by managing shoot height (through pruning, proper care, fertilization or irrigation), fruit production can be optimized, both in terms of quantity (number of fruits) and quality (fruit weight). This information can be extremely valuable for optimizing agricultural practices and increasing yields.

In the case of raspberry and blackberry crops, from the analysis of the mentioned correlations, the growth parameters that are most related to crop yield are plant height and stem diameter. According to the correlations, on the one hand plant height has a very strong correlation ($r = 0.99$) with the number of fruits and fruit weight. This suggests that taller plants tend to have a higher yield, as they are able to support more fruits and heavier fruits. Vertical plant growth is an important indicator of vigorous development and the ability to support a higher production. On the other hand, the very strong correlation (r = 0.96) between stem diameter and the number of fruits and fruit weight indicates that a larger stem diameter is associated with higher fruit production. Thick stems are generally a sign of a healthy and robust plant, and are able to support more and larger fruits, which directly contributes to the yield.

Therefore, both plant height and stem diameter are growth parameters that significantly influence the yield of raspberry and blackberry crops, having very strong correlations with the number and weight of fruits. Therefore, optimizing the vertical development of plants should be a priority in crop management to maximize fruit production.

Average dimensions for ripe fruits of the two studied berry varieties are presented in Figure 4.

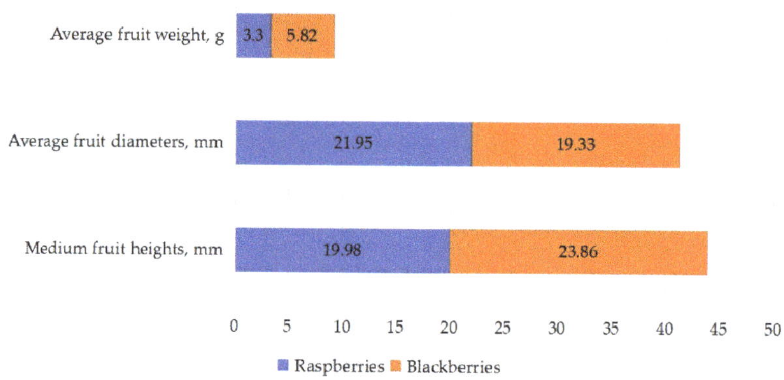


**Figure 4.** Average sizes of harvested ripe fruits.

The weight of the fruits of the two varieties indicates the value of 5.82 g/fruit for blackberries and 3.3 g/fruit for raspberries. If in the case of mass, the blackberry is heavier than the raspberry, in the case of diameter, it is larger for raspberries at 21.95 mm than for blackberries at 19.33 mm. In general, raspberry fruits have a smaller diameter than blackberry fruits, but there is some variability depending on the type of variety and growing conditions. The "*Delniwa*" raspberry variety is a hybrid variety specially developed to produce large fruits if the growing conditions are provided precisely with its cultivation technology. Regarding the heights of the fruits, because blackberries have a more elongated shape, they are taller (23.86 mm) than raspberry heights (19.98 mm).

### *3.2. Statistical Models for the Cumulative Production of a Raspberry and Blackberry Plant, Depending on Vegetative and Productive Indices, During the First Year of Vegetation*

The data found in the experiments described in Table 5 were used to obtain statistical mathematical models by linearization and interpolation.

In this paper, we chose to analyze the experimental data using both linear and polynomial (2nd, 3rd and 4th degree) and hyperbolic models to provide a complete perspective on the dynamics of raspberry and blackberry crops. Linear models, although useful for a first understanding of the data, failed to capture the significant nonlinear variations observed in fruit production and plant growth. In contrast, polynomial models provided a more accurate approximation, allowing the identification of seasonal trends and production tipping points.

#### 3.2.1. Application of the Linearized Model on Experimental Data of Raspberry and Blackberry Crops

Based on the experimental results presented in Table 5, the logarithmic equation (Equation (1)) was applied, resulting in the linearized model coefficients, the model interpretation and the correlation coefficient (r) for each parameter and crop, shown in Table 7.

$$\ln(y) = a \cdot x + b \quad (1)$$

where ln(y) is the natural logarithm of the parameter, y is the measured parameter, the dependent variable, (e.g., plant height, stem diameter, fruit weight, number of fruits); x is the month, the independent variable (June, July, August, September, October) used for linearization; the coefficient a (slope) shows the rate of increase or decrease of the parameter per month, of y as a function of x; the coefficient b (intercept) represents the initial value of the parameter, the value of ln(y) when x = 0.

**Table 7.** The coefficients of the equations, the linearized model, the interpretation and the value of r (correlation coefficient) for each parameter and crop.

| Parameter | Culture | Coefficient A | Coefficient B | Interpretation | R-Value |
|---|---|---|---|---|---|
| Plant height | Raspberry | 0.036 | 7.136 | Height grows by ~3.6% per month. | 0.992 |
| | Blackberry | 0.072 | 7.892 | Height grows by ~7.2% per month. | 0.998 |
| Stem diameter | Raspberry | 0.042 | 1.956 | Diameter is increasing by ~4.2% per month. | 0.987 |
| | Blackberry | 0.038 | 2.412 | Diameter is increasing by ~3.8% per month. | 0.991 |
| Fruit weight | Raspberry | −0.512 | 6.452 | Weight drops significantly in August. | 0.945 |
| | Blackberry | −1.204 | 8.912 | Weight drops sharply in August. | 0.932 |
| Number of fruits | Raspberry | −0.128 | 4.912 | The number of fruits decreases in August, then increases slightly in September. | 0.921 |
| | Blackberry | −0.956 | 6.412 | The number of fruits drops sharply in August. | 0.912 |

Table 7 shows that the linearized models provide a good approximation of the data, with correlation coefficients (r) close to 1 for most parameters. Plant height and stem diameter show a constant increase throughout the season, while fruit weight and number decrease significantly in August. These results suggest that the linearized models are suitable for analyzing plant growth but may be insufficient to describe sudden variations in fruit production.

Next, to obtain the real values of the parameters from the linearized models, we applied the exponential to both sides of the equation; this transformation is necessary to interpret the results in real units of measurement (m, mm, g). The values calculated with the exponential were very close to the experimental data, which confirms that the linearized model is accurate.

By applying the linearized model and exponential to experimental data of raspberry and blackberry crops, we have obtained a clear understanding of their growth dynamics and productivity. Raspberries and blackberries have constant growth, but different rhythms. Raspberries produce fruit consistently, with a peak in June–July and a slight recovery in September, while blackberries have high production in early summer but drop dramatically in August. Therefore, this model can help farmers optimize harvest and resources, adapting irrigation and fertilization to prevent yield decline.

### 3.2.2. Application of Polynomial and Hyperbolic Models on Experimental Data of Raspberry and Blackberry Crops

The following are 2nd, 3rd and 4th degree interpolations of the production/plant at the monthly harvest of ripe fruits, depending on certain vegetative and productive indices, such as average plant height, average stem diameter and number of ripe fruits harvested.

Polynomial interpolations of the production/raspberry and blackberry plant (P) during the fruiting, ripening and harvesting period (approximately five months for raspberries and three months for blackberries) depend on three vegetative and productive indices ($I_{vr}$), namely, average plant height, average stem diameter and number of fruits harvested during the first vegetative year.

The general form of the interpolation polynomial (degrees one to four) is

$$P(I_{vr}) = c_0 + c_1 I_{vr} + c_2 I_{vr}^2 + c_3 I_{vr}^3 + c_4 I_{vr}^4 \tag{2}$$

The general form of a hyperbola is

$$P(I_{vr}) = \frac{c_0}{I_{vr}} + c_1 \tag{3}$$

The coefficients of the polynomials (1) are given in Table 8, and those of the hyperbola (2) are given in Table 9.

**Table 8.** The interpolation equations corresponding to the production of raspberry and blackberry/plant.

| Fruit | Index | Coefficients of Interpolation Polynomials | | | | | Error, % |
|---|---|---|---|---|---|---|---|
| | | $c_0$ | $c_1$ | $c_2$ | $c_3$ | $c_4$ | |
| Raspberry | Shoot height | $1.745 \times 10^3$ | $-1.128 \times 10^3$ | 0 | 0 | 0 | 88.74 |
| | | $1.947 \times 10^4$ | $-2.646 \times 10^4$ | $9.035 \times 10^3$ | 0 | 0 | 82.428 |
| | | $3.688 \times 10^5$ | $-7.783 \times 10^5$ | $5.48 \times 10^5$ | $-1.286 \times 10^5$ | 0 | 80.525 |
| | | $-9.414 \times 10^7$ | $2.687 \times 10^8$ | $-2.874 \times 10^8$ | $1.366 \times 10^8$ | $-2.431 \times 10^7$ | $2.1 \times 10^{-7}$ |
| | Stem diameter | 835.074 | −78.543 | 0 | 0 | 0 | 86.493 |
| | | $1.638 \times 10^3$ | −273.881 | 11.753 | 0 | 0 | 85.838 |
| | | $1.753 \times 10^4$ | $-6.052 \times 10^3$ | 706.881 | −27.691 | 0 | 85.41 |
| | | $-6.575 \times 10^6$ | $3.117 \times 10^6$ | $-5.514 \times 10^5$ | $4.318 \times 10^4$ | $-1.263 \times 10^3$ | $1.239 \times 10^{-8}$ |
| | Number of fruits | 17.499 | 1.883 | 0 | 0 | 0 | 101.44 |
| | | −202.003 | 8.982 | −0.049 | 0 | 0 | 87.88 |
| | | 875.142 | −47.373 | 0.797 | $-3.822 \times 10^{-3}$ | 0 | 62.77 |
| | | $1.009 \times 10^4$ | −643.95 | 13.826 | −0.122 | $3.779 \times 10^{-4}$ | $2.893 \times 10^{-11}$ |
| Blackberries | Shoot height | $3.249 \times 10^3$ | −817.219 | 0 | 0 | 0 | 184.8 |
| | | $-1.349 \times 10^4$ | $9.504 \times 10^3$ | $-1.556 \times 10^3$ | 0 | 0 | 131.577 |
| | | $-2.546 \times 10^5$ | $2.326 \times 10^5$ | $-6.957 \times 10^4$ | $6.84 \times 10^3$ | 0 | 7.045 |
| | | $-1.488 \times 10^6$ | $1.726 \times 10^6$ | $7.431 \times 10^5$ | $1.409 \times 10^5$ | $-9.952 \times 10^3$ | $6.183 \times 10^{-10}$ |
| | Stem diameter | $5.213 \times 10^3$ | −375.687 | 0 | 0 | 0 | 154.896 |
| | | $1.124 \times 10^4$ | $-1.33 \times 10^3$ | 37.449 | 0 | 0 | 153.968 |
| | | $-7.146 \times 10^5$ | $1.706 \times 10^5$ | $1.348 \times 10^4$ | 353.073 | 0 | 92.029 |
| | | $-9.006 \times 10^6$ | $2.796 \times 10^6$ | $3.245 \times 10^5$ | $1.669 \times 10^4$ | −320.865 | $2.55 \times 10^{-9}$ |
| | Number of fruits | 36.369 | 3.949 | 0 | 0 | 0 | 55.402 |
| | | −20.425 | 7.154 | $-9.964 \times 10^{-3}$ | 0 | 0 | 14.353 |
| | | −0.022 | 2.347 | 0.041 | $-1.13 \times 10^{-4}$ | 0 | $2.031 \times 10^{-13}$ |
| | | −0.02 | 2.141 | 0.055 | $-2.42 \times 10^{-4}$ | $2.728 \times 10^{-7}$ | $1.418 \times 10^{-3}$ |

**Table 9.** Hyperbolic interpolation equations corresponding to the production of raspberry and blackberry/plant.

| Fruit | Index | Hyperbola Coefficients | |
|---|---|---|---|
| | | $c_0$ | $c_1$ |
| Raspberry | Shoot height | $2.236 \times 10^3$ | $-1.434 \times 10^3$ |
| | Stem diameter | $5.329 \times 10^3$ | $-466.178$ |
| | Number of fruits | $-7.364 \times 10^3$ | $278.963$ |
| Blackberries | Shoot height | $8.025 \times 10^3$ | $-1.946 \times 10^3$ |
| | Stem diameter | $6.023 \times 10^4$ | $-4.338 \times 10^3$ |
| | Number of fruits | $-6.838$ | $683.594$ |

The graphical representations of the polynomial interpolations of fruit production/plant (raspberry, blackberry) at harvest depending on plant height, stem diameter and number of ripe fruits harvested are shown in Figure 5.

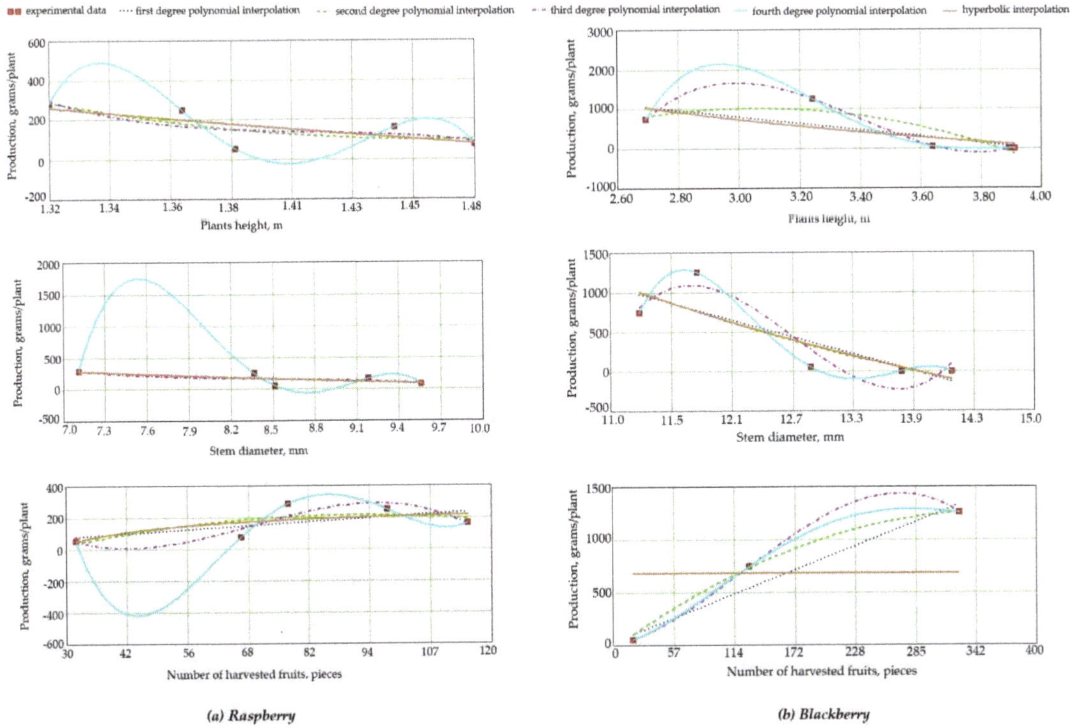


**Figure 5.** Graphical representations of the interpolations of production/plant for raspberry (**a**) and blackberry (**b**) according to vegetative and productive indices.

The interpolation formulas can be used to forecast raspberry and blackberry production at the end of the vegetation period for each of the studied plants. The forecast will be made, however, only within the limits of the interval of each vegetation and reproduction index given by their minimum and maximum values used in the experiments.

A first conclusion shows that interpolation polynomials of degrees one and two are the most usable because they have small variations between the experimental points. Polynomial curves of degree three and four show relatively large variations between the experimental points and therefore their use presents the risk of producing large errors, although, for example, restricting ourselves only to the set of experimental points, the polynomial of degree four passes exactly through the experimental points.

In general, the interpolation curves show a decreasing trend in productivity per plant (at the final stage of vegetation/harvest), depending on the height of the shoots and the diameter of the stem, and the number of ripe fruits harvested. For this reason, a polynomial Formula (1) and a hyperbolic interpolation Formula (2) were considered, which prove adequate with some exceptions in which the experimental data do not detect the above-mentioned trend.

The statistical mathematical models provided by Formulas (1) and (2), whose generic coefficients, for each specific case, are given in Tables 4 and 5, can be used for interpolation (calculation of any productivity for a raspberry or blackberry plant, respectively, only in the experimental interval considered for the values of plant height, stem diameter and number of ripe fruits).

## 4. Discussion

In the context of climate forecasts, precipitation shows the provision of water to crops to obtain optimal production from a technical and economic point of view. Precipitation records aim to establish the level to which plants can be water-stressed in certain vegetation phases, with a beneficial effect on irrigation norms and consumption applied to watering to obtain high and quality productions.

The dominant effects of environmental factors (e.g., temperature, light, humidity, precipitation, solar radiation, carbon dioxide and others) on raspberry and blackberry crops observed in the present study were consistent with those reported in other research works [5,39].

The effects of temperature under controlled environmental conditions after the flowering period in raspberries on the chemical composition of the fruits were studied in this paper [43]. The results showed that the weight of the berries decreased significantly with increasing temperature (12, 18 and 24 °C) and with the progress of the harvest period. Also, the temperature of raspberry fruit growth has significant and contrasting effects on the concentration of a series of potentially bioactive compounds in raspberries [43].

Raspberries react negatively to a lack of moisture during fruit formation and growth, which can lead to a reduction in fruit weight, quantity and quality, a fact also observed from the values observed and monitored in our study, especially those in August. The requirements of the raspberry crop in terms of soil moisture are necessary due to the superficial root system. The lack of precipitation during the periods of shoot growth and fruit formation is visible and leads to a slowdown in plant development, implicitly of the fruits, and therefore requires irrigation of the crop [17].

For blackberries, the values of the growth parameters obtained in our study, namely, average height 3.47 m, average diameter 12.8 mm, average fruit weight 5.82 g and productivity per plant 2 kg, are higher for the same variety "*Thornfree*" than the values obtained in the work [20], where it is shown that for the same variety grown in a Mediterranean climate in a region of Turkey, the values were the following: for shoot height 2.57 m, stem diameter 20.8 mm, average fruit weight 3.9 g and yield per plant 3 kg.

Correlation analyses between vegetative and productive performances (number, height and thickness of shoots with yield) over three years are presented in the paper [18] for the raspberry variety "*Shopska Alena*" and in the paper [16] for the variety "*Magdalena*", both varieties being cultivated in Bulgaria. Compared to our study conducted in Romania, where the vegetative indices in the first year of vegetation for the raspberry variety "*Delniwa*" were average height (1.4 m), average stem diameter (8.5 mm) and productivity 820 g/plant, for the first variety, cultivated in the neighboring country, the values recorded in the first year were average plant height (1.35 m), average stem thickness (7.48 mm) and yield of 0.51 kg/1 m$^2$; and for the second variety, the values recorded in the first year were average plant height (1.39–1.83 m), average stem thickness (6.44–6.47 mm) and yield of 0.82–0.94 kg/1 m$^2$. Therefore, the values of the vegetative and productive parameters are very close, although the raspberry varieties differ; and the soil conditions and climate in which the two crops developed are very similar.

Regarding productivity per plant obtained in the present study, it was 820 g/plant in the first year of fruiting for raspberries and 2050 g/plant in the first year of fruiting for blackberries. Comparatively, in the study [34], the yield for raspberries under ambient conditions was 2.5 kg/plant, while the yield at controlled temperatures (12–24 °C) stabilized at approximately 1.6 kg/plant. In the study [8], the yield for blackberries varied between 0.56 and 1.1 kg/plant for the crop grown in the open field and between 0.5 and 1.98 kg/plant for the crop grown in the tunnel during a one-year vegetation period.

Therefore, according to [18], the profitability of raspberry fruit production is determined by the correct choice of variety, appropriate agroclimatic conditions of the habitat (temperature, wind, precipitation, pests) and modern cultivation technology.

The average weight of raspberry fruits is a criterion for evaluating quality varieties. The results obtained regarding the size of the fruits in the work [17] highlighted that the following raspberry varieties have the largest fruit sizes: "*Hybrid Bulgarian*" 3.5 g, "*Rubin September*" 3.2 g, "*Delbard Magnific*" 2.9 g and "*Lazarevscaia*" 2.7 g. Small fruit varieties from 1.3 to 1.6 g were "*June, Meteor, Pathfinder, Red Wadenswil, Indian Summer, Kuthbert, St. Walfred*". In the case of our study, the fruit weight of the two varieties indicates the value of 5.82 g/fruit for blackberries and 3.3 g/fruit for raspberries. The raspberry variety "*Delniwa*" is a more recent variety, developed to provide good production and medium to large quality fruits. This variety comes from a Polish selection and is valued for its characteristics such as disease resistance and high yield [22]. The diameter of the "*Delniwa*" variety cultivated in the Prishtina region (Kosovo) was 24.79 mm, in a crop in the 2nd year of vegetation, compared to our study where the diameter was 21.95 mm, in a crop in the first year of vegetation [19].

Similar to the values obtained in this work for blackberry are those in the work [21] in which the same variety, "*Thornfree*", but cultivated in Serbia, in a mild climate, where the fruit weight varied between 4.65 and 5.32 g, fruit heights between 21.52 and 23.69 mm and diameters between 17.52 and 19.31 mm.

The results presented demonstrate that variations of environmental factors, especially precipitation and temperature, influence the growth of shoot height and stem diameter and flower formation, implicitly, the formation of fruits in raspberries and blackberries grown in a natural environment of temperate continental climate.

Polynomial representations are a primary and common form of statistical models with practical applications both in scientific research and in agricultural management. Using such formulas, one can calculate the production and yield of fruits per hectare. Using the same formulas, vegetation processes and adequate growth of raspberry and blackberry crops can be optimized in suitable areas, considering their use for marketing and/or human consumption.

## 5. Conclusions

The results demonstrate that in the studied plants, the raspberry variety "*Delniwa*" and the blackberry variety "*Thornfree*", there is a functional link between the growth in height of the shoots and the increase in the diameter of the stems to achieve a large number of fruits harvested from a raspberry and blackberry plant, respectively, implicitly obtaining high productivity for the two selected varieties from the first year of vegetation. In addition, the results suggest that the effects of environmental factors (temperature, precipitation, air humidity, solar radiation, evapotranspiration, soil temperature and humidity, soil salinity) and the interaction between them on vegetative growth, plant development and fruit production offer a better understanding of the influence of climate on the productivity and quality of raspberry and blackberry fruits. The results of the study confirm that the varieties *"Delniwa"* and *"Thornfree"* are suitable for cultivation in the pedoclimatic conditions of Bucharest, given their high yield. However, for a broader understanding, it is recommended to include more varieties in future studies to evaluate a wider range of adaptability and productivity and to support farmers growing berries in the Bucharest area, including for consumer satisfaction.

**Author Contributions:** Conceptualization, A.P., G.A.C. and N.A.V.; methodology, A.P. and G.A.C.; software, A.P. and G.A.C.; validation, A.P., G.A.C. and V.V.; formal analysis, V.V.; investigation, A.P. and N.A.V.; resources, G.A.C. and V.V.; data curation, V.V.; writing—original draft preparation, A.P. and N.A.V.; writing—review and editing, A.P., G.A.C. and N.A.V.; visualization, A.P., N.A.V. and G.A.C.; supervision, V.V.; project administration, G.A.C. and V.V.; funding acquisition G.A.C. and V.V. All authors have read and agreed to the published version of the manuscript.

**Funding:** This research was supported by the project Establishment and Operationalization of a Competence Center for Soil Health and Food Safety—CeSoH, Contract no.: 760005/2022, specific project no. 5, with the title: Improving soil conservation and resilience by boosting biodiversity and functional security of organic food products, Code 2, financed through PNRR-III-C9-2022—I5 (PNRR-National Recovery and Resilience Plan, C9 Support for the private sector, research, development and innovation, I5 Establishment and operationalization of Competence Centers).

**Data Availability Statement:** The original contributions presented in the study are included in the article; further inquiries can be directed to the corresponding authors.

**Conflicts of Interest:** The authors declare no conflicts of interest.

## References

1. Sadojevic, M. Growth and Development of Blackberry Cultivars. Master's Thesis, Norwegian University of Life Sciences (NMBU), Ås, Norway, 2024.
2. Shahidi, F.; Naczk, M. Biosynthesis, classification, and nomenclature of phenolics in food and nutraceuticals. In *Phenolics in Food and Nutraceuticals;* CRC Press LLC: Boca Raton, FL, USA, 2004; pp. 1–16. [CrossRef]
3. Krzepiłko, A.; Prażak, R.; Święciło, A. Chemical Composition, Antioxidant and Antimicrobial Activity of Raspberry, Blackberry and Raspberry-Blackberry Hybrid Leaf Buds. *Molecules* **2021**, *26*, 327. [CrossRef] [PubMed]
4. Rao, A.V.; Snyder, D.M. Raspberries and Human Health: A Review. *J. Agric. Food Chem.* **2010**, *58*, 3871–3883. [CrossRef] [PubMed]
5. Dickson, R.W.; Machesney, L.M.; Henderson, E.O. Forcing Temperature Effects on Plant Development for Containerized Blackberry Grown Using Long-Cane Techniques. *Horticultura* **2023**, *9*, 1090. [CrossRef]
6. FAO. Available online: https://www.fao.org/faostat/en/#data/QCL/visualize (accessed on 10 February 2025).
7. Pritts, M.P. Growing blackberries in a cold climate using high tunnels. *Acta Hortic.* **2016**, *1133*, 263–269. [CrossRef]
8. Thompson, E.; Strik, B.C. High Tunnel versus Open Field: Management of Primocane-fruiting Blackberry Using Pruning and Tipping to Increase Yield and Extend the Fruiting Season. *Hortscience* **2009**, *44*, 1581–1587. [CrossRef]
9. Gallagher, E.J.; Mudge, K.W.; Pritts, M.P.; DeGloria, S.D. Growth and development of 'Illini Hardy' blackberry (*Rubus* subgenus *Eubatus* Focke) under shaded systems. *Agrofor. Syst.* **2015**, *89*, 1–17. [CrossRef]
10. Duan, Y.; Yang, H.; Yang, H.; Wei, Z.; Che, J.; Wu, W.; Lyu, L.; Li, W. Physiological and Morphological Responses of Blackberry Seedlings to Different Nitrogen Forms. *Plants* **2023**, *12*, 1480. [CrossRef]
11. Strik, B.C.; Bryla, D.R. Uptake and Partitioning of Nutrients in Blackberry and Raspberry and Evaluating Plant Nutrient Status for Accurate Assessment of Fertilizer Requirements. *HortTechnology* **2015**, *25*, 452–459. [CrossRef]
12. Lu, Q.; Miles, C.; Tao, H.; DeVetter, L.W. Reduced Nitrogen Fertilizer Rates Maintained Raspberry Growth in an Established Field. *Agronomy* **2022**, *12*, 672. [CrossRef]
13. Sønsteby, A.; Heide, O.M. Temperature and Daylength Effects on Growth and Floral Initiation in Biennial-Fruiting Blackberry. *Horticulturae* **2023**, *9*, 1285. [CrossRef]
14. Carew, J.G.; Trudi, G.; White, J.; Wainwright, H.; Brennan, R.; Battey, N.H. The control of the annual growth cycle in raspberry. *J. Hortic. Sci. Biotechnol.* **2000**, *75*, 495–503. [CrossRef]

15. Dale, A. How Climate Change Could Influence Breeding and Modern Production Systems in Berry Crops. *Acta Hortic.* **2009**, *838*, 161–168. [CrossRef]
16. Atanasova, S.; Georgieva, M.; Georgiev, D. Interaction between vegetative and reproductive manifestations in raspberry candidate cultivar 'Magdalena'. *Bulg. J. Agric. Sci.* **2022**, *28*, 1099–1104.
17. Sava, P. Research on Factors Affecting Raspberry Plant Growth. *Horticulture* **2013**, *57*, 105–108.
18. Georgiev, D.; Georgieva, M.; Hristova, D. Dependencies Between Vegetative and Reproductive Performances in 'Shopska Alena' Raspberry Cultivar. *Horticulture* **2023**, *LXVII*, 76–79.
19. Lepaja, K.; Lepaja, L.; Kullaj, E. Fruit quality parameters of five primocane raspberry cultivars in Kosovo. In Proceedings of the IV International Symposium on Horticulture in Europe—SHE2021, Stuttgart, Germany, 8–12 March 2021; p. 1327. [CrossRef]
20. Özdemir, E.; Ayanoğlu, H.; Gündüz, K.; Bayazit, S. Determination of Vegetative and Fruit Characteristics of Some Thornless Blackberry Genotypes in Hatay (Turkey). *Int. J. Fruit Sci.* **2005**, *5*, 77–83. [CrossRef]
21. Milošević, T.; Milošević, N.; Glišić, I.; Mladenović, J. Fruit quality attributes of blackberry grown under limited environmental conditions. *Plant Soil Environ.* **2012**, *58*, 322–327. [CrossRef]
22. Maj, G.; Buczyński, K.; Klimek, K.E.; Kapłan, M. Evaluation of Growth and Energy Parameters of One-Year-Old Raspberry Shoots, Depending on the Variety. *Energies* **2024**, *17*, 3153. [CrossRef]
23. Google Earth. Available online: https://earth.google.com/web/@44.50106347,26.06972411,104.04988003a,791.59168963d,35y,359.99982591h,0t,0r/data=CgRCAggBMikKJwolCiExQ004anBrWDVxbkhCaUhYczhDb2Y3d0FUTzFETzZPSkUgAToDCgEwQgIIAEoICJDjsKcDEAE (accessed on 4 February 2025).
24. Cristian, I. Adaptarea la Schimbările Climatice, (Raport de Cercetare). 2022. Available online: https://www.platformademediu.ro/CAP5_Adaptarea_la_schimbarile_climatice_in_municipiul_Bucuresti.pdf (accessed on 10 February 2025).
25. Strik, B.C. Flowering and Fruiting on Command in Berry Crops. In Proceedings of the XXVIII International Horticultural Congress on Science and Horticulture for People (IHC2010): International Symposium, Lisbon, Portugal, 14 December 2012; p. 926. [CrossRef]
26. Balan, V.; Sava, P.; Calalb, T.; Ciorchină, N.; Cumpanici, A.; Dodica, D.; Roşca, I.; Todiraș, V.; Zbancă, A. *Culture of Fruit Shrubs and Strawberries Chișinău;* HEKS Publishing House EPER: Chisinau, Moldova, 2017; Within the Project "Improving Productivity and Market Access of Fruit Producers in the Center and North of the Republic of Moldova (AMIB)". Available online: https://agrobiznes.md/wp-content/uploads/2018/03/Manual-didactic-Cultura-arbustilor-fructiferi-si-a-capsunului.pdf (accessed on 4 February 2025).
27. Wu, H.; Li, J.; Tian, L.; Zhao, F.; Yin, J.; Shao, Y. pH-Sensitive Phosphorescence in Penicillamine-Coated $Au_{22}$ Nanoclusters: Theoretical and Experimental Insights. *Chem. Eng. J.* **2024**, *495*, 153608. [CrossRef]
28. Yang, Z.; Xiong, J.; Zhao, X.; Meng, X.; Wang, S.; Li, R.; Wang, Y.; Chen, M.; He, N.; Yang, Y.; et al. Column-Hemispherical Penetration Grouting Mechanism for Newtonian Fluid Considering the Tortuosity of Porous Media. *Processes* **2023**, *11*, 1737. [CrossRef]
29. Blake, G.R.; Hartge, K.H. Bulk Density. In *Methods of Soil Analysis: Part 1 Physical and Mineralogical Methods*, 2nd ed.; American Society of Agronomy: Madison, WI, USA, 1986; Chapter 13. [CrossRef]
30. Yang, Z.; Liu, Y.; Chen, M.; Wang, X.; Ye, C.; Li, X.; Chen, W.; Yang, Y.; Wang, B.; Li, C.; et al. Influence of Coupling Effects between Gravel Soil Porosity and Cement Grout Weight on Diffusion Laws and Morphologies of Penetration Grouting. *Appl. Sci.* **2022**, *12*, 7601. [CrossRef]
31. Yao, Y.; An, L.K.; Gopi, K.; Chen, S. Role of soil in improving process performance and methane yield of anaerobic digestion with corn straw as substrate. *Energy* **2018**, *151*, 998–1006. [CrossRef]
32. Faloye, O.T.; Ajayi, A.E.; Kamchoom, V.; Akintola, O.A.; Oguntunde, P.G. Evaluating Impacts of Biochar and Inorganic Fertilizer Applications on Soil Quality and Maize Yield Using Principal Component Analysis. *Agronomy* **2024**, *14*, 1761. [CrossRef]
33. Yang, Z.; Chen, M.; Zhang, J.; Ding, P.; He, N.; Yang, Y. Effect of initial water content on soil failure mechanism of loess mudflow disasters. *Front. Ecol. Evol.* **2023**, *11*, 1141155. [CrossRef]
34. Menzel, A.; Sparks, T. Temperature and Plant Development: Phenology and Seasonality. In *Plant Growth and Climate Change;* Wiley: Hoboken, NJ, USA, 2006; Chapter 4; pp. 70–95. [CrossRef]
35. Renquist, A.R.; Hughes, H.G.; Rogoyski, M.K. Combined High Temperature and Ultraviolet Radiation Injuiy of Red Raspberry Fruit. *HortScience* **1989**, *24*, 597–599. [CrossRef]
36. Morecroft, M.D.; Paterson, J.S. Effects of temperature and precipitation changes on plant communities. In *Plant Growth and Climate Change;* Wiley: Hoboken, NJ, USA, 2006; Chapter 7; pp. 146–164. [CrossRef]
37. Körner, C. Significance of temperature in plant life. In *Plant Growth and Climate Change;* Wiley: Hoboken, NJ, USA, 2006; Chapter 3; pp. 48–69. [CrossRef]
38. Pereira, J.S.; Chaves, M.M.; Caldeira, M.C.; Correia, A.V. Water Availability and Productivity. In *Plant Growth and Climate Change;* Wiley: Hoboken, NJ, USA, 2006; Chapter 6; pp. 118–145. [CrossRef]

39. Grace, J.; Zhang, R. Predicting the effect of climate change on global plant productivity and the carbon cycle. In *Plant Growth and Climate Change*; Wiley: Hoboken, NJ, USA, 2006; Chapter 9; pp. 187–208. [CrossRef]
40. Milošević, T.M.; Glišić, I.P.; Glišić, I.S.; Milošević, N.T. Cane properties, yield, berry quality attributes and leaf nutrient composition of blackberry as affected by different fertilization regimes. *Sci. Hortic.* **2018**, *227*, 48–56. [CrossRef]
41. Deng, Y.; Qiu, l.; Shao, Y.; Yao, Y. Process Modeling and Optimization of Anaerobic Co-Digestion of Peanut Hulls and Swine Manure Using Response Surface Methodology. *Energy Fuels* **2019**, *33*, 11021–11033. [CrossRef]
42. Zhiquan, Y.; Xiangdong, N.; Kepeng, H.; Wei, L.; Yanhui, G. Relationships betweenWater-Cement Ratio and Rheological Characteristics of Power-Law Cement Grouts. *Electron. J. Geotech. Eng.* **2015**, *20*, 5825–5831.
43. Remberg, S.F.; Sønsteby, A.; Aaby, K.; Heide, O.M. Influence of Postflowering Temperature on Fruit Size and Chemical Composition of Glen Ample Raspberry (*Rubus idaeus* L.). *J. Agric. Food Chem.* **2010**, *58*, 9120–9128. [CrossRef]

*horticulturae*

MDPI

*Article*

# Isolation and Characterization of *Pseudomonas* sp. HX1, *Streptomyces luteogriseus* HR40, and *Streptomyces flavofungini* HR77 as Promising Biocontrol Agents Against Verticillium Wilt in Hops Affected by *Verticillium nonalfalfae*

**Seyedehtannaz Ghoreshizadeh [1], Carla Calvo-Peña [1], Marina Ruiz-Muñoz [1], Maja Dobrajc [2], Sebastjan Radišek [2], Juan José R. Coque [1,*] and Rebeca Cobos [1,*]**

1 Instituto de Investigación de la Viña y el Vino, Escuela de Ingeniería Agraria, Universidad de León, 24009 León, Spain; sghore00@estudiantes.unileon.es (S.G.); ccalp@unileon.es (C.C.-P.); mruim@unileon.es (M.R.-M.)

2 Slovenian Institute of Hop Research and Brewing, 3310 Žalec, Slovenia; maja.dobrajc@ihps.si (M.D.); sebastjan.radisek@ihps.si (S.R.)

* Correspondence: jjrubc@unileon.es (J.J.R.C.); rebeca.cobos@unileon.es (R.C.)

**Abstract:** *Verticillium* spp. are soil-borne fungal pathogens capable of infecting over 400 plant species via their root systems, with each species demonstrating a distinct host range. The primary host of *Verticillium nonalfalfae* is the hop plant (*Humulus lupulus*), which becomes affected by Verticillium wilt and, in severe cases, dieback. A dual screening approach was employed to isolate Streptomycetes as potential biocontrol agents from the hop root systems alongside other bacterial species from hop xylem. Three *Pseudomonas* strains from the xylem and three *Streptomyces* strains from the roots were selected based on their in vitro antifungal activity against *Verticillium dahliae* and/or *V. nonalfalfae*. Their potential for controlling Verticillium wilt was further assessed through in planta assays conducted under greenhouse conditions. The results demonstrated that *Pseudomonas* sp. HX1, *Streptomyces luteogriseus* HR40, and *Streptomyces flavofungini* HR77 significantly reduced the disease severity index by 32.56% compared to artificially infected control plants. In contrast, the commercial biocontrol product Serenade® ASO, containing *Bacillus amyloliquefaciens* QST 713, achieved a reduction of 13.96%. These findings underscore the potential of the selected bacterial strains as promising candidates for the biological control of Verticillium wilt in hops affected by *V. nonalfalfae*.

**Keywords:** antifungal activity; biocontrol; hops; *Humulus lupulus*; rhizosphere; soil-borne pathogen; *Streptomyces*; *Verticillium nonalfalfae*; Verticillium wilt

## 1. Introduction

The hop plant (*Humulus lupulus*), a dioecious climbing plant of the *Cannabaceae* family, is primarily cultivated for its female inflorescences (hop cones). These cones are rich in secondary metabolites, including bitter acids and essential oils, which are integral to the brewing industry [1]. Hops play a vital role in beer production by imparting bitterness, flavor, and aromatic qualities. Furthermore, they are recognized for their therapeutic applications, such as managing insomnia, anxiety, and digestive disorders, due to their sedative properties [2]. Beyond their use in brewing, hops are utilized in various industries, including the manufacture of natural dyes, cosmetics, and food preservatives [3]. Hops are predominantly cultivated in temperate regions worldwide, with major production hubs in Europe, the United States, and Asia [4].

Verticillium wilt in hops is a vascular disease caused by the soil-borne fungal pathogens *Verticillium nonalfalfae* (formerly *V. albo-atrum sensu lato*) and *V. dahliae*. Of these, *V. nonalfalfae* is the most commonly isolated from hops and accounts for the majority of disease outbreaks. In contrast, infections caused by *V. dahliae* are relatively infrequent and typically less severe [5]. *Verticillium* spp. infect plants through the roots and subsequently disseminate via the vascular (xylem) systems. This infection results in the blockage of vascular vessels, which manifests as symptoms such as chlorosis, leaf curling, wilting, necrosis of lower leaves, and bine swelling. Upon dissection, infected vascular tissues exhibit medium-to-dark-brown discoloration [6,7]. By the time these symptoms become visible, the fungal pathogen has colonized all plant tissues, rendering effective chemical treatments unavailable. Consequently, implementing preventive measures to restrict fungal invasion is essential for managing Verticillium wilt.

Biocontrol agents (BCAs) have emerged as a sustainable and environmentally alternative to chemical fungicides, offering innovative solutions for plant disease management. BCAs consist of beneficial microorganisms that mitigate plant pathogens through diverse mechanisms, including antibiosis, competition for nutrients and physical space, parasitism, and the induction of systemic resistance in host plants [8]. The rhizosphere—a narrow soil region influenced by root exudates and microbial activity—serves as a rich reservoir of potential BCAs due to its remarkable microbial diversity [9]. In recent years, numerous promising BCAs have been identified for the management of Verticillium wilt in various crops, particularly olive trees. These BCAs encompass both fungal [10–14] and bacterial strains [15–19], including different bacterial strains isolated from the olive root endosphere [15,18] or rhizosphere [16,17,19]. Additionally, certain endophytic strains, such as *Pseudomonas aurantiaca* and other *Pseudomonas* spp., have demonstrated efficacy in controlling Verticillium wilt in cotton [20]. However, to the best of our knowledge, there is currently no information available regarding the use of BCAs for the management of Verticillium wilt in hops. Among the BCAs studied, *Streptomyces* spp. stand out as promising candidates for the biocontrol of soil-borne fungal pathogens [21]. The genus *Streptomyces*, comprising 743 validly named species (https://lpsn.dsmz.de/genus/streptomyces, accessed on 24 April 2025), is one of the most taxonomically diverse bacterial genera. *Streptomyces* spp. are renowned for their prolific production of antibiotics [22] and their ability to suppress a wide range of plant diseases, including those caused by soil-borne fungi [23,24]. Furthermore, *Streptomyces* spp. are prevalent across various soil types, representing approximately 10% of the total soil microbiome [25]. Several *Streptomyces* strains have demonstrated antifungal (AF) activity against *Verticillium* wilt in various crops such as cotton, potato, strawberry, tomato, and olive [19,26]. However, no data are currently available regarding the biocontrol potential of *Streptomyces* spp. against *Verticillium nonalfalfae*. Another promising strategy for managing soil-borne fungal pathogens capable of colonizing the xylem of hop plants is the isolation and characterization of bacterial endophytes with AF activity. Endophytic bacteria, which reside within plant tissues, are efficient colonizers that can enhance plant immune responses and establish symbiotic relationships with their host plants without inducing disease symptoms [27,28].

This study, therefore, aimed to evaluate the AF activity of putative BCAs isolated from the root systems of hop plants, including both rhizosphere-associated and endophytic strains, against *Verticillium nonalfalfae*. Additionally, the ability of these strains to suppress pathogen growth was investigated through both soil and in planta analyses. These findings have the potential to contribute to the development of sustainable alternatives to chemical fungicides, thereby promoting the health and productivity of hop plants.

## 2. Materials and Methods

### 2.1. Isolation of Culturable Streptomycetes Strains from the Rhizosphere of Hop Plants

Bacterial strains were isolated from the rhizosphere of hop plants (Nugget cultivar) cultivated under conventional agronomic practices in a field located in León, Spain (5°35′26.45″ W, 42°35′2.524″ N). Rhizosphere soil samples were collected from five individual plants by carefully excavating the surrounding root systems. Soil in direct contact with the roots was extracted using a sterile spatula, transferred into sterile 50 mL Falcon tubes, and stored in an icebox prior to preservation at 4 °C until further processing. Culturable Streptomycetes were isolated on starch casein agar (SCA) [29] and International *Streptomyces* Project 2 (ISP2) agar medium [30] following established protocols [19]. Plates were incubated at 30 °C for 5–7 days. Isolates were selected based on distinct morphological and cultural characteristics, including colony morphology, presence or absence of aerial mycelia, spore mass coloration, reverse colony pigmentation, and the production of diffusible pigments. Selected isolates were routinely cultured and maintained on MEY (Maltose Yeast Extract, Condalab, Madrid, Spain) medium [31] at 4 °C. Spore-producing isolates were preserved as spore suspensions in 40% glycerol at −20 °C.

### 2.2. Isolation of Culturable Streptomycetes from the Endosphere of Hop Roots

Endophytic *Streptomyces* strains were isolated from hop root tissues using the methodology described by Álvarez-Pérez et al. (2017) [32]. Sterilized root tissues were mechanically disrupted with a mortar and pestle, following by the addition of 20 mL of 0.9% ($w/v$) NaCl solution. The resulting samples were incubated at 28 °C with vigorous agitation (220 rpm) to facilitate the extraction of endophytic microorganisms. Serial tenfold dilutions were plated on SCA and ISP2 agar plates, which were incubated under the previously outlined conditions.

### 2.3. Isolation of Culturable Bacterial Strains from the Xylem of Hop Plants

Samples were collected from hop plants (variety Celeia) displaying visual symptoms of Verticillium wilt in a *V. nonalfalfae*-infested field located in the Savinja Valley, Slovenia (GPS coordinates: Y 50°61′65.424/X 12°39′75.556). Each sample consisted of 30 cm sections of bines taken from the basal part of symptomatic plants. Surface sterilization was performed using 2% sodium hypochlorite for 3 min to eliminate epiphytic fungi and bacteria, followed by thorough rinsing with sterile distilled water. The epidermis of the bines was removed with a sterile scalpel, and four 1 cm pieces of xylem tissue per bine were excised and placed onto potato dextrose agar (PDA) plates (Condalab). After 4 days of incubation at 25 °C in the dark, bacterial colonies that inhibited the growth of *V. nonalfalfae* from the xylem tissue were transferred to tryptic soy agar (TSA) plates (Condalab). All isolates were preserved in Luria-Bertani (LB) broth (Condalab) containing 20% glycerol and stored at −80 °C.

### 2.4. In Vitro Selection of Isolates Based on Antifungal Activity in Plate Assays

The AF activity of the isolated bacterial strains was assessed using in vitro plate assays, as described by Calvo-Peña et al. (2023) and Álvarez-Pérez et al. (2017) [19,32]. Briefly, bacterial isolates were inoculated onto MEY agar plates within a 1.0 cm$^2$ area, with four isolates per plate positioned 1 cm from the edge. An agar plug (0.5 cm in diameter) containing the *V. nonalfalfae* (hop pathotype PV1, isolate T2) [33] was placed at the center of each plate. The plates were incubated at 25 °C for up to 12 days, and growth inhibition zones were measured. To quantify the AF activity of the most effective isolates, individual bacterial strains were inoculated on MEY agar plates in a circular pattern, 1 cm from the plate's edge. A *V. nonalfalfae* T2 agar plug was placed at the center, and the plates were incubated at 25 °C for up to 10 days. The inhibition index (*I* index) was calculated using the formula *I* index (%) = [(Rc − R)/Rc] × 100, where R represents the radius of the fungal colony in the presence of the bacterial isolate and Rc represents the maximum radius of the fungal colony in the control condition. Additionally, AF activity was evaluated against *V. dahliae* due to its significance as a pathogen affecting hop plants and other crops. All assays were performed in triplicate for each bacterial strain tested.

### 2.5. Preliminary Identification of Bacterial Strains by 16S rRNA and rpoD Gene Sequencing

Bacterial strains exhibiting the highest AF activity were identified through partial sequencing of the 16S rRNA gene. Genomic DNA was extracted following the protocol outlined by Hopwood et al. (1985) [34]. Amplification of the *16S rRNA* gene was performed using universal primers 27F and 1492R [35]. The resulting sequences were compared against reference sequences of type strains available in the EzTaxon-e database [36] (http://www.ezbiocloud.net/eztaxon/identify, accessed on 9 January 2025). Sequence alignments were conducted using MEGA11 software (http://www.megasoftware.net/, accessed on 10 January 2025). For species-level identification of *Pseudomonas* isolates, an internal fragment of the *rpoD* gene was sequenced in accordance with the methodology recommended by Girard et al. (2020) [37].

### 2.6. Identification of Streptomyces Isolates Through Multilocus Sequence Analysis (MLSA)

MLSA was employed to identify *Streptomyces* species based on five housekeeping genes: *atpD* (ATP synthase F1, β-subunit), *gyrB* (DNA gyrase B subunit), *recA* (recombinase A), *rpoB* (RNA polymerase, β-subunit), and *trpB* (tryptophan synthase, β-subunit) [38]. Gene amplification was performed using primers and conditions specified by Guo et al. (2008) [39] and Rong et al. (2009) [40]. A phylogenetic tree was constructed utilizing concatenated sequences from the five housekeeping genes. Sequences were manually trimmed to uniform lengths prior to alignment, which was carried out using MEGA11 software [41]. Reference sequences of type strains were retrieved from the *Streptomyces* Sequence Database maintained by the Buckley Lab at Cornell University (https://blogs.cornell.edu/buckley/streptomyces-sequence-database/, accessed on 19 January 2025). Phylogenetic relationships were inferred through the maximum likelihood method, applying the Kimura two-parameter (K2P) model to estimate nucleotide sequence divergence [42]. MLSA evolutionary distances were calculated using MEGA11 software, also employing the K2P model to estimate nucleotide sequence divergence [42]. As defined by Rong and Huang (2012) [38], strain pairs demonstrating an MLSA evolutionary distance of $\leq$0.007 were considered conspecific.

### 2.7. Genome Sequencing and Biosynthetic Gene Cluster Analysis

The genomes of strains HR31 and HR77 were sequenced using next-generation sequencing on the PacBio Revio platform provided by Macrogen Inc. (Seoul, Republic of Korea). De novo assembly was performed using Canu version 2.2 [43], and assembly quality assessed via QUAST-LG version 5.3.0 [44]. Genome annotation was conducted using the RAST-SEED webserver [45] (https://rast.nmpdr.org/, accessed on 12 February 2025). Species identification was further refined through digital DNA:DNA hybridization (dDDH) utilizing the Genome-To-Genome Distance Calculator (GGDC) version 2.1 [46], in conjunction with the Type (Strain) Genome Server (TYGS) v1.0 (https://tygs.dsmz.de/, accessed on 13 February 2025), a bioinformatics platform for whole-genome taxonomic analysis [47]. Biosynthetic gene cluster (BGC) identification in the genome sequences of HR31 and HR77 was performed using antiSMASH version 7.1.0 [48], with annotation results manually inspected for confirmation and refinement.

The genome sequences of strains HR31 and HR77 were deposited in the GenBank database under the accession numbers SAMN47195979 and SAMN47195978, respectively, with the associated BioProject accession number PRJNA1230872.

### 2.8. Plant Assays to Evaluate the Effectiveness of Selected Bacterial Isolates Against Verticillium Wilt Caused by V. nonalfalfae

The efficacy of selected bacterial isolates (HX1, HX2, HX3) and *Streptomyces* isolates (HR40, HR77, HR31) against Verticillium wilt caused by *V. nonalfalfae* was assessed through a pot experiment conducted under controlled conditions. One-year-old hop plants (variety Celeia, susceptible to Verticillium wilt) were used as test subjects. The soil was artificially contaminated with *V. nonalfalfae* (hop pathotype PV1, isolate T2). The experiment setup included the commercial bacterial fungicide Serenade® ASO (Bayer Crop Science, Leverkusen, Germany), containing *Bacillus amyloliquefaciens* (QST 713), as well as a negative control consisting of plants cultivated in infected soil without BCA treatment.

The *V. nonalfalfae*-contaminated soil was prepared by autoclaving field soil at 121 °C for 30 min, followed by inoculation with a mixture of conidial and mycelium suspension ($1 \times 10^7$ conidia mL$^{-1}$ and $1 \times 10^2$ mycelium fragments mL$^{-1}$) at a ratio of 100 mL per liter of soil. This resulted in a final concentration of $1 \times 10^6$ conidia g$^{-1}$ of soil and 10 mycelium fragments g$^{-1}$ of soil. To enhance contamination, pieces of *V. nonalfalfae*-infected hop bines (3–5 cm in length) were added at a dose of 100 mL per liter of soil.

*Streptomyces* isolates were cultivated in MEY or ISP2 liquid media in baffled flasks shaken at 120 rpm at 30 °C for 7 days. Similarly, HX1, HX2, and HX3 isolates were cultured in LB broth under identical conditions. Following incubation, the liquid cultures were centrifuged at 6000 rpm for 10 min in 50 mL tubes. The supernatant was discarded, and the resulting pellets were resuspended in an equal volume of sterile distilled water. For Serenade® ASO, a 1% solution was prepared using sterile distilled water. To estimate microbial concentrations, 1 mL of each microbial suspension was serially diluted tenfold in sterile water. Aliquots of 0.1 mL from each dilution were inoculated onto TSA (for HX1, HX2, HX3) and MEY agar (for *Streptomyces* strains) and incubated at 28 °C for 3 days. For inoculation, 10 plants per treatment were uprooted, and their roots cleaned with tap and distilled water. The roots were subsequently dipped into microbial suspensions for 15 min. Control plants were treated with sterile distilled water. The treated plants were transplanted into 2 L pots filled with *V. nonalfalfae*-contaminated soil, except for the negative control, which received sterile soil. To promote colonization by *Streptomyces* strains, five agar plugs containing sporulated mycelium were placed near the root systems in each pot.

Plants were maintained in a walk-in growth chamber (RK-13300, Kambič d.o.o., Semič, Slovenia) under controlled environmental conditions, with a 13 h light/11 h dark photoperiod provided by fluorescent light, temperatures set at 22 °C during the light period and 15 °C during the dark period, and 70% relative humidity. Upon the appearance of the first symptoms, plants were assessed weekly for the severity of foliar symptoms over a four-week period using a scale from 0 to 5 [49], where 0 = no leaf symptoms, 1 = 1–20% of leaf area wilted, 2 = 21–40%, 3 = 41–60%, 4 = 61–80%, and 5 = 81–100%. Following the final foliar assessment, the presence of the fungal pathogen was confirmed by its re-isolation from roots and stem tissues. Plants in which fungal infection was not confirmed were excluded from scoring. The disease severity index (DSI) for each treatment was determined as the mean wilt score of infected plants at each time point of assessment. The Area Under the Disease Progress Curve (AUDPC) was calculated based on the DSI values recorded during the four-week evaluation period using the following formula:

$$\text{AUDPC} = \sum_{i=1}^{n-1} \frac{(\text{DSI}_i + \text{DSI}_{i+1})}{2} \times (t_{i+1} - t_i)$$

where $n$ represents the total number of plants evaluated and $t$ corresponds to the time (in days post-inoculation) at the $i$th assessment. The AUDPC provides a cumulative measure of disease progression, taking into account both the severity of the symptoms and their duration over the evaluation period. This approach facilitates comparisons of disease progression across treatments and aids in assessing the effectiveness of the bacterial isolates in reducing disease severity [50].

### 2.9. Confrontation Assays

Potential negative interactions among the selected *Streptomyces* strains, as well as between *Streptomyces* and *Pseudomonas* strains, were evaluated using pairwise bioassays following the methodology described by Schrey et al. (2012) [51]. *Streptomyces* cultures were grown in ISP2 medium at 28 °C for three days, while *Pseudomonas* spp. were cultivated in Nutrient Broth (Condalab) at 30 °C for one day. For the bioassay, a 40 µL aliquot of the tester strain suspension was applied in a linear pattern along the lower section of a MEY agar plate. Once the tester strain had developed—marked by the onset of sporulation in *Streptomyces* strains—three parallel lines of the receiver strains were inoculated perpendicularly to the tester line. Each strain combination was evaluated in triplicate. The effect of the tester strain on the growth of the receiver strain, including substrate mycelium formation and sporulation (specifically for *Streptomyces* strains), was assessed at the onset of sporulation in the control cultures.

### 2.10. Analysis of AF Activity in Small-Scale Soil Tests

To assess AF activity in soil, small-scale experiments were conducted using 10 mL Falcon tubes, each containing 1 g of sterile soil, as previously described [19]. Spores of the tested strains were obtained by culturing on MEY agar plates at 28 °C for 6–8 days. The spores were collected in a 20% glycerol solution and subsequently quantified through tenfold serial dilutions followed by plating on MEY agar plates. The soil samples were inoculated with 100 µL of a bacterial suspension supplemented with 1% starch and 0.5% casein, resulting in a final concentration of $10^6$ CFU g of soil$^{-1}$. Homogenization was achieved through vortex mixing and manual stirring with a spatula to ensure uniform inoculum distribution. The samples were incubated at 30 °C for 48 h. Subsequently, $2 \times 10^6$ spores g of soil$^{-1}$ of *V. nonalfalfae* were added to each tube, followed by homogenization as described above. The study included a non-inoculated negative control and a positive control inoculated only with the fungal pathogen.

The tubes were incubated at 25 °C, and soil samples (100 mg) were collected 7 and 14 days after bacterial inoculation. Tenfold serial dilutions of the collected samples were prepared in sterile water, and 0.1 mL aliquots of each dilution were plated on PDA plates supplemented with 250 µg mL$^{-1}$ chloramphenicol (Sigma-Aldrich, St. Louis, MO, USA) to quantify fungal growth. Additionally, SCA plates supplemented with 100 µg mL$^{-1}$ pimaricin (VGP Pharmachem, Vic, Spain) and 50 µg mL$^{-1}$ nalidixic acid (Sigma-Aldrich) were used to quantify the presence of the tested actinobacteria.

### 2.11. Soil Physicochemical Properties

Soil samples were collected from a depth of 20–30 cm, after the removal of the top 5 cm of superficial soil. Five samples were taken and combined for analysis. Soil texture and mechanical analyses were performed according to the standard sedimentation method, involving particle size distribution through a combination of sieving and sedimentation techniques (ISO 11277:2020) [52]. Organic carbon content was determined using the Walkley–Black wet oxidation method. The pH and electrical conductivity (EC) were measured in the supernatants of soil/water (1:2.5) suspensions after 25 min of shaking and 5 min of soil settlement, using a Crison micropH 2001 pH meter and a Crison 522 Conductivity Meter, respectively (Crison Instruments S.A., Alella, Spain). Soil nitrogen was determined by the Kjeldahl method [53], which quantifies all soil nitrogen except that in nitrates [54]. Available soil phosphorus was determined by ultraviolet–visible spectroscopy following treatment with the Olsen–Watanabe extractant [55]. Exchangeable potassium was extracted with successive aliquots of 1 M ammonium acetate and quantified by atomic absorption spectrometry (AAS) with a Unicam Model SOLAAR 969 (Unicam Ltd., Cambridge, UK). Micronutrients (Fe, Cu, Mn, and Zn) were determined using AAS in an air–acetylene flame [56]. Magnesium was analyzed using the $CaCl_2$ method [57] determined by AAS. Available boron was measured according to standard NF X31-122 [58], and humus content according to the ISO 14235:1999 [59]. Total carbonates and active carbonates were quantified using the Bernard calcimeter method, with active carbonates extracted using 0.2 N ammonium oxalate (Sigma-Aldrich).

### 2.12. Statystical Analysis

Data on Verticillium wilt severity in the in planta assays and the AF activity of the selected bacterial strains in the small-scale soil tests were statistically analyzed using R software version 4.4.3 [60]. The assumption of normality was assessed using the Shapiro–Wilk test (via the "stats" package), while homogeneity of variances was evaluated using Levene's test (via the "car" package) [61]. Based on the outcomes of these preliminary tests, the non-parametric Kruskal–Wallis test (from the "stats" package) was employed to evaluate significant differences between treatment and control groups. When significant effects were detected ($p < 0.05$), post hoc pairwise comparisons were conducted using Dunn's test with Bonferroni correction (via the "FSA" package version 0.9.5) [62] to control for Type I error. Statistical significance was defined as an adjusted $p$-value < 0.05. To aid in the visual interpretation of significant differences among treatments, grouping letters were assigned according to adjusted $p$-values using the "multcompView" package version 0.1–8 [63].

## 3. Results

### 3.1. Isolation of Culturable Streptomycetes from the Root System and Other Endophytic Bacterial Strains from the Xylem of Hop Plants and Their Antifungal Activity

A total of 100 putative Streptomycetes strains were isolated from the hop rhizosphere (designated HR1–HR100) based on their distinct morphological and cultural characteristics, including the color of substrate mycelium, aerial mycelium, exospores, and pigment production. An in vitro bioassay-based screening revealed that 37% of these isolates exhibited AF activity against *V. nonalfalfae*, while 35% demonstrated activity against *V. dahliae*.

Similarly, 85 endophytic strains (designated HE1–HE85) were isolated from the hop endosphere. Among these, 13 isolates (15.29%) showed AF activity against *V. dahliae*, whereas 16 isolates (18.8%) exhibited AF activity against *V. nonalfalfae*.

In the case of bacterial isolation from the xylem of hop plants infected with *V. nonalfalfae*, three isolates (HX1, HX2, and HX3, named after "Hop Xylem") were obtained. These strains clearly inhibited the growth of *V. nonalfalfae* developing on agar plates from xylem tissue fragments of hop infected bines. After selection, their AF activity on *V. nonalfalfae* was confirmed by in vitro bioassay tests.

The six Streptomycete strains demonstrating the highest AF activity in in vitro assays, along with the three strains isolated from the xylem, were individually tested to calculate their inhibition index (I index) against both *V. dahliae* and *V. nonalfalfae* (Figure 1). The I index values for *V. dahliae* ranged from 12.5% (strain HX1) to 100% (strain HX3). In the case of *V. nonalfalfae*, the I index values varied between 15% (strain HX1) and 100% (strains HR40, HR77, and HR31) (Figure 1).

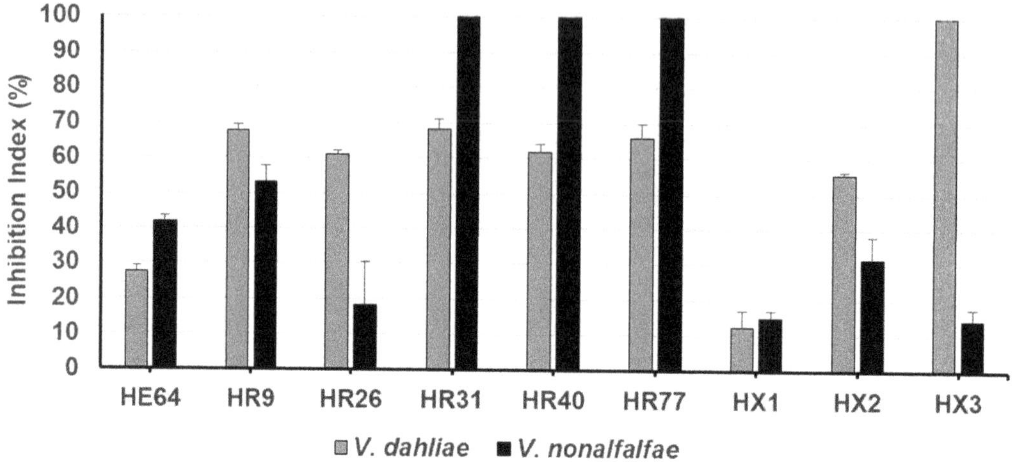


**Figure 1.** Antifungal activity of promising strains isolated from root endosphere (HE), hop rhizosphere (HR), and xylem of hop plants (HX) against *V. dahliae* and *V. nonalfalfae*. Values presented are means of two independent experiments, each conducted in triplicate. Error bars indicate standard deviation (SD) of mean.

### *3.2. Molecular Identification of the Selected Isolates by Partial Sequencing of 16S rRNA and rpoD Genes and Multilocus Sequence Analysis (MLSA)*

Partial sequencing of the 16S rRNA gene of the six selected Streptomycete strains isolated from hop roots identified them as members of the *Streptomyces* genus. Given the high genetic complexity of the *Streptomyces* genus and the limited resolution of 16S rRNA sequencing for species differentiation, an MLSA was conducted using five housekeeping genes (*atpD*, *gyrB*, *recA*, *rpoB*, and *trpB*). The GenBank accession numbers of the DNA sequences corresponding to these housekeeping genes are provided in Table 1.

**Table 1.** GenBank accession numbers of DNA sequences corresponding to partially sequenced genes used in MLSA analysis of different *Streptomyces* sp. strains characterized.

| Strain | GenBank Accession Numbers | | | | | |
|---|---|---|---|---|---|---|
| | **16S rRNA** | ***atpD*** | ***gyrB*** | ***recA*** | ***rpoB*** | ***trpB*** |
| HE64 | PQ638391 | PQ654892 | PQ654898 | PQ654886 | PQ654904 | PQ654910 |
| HR9 | PQ638389 | PQ654890 | PQ654896 | PQ654884 | PQ654902 | PQ654908 |
| HR26 | PQ638386 | PQ654887 | PQ654893 | PQ654881 | PQ654899 | PQ654905 |
| HR31 | PQ638390 | PQ654891 | PQ654897 | PQ654885 | PQ654903 | PQ654909 |
| HR40 | PQ638387 | PQ654888 | PQ654894 | PQ654882 | PQ654900 | PQ654906 |
| HR77 | PQ638388 | PQ654889 | PQ654895 | PQ654883 | PQ654901 | PQ654907 |

Once the partial sequences of the housekeeping genes were obtained, a phylogenetic tree was generated (Supplementary Material, Figure S1) and the corresponding MLSA evolutionary distances were calculated (Table 2).

**Table 2.** Identification of selected isolates by MLSA analysis and evolutionary distances from phylogenetically nearest strains.

| Strain | MLSA (Kimura 2-Parameter) Distance * | | | | | |
|---|---|---|---|---|---|---|
| | HE64 | HR9 | HR26 | HR31 | HR40 | HR77 |
| *Streptomyces* sp. HE64 | | | | | | |
| *Streptomyces* sp. HR9 | 0.1032 | | | | | |
| *Streptomyces* sp. HR26 | 0.1033 | 0.1184 | | | | |
| *Streptomyces* sp. HR31 | 0.1105 | 0.1050 | 0.1129 | | | |
| *Streptomyces* sp. HR40 | 0.0663 | 0.1152 | 0.1053 | 0.1123 | | |
| *Streptomyces* sp. HR77 | 0.0869 | 0.1170 | 0.0829 | 0.1144 | 0.0910 | |
| *Streptomyces flavofungini* NRRL B12307 | 0.1092 | 0.0958 | 0.0919 | 0.1134 | 0.1144 | 0.0160 |
| *Streptomyces luteogriseus* NRRL B12422 | 0.0672 | 0.1157 | 0.1048 | 0.1133 | 0.0040 | 0.0901 |
| *Streptomyces rectiviolaceus* NRRL B16374 | 0.1171 | 0.1214 | 0.0646 | 0.1152 | 0.1125 | 0.0968 |
| *Streptomyces robefuscus* NRRL B5743 | 0.1034 | 0.0020 | 0.1205 | 0.1164 | 0.1147 | 0.1231 |
| *Streptomyces rubiginosohelvolus* CGMCC 4.0127 | 0.0329 | 0.1069 | 0.1047 | 0.1086 | 0.0725 | 0.0860 |
| *Streptomyces sirii* CCTCC AA2024094 | 0.1259 | 0.1156 | 0.1340 | 0.0020 | 0.1276 | 0.1348 |

* Strain pairs having ≤0.007 MLSA evolutionary distance (highlighted by shading) were considered conspecific based on guideline empirically determined by Rong and Huang in 2012 [38].

The analysis revealed that the endophytic isolate HE64 is a putative new species closely related to *Streptomyces rubiginosohelvolus* (Figure 2A) with an MLSA evolutionary distance of 0.0329. HR9 was identified as *Streptomyces robefuscus* (Figure 2B) with an MLSA evolutionary distance of 0.002 (less than 0.007). Isolate HR26 is a putative new species in the *Streptomyces* genus closely related to *Streptomyces rectiviolaceus* (Figure 2C), with an MLSA evolutionary distance of 0.0646, whereas HR31 was identified as *Streptomyces sirii* (Figure 2D), exhibiting an MLSA evolutionary distance of 0.002. The HR77 strain is closely related to *Streptomyces flavofungini* (Figure 2E) with an MLSA evolutionary distance of 0.016, suggesting it could represent a new species within the genus. Finally, HR40 was assigned to *Streptomyces luteogriseus* (Figure 2F), with an MLSA evolutionary distance of 0.004.

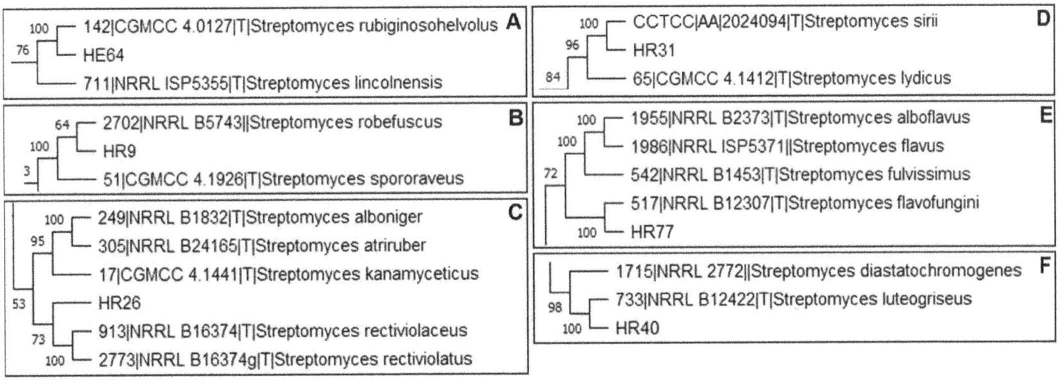


**Figure 2.** Partial details of the Neighbor-Joining tree generated from the MLSA analysis. The figure illustrates the positions of the selected strains: *Streptomyces* sp. HE64 (**A**), *Streptomyces* sp. HR9 (**B**), *Streptomyces* sp. HR26 (**C**), *Streptomyces sirii* HR31 (**D**), *Streptomyces flavofungini* HR77 (**E**), and *Streptomyces luteogriseus* HR40 (**F**). For the complete tree, refer to Supplementary Materials, Figure S1.

Partial sequencing of the 16S rRNA gene confirmed that the three bacterial strains isolated from hop xylem belonged to the *Pseudomonas* genus. However, only the HX3 strain could be identified at the species level. Specifically, isolate HX3 was identified as *P. putida*, as the *rpoD* partial sequence showed a 99.83% homology to the *P. putida* JCM 20005 strain

(GenBank accession number LC752193.1). This value exceeds the 98.0% cutoff defined for differentiating *Pseudomonas* spp. strains at the species level using this molecular marker [37]. HX1 and HX2 are closely related to *P. fluorescens*, with the *rpoD* partial sequence showing 97.06% and 97.03% homology to *P. fluorescens* YK-310 strain (GenBank accession number CP071797.1), respectively.

Since the analysis of the AF activity of the isolates HR40, HR77, and HR31 confirmed its very high effectiveness in inhibiting the in vitro growth of *V. nonalfalfae* (I index values of 100 were obtained for all three strains, as previously shown in Figure 1), we decided to analyze their genomes to elucidate the potential production of AF compounds.

### 3.3. Genome Analysis of Streptomyces Flavofungini HR77 Strain

The MLSA analysis previously established that strain HR77 shared close similarity with *Streptomyces flavofungini*, exhibiting a genetic distance of 0.016. Since the species delimitation threshold within the *Streptomyces* genus based on MLSA is defined as 0.007 [38], this initial finding suggested that strain HR77 could potentially represent a novel species. Furthermore, due to the limited genomic information available for *S. flavofungini*, whole-genome sequencing was conducted to explore potential AF compounds biosynthesized by this strain.

For strain HR77, de novo genome assembly yielded a single contig comprising 9,363,350 bp with a GC content of 72.17%, differing marginally by 0.09% from the type strain *S. flavofungini* JCM 4753. Species identification was further corroborated through TYGS analysis, which revealed a d4DDH value of 95.3%, surpassing the 70% threshold for species delineation [64]. These results underscore the significance of genome-wide methodologies in accurate taxonomic classification. Additionally, quality assessment using QUAST indicated that 95.66% of the HR77 genome aligned to the type strain reference genome, with an average sequencing depth of 71×, indicating high-quality genome coverage.

The antiSMASH analysis identified a total of 29 putative biosynthetic gene clusters (BGCs) in *S. flavofungini* HR77 (Supplementary Material, Table S1). These included six polyketide synthase (PKS) clusters (PKS I and PKS II), four nonribosomal peptide synthetase (NRPS) clusters, five terpene clusters, two siderophore clusters, two ribosomally synthesized and post-translationally modified peptide (RiPP-like) clusters, and ten hybrid clusters. Among these, ten exhibited a high degree of similarity ($\geq$80%) to known clusters, with four clusters—responsible for the biosynthesis of albaflavenone, ectoine, geosmin, and isorenieratene—demonstrating 100% similarity. The siderophores griseobactin and coelichelin showed similarities of 92% and 81%, respectively. Additionally, one BGC exhibited 86% similarity to the cluster involved in the biosynthesis of undecylprodigiosin, an antimicrobial compound known to disrupt bacterial membranes [65,66].

In relation to the putative AF compounds produced by HR77, the genome was found to harbor a PKS cluster exhibiting 90% similarity to spectinabilin, a polyketide noted for its diverse derivatives with nematicidal and AF properties [67]. Moreover, several BGCs displayed lower similarity to entries in current databases yet were associated with reported AF activity. Notably, an NRPS-like cluster exhibited 60% similarity to desertomycin, a compound with established known AF properties [68]. Another BGC showed 25% similarity to the cluster responsible for producing 67-121C, a macrolide polyene complex with documented AF activity [69]. Additionally, two clusters displayed low similarity—24% and 13%—to kutzneride 2, a compound also recognized for its AF activity [70].

### 3.4. Genome Analysis of Streptomyces luteogriseus DSM 40483 Type Strain

The MLSA analysis confirmed the classification of strain HR40 within the *Streptomyces luteogriseus* species. A search of available genomic data revealed the existence of 23 fully sequenced genomes of this species in the NCBI database (accessed 28 February 2025). Consequently, sequencing the genome of strain HR40 was deemed unnecessary, as the existing datasets were considered sufficient to capture the genetic diversity of the species. Instead, an antiSMASH analysis was performed on the genome of the *S. luteogriseus* DSM 40483 type strain. This analysis identified 30 putative BGCs, including eight terpene clusters, alongside a diverse array of PKS, NRPS, RiPP-like, and siderophore-related pathways (Supplementary Material, Table S2). Of these, ten BGCs displayed high similarity to known clusters. These included those responsible for synthesizing albaflavenone, citrulassin D, coelichelin, desferrioxamine B/E, ectoine, flaviolin, geosmin, and informatipeptin, all showing 100% similarity. Additionally, clusters responsible for the biosynthesis of hopene and a spore pigment demonstrated 92% and 83% similarity, respectively. Notably, one BGC exhibited 88% similarity to oligomycin, a macrolide antibiotic with proven AF activity [71–73].

### 3.5. Genome Analysis of Streptomyces sirii HR31 Strain

As previously noted, initial MLSA analysis suggested that strain HR31 could represent a new species, leading to the decision to sequence its genome. However, during the preparation of this manuscript, a new species, *S. sirii*, was described [74].

Whole-genome sequencing of strain HR31 was performed, resulting in a de novo assembly comprising a single contig of 8,600,360 bp with a GC content of 71.46%. Digital DNA-DNA hybridization analysis revealed a d4DDH value of 94.2% and a G + C content difference of 0.02% in comparison to the *S. sirii* BP-8 type strain. Alignment with the reference genome indicated genome coverage of 95.80%, with an average sequencing depth of 38×. These findings unequivocally identified strain HR31 as *S. sirii*.

AntiSMASH analysis of the genome of *S. sirii* HR31 identified 37 putative BGCs (Supplementary Material, Table S3). These included seven PKS clusters (encompassing both PKS I and PKS II), five NRPS clusters, six terpene clusters, three siderophore clusters, three RiPP-like clusters, and thirteen hybrid clusters. Among these, eight BGCs demonstrated high similarity (≥80%) to known clusters, including ectoine (100%), isomigrastatin (100%), SapB (100%), isorenieratene (85%), a spore pigment cluster (83%), lagmysin (80%), and two siderophores—legonoxamine A-B/desferrioxamine B and ethylenediaminesuccinic acid hydroxyarginine (EDHA)—both with 100% similarity. Despite these findings, none of these metabolites have been reported to exhibit direct AF activity, although *S. sirii* HR31 demonstrated strong AF activity against *Verticillium* spp.

### 3.6. In Planta Efficacy of Selected Bacterial Isolates Against Verticillium Wilt Caused by V. nonalfalfae

A pot trial was conducted under controlled growing chamber conditions to assess the efficacy of three bacterial strains isolated from hop xylem (HX1, HX2, HX3) and three *Streptomyces* sp. strains isolated from the root system (HR40, HR77, HR31) against the Verticillium wilt pathogen *V. nonalfalfae*. The commercial bacterial fungicide Serenade® ASO (1% dilution) served as the standard treatment, while sterile water was used as the negative control. Quantification of the prepared microbial inocula revealed varying concentrations (Supplementary Material, Table S4), which were attributed to differences in biomass production processes and organism types. The results indicated a 100% disease incidence across all microbial treatments and the negative control (plants infected but untreated with BCAs), with typical symptoms of Verticillium wilt observed in all treatments. However, at the end of the experiment (42 days post-pathogen inoculation), significantly reduced disease severity was observed in plants treated with *Pseudomonas* sp. isolate HX1 (DSI 2.9) and the *Streptomyces* isolates HR40 (DSI 2.9) and HR77 (DSI 2.9) compared to the negative control (DSI 4.3) (Figure 3A,B). Additionally, plants treated with the commercial fungicide

Serenade® ASO displayed a lower DSI value (3.7) in comparison to the control plants. However, no statistically significant differences were detected between the Serenade® ASO treatment and the negative control (Figure 3B).

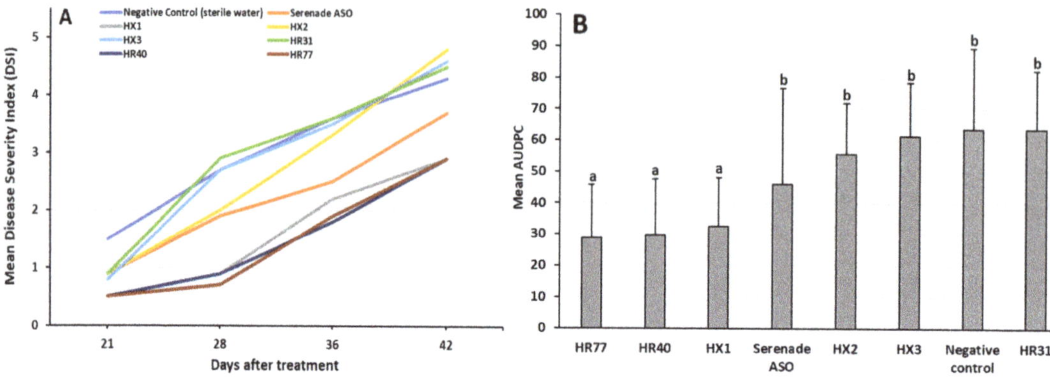


**Figure 3.** (**A**) The progression of the disease severity index (DSI) over time in *Humulus lupulus* cv. Celeia plants treated with selected *Pseudomonas* sp. and *Streptomyces* sp. isolates. (**B**) The Area Under the Disease Progress Curve (AUDPC) corresponding to each treatment. Error bars represent the standard deviation (SD) of the mean. Different letters above the bars denote statistically significant differences between treatments (adjusted *p*-value < 0.05). The results are presented as the means of two independent experiments.

These findings demonstrate that treatments incorporating *Pseudomonas* sp. HX1, *Streptomyces* sp. HR40, and *Streptomyces* sp. HR77 yielded a substantial 32.56% reduction in DSI values. In contrast, the commercial treatment with Serenade® ASO achieved a non-significant reduction of 13.96% in DSI values (Figure 3A) when compared to the control plants.

### 3.7. Strain-Specific Patterns of Inhibition in Interaction Bioassays

Based on the findings from the in planta experiments, the strains *Pseudomonas* sp. HX1, *S. luteogriseus* HR40, and *Streptomyces* sp. HR77 demonstrated the highest efficacy in reducing the incidence of Verticillium wilt caused by *V. nonalfalfae* in hop plants. The potential for negative interactions (cross-inhibition) between these strains was assessed to enable future field trials on commercially cultivated plots using these three promising BCAs. Consequently, co-cultured bioassays were performed. As depicted in Figure 4, no evidence of cross-inhibition was observed between the HX1 strain and the HR40 and HR77 strains. However, partial growth inhibition was detected when HR40 served as the "receiver strain" and HR77 as the "tester strain", with an inhibition index of 24.4% (±0.17). This level of negative interaction is considered acceptable for their combined application in field conditions. Overall, these results exclude the presence of any quantitatively significant cross-inhibition between these three strains, thereby supporting their joint application in future field trials to validate their efficacy under real cultivation conditions.

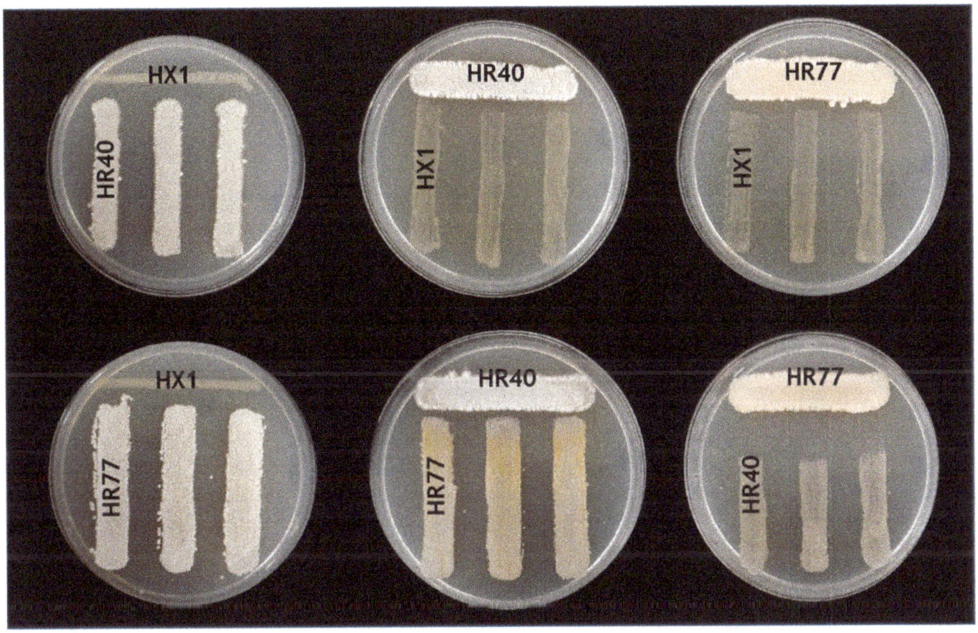


**Figure 4.** Analysis of cross-interactions between HX1, HR40, and HR77 strains. Co-culture bioassays were conducted on Petri dishes, with each interaction replicated three times ($n$ = 9).

### *3.8. Analysis of the Antifungal Activity of the Selected Strains in Small-Scale Soil Tests*

Building upon the results from the preceding co-culture bioassays and considering the low level of negative interaction (<25.0%) observed between strains HR40 and HR77, small-scale soil tests were conducted to evaluate their efficacy. These tests were conducted to evaluate the capacity of the strains, both individually and in combination, to inhibit the development of *V. nonalfalfae*, and to rule out any potential adverse effects on the soil that may occur. Sterile soil samples were inoculated with the HR40 and HR77 strains, either individually or in combination, followed by inoculation with *V. nonalfalfae*. The presence of the pathogen was quantified 7 and 14 days post-inoculation (Figure 5). The soil used in this study was classified as loamy according to the USDA standard classification, with a pH of 7.98, an organic matter content of 6.59%, and a total nitrogen content of 0.32%. These were identified as the most characteristic attributes of the soil, with no additional notable data. A complete description of the physical–chemical properties of the soil is provided in Supplementary Material, Table S5.

As illustrated in Figure 5, at 7 days post-inoculation, the HR40 strain significantly reduced the viable cells of *V. nonalfalfae* in the soil by 99.81%, achieving a three-order-of-magnitude reduction in pathogen levels. The HR77 strain yielded a reduction of 99.58%. When the two strains were applied in combination, a significant difference in pathogen viability was observed, with the combined treatment achieving a reduction of 99.96%. At 14 days post-inoculation, analysis revealed that the HR40 strain reduced pathogen survival by 98.41%, while the reduction achieved by HR77 was quantified at 94.53%. Joint application of both strains resulted in a significant reduction in pathogen survival, estimated at 99.97%.

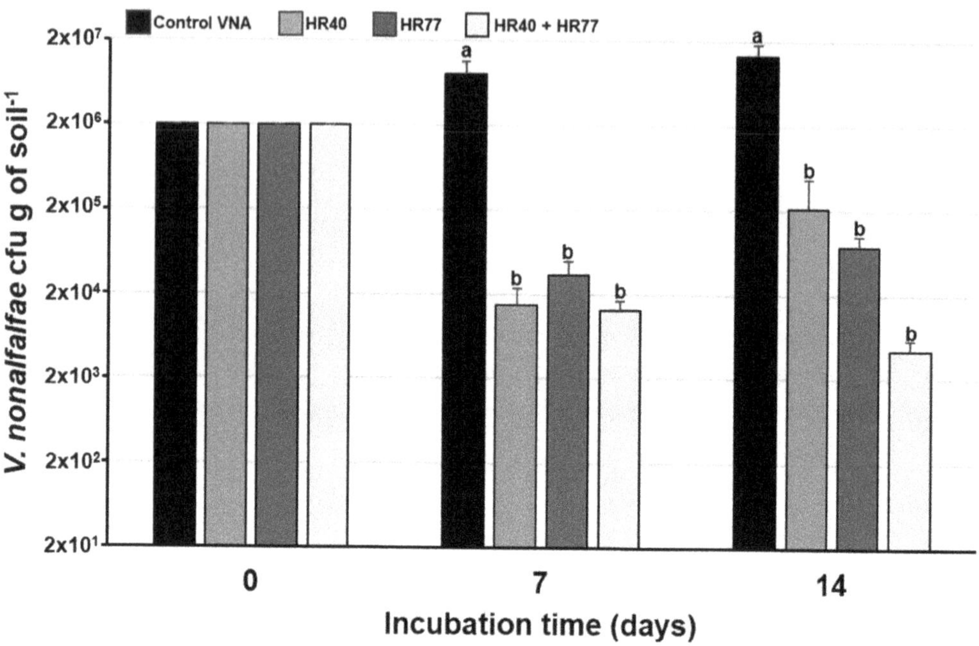


**Figure 5.** A small-scale in vitro soil assay to assess the AF activity of the selected strains HR40 and HR77 against *Verticillium nonalfalfae* (VNA). The values shown represent the means of two independent experiments conducted in triplicate. Error bars indicate the standard deviation (SD) of the mean. Different letters above the bars indicate statistically significant differences between treatments (adjusted $p$-value < 0.05).

## 4. Discussion

To the best of our knowledge, no BCAs have been identified as exhibiting AF activity against *Verticillium* wilt in hops caused by *V. nonalfalfae*. Consequently, this study aims to isolate and select BCAs to combat this destructive disease in hops. To identify potential BCAs, we employed two distinct methodologies. First, bacterial strains from the genus *Streptomyces* were isolated from the endosphere and rhizosphere of the hop plant's root system. *Streptomyces* spp. are widely recognized as promising candidates for the biocontrol of soil-borne fungal pathogens [21,75], with several *Streptomyces* strains previously demonstrating AF activity against *V. dahlia*, which affects a variety of crops [19,26].

Given that soil-borne fungal pathogens such as *Verticillium* spp. partially complete their life cycle in soil and infect plants via the root system, the isolation of BCAs from the root system represents a logical approach. The identification of *Streptomyces* strains in the rhizosphere of hop plants indicates their adaptability to proliferate within soil environments, from which they are subsequently recruited by the plant into the rhizosphere. These strains have the potential to function as a primary barrier, mitigating fungal infection within the plant. Notably, highly effective BCAs targeting Verticillium wilt in olive trees have been isolated from the rhizosphere [16,17,19] or the endosphere of roots [15,18]. Endophytic BCAs, capable of colonizing the interior of the root system, could offer a secondary line of defense if the pathogen breaches the root system. By exploring this dual approach, both rhizosphere-associated and endophytic *Streptomyces* strains exhibiting AF activity against *V. nonalfalfae*—and also *V. dahliae*—were successfully isolated from the hop plant's root system. These findings underscore the significance of both rhizosphere and endosphere environments for the selection of BCAs against various fungal phytopathogens, such as those affecting olive [19] and grapevine [32], among other crops.

A secondary protective strategy involves the isolation of bacterial BCAs from the xylem of plants infected with *V. nonalfalfae*. The capacity of *Verticillium* spp. to infiltrate the xylem of host plants it infects is well established, utilizing this pathway to disseminate towards the plant's upper tissues, including leaves and branches [76]. Endophytic BCAs capable of colonizing the xylem could serve as a tertiary defensive barrier.

Following the selection of the most promising BCAs, it is imperative to identify them at the genus level, or preferably at the species level, as this information could provide insights into the potential AF compounds they produce. MLSA serves as a robust tool for differentiating novel strains within established species of the genus *Streptomyces* and for identifying putative new species within this complex genus. In both this and previous studies, MLSA has proven effective in delineating species-level distinctions and detecting novel species among various rhizosphere and/or endophytic strains isolated from the root environments of olive [19] and grapevine plants [32]. The case of the isolate HR31 is particularly illustrative. Initial MLSA analysis positioned this strain within a subclade containing *Streptomyces lydicus* and *Streptomyces chattanoogensis*, with MLSA evolutionary distances of 0.0837 and 0.0841, respectively. These results indicated that HR31 represented a new *Streptomyces* species. However, in December 2024, a novel species, *Streptomyces sirii*, was identified and characterized from the rhizosphere of bamboo plants [74]. Subsequent MLSA analysis confirmed that the HR31 isolate belonged to this newly described species, as evidenced by an MLSA evolutionary distance of 0.002 (Table 2).

The fact that the three endophytes isolated from the xylem of hop plants belong to the genus *Pseudomonas* suggests a strong capacity of different species within this genus to penetrate the xylem. Indeed, species of *Pseudomonas* spp. have been widely reported as endophytes of different plants [77]. They can colonize plant tissues without eliciting disease symptoms in host plants [27]. The AF activity of different *Pseudomonas* spp. is well documented and is exerted through various mechanisms, such as the production of diffusible compounds [78], volatile organic compounds [79], or siderophores [80]. However, as of now we do not have any direct evidence of the AF mechanism exerted by the *Pseudomonas* sp. HX1 strain.

For the *Streptomyces* strains under investigation, an in silico analysis using antiSMASH enabled the identification of numerous BGCs within their genomes, some of which are implicated in the biosynthesis of metabolites with AF activity. Specifically, for the *Streptomyces flavofungini* HR77 strain, this analysis revealed a BGC with 90% similarity to the AF compound spectinabilin, positioning this compound (or a structurally similar one) as the primary candidate responsible for its AF activity. Moreover, the HR77 genome harbors BGCs with low similarity to known pathways associated with AF compounds, such as desertomycin, kutzneride 2, or the macrolide polyene 67-121C. This finding suggests that HR77 may synthesize novel bioactive compounds. Hybrid PKS-NRPS clusters are particularly noteworthy due to their structural diversity, which frequently gives rise to unique biological activities. The discovery of such clusters highlights the necessity for further functional characterization, as they may encode metabolites with AF or other antimicrobial properties yet to be explored. Additionally, while *S. flavofungini* is known to produce flavofungin [81], the corresponding biosynthetic cluster remains uncharacterized in publicly available databases. This raises the possibility that some of the unclassified BGCs in HR77 are involved in the biosynthesis of flavofungin or structurally related compounds. Investigating these pathways could yield valuable insights into the diversity of secondary metabolites produced by this species and their potential applications.

The case of the *Streptomyces sirii* HR31 strain presents a greater degree of complexity, as none of the metabolites encoded by the BGCs identified in its genome appear to be associated with the biosynthesis of compounds with established AF activity. Nonetheless, *S. sirii* HR31 has been shown to exhibit significant AF activity against *Verticillium* spp., implicating the potential involvement of bioactive metabolites. This AF activity may result from synergistic interactions among various metabolites or from as-yet-unknown mechanisms. Furthermore, HR31 possesses BGCs with low similarity to known AF clusters, such as auroramycin (8%) [82] and blasticidin S (7%) [83]. Although these clusters exhibit only limited homology to established pathways, their presence in HR31 underscores its potential as a source of novel AF compounds.

Meanwhile the antiSMASH analysis of the genome of the type strain *Streptomyces luteogriseus* DSM 40483 suggests that oligomycin is the most likely candidate metabolite responsible for the AF activity exhibited by strain HR40 against *V. nonalfalfae*.

In planta analyses conducted to evaluate the effectiveness of the selected bacterial strains demonstrated that treatments utilizing *Pseudomonas* sp. HX1, *Streptomyces* sp. HR40, and *Streptomyces* sp. HR77 resulted in a significant reduction of 32.56% in DSI values. In comparison, the commercial treatment with Serenade® ASO achieved only a 13.96% reduction in DSI values. These findings indicate that the three bacterial isolates may alleviate the pathogen's detrimental effects on the plant, either by decreasing its infectivity in the soil or by limiting its virulence once the plant has been infected. Furthermore, the data suggest that isolates HR40 and HR77 could be more effective than the other treatments tested, as their application concentrations were two and three orders of magnitude lower than those of the commercial product Serenade® ASO and the *Pseudomonas* sp. HX1 strain, respectively (Supplementary Material, Table S4).

As the selected bacterial strains were applied by immersing the plant root system in a bacterial suspension, the positive effects observed in the aforementioned strains may indicate a notable ability of these microorganisms to effectively colonize, or at least establish themselves within, the rhizosphere environment of the treated plants. For *Pseudomonas* sp. HX1, this capability may extend to penetration into the plant xylem. However, the absence of specific molecular markers for these strains has hindered the verification of this assumption, presenting a research gap to be addressed in future studies. Conversely, the lack of a protective effect observed for strains HR31, HX2, and HX3 may be attributable to inadequate colonization of the plant root environment under the tested conditions.

Although many microorganisms identified as potential BCAs exhibit strong AF activity in in vitro assays, their efficacy under soil conditions remains uncertain. Small-scale soil tests are essential for evaluating the potential effectiveness of the selected BCAs when present in a soil sample. Results from such experiments demonstrated that strains HR40 and HR77 exhibit significant AF activity against *V. nonalfalfae* while persisting and proliferating in soil, at least in environments with comparable physicochemical properties. Nevertheless, certain soil characteristics, such as highly acidic or alkaline pH levels, could adversely affect the survival of these strains and thereby diminish their AF activity. Consequently, further studies are needed to assess their behavior across a broader range of soil types. Interestingly, the negative effect observed during in vitro tests, wherein strain HR77 appeared to impact the activity of strain HR40, was not evident in the small-scale soil tests. If this effect does occur in soil, it does not appear to be sufficient to counteract the potential synergistic effects of the AF compounds produced by strains HR40 and HR77 in such assays.

Finally, we would like to mention that the promising results obtained from the in planta analysis, which confirmed the efficacy of strains *Pseudomonas* sp. HX1, *Streptomyces luteogriseus* HR40, and *Streptomyces flavofungini* HR77 in controlling Verticillium wilt in hops affected by *V. nonalfalfae*, underscore the need for further characterization of these isolates. A particularly valuable research direction involves identifying the compounds responsible for their AF activity. This could be accomplished through the purification of these compounds via Vacuum Flash Chromatography, followed by their identification using liquid chromatography–mass spectrometry (LC-MS) analysis, nuclear magnetic resonance (NMR) spectroscopy, or other complementary analytical techniques.

We are currently conducting a long-term field assay in which the three selected isolates are being applied to commercial hop plots as a biofertilizer. The development of specific molecular markers for each strain will be critical in evaluating their capacity for survival and establishment in agricultural soils, as well as in quantifying their presence and potential for colonization of the rhizosphere environment or the interior of the plant. This is particularly significant for the endophytic strain *Pseudomonas* sp. HX1.

Additionally, long-term metataxonomic studies will be necessary to assess the impact of repeated application of these isolates over successive years on soil microbiota. The primary objective of this research is to determine whether continued application can foster the development of soil characteristics capable of suppressing this highly detrimental pathology in hops.

## 5. Conclusions

The endosphere and rhizosphere of hop plants have been identified as favorable microenvironments for isolating *Streptomyces* strains exhibiting AF activity against *V. dahliae* and *V. nonalfalfae*. An in vitro screening strategy employing bioassay techniques facilitated the selection of six potential BCAs to combat Verticillium wilt caused by *V. nonalfalfae*. MLSA, based on the partial sequencing of five housekeeping genes, has proven to be a reliable method of identifying species within the genus *Streptomyces* as well as detecting isolates that may represent new species. Additionally, various species within the genus *Pseudomonas* were found to colonize the xylem of hop plants, some of which demonstrated antagonistic activity against *V. nonalfalfae*. In planta analyses evaluating the efficacy of the selected BCAs for controlling Verticillium wilt identified *Pseudomonas* sp. HX1, *Streptomyces luteogriseus* HR40, and *Streptomyces flavofungini* HR77 as the most promising candidates. These isolates significantly reduced disease progression in hop plants.

**Supplementary Materials:** The following supporting information can be downloaded at: https://www.mdpi.com/article/10.3390/horticulturae11050459/s1, Figure S1: *Streptomyces* phylogenetic tree inferred from concatenated partial sequences of the housekeeping genes (*atpD*, *gyrB*, *recA*, *rpoB*, and *trpB*) of rhizosphere isolates HR9, HR26, HR31, HR40, and HR77, and endophytic strain HE64 with type strains obtained from the ARS Microbial Genomic Sequence Database server; Table S1: Predicted Biosynthetic Gene Clusters (BGCs) in the genome of *S. flavofungini* HR77 using antiSMASH; Table S2: Predicted biosynthetic gene clusters (BGCs) in the genome of *S. luteogriseus* DSM 40483 using antiSMASH; Table S3: Predicted Biosynthetic Gene Clusters (BGCs) in the genome of *S. sirii* HR31 using antiSMASH; Table S4: Concentrations of the different inocula used in the in planta tests to evaluate the efficacy of the different selected bacterial isolates (HX1, HX2, HX3, HR31, HR40 and HR77) to reduce the incidence of Verticillium wilt in hops caused by *V. nonalfalfae*; Table S5: Physicochemical properties of the soil used in the small-scale soil test.

**Author Contributions:** Conceptualization, S.R., R.C. and J.J.R.C.; methodology, S.G., M.D. and C.C.-P.; investigation, S.G., M.D. and C.C.-P.; bioinformatics analysis, M.R.-M.; writing—original draft preparation, S.R., R.C., M.R.-M. and J.J.R.C.; writing—review and editing, S.R., R.C., M.R.-M. and J.J.R.C.; funding acquisition, S.R. and J.J.R.C. All authors have read and agreed to the published version of the manuscript.

**Funding:** This work was financed through a PRIMA grant (Section 2-2021) and is part of the project PCI2022-132966, funded by the Ministry of Science and Innovation (MCIN), the State Investigation Agency (AEI) (DOI/10.13039/501100011033), and project N4-0270, funded by the Slovenian Research and Innovation Agency (ARIS), and the European Union "NextGenerationEU"/Recovery Plant, Transformation, and Resilience (PRTR). Marina Ruiz-Muñoz was supported by a postdoctoral contract financed by the Ministry of Science and Innovation (MCIN), the State Investigation Agency (AEI) (DOI/10.13039/501100011033), and the European Union "NextGenerationEU"/Recovery Plant, Transformation, and Resilience (PRTR). Carla Calvo-Peña was first supported by a predoctoral contract from the Junta de Castilla y León and the European Social Fund (EDU/601/2020) and later by a postdoctoral contract financed by the Ministry of Science and Innovation (MCIN), the State Investigation Agency (AEI) (DOI/10.13039/501100011033), and the European Union "NextGenerationEU"/Recovery Plant, Transformation, and Resilience (PRTR).

**Data Availability Statement:** The data are contained within the article.

**Acknowledgments:** We thank Rebeca Otero-Suárez and Paola Callado Tejerina for their technical support.

**Conflicts of Interest:** The authors declare no conflicts of interest.

## References

1. González-Salitre, L.; Guillermo González-Olivares, L.; Antobelli Basilio-Cortes, U. *Humulus lupulus* L. a Potential Precursor to Human Health: High Hops Craft Beer. *Food Chem.* **2023**, *405*, 134959. [CrossRef] [PubMed]
2. Gerhäuser, C. Broad Spectrum Antiinfective Potential of Xanthohumol from Hop (*Humulus lupulus* L.) in Comparison with Activities of Other Hop Constituents and Xanthohumol Metabolites. *Mol. Nutr. Food Res.* **2005**, *49*, 827–831. [CrossRef]
3. Stevens, J.F.; Page, J.E. Xanthohumol and Related Prenylflavonoids from Hops and Beer: To Your Good Health! *Phytochemistry* **2004**, *65*, 1317–1330. [CrossRef]
4. *International Hop Growers' Convention Economic Commission—Summary Reports*; IHGC-Economic Commission: Lublin, Poland, 2024. Available online: https://www.ihgc.org/activities/lublin-poland/ (accessed on 20 April 2025).
5. Neve, R.A. *Hops*; Springer: Dordrecht, The Netherlands, 1991; ISBN 978-94-010-5375-4.
6. Radisek, S. Verticillium Wilt. In *Compendium of Hop Diseases and Arthropod Pests*; Mahaffee, W.F., Pethybridge, S.J., Gent, D.H., Eds.; APS Press the American Phytopathological Society: St. Paul, MN, USA, 2009; pp. 33–36.
7. Gent, D.H.; Woods, J.L.; Putnam, M.L. New Outbreaks of Verticillium Wilt on Hop in Oregon Caused by Nonlethal *Verticillium albo-atrum*. *Plant Health Prog.* **2012**, *13*, 14. [CrossRef]
8. Compant, S.; Duffy, B.; Nowak, J.; Clément, C.; Barka, E.A. Use of Plant Growth-Promoting Bacteria for Biocontrol of Plant Diseases: Principles, Mechanisms of Action, and Future Prospects. *Appl. Environ. Microbiol.* **2005**, *71*, 4951–4959. [CrossRef]
9. Lugtenberg, B.; Kamilova, F. Plant-Growth-Promoting Rhizobacteria. *Annu. Rev. Microbiol.* **2009**, *63*, 541–556. [CrossRef]
10. Carrero-Carrón, I.; Trapero-Casas, J.L.; Olivares-García, C.; Monte, E.; Hermosa, R.; Jiménez-Díaz, R.M. *Trichoderma asperellum* Is Effective for Biocontrol of Verticillium Wilt in Olive Caused by the Defoliating Pathotype of *Verticillium dahliae*. *Crop Prot.* **2016**, *88*, 45–52. [CrossRef]
11. Varo, A.; Raya-Ortega, M.C.; Trapero, A. Selection and Evaluation of Micro-Organisms for Biocontrol of *Verticillium dahliae* in Olive. *J. Appl. Microbiol.* **2016**, *121*, 767–777. [CrossRef]
12. Lozano-Tovar, M.D.; Garrido-Jurado, I.; Quesada-Moraga, E.; Raya-Ortega, M.C.; Trapero-Casas, A. *Metarhizium brunneum* and *Beauveria bassiana* Release Secondary Metabolites with Antagonistic Activity against *Verticillium dahliae* and *Phytophthora megasperma* Olive Pathogens. *Crop Prot.* **2017**, *100*, 186–195. [CrossRef]
13. Mulero-Aparicio, A.; Agustí-Brisach, C.; Varo, Á.; López-Escudero, F.J.; Trapero, A. A Non-Pathogenic Strain of *Fusarium oxysporum* as a Potential Biocontrol Agent against Verticillium Wilt of Olive. *Biol. Control* **2019**, *139*, 104045. [CrossRef]
14. Mulero-Aparicio, A.; Cernava, T.; Turrà, D.; Schaefer, A.; Di Pietro, A.; López-Escudero, F.J.; Trapero, A.; Berg, G. The Role of Volatile Organic Compounds and Rhizosphere Competence in Mode of Action of the Non-Pathogenic *Fusarium oxysporum* FO12 Toward Verticillium Wilt. *Front. Microbiol.* **2019**, *10*, 468375. [CrossRef] [PubMed]
15. Markakis, E.A.; Tjamos, S.E.; Antoniou, P.P.; Paplomatas, E.J.; Tjamos, E.C. Biological Control of Verticillium Wilt of Olive by *Paenibacillus alvei*, Strain K165. *BioControl* **2016**, *61*, 293–303. [CrossRef]
16. Cabanás, C.G.L.; Sesmero, R.; Valverde-Corredor, A.; Javier López-Escudero, F.; Mercado-Blanco, J. A Split-Root System to Assess Biocontrol Effectiveness and Defense-Related Genetic Responses in above-Ground Tissues during the Tripartite Interaction *Verticillium dahliae*-Olive-*Pseudomonas fluorescens* PICF7 in Roots. *Plant Soil.* **2017**, *417*, 433–452. [CrossRef]
17. Gómez-Lama Cabanás, C.; Legarda, G.; Ruano-Rosa, D.; Pizarro-Tobías, P.; Valverde-Corredor, A.; Niqui, J.L.; Triviño, J.C.; Roca, A.; Mercado-Blanco, J. Indigenous *Pseudomonas* spp. Strains from the Olive (*Olea europaea* L.) Rhizosphere as Effective Biocontrol Agents against Verticillium dahliae: From the Host Roots to the Bacterial Genomes. *Front. Microbiol.* **2018**, *9*, 297478. [CrossRef]

18. Cheffi Azabou, M.; Gharbi, Y.; Medhioub, I.; Ennouri, K.; Barham, H.; Tounsi, S.; Triki, M.A. The Endophytic Strain *Bacillus Velezensis* OEE1: An Efficient Biocontrol Agent against Verticillium Wilt of Olive and a Potential Plant Growth Promoting Bacteria. *Biol. Control* **2020**, *142*, 104168. [CrossRef]
19. Calvo-Peña, C.; Cobos, R.; Sánchez-López, J.M.; Ibañez, A.; Coque, J.J.R. Albocycline Is the Main Bioactive Antifungal Compound Produced by *Streptomyces* sp. OR6 against *Verticillium dahliae*. *Plants* **2023**, *12*, 3612. [CrossRef]
20. Zeng, Q.; Man, X.; Dai, Y.; Liu, H. *Pseudomonas* spp. Enriched in Endophytic Community of Healthy Cotton Plants Inhibit Cotton Verticillium Wilt. *Front. Microbiol.* **2022**, *13*, 906732. [CrossRef] [PubMed]
21. Bubici, G. *Streptomyces* spp. as Biocontrol Agents against Fusarium Species. *CABI Rev.* **2018**, *13*, 1–15. [CrossRef]
22. Labeda, D.P.; Goodfellow, M.; Brown, R.; Ward, A.C.; Lanoot, B.; Vanncanneyt, M.; Swings, J.; Kim, S.B.; Liu, Z.; Chun, J.; et al. Phylogenetic Study of the Species within the Family Streptomycetaceae. *Antonie Van Leeuwenhoek* **2012**, *101*, 73–104. [CrossRef]
23. van der Meij, A.; Worsley, S.F.; Hutchings, M.I.; van Wezel, G.P. Chemical Ecology of Antibiotic Production by Actinomycetes. *FEMS Microbiol. Rev.* **2017**, *41*, 392–416. [CrossRef]
24. Vurukonda, S.S.K.P.; Giovanardi, D.; Stefani, E. Plant Growth Promoting and Biocontrol Activity of *Streptomyces* spp. as *Endophytes*. *Int. J. Mol. Sci.* **2018**, *19*, 952. [CrossRef]
25. Janssen, P.H. Identifying the Dominant Soil Bacterial Taxa in Libraries of 16S RRNA and 16S RRNA Genes. *Appl. Environ. Microbiol.* **2006**, *72*, 1719–1728. [CrossRef] [PubMed]
26. Deketelaere, S.; Tyvaert, L.; França, S.C.; Hofte, M. Desirable Traits of a Good Biocontrol Agent against Verticillium Wilt. *Front. Microbiol.* **2017**, *8*, 1186. [CrossRef] [PubMed]
27. Bacon, C.W.; Hinton, D.M. Bacterial Endophytes: The Endophytic Niche, Its Occupants, and Its Utility. In *Plant-Associated Bacteria*; Springer: Berling, Germany, 2007; pp. 155–194. [CrossRef]
28. Ali, M.A.; Ahmed, T.; Ibrahim, E.; Rizwan, M.; Chong, K.P.; Yong, J.W.H. A Review on Mechanisms and Prospects of Endophytic Bacteria in Biocontrol of Plant Pathogenic Fungi and Their Plant Growth-Promoting Activities. *Heliyon* **2024**, *10*, e31573. [CrossRef]
29. Küster, E.; Williams, S.T. Selection of Media for Isolation of Streptomycetes. *Nature* **1964**, *202*, 928–929. [CrossRef] [PubMed]
30. Shirling, E.B.; Gottlieb, D. Methods for Characterization of *Streptomyces* Species. *Int. J. Syst. Bacteriol.* **1966**, *16*, 313–340. [CrossRef]
31. Kieser, T.; Bibb, M.J.; Buttner, M.J.; Chater, K.F.; Hopwood, D.A. *Practical Streptomyces Genetics*; John Innes Foundation: Norwich, UK, 2000; ISBN 0708406238.
32. Álvarez-Pérez, J.M.; González-García, S.; Cobos, R.; Olego, M.Á.; Ibañez, A.; Díez-Galán, A.; Garzón-Jimeno, E.; Coque, J.J.R. Use of Endophytic and Rhizosphere Actinobacteria from Grapevine Plants to Reduce Nursery Fungal Graft Infections That Lead to Young Grapevine Decline. *Appl. Environ. Microbiol.* **2017**, *83*, e01564-17. [CrossRef]
33. Radišek, S.; Jakše, J.; Javornik, B. Genetic Variability and Virulence among *Verticillium albo-atrum* Isolates from Hop. *Eur. J. Plant Pathol.* **2006**, *116*, 301–314. [CrossRef]
34. Hopwood, D.A.; Bibb, M.J.; Chater, K.F.; Kieser, T.; Bruton, C.J.; Kieser, H.M.; Lydiate, D.J.; Smith, C.P.; Ward, J.M.; Schrempf, H. *Genetic Manipulation of Streptomyces: A Laboratory Manual*; The John Innes Institute: Norwich, UK, 1985.
35. Lane, D.J. 16S/23S RRNA Sequencing. In *Nucleic Acid Techniques in Bacterial Systematics*; Stackebrandt, E., Goodfellow, M., Eds.; John Wiley and Sons: New York, NY, USA, 1991.
36. Kim, O.S.; Cho, Y.J.; Lee, K.; Yoon, S.H.; Kim, M.; Na, H.; Park, S.C.; Jeon, Y.S.; Lee, J.H.; Yi, H.; et al. Introducing EzTaxon-e: A Prokaryotic 16s RRNA Gene Sequence Database with Phylotypes That Represent Uncultured Species. *Int. J. Syst. Evol. Microbiol.* **2012**, *62*, 716–721. [CrossRef]
37. Girard, L.; Lood, C.; Rokni-Zadeh, H.; van Noort, V.; Lavigne, R.; De Mot, R. Reliable Identification of Environmental *Pseudomonas* Isolates Using the *rpoD* Gene. *Microorganisms* **2020**, *8*, 1166. [CrossRef]
38. Rong, X.; Huang, Y. Taxonomic Evaluation of the *Streptomyces hygroscopicus* Clade Using Multilocus Sequence Analysis and DNA–DNA Hybridization, Validating the MLSA Scheme for Systematics of the Whole Genus. *Syst. Appl. Microbiol.* **2012**, *35*, 7–18. [CrossRef] [PubMed]
39. Guo, Y.P.; Zheng, W.; Rong, X.Y.; Huang, Y. A Multilocus Phylogeny of the *Streptomyces griseus* 16S RRNA Gene Clade: Use of Multilocus Sequence Analysis for Streptomycete Systematics. *Int. J. Syst. Evol. Microbiol.* **2008**, *58*, 149–159. [CrossRef] [PubMed]
40. Rong, X.; Guo, Y.; Huang, Y. Proposal to Reclassify the *Streptomyces albidoflavus* Clade on the Basis of Multilocus Sequence Analysis and DNA–DNA Hybridization, and Taxonomic Elucidation of *Streptomyces griseus* Subsp. *solvifaciens*. *Syst. Appl. Microbiol.* **2009**, *32*, 314–322. [CrossRef]
41. Tamura, K.; Stecher, G.; Kumar, S. MEGA11: Molecular Evolutionary Genetics Analysis Version 11. *Mol. Biol. Evol.* **2021**, *38*, 3022–3027. [CrossRef] [PubMed]
42. Kimura, M. A Simple Method for Estimating Evolutionary Rates of Base Substitutions through Comparative Studies of Nucleotide Sequences. *J. Mol. Evol.* **1980**, *16*, 111–120. [CrossRef]
43. Koren, S.; Walenz, B.P.; Berlin, K.; Miller, J.R.; Bergman, N.H.; Phillippy, A.M. Canu: Scalable and Accurate Long-Read Assembly via Adaptive k-Mer Weighting and Repeat Separation. *Genome Res.* **2017**, *27*, 722–736. [CrossRef]

44. Mikheenko, A.; Prjibelski, A.; Saveliev, V.; Antipov, D.; Gurevich, A. Versatile Genome Assembly Evaluation with QUAST-LG. *Bioinformatics* **2018**, *34*, i142–i150. [CrossRef]
45. Overbeek, R.; Olson, R.; Pusch, G.D.; Olsen, G.J.; Davis, J.J.; Disz, T.; Edwards, R.A.; Gerdes, S.; Parrello, B.; Shukla, M.; et al. The SEED and the Rapid Annotation of Microbial Genomes Using Subsystems Technology (RAST). *Nucleic Acids Res.* **2014**, *42*, D206–D214. [CrossRef]
46. Meier-Kolthoff, J.P.; Auch, A.F.; Klenk, H.-P.; Göker, M. Genome Sequence-Based Species Delimitation with Confidence Intervals and Improved Distance Functions. *BMC Bioinform.* **2013**, *14*, 60. [CrossRef]
47. Meier-Kolthoff, J.P.; Göker, M. TYGS Is an Automated High-Throughput Platform for State-of-the-Art Genome-Based Taxonomy. *Nat. Commun.* **2019**, *10*, 2182. [CrossRef]
48. Blin, K.; Shaw, S.; Augustijn, H.E.; Reitz, Z.L.; Biermann, F.; Alanjary, M.; Fetter, A.; Terlouw, B.R.; Metcalf, W.W.; Helfrich, E.J.N.; et al. AntiSMASH 7.0: New and Improved Predictions for Detection, Regulation, Chemical Structures and Visualisation. *Nucleic Acids Res.* **2023**, *51*, W46–W50. [CrossRef]
49. Flajšman, M.; Radišek, S.; Javornik, B. Pathogenicity Assay of *Verticillium nonalfalfae* on Hop Plants. *Bio-Protocol* **2017**, *7*, e2171. [CrossRef]
50. Brglez, A.; Devetak, Z.; Ogris, N.; Radišek, S.; Piškur, B. An Outbreak of *Verticillium dahliae* on Sycamore Maple in a Forest Stand in Slovenia. *J. Plant Pathol.* **2024**, *106*, 609–621. [CrossRef]
51. Schrey, S.D.; Erkenbrack, E.; Früh, E.; Fengler, S.; Hommel, K.; Horlacher, N.; Schulz, D.; Ecke, M.; Kulik, A.; Fiedler, H.P.; et al. Production of Fungal and Bacterial Growth Modulating Secondary Metabolites Is Widespread among Mycorrhiza-Associated Streptomycetes. *BMC Microbiol.* **2012**, *12*, 164. [CrossRef] [PubMed]
52. International Organization for Standardization. ISO 11277:2020—Soil Quality—Determination of Particle Size Distribution in Mineral Soil Material—Method by Sieving and Sedimentation. 2020. Available online: https://www.iso.org/standard/69496.html (accessed on 20 April 2025).
53. Tan, K.H. Soil Sampling, Preparation, and Analysis. In *Soil Sampling, Preparation, and Analysis*; CRC Press: Boca Raton, FL, USA, 2005. [CrossRef]
54. Batjes, N.H. Total Carbon and Nitrogen in the Soils of the World. *Eur. J. Soil Sci.* **1996**, *47*, 151–163. [CrossRef]
55. Olsen, S.R.; Dean, L.A. Phosphorus. In *Methods of Soil Analysis, Part 2: Chemical and Microbiological Properties*; Wiley: Hoboken, NJ, USA, 2016; pp. 1035–1049.
56. Ostrowska, A.; Gawliński, S.; Szczubiałka, Z. *Methods of Analysis and Assessment of Soil and Plant Properties*; Instytut Ochrony Środowiska: Warsaw, Poland, 1991.
57. Houba, V.J.G.; Temminghoff, E.J.M.; Gaikhorst, G.A.; van Vark, W. Soil Analysis Procedures Using 0.01 M Calcium Chloride as Extraction Reagent. *Commun. Soil Sci. Plant Anal.* **2000**, *31*, 1299–1396. [CrossRef]
58. Association Française de Normalisation. NF X31-122:1999—Soil Quality—Measurement of the Aerobic Biological Activity of Soil—Respirometric Method with Oxygen Measurement. 1999. Available online: https://www.dinmedia.de/en/standard/nf-x31-122/23265586 (accessed on 20 April 2025).
59. Slovenian Institute for Standardization. SIST ISO 14235:1999—Soil Quality—Determination of Organic Carbon by Sulfochromic Oxidation. 1999. Available online: https://standards.iteh.ai/catalog/standards/sist/215eea8b-6056-4b77-942d-e4907c003860/sist-iso-14235-1999 (accessed on 20 April 2025).
60. R Core Team, R. *A Language and Environment for Statistical Computing*; R Foundation for Statistical Computing: Vienna, Austria, 2022; Available online: https://www.R-project.org/ (accessed on 20 April 2025).
61. Fox, J.; Weisberg, S. *An R Companion to Applied Regression*, 3rd ed.; Sage Publications: New York, NY, USA, 2019; Available online: https://www.john-fox.ca/Companion/ (accessed on 11 April 2025).
62. Ogle, D.H.; Doll, J.C.; Wheeler, A.P.; Dinno, A. *Simple Fisheries Stock Assessment Methods*; R Package FSA Version 0.9.6; CRAN: Vienna, Austria, 2025. [CrossRef]
63. Spencer, G.; Hans-Peter, P.; Luciano, S.; Sundar, D.-R. *MultcompView: Visualizations of Paired Comparisons*; Version 0.1-10; CRAN: Vienna, Austria, 2006. [CrossRef]
64. Wayne, L.G.; Brenner, D.J.; Colwell, R.R.; Grimont, P.A.D.; Kandler, O.; Krichevsky, M.I.; Moore, L.H.; Moore, W.E.C.; Murray, R.G.E.; Stackebrandt, E.; et al. Report of the Ad Hoc Committee on Reconciliation of Approaches to Bacterial Systematics. *Int. J. Syst. Evol. Microbiol.* **1987**, *37*, 463–464. [CrossRef]
65. Meschke, H.; Walter, S.; Schrempf, H. Characterization and Localization of Prodiginines from *Streptomyces lividans* Suppressing *Verticillium dahliae* in the Absence or Presence of *Arabidopsis thaliana*. *Environ. Microbiol.* **2012**, *14*, 940–952. [CrossRef]
66. Ramarajan, M.; Devilla, R.; Dow, L.; Walsh, N.; Mead, O.; Zakeel, M.C.M.; Gallart, M.; Richardson, A.E.; Thatcher, L.F. Genomic and Untargeted Metabolomic Analysis of Secondary Metabolites in the *Streptomyces griseoaurantiacus* Strain MH191 Shows Media-Based Dependency for the Production of Bioactive Compounds with Potential Antifungal Activity. *J. Agric. Food Chem.* **2024**, *72*, 24432–24448. [CrossRef]

67. Liu, S.; Xu, M.; Zhang, H.; Qi, H.; Zhang, J.; Liu, C.; Wang, J.; Xiang, W.; Wang, X. New Cytotoxic Spectinabilin Derivative from Ant-Associated *Streptomyces* sp. 1H-GS5. *J. Antibiot.* **2016**, *69*, 128–131. [CrossRef]
68. Li Xia, F.; Zheng Yan, G.; Wen Jun, W. Isolation and Characterization of *Streptomyces alboflavus* SC11 Producing Desertomycin A. *Afr. J. Microbiol. Res.* **2013**, *7*, 1246–1252. [CrossRef]
69. Sheehan, J.; Murphy, C.D.; Caffrey, P. New Insights into Polyene Macrolide Biosynthesis in *Couchioplanes caeruleus*. *Mol. Biosyst.* **2017**, *13*, 866–873. [CrossRef]
70. Zhang, D.; Lu, Y.; Chen, H.; Wu, C.; Zhang, H.; Chen, L.; Chen, X. Antifungal Peptides Produced by Actinomycetes and Their Biological Activities against Plant Diseases. *J. Antibiot.* **2020**, *73*, 265–282. [CrossRef]
71. Khebizi, N.; Boudjella, H.; Bijani, C.; Bouras, N.; Klenk, H.-P.; Pont, F.; Mathieu, F.; Sabaou, N. Oligomycins A and E, Major Bioactive Secondary Metabolites Produced by *Streptomyces* sp. Strain HG29 Isolated from a Saharan Soil. *J. Mycol. Med.* **2018**, *28*, 150–160. [CrossRef] [PubMed]
72. Lardy, H.A.; Johnson, D.; McMurray, W.C. Antibiotics as Tools for Metabolic Studies. I. A Survey of Toxic Antibiotics in Respiratory, Phosphorylative and Glycolytic Systems. *Arch. Biochem. Biophys.* **1958**, *78*, 587–597. [CrossRef]
73. Xiao, L.; Niu, H.-J.; Qu, T.-L.; Zhang, X.-F.; Du, F.-Y. *Streptomyces* sp. FX13 Inhibits Fungicide-Resistant *Botrytis Cinerea* in Vitro and in Vivo by Producing Oligomycin A. *Pestic. Biochem. Physiol.* **2021**, *175*, 104834. [CrossRef] [PubMed]
74. Zakalyukina, Y.V.; Alferova, V.A.; Nikandrova, A.A.; Kiriy, A.R.; Chernyshova, A.P.; Kabilov, M.R.; Baturina, O.A.; Biryukov, M.V.; Sergiev, P.V.; Lukianov, D.A. Genomic and Phenotypic Characterization of *Streptomyces sirii* sp. Nov., Amicetin-Producing Actinobacteria Isolated from Bamboo Rhizospheric Soil. *Microorganisms* **2024**, *12*, 2628. [CrossRef]
75. Coque, J.J.R.; Álvarez-Pérez, J.M.; Cobos, R.; González-García, S.; Ibáñez, A.M.; Diez Galán, A.; Calvo-Peña, C. Advances in the Control of Phytopathogenic Fungi That Infect Crops through Their Root System. *Adv. Appl. Microbiol.* **2020**, *111*, 123–170. [CrossRef]
76. López-Escudero, F.J.; Mercado-Blanco, J. Verticillium Wilt of Olive: A Case Study to Implement an Integrated Strategy to Control a Soil-Borne Pathogen. *Plant Soil.* **2011**, *344*, 1–50. [CrossRef]
77. Lalucat, J.; Mulet, M.; Gomila, M.; García-Valdés, E. Genomics in Bacterial Taxonomy: Impact on the Genus *Pseudomonas*. *Genes* **2020**, *11*, 139. [CrossRef]
78. Lu, J.; Huang, Y.; Liu, R.; Liang, Y.; Zhang, H.; Shen, N.; Yang, D.; Jiang, M. Antimicrobial Mechanisms and Antifungal Activity of Compounds Generated by Banana Rhizosphere *Pseudomonas aeruginosa* Gxun-2 against *Fusarium oxysporum* f. sp. cubense. *Front. Microbiol.* **2024**, *15*, 1456847. [CrossRef] [PubMed]
79. Ni, H.; Kong, W.-L.; Zhang, Y.; Wu, X.-Q. Effects of Volatile Organic Compounds Produced by *Pseudomonas aurantiaca* ST-TJ4 against *Verticillium dahliae*. *J. Fungi* **2022**, *8*, 697. [CrossRef] [PubMed]
80. Rane, M.R.; Sarode, P.D.; Chaudhari, B.L.; Chincholkar, S.B. Exploring Antagonistic Metabolites of Established Biocontrol Agent of Marine Origin. *Appl. Biochem. Biotechnol.* **2008**, *151*, 665–675. [CrossRef]
81. Uri, J.; Békési, I. Flavofungin, a New Crystalline Antifungal Antibiotic: Origin and Biological Properties. *Nature* **1958**, *181*, 908. [CrossRef] [PubMed]
82. Lim, Y.H.; Wong, F.T.; Yeo, W.L.; Ching, K.C.; Lim, Y.W.; Heng, E.; Chen, S.; Tsai, D.; Lauderdale, T.; Shia, K.; et al. Auroramycin: A Potent Antibiotic from *Streptomyces roseosporus* by CRISPR-Cas9 Activation. *ChemBioChem* **2018**, *19*, 1716–1719. [CrossRef]
83. Cone, M.C.; Yin, X.; Grochowski, L.L.; Parker, M.R.; Zabriskie, T.M. The Blasticidin S Biosynthesis Gene Cluster from *Streptomyces griseochromogenes*: Sequence Analysis, Organization, and Initial Characterization. *ChemBioChem* **2003**, *4*, 821–828. [CrossRef]

*horticulturae*

MDPI

*Article*

# Orthodox vs. Recalcitrant? Germination and Early Growth of *Phoenix* Species (Arecaceae) Stored for up to Ten Years

**Concepción Obón [1,*], Sofía Pardo-Pina [1], Dennis Johnson [2] and Diego Rivera [3,*]**

1 CIAGRO, Escuela Politécnica Superior de Orihuela, Universidad Miguel Hernández, Carretera de Beniel, km 3,2, 03312 Orihuela, Spain; spardo@umh.es
2 Independent Researcher, Cincinnati, OH 45208, USA; djohn37@aol.com
3 Departamento de Biología Vegetal, Facultad de Biología, Universidad de Murcia, 30100 Murcia, Spain
* Correspondence: cobon@umh.es (C.O.); drivera@um.es (D.R.)

**Abstract:** This study investigated seed storage behavior and seedling development patterns in the genus *Phoenix* L. (*Arecaceae*), addressing the knowledge gap regarding orthodox versus recalcitrant characteristics in these ecologically and economically significant palms. We examined the germination capacity and subsequent growth in 31 seed samples from various *Phoenix* species stored for up to 10 years at approximately 5 °C, at the Germplasm Bank at the Escuela Politécnica Superior de Orihuela, comprising 465 seeds monitored over a one-year period. The seed germination trials involved planting seeds in pots placed in an open-air greenhouse after ambient temperatures consistently exceeded 20 °C, typically after mid-June. *Phoenix dactylifera*, *P. canariensis*, *P. theophrasti*, the hybrid *P. dactylifera* × *P. canariensis*, and *P.* × "Palmeri" demonstrated orthodox seed storage behavior, maintaining viability for up to nine years. Conversely, *P. sylvestris*, *P. pusilla*, *P. rupicola*, and *P. loureiroi* consistently failed to germinate despite previous germination success, suggesting potential recalcitrant characteristics. Statistical analyses revealed that species identity and geographic origin exerted greater influence on germination success than seed age. Seedling development exhibited a conserved seasonal pattern across all species, with synchronized leaf emergence in September and March–July, followed by winter dormancy. Significant intraspecific variation was observed, particularly within *P. dactylifera* varieties, in both leaf production and final leaf length. These findings provide valuable insights into germplasm preservation and cultivation strategies, demonstrating that while some *Phoenix* species are suitable for long-term seed banking, others may require alternative conservation approaches. The observed species-specific and variety-specific differences offer important selection criteria for horticultural applications and conservation efforts.

**Keywords:** conservation; development; dormancy; germplasm bank; seed biology; seedling

## 1. Introduction

The genus *Phoenix* L. Sp. Pl. 2, 1753, comprises a group of palms belonging to the order Arecales. It is notable for its pinnate leaves, unlike the palmate leaves characteristic of other genera in the subfamily Coryphoideae.

Currently, between 14 and 20 species are recognized, distributed from the Macaronesian Archipelago and Cape Verde Islands, North Africa, to the Middle East, India, China, Vietnam, southern Europe, Thailand, and islands of the Indian Ocean [1–3] (Figure 1).

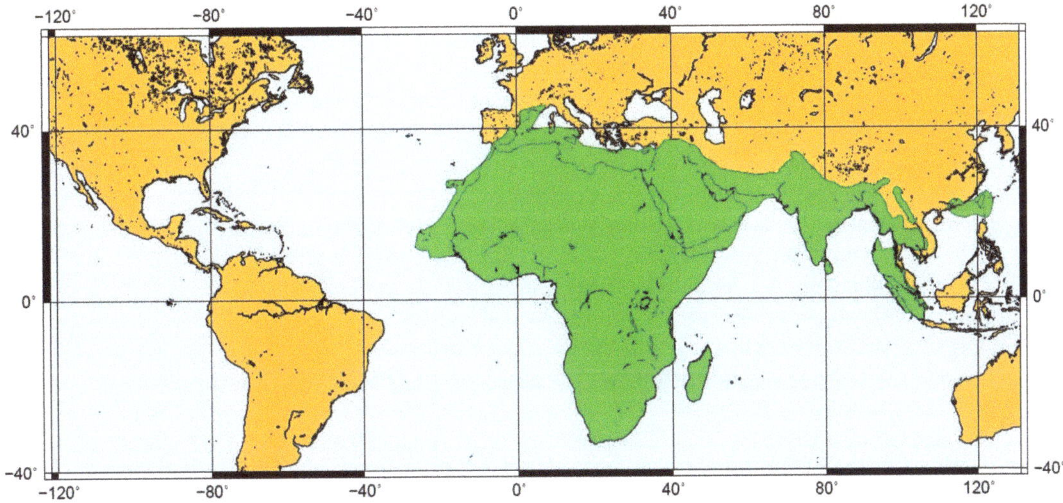


**Figure 1.** Primary distribution of the *Phoenix* genus in green. Image by Francisco Alcaraz, with permission.

Seedling date palms are prevalent in traditional cultivation regions and naturalized populations where intentional introduction occurred, notably in Spain and the Americas. These populations, particularly in Spain, Peru, and Mexico, represent valuable genetic repositories. Growing in diverse climatic conditions, these specimens may possess advantageous traits that, through biotechnological applications, could enhance elite cultivars' resistance, hardiness, salt tolerance, and fruit characteristics, thereby advancing commercial production [4].

Seed storage behavior is a crucial biological trait that significantly influences plant conservation strategies. Seeds are primarily classified into two categories based on their post-harvest physiology and storage potential: orthodox and recalcitrant. This classification reflects adaptations to diverse environmental pressures and has profound implications for ex situ conservation efforts [5].

Orthodox seeds can endure substantial desiccation, typically to 5–7% moisture content, and can be stored at low temperatures for extended periods, often lasting decades or more. Their longevity increases predictably with decreasing storage temperature and moisture content. Examples include most annual crops like wheat, rice, and corn, as well as many temperate species and plants adapted to seasonal dry periods [6,7].

In contrast, recalcitrant seeds are intolerant of desiccation, losing viability rapidly when their moisture content falls below a relatively high threshold, usually 20–30%. They cannot survive conventional seed bank storage conditions and remain metabolically active after shedding. Examples include many tropical tree species such as mango, cacao, and avocado, as well as aquatic plants and several economically important tree crops [6,8].

There is also an intermediate category of seeds that can tolerate some desiccation but not to the extent of truly orthodox seeds, denominated here "non-orthodox", representing a physiological continuum rather than strict categories [9,10].

The classification of seeds as orthodox or recalcitrant represents a fundamental criterion for germplasm preservation protocols [11,12]. About 75 to 80% of angiosperm species produce orthodox seeds that can survive drying and prolonged storage at −20 °C. By contrast, 5 to 10% of angiosperm species produce recalcitrant seeds that do not survive desiccation and are killed by freezing when ice crystals form [13]. While this functional dichotomy guides gene bank practices, the seed biology literature typically frames these categories in terms of desiccation tolerance: orthodox seeds survive significant dehydration, whereas recalcitrant seeds exhibit marked sensitivity to moisture reduction [12].

The distribution of seed storage behavior shows distinct patterns across plant taxonomy and biogeography:

- Orthodox seeds predominate in temperate regions, in arid environments, and among herbaceous plants.
- Recalcitrance is more common in tropical and subtropical regions, particularly in moist broadleaf forests [14,15].
- Within some families, such as Dipterocarpaceae, recalcitrance is the norm.
- Other families, like Fabaceae, exhibit a mix of both behaviors.
- Primitive seed plants often display recalcitrant characteristics.
- Some plant families show strong phylogenetic conservation of seed storage behavior.

This distribution suggests that recalcitrance may be an ancestral trait in seed plants, with orthodoxy evolving as an adaptation to seasonal or unpredictable environments where dormancy confers survival advantages.

The physiological differences between orthodox and recalcitrant seeds include the following:

1. Cellular and biochemical adaptations:
    - Orthodox seeds accumulate protective proteins (LEA proteins), sugars (especially oligosaccharides), and antioxidants during maturation and undergo cytoplasmic vitrification during drying.
    - Recalcitrant seeds lack sufficient protective mechanisms against desiccation damage and maintain high metabolic activity [9].
2. Water relations:
    - Orthodox seeds can survive equilibration with very dry air, with cellular water tightly bound [9].
    - Recalcitrant seeds maintain a high proportion of free water and experience lethal membrane damage upon drying [9].
3. Developmental pathways:
    - Orthodox seeds typically undergo a programmed drying phase during maturation.
    - Recalcitrant seeds are shed at high moisture content and remain metabolically active [16].

Genetic basis: Recent genomic studies have identified gene expression patterns associated with desiccation tolerance, including transcription factors that regulate LEA protein synthesis, antioxidant systems, and sugar metabolism pathways.

Ecological significance: Seed storage behavior reflects ecological adaptation and reproductive strategy as follows:

- Orthodox seeds are adapted to unpredictable or seasonal environments where dormancy increases survival probability and are often smaller with higher production numbers [17].
- Recalcitrant seeds are adapted to stable, humid environments where immediate germination is advantageous and are often larger with greater nutrient reserves.
- Seed sizes frequently correlate with storage behavior, with larger seeds tending toward recalcitrance [18,19].
- Dispersal mechanisms often align with storage behavior, with wind-dispersed seeds typically being orthodox.

- Pioneer species generally produce orthodox seeds, while climax forest species often produce recalcitrant seeds.

These ecological correlations reflect the tradeoffs between seed longevity, establishment success, and reproductive investment.

Implications for ex situ conservation: The physiological differences between orthodox and recalcitrant seeds create distinct challenges for conservation as follows:

Orthodox seeds:

- Well suited for conventional seed banking (dried and stored at −18 °C).
- Predictable longevity under proper storage conditions.
- Cost-effective conservation of genetic diversity.
- Large sample sizes can be stored in limited space.

Recalcitrant seeds:

- Conventional seed banking is ineffective.
- Alternative approaches are required, such as cryopreservation of embryonic axes or embryos, in vitro culture of embryos or embryonic axes, field gene banks (living collections), pollen storage and artificial pollination, and synthetic seed technology.

These limitations have significant implications for conservation priorities, particularly for tropical tree species with recalcitrant seeds that are disproportionately threatened by habitat loss and climate change.

Current research focuses on the following:

1. Identifying the molecular mechanisms underlying desiccation tolerance.
2. Understanding the genetic basis of seed storage behavior to potentially engineer greater storage tolerance.
3. Exploring the continuum of seed storage behaviors beyond the simple orthodox–recalcitrant dichotomy [10].
4. Predicting seed storage behavior based on phylogenetic and ecological information [15,17–20].
5. Developing improved techniques for preserving recalcitrant germplasm [21,22].

The distinction between orthodox and recalcitrant seeds represents a fundamental aspect of plant reproductive biology with critical implications for conservation. As habitat loss accelerates and climate change threatens natural ecosystems, effective ex situ conservation strategies must account for these physiological differences. The continued development of technologies for preserving recalcitrant germplasm remains a priority for global plant conservation efforts, particularly for tropical forest species that are both highly threatened and likely to possess recalcitrant seeds [5,23].

Seed longevity is a critical parameter in germplasm preservation facilities, guiding storage protocols, regeneration schedules, and viability testing. Although its biological importance was recognized in the early 20th century when Nicolai I. Vavilov identified the need for systematic conservation of threatened phytogenetic resources, research has since revealed significant variation in longevity both and within species [24]. Multiple factors influence ex situ seed longevity, including genetics, developmental stage at collection, processing methods, and environmental storage conditions—particularly relative humidity and temperature [25,26].

Cryopreservation, under specific parameters, facilitates the long-term conservation of plant germplasm. A remarkable demonstration of this phenomenon was observed in *Silene stenophylla* tissue specimens from Siberia, dating to the Late Pleistocene epoch (approximately 30,000 years before present), which were preserved in permafrost conditions. Through the application of in vitro culture techniques and micropropagation protocols, these ancient tissues were successfully regenerated into viable specimens capable of sexual reproduction [27].

Anhydrobiotic seeds of diverse plant taxa exhibit remarkable tolerance to adverse environmental conditions when maintaining a desiccated state. Under these conditions, their metabolic processes are downregulated to a minimal threshold of activity—a state of quiescence—while maintaining germination viability over extended temporal periods [28,29]. During ontogeny, orthodox seeds develop desiccation tolerance mechanisms through a complex maturation process. This developmental progression is characterized by the gradual acquisition of longevity—defined as the cumulative duration throughout which desiccated seeds retain their biological viability [9]. Francis E. Lloyd investigated seed ontogeny and histochemical changes that occur during germination in cultivars of *P. dactylifera* [30].

Various *Phoenix* species exhibit seed characteristics that fall between orthodox and recalcitrant, which we designate here as "non-orthodox". Studies of *Phoenix reclinata* demonstrate that although both embryonic and endospermic water content decrease during seed maturation, they maintain relatively high levels—a characteristic frequently associated with, although not exclusive to, non-orthodox seeds. The embryonic cellular ultrastructure and minimal water absorption requirements for germination initiation further support the non-orthodox classification of these seeds. Current evidence suggests that *P. reclinata* seeds display intermediate post-harvest behavior, placing them between orthodox and recalcitrant categories in terms of their storage characteristics [31].

Desiccation tolerance in *Phoenix* species appears to be associated with a form of seed dormancy, which can be alleviated through acid treatments or immersion in water at varying temperatures and durations. This phenomenon has been specifically investigated in *P. dactylifera* [32,33].

Very long-term preservation of *Phoenix dactylifera* seeds has a recent history. In 1963, archaeological excavations in southern Israel yielded a remarkable discovery: an undisturbed clay vessel containing several date palm seeds. The vessel, recovered from arid soil conditions, underwent radiocarbon dating analysis, which placed its chronological context between 155 BCE and 64 CE. The seeds were subsequently preserved at Bar-Ilan University in Tel Aviv. Four decades later, in 2003, agricultural specialist Elaine Solowey initiated germination trials with three of these ancient seeds. Notably, one seed successfully germinated, and by 2011, eight years post-germination, the specimen flowered [34,35]. This extraordinary case demonstrates the exceptional longevity potential of some *Phoenix* seeds under specific preservation conditions.

In 2007, the Valencian Regional Government commissioned a comprehensive study on date palm and wild palm populations within its territory. This led to the establishment of the Spanish Palm Germplasm Bank, funded by the National Institute for Agricultural Research (INIA). The bank was created to facilitate the collection, propagation, preservation, and characterization of *Phoenix* species and other related palms. Currently, the Palm Germplasm Bank houses over 600 living accessions, representing 19 species and subspecies and 4 interspecific hybrids [36]. As of 2024, the National *Phoenix* Collection has documented 1675 accessions, illustrating the extensive range of palm genetic resources stored in this repository (Figure 2). A key feature of the collection is that over 90% of the accessions were obtained from seeds, which are desiccated and stored at 4 to 5 °C. Consequently, evaluating the germination capacity of these seeds after long-term storage is crucial to determine the viability of this seed collection as a genetic resource.

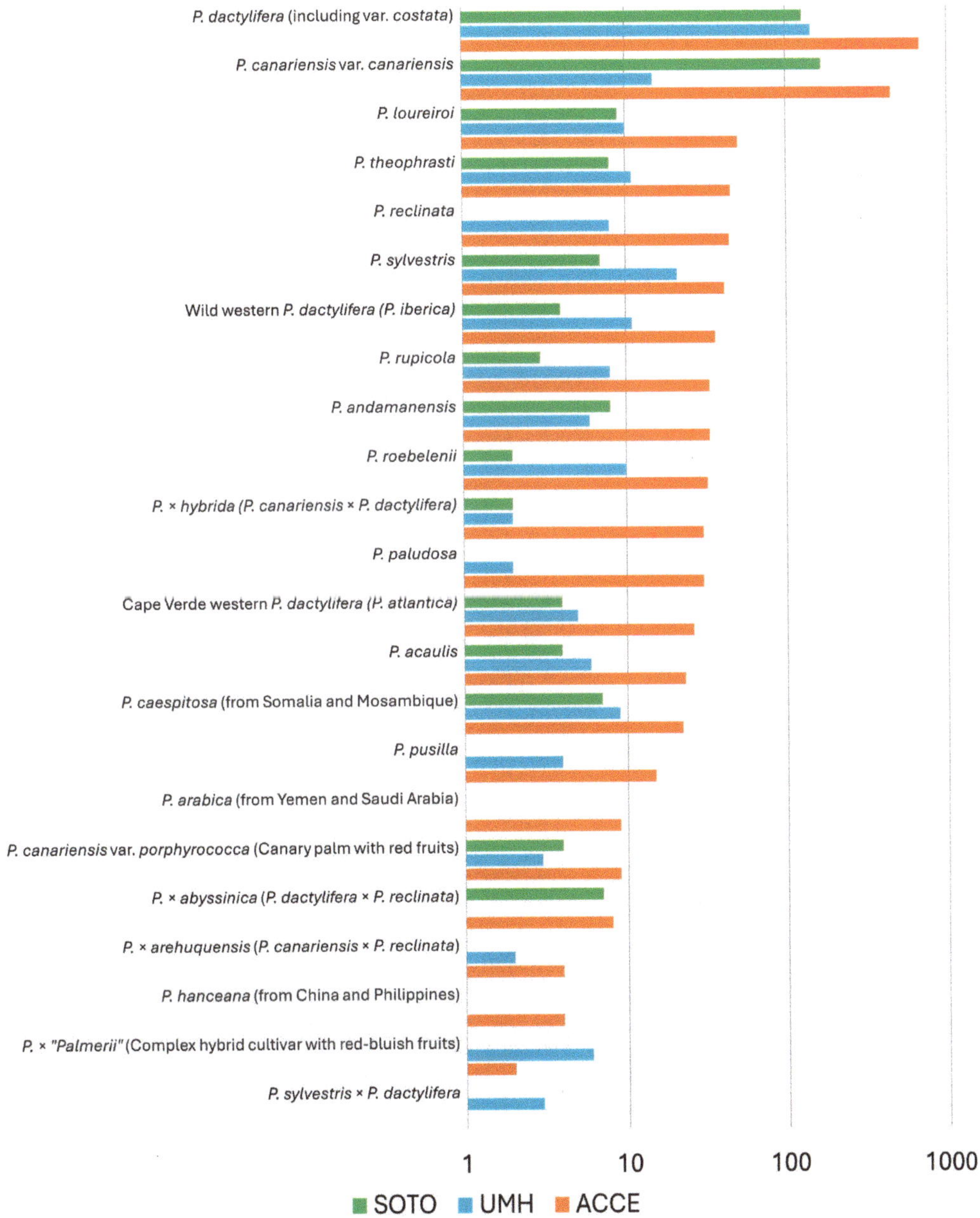


**Figure 2.** Number of accessions for each *Phoenix* taxon maintained in ex situ conservation at three sites in Orihuela, Spain, as per 2024. The *x*-axis indicates the number of accessions, while the *y*-axis lists the different *Phoenix* taxa. Bars represent the following: ACCE (green)—accessions conserved as seeds in the seed bank; SOTO (blue)—accessions cultivated as living palm trees in the 16-hectare Palmetum gardens of Orihuela; and UMH (orange)—accessions grown in the experimental fields (EPSO) of the Miguel Hernández University at Desamparados. A given taxon may be present in one, two, or all three locations, depending on germination success. Overall, most accessions are preserved in seed form, with approximately 50% also represented as living specimens in the field collections. Graph by Diego Rivera.

This extensive seed collection serves as a valuable resource for future studies on the morphological, genetic, and biochemical characteristics of the *Phoenix* genus, with significant potential for applications in horticulture and conservation.

This study investigated seed samples from the National *Phoenix* Collection to determine seed storage behavior (orthodox or recalcitrant) and analyze growth rates and developmental patterns across species during the first year. The specific objectives included (1) evaluating germination capacity after extended standardized storage, (2) assessing the influence of seed age and species on the percentage of germination, (3) analyzing seasonal patterns of growth rates, and (4) comparing vegetative development among species. This systematic approach contributes to a comprehensive understanding of *Phoenix* seed biology and development, informing both conservation and cultivation practices.

## 2. Materials and Methods

### 2.1. Plant Material

The present study was conducted using seed material from the *Phoenix* Germplasm Bank located at the Miguel Hernández University (UMH), specifically at the Escuela Politécnica Superior de Orihuela (EPSO), which is part of the global germplasm bank network and is supported by the Miguel Hernández University of Elche, the Municipality of Orihuela, and the National Institute of Agricultural Research (INIA).

Seeds from multiple *Phoenix* species and varieties were collected and maintained in controlled storage conditions, refrigerated chambers (Liebherr, Ochsenhausen, Germany) at 5 °C with silica gel with humidity indicator (orange, 2.5–6 mm, Scharlab S.L., Barcelona, Spain). This study utilized 31 seed samples (Table 1), representing a broad spectrum of *Phoenix* taxa. The sampling strategy prioritized Spanish localities, particularly the Valencia and Murcia regions, while also incorporating samples from the northern Iberian Peninsula (Asturias and Galicia). To achieve a comprehensive representation of *Phoenix* diversity, seeds from other Mediterranean countries and Southeast Asia were also included. The collection comprised the following taxa: *Phoenix dactylifera* L. (including cultivars "Negres", "Candios de mala clase", and "Rojos"), *P. dactylifera* var. *costata* Becc., *P. canariensis* H. Wildpret var. *porphyrococca* Vasc. & Franco, *P. loureiroi* Kunth, *P.* × "Palmeri" (identified as a hybrid through genetic analysis by Carreño [3]), *P. rupicola* T. Anderson, *P. sylvestris* (L.) Roxb., and *P. theophrasti* Greuter "Datça." The samples were acquired between 2008 and 2015 from diverse sources, including commercial suppliers, germplasm repositories, and field collections conducted across Spain and other relevant geographical locations. Seed ages ranged from 633 to 3440 days since collection.

**Table 1.** *Phoenix* accessions studied with their codes.

| Repository Code | Species and Variety | Accessions Data (Date and Locality of Seed Collection) | Storage (Days) ** | Previous Germination Results |
|---|---|---|---|---|
| | *Phoenix canariensis* H. Wildpret | | 1870 * | |
| E160-1677 | *P. canariensis* var. *porphyrococca* Vasc. & Franco | 24-11-2009; Espinardo (Murcia, Spain) | 2771 | 30-11-2009, – |
| E160-1036 | *P. canariensis* var. *canariensis* | 26-08-2013; Colombres (Asturias, Spain) | 1400 | 17-09-2013, – |
| E160-1478 | *P. canariensis* var. *porphyrococca* | 28-08-2013; Cambados (Pontevedra, Spain) | 1398 | 20-06-2014, + |
| E160-1488 | *P. canariensis* var. *porphyrococca* | 29-10-2010; Valencia (Spain) | 2432 | 20-06-2011, + |
| E160-1490 | *P. canariensis* var. *macrocarpa* H. Wildpret | 20-12-2014; Granada (Spain) | 919 | 30-06-2015, + |
| E160-1496 | *P. canariensis* var. *canariensis* | 27-08-2013; Cudillero (Asturias, Spain) | 1399 | 10-07-2014, + |
| E160-1676 | *P. canariensis* var. *canariensis* | 24-11-2009; Espinardo (Murcia, Spain) | 2771 | 30-11-2009, – |

**Table 1.** *Cont.*

| Repository Code | Species and Variety | Accessions Data (Date and Locality of Seed Collection) | Storage (Days) ** | Previous Germination Results |
|---|---|---|---|---|
| | *Phoenix dactylifera* L. | | *2591* * | |
| E160-1476 | *P. dactylifera* var. *dactylifera* | 28-01-2008; Fortuna (Murcia, Spain) | 3437 | none |
| E160-1477 | *P. dactylifera* var. *dactylifera* | 28-01-2008; Fortuna (Murcia, Spain) | 3437 | none |
| E160-1479 | *P. dactylifera* var. *dactylifera* | 25-09-2011; Elche (Alicante, Spain) | 2101 | 30-06-2012, + |
| E160-1480 | *P. dactylifera* var. *costata* Becc. | 16-01-2011; Santomera (Murcia, Spain) | 2353 | 30-6-2011, + |
| E160-1481 | *P. dactylifera* var. *costata* | 10-11-2010; Santomera (Murcia, Spain) | 2420 | 30-6-2011, + |
| E160-1483 | *P. dactylifera* var. *dactylifera* | 03-12-2008; Ojós (Murcia, Spain) | 3127 | 20-07-2009, + |
| E160-1484 | *P. dactylifera* var. *dactylifera* | 04-02-2013; Elche (Alicante, Spain) | 1603 | 04-02-2013, – |
| E160-1485 | *P. dactylifera* var. *dactylifera* | 11-11-2011; Orihuela (Alicante, Spain) | 2054 | 29-12-2011, – |
| E160-1486 | *P. dactylifera* var. *dactylifera* | 12-10-2010; Albatera (Alicante, Spain) | 2449 | 30-06-2011, + |
| E160-1489 | *P. dactylifera* var. *dactylifera* "Negres" | 25-01-2008; Elche (Alicante, Spain) | 3440 | 30-06-2009, – |
| E160-1491 | *P. dactylifera* var. *dactylifera* "Rojos" | 02-10-2015; Elche (Alicante, Spain) | 633 | 20-06-2016, + |
| E160-1493 | *P. dactylifera* var. *dactylifera* "Negres" | 25-01-2008; Elche (Alicante, Spain) | 3440 | 10-07-2016, + |
| E160-1495 | *P. dactylifera* var. *dactylifera* "Negres" | 25-01-2008; Elche (Alicante, Spain) | 3440 | 10-07-2016, + |
| E160-1497 | *P. dactylifera* var. *dactylifera* "Candios de mala clase" | 23-01-2011; Elche (Alicante, Spain) | 2346 | 20-06-2011, + |
| | *Phoenix* × "Palmeri" | | *1710* * | |
| E160-1482 | *Phoenix* × "Palmeri" | 18-06-2012; B & T World Seeds [1] | 1834 | 10-07-2014, + |
| E160-1487 | *Phoenix* × "Palmeri" | 20-02-2013; KPR [2] | 1587 | 10-07-2014, + |
| | *Phoenix theophrasti* Greuter | | | |
| E160-1492 | *P. theophrasti* "Datça" | 06-07-2013; Palmiye Merkezi [3] | 1451 | 30-9-2013, + |
| | *Phoenix* × *hybrida* André (*Phoenix dactylifera* L. × *Phoenix canariensis* H. Wildpret) | | | |
| E160-1494 | *P.* × *hybrida* | 16-01-2011; El Siscar (Murcia, Spain) | 2353 | 10-07-2011, + |
| | *Phoenix pusilla* Gaertn. | | | |
| E160-327 | *P. pusilla* | 22-11-2009; Kenibreed [4] | 2773 | 24-12-2011, – |
| | *Phoenix sylvestris* (L.) Roxb. | | *2773* * | |
| E160-328 | *P. sylvestris* | 22-11-2009; Kenibreed [4] | 2773 | 30-06-2010, + |
| E160-329 | *P. sylvestris* | 22-11-2009; Kenibreed [4] | 2773 | 30-06-2010, + |
| | *Phoenix loureiroi* Kunth | | *1650* * | |
| E160-712 | *P. loureiroi* | 25-12-2012; Kenibreed [4] | 1644 | 30-06-2013, + |
| E160-714 | *P. loureiroi* | 12-12-2012; Kenibreed [4] | 1657 | 30-06-2013, + |
| | *Phoenix rupicola* T. Anderson | | | |
| E160-713 | *P. rupicola* | 25-12-2012; Kenibreed [4] | 1644 | 30-06-2013, + |

Note: Seed sample details including mean ages in *italics* (*) of the group of samples from each species, identified by genus, species, and variety name. Samples included in the experiment are identified with their accession codes. (**) The column "Age (days)" records the number of days since seed collection until the experiment began. Germination results are displayed as "+" (successful germination) and "–" (no germination). Commercial seed sources: [1] B & T World Seeds (Aigues-Vives, France), [2] KPR Gardeners Club (Slovakia), [3] Palmiye Merkezi (Koycegiz, Mugla, Turkiye), and [4] Kenibreed Plant (West Bengal, India).

Seed samples were selected to prioritize species of research interest while ensuring sufficient seed quantities for germination experiments, minimizing depletion of the germplasm collection. Thirty-one samples, each containing 15 seeds, were chosen for this study (Table 1). These samples were stored in refrigerated chambers at 5 °C with silica gel desiccant to control humidity. Germination trials were conducted by planting seeds in pots, which were then placed in an open-air greenhouse after the ambient air temperatures consistently exceeded 20 °C, typically after mid-June in southeastern Spain. Germination of tropical *Phoenix* species presented greater challenges compared with arid-adapted species, even with fresh seed stock. Following selection, a standardized photographic protocol was implemented to document the dorsal and ventral surfaces of each seed sample (Supplementary Table S1). Images were captured using a Panasonic DMC-FZ50 camera under consistent lighting conditions. Millimeter paper and a 30 cm ruler were included in each photograph for accurate scale reference.

### 2.2. Planting Seeds

The germination protocol was selected based on the standard procedure routinely used by the *Phoenix* germplasm bank, which has successfully applied it to over 1500 seed-derived accessions representing the full diversity of the genus (Figure 1). Uniform conditions were preferred to ensure consistency across the samples analyzed in this study.

On 26 June 2017, the planting phase of the study was initiated. Fifteen seeds from each sample were directly sown into standardized black containers with the shape of an inverted pyramidal frustum, with commercial potting soil, each with a 3 L volume capacity and dimensions of 15 cm × 20 cm, without any pre-treatment or hydration. To maintain experimental uniformity, these containers were utilized across all treatments. Each container was meticulously labeled to record critical metadata, including the provenance, collection date, collector identification, and pertinent additional information, thereby ensuring accurate and comprehensive data collection and facilitating subsequent interpretation of results. This systematic methodological approach was implemented to ensure rigorous sample management throughout the initial phase of the study. Subsequently, on 2 April 2018, to encourage greater root development and ensure optimal adaptation for future field planting, plants were transplanted from the initial 3L pots to larger 13L pots.

### 2.3. Data Collection

Data collection commenced on 17 July 2017, with the initial observation of seed germination. The initial recorded data consisted of the total number of germinated seeds per pot. Subsequently, seedling growth was monitored using a 40 cm ruler. To ensure consistent longitudinal data, the first three germinated seedlings in each pot were measured at 15-day intervals from the onset of germination until 19 July 2018. Growth measurements were continued until the development of five leaves per seedling, with all data recorded in Excel.

Concurrent with growth measurements, environmental parameters were monitored. Air temperature, both internal and external to the greenhouse, was recorded using two "Six-Bellani" thermometers. The soil temperature within the pots was measured using a digital food thermometer (model "5989"), equipped with a 120 mm stainless steel probe and a measurement range of −50 °C to +150 °C, to investigate potential correlations between soil temperature and growth rates. These measurements were also taken at 15-day intervals, and comparative temperature data were obtained from the AEMET Orihuela observatory [37].

To evaluate early seedling development, a common method was employed: the assessment of plant growth rate during the first year through the measurement of plant length. Specifically, the length of the first leaf was recorded for three individuals per sample, as this metric serves as a reliable indicator of growth rate. All individuals were meticulously labeled to ensure accurate tracking and data association.

### 2.4. Data Analysis

The relationship between seed age and germination percentage was analyzed using Pearson's and Spearman's correlation coefficients. Population means were compared using Welch's *t*-test to accommodate potential heteroscedasticity. Species-specific germination patterns were evaluated using one-way ANOVA followed by post hoc analysis when appropriate. Statistical significance was established at $p < 0.05$. Effect sizes were calculated using

Cohen's d to quantify the magnitude of differences between groups. All statistical analyses were performed using standard statistical software in Python 3.11 and Claude 3.7 Sonnet, released in February 2025, analysis [38,39].

We also implemented a Bayesian framework to estimate the correlation coefficient (ρ) between seed age and germination rate. The bivariate relationship was modeled using a bivariate normal distribution with an uninformative prior on the correlation parameter.

For the prior distribution, we employed a uniform prior for ρ over the interval $[-1, 1]$, indicating no prior belief about the direction or strength of the correlation.

Posterior distributions were computed using Markov Chain Monte Carlo (MCMC) sampling with the No-U-Turn Sampler (NUTS) algorithm. We ran four independent chains, each with 2000 iterations, discarding the first 1000 iterations as warm-up.

Convergence was verified using the potential scale reduction factor ($\hat{R} < 1.05$) and effective sample size (>400 for all parameters).

From the posterior distribution, we calculated the mean posterior correlation coefficient, 95% credible intervals, the probability of a positive correlation ($P(\rho > 0)$), and the probability of a practically significant correlation ($P(\rho > 0.3)$).

The same Bayesian framework was applied separately to data subsets for *Phoenix dactylifera* and *Phoenix canariensis* to assess species-specific correlations.

This methodology provides a comprehensive probabilistic assessment of the relationship between seed age and germination, allowing for direct statements about the probability of correlations of different magnitudes rather than relying solely on significance testing.

## 3. Results

### 3.1. *Phoenix Seed Age and Germination*

Our results, as depicted in Figure 3, revealed an irregular pattern in seed germination.

Multiple variables influenced germination irregularity. Seed water content was essential during germination, while seed quality significantly affected germination rates. For instance, *P. canariensis* seeds from diverse origins demonstrated inconsistent behavior, with seeds in pot 27 exhibiting decomposition without germination. Seed quality typically peaked at harvest maturity before declining. Environmental conditions (temperature, moisture, and light) and physiological factors (dormancy status and nutrient reserves) further modulated germination patterns, with stronger dormancy or inadequate reserves potentially causing delayed or reduced germination.

In summary, the irregular germination pattern highlighted in our results was likely influenced by a combination of seed age, botanical source, seed quality, environmental factors, and physiological states. A comprehensive analysis considering these factors, and in particular age and botanical source, is essential to understand and address the observed variability in seed germination considering that we paid attention to work under uniform environmental factors and physiological states.

#### 3.1.1. Seed Age and Germination Percentage: Pearson Correlation Coefficient

The Pearson correlation ($r \approx 0.215$, $p = 0.246$) between the seed age and germination percentage revealed a weak, non-significant positive relationship. Seeds averaged 2129 days (~5.8 years) in age with 49% mean germination rate, with substantial variation in both metrics (age SD ≈ 834 days; germination SD ≈ 37%). While older seeds demonstrated a slight tendency toward higher germination, this trend lacked statistical significance. Other factors appeared more influential, particularly the species of origin, as evidenced by *P. dactylifera* seeds both being older and exhibiting higher germination percentages (Table 1).

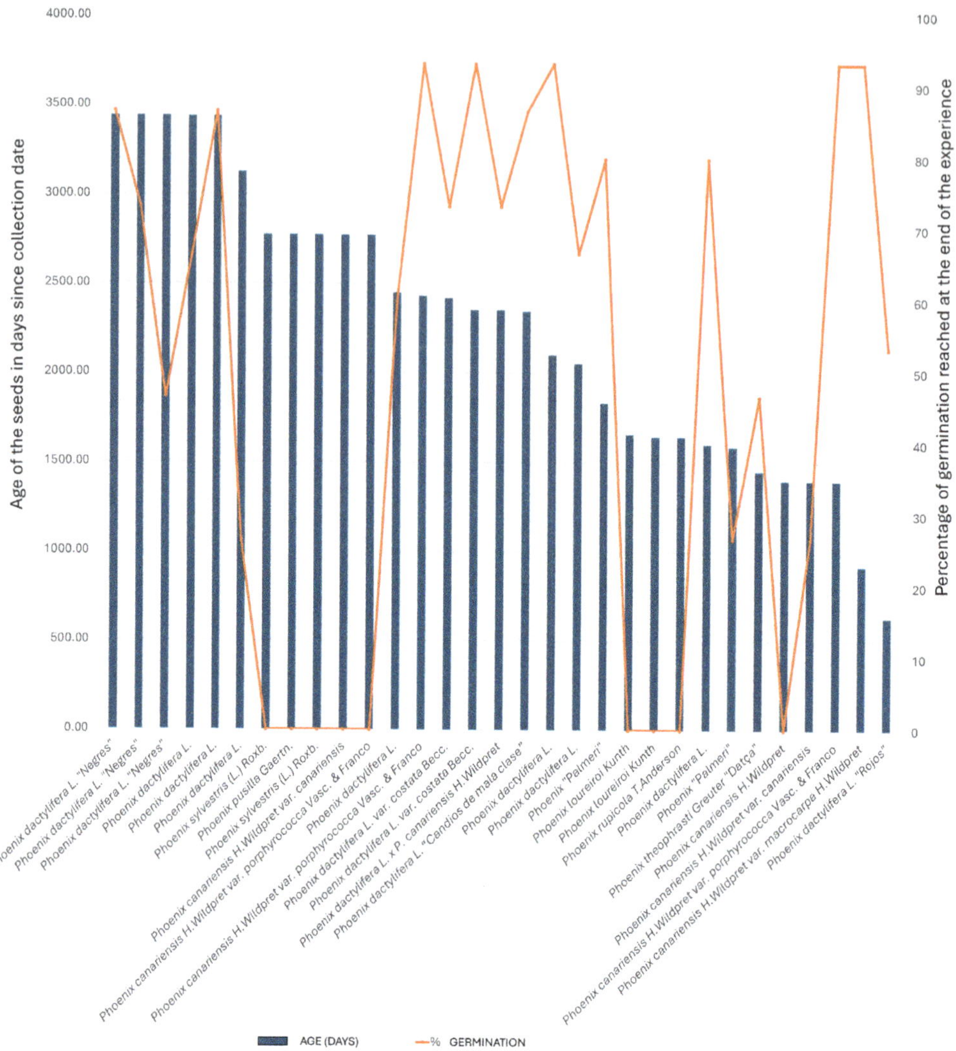


**Figure 3.** Comparison of seed age (blue bars) and germination percentage (orange line) across *Phoenix* plant species and varieties. The $x$-axis lists different *Phoenix* types described in Table 1 and Supplementary Table S1, while the left $y$-axis represents the seed age in days, and the right $y$-axis represents the germination percentage. The chart highlights variations in germination success relative to seed age and botanical origin. The names of local varieties or ethnovarieties are shown in quotation marks ("").

### 3.1.2. Seed Age and Germination Percentage: Spearman Correlation Coefficient

The Spearman correlation ($\rho$ = 0.184, $p$ = 0.323) indicated a weak, non-significant monotonic relationship between the seed age and germination percentage. This result, slightly weaker than the Pearson correlation (r = 0.215, $p$ = 0.246), suggested the relationship was neither strongly linear nor monotonic. Rank transformation failed to reveal patterns obscured in raw values. These findings confirmed that seed age was not a reliable predictor of germination percentage, with other factors likely exerting greater influence on germination outcomes.

3.1.3. Seed Age and Germination Percentage: *t*-Test

A *t*-test analysis was conducted by dividing samples at the median seed age (2346 days) into older ($n$ = 16) and younger ($n$ = 15) groups. Mean germination percentages were comparable between older (48.13%) and younger (50.89%) seeds, with a difference of −2.76%. Welch's two-sample *t*-test revealed no significant difference (t = −0.187, $p$ = 0.853) with a negligible effect size (Cohen's d = −0.068). This analysis corroborated the correlation findings, confirming that seed age did not significantly influence germination rates in this dataset.

*3.2. Phoenix Seed Age, Species, and Germination*

3.2.1. Overall Species Comparison

*Phoenix dactylifera* had the highest number of samples with 12, a mean germination rate of 70.14%, and a wide range from 26.67% to 93.33%. The average seed age was approximately 2437 days or about 6.7 years.

*Phoenix canariensis* had seven samples with a mean germination rate of around 42.86%. The germination rate varied significantly, from 0% to 93.33%, and the seeds were younger on average, at approximately 1724 days or about 4.7 years.

*Phoenix sylvestris*, *P. loureiroi*, and *P. rupicola* all showed 0% germination. These seeds ranged from relatively young to older, with ages exceeding 1600 days.

*Phoenix* × "Palmeri", likely a hybrid, exhibited variable germination rates ranging from 26.67% to 80%. The seeds were relatively young, averaging around 1710 days.

*Phoenix theophrasti* had a single sample, with a germination rate of 46.67%. The seed age was 1451 days.

Statistical Comparisons:

A *t*-test comparing the germination rates of *P. dactylifera* and *P. canariensis* revealed no statistically significant difference, despite *P. dactylifera*'s higher mean germination rate. This suggested that the observed difference may be due to chance or the limited sample size.

Age–Germination Correlation by Species:

For species with sufficient samples, neither *P. dactylifera* nor *P. canariensis* showed a significant correlation between the seed age and germination rate.

Practical Implications:

*P. dactylifera* demonstrated the highest overall germination success, maintaining good viability even with older seeds.

Some species, such as *P. sylvestris*, *P. pusilla*, *P. loureiroi*, and *P. rupicola*, showed no germination. This suggested either inherently lower seed viability, poor tolerance to long-term storage, or specific germination requirements that were not met in the study conditions.

Varieties within the same species, such as different varieties of *P. canariensis*, exhibited variable germination rates, indicating that genetic factors played an important role in seed viability.

3.2.2. Bayesian Approach to Correlation Analysis

A Bayesian approach to correlation analysis offered several advantages for our seed germination data: The analysis provided the full posterior distribution of correlation values rather than a single coefficient, offering a more comprehensive view. We could make direct probability statements, such as "there is an X% probability that the correlation is positive", instead of relying on *p*-values. Credible intervals were intuitive; a 95% credible interval directly indicated the range containing 95% of the plausible correlation values. If we had prior information about seed germination patterns, it could be formally incorporated into the analysis.

The key findings from Bayesian analysis were as follows: For the overall dataset, the mean posterior correlation was approximately 0.21, with a 95% credible interval ranging from about −0.15 to 0.55. The probability of a positive correlation was around 85%, while the probability of a practically significant correlation ($\rho > 0.3$) was about 30%. This meant there was an 85% chance that the seed age and germination were positively correlated but only a 30% chance that this correlation was strong enough to be practically meaningful.

Species-Specific Analysis: For *Phoenix dactylifera*, the mean posterior correlation was approximately 0.18, with a 95% credible interval ranging from −0.35 to 0.67. The probability of a positive correlation was around 75%.

For *Phoenix canariensis*, the mean posterior correlation was about 0.30, with a 95% credible interval ranging from −0.54 to 0.89. The probability of a positive correlation was around 71%. These species-specific results suggested that both species showed more evidence for positive than negative correlations. *P. canariensis* may have had a slightly stronger correlation between seed age and germination. However, the wide credible intervals for both species indicated substantial uncertainty.

This Bayesian analysis complemented frequentist approaches, such as Pearson, Spearman, and *t*-tests, by providing a more nuanced understanding of the relationship between seed age and germination in *Phoenix* species.

#### 3.2.3. Relationship Between the Year of Collection and the Germination Percentage of the Species

When examining the relationship between the year of collection and the germination percentage of the species, it was evident that the highest germination rates were observed in samples of *Phoenix dactylifera* from Spain, which belonged to the Western Group of this species, and in certain samples of *P. canariensis*. Remarkably, some *P. canariensis* samples, including both the typical variety and var. *porphyrococca*, failed to germinate throughout the observation period. Conversely, a hybrid between these two species, represented by a single sample, demonstrated over 70% germination (Figures 4 and 5).

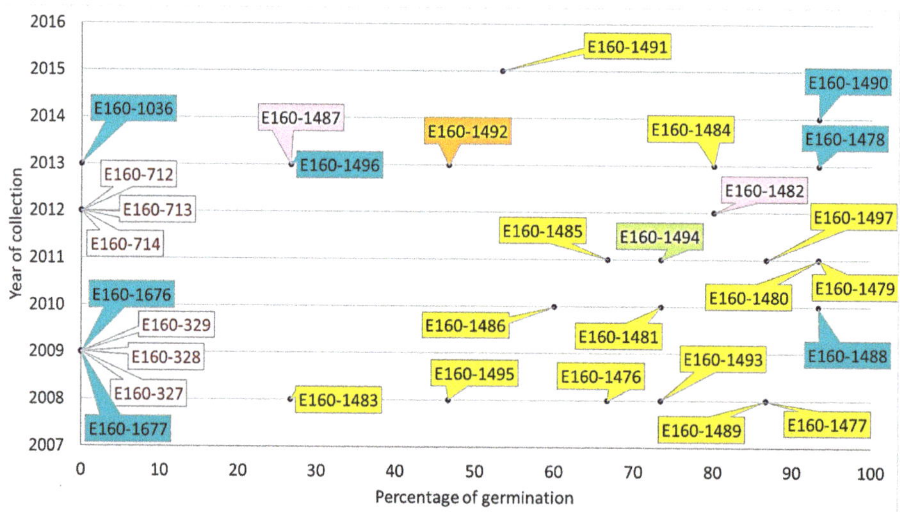


**Figure 4.** Germination percentage and year of collection across *Phoenix* samples. Color code: yellow—*Phoenix dactylifera*; blue—*Phoenix canariensis*; purple—*Phoenix* "Palmeri"; green—*Phoenix × hybrida* (*P. canariensis × P. dactylifera*); orange—*Phoenix theophrasti*; white—tropical *Phoenix* species of the Indian subcontinent. Note: Codes as in Table 1 and Supplementary Table S1.

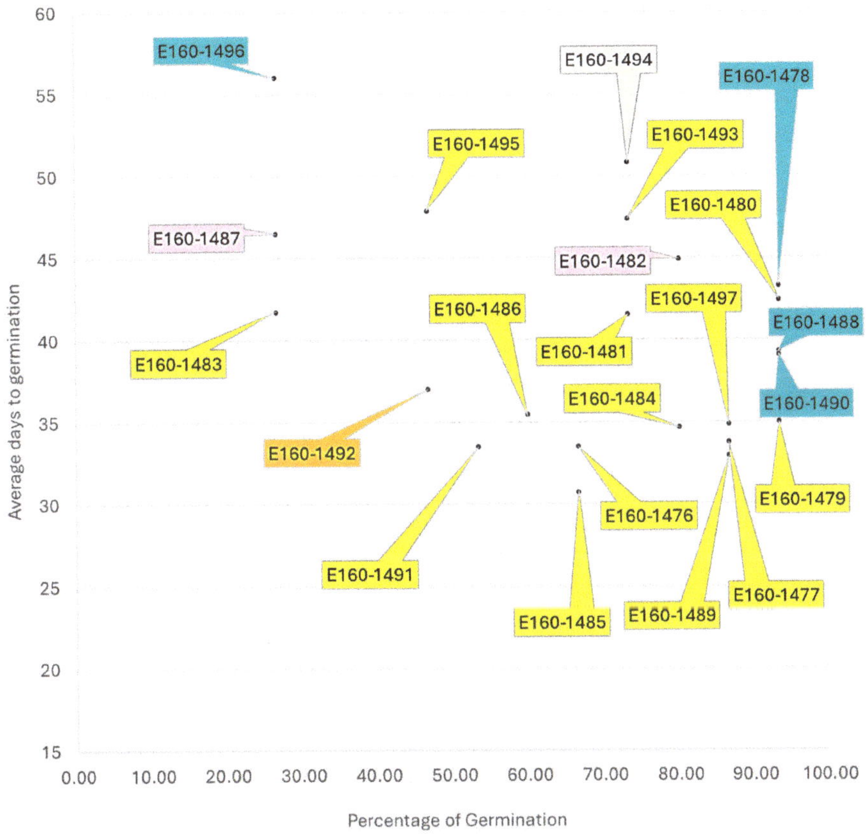


**Figure 5.** Germination percentage and average days to germination across *Phoenix* species. Color code: yellow—*Phoenix dactylifera*; blue—*Phoenix canariensis*; purple—*Phoenix* "Palmeri"; green—*Phoenix* × *hybrida* (*P. canariensis* × *P. dactylifera*); orange—*Phoenix theophrasti*. Note: Codes as in Table 1 and Supplementary Table S1.

It is noteworthy that the *P. canariensis* samples from the Espinardo Campus (E160-1676 and E160-1677) previously did not germinate in the year they were first planted (Table 1). This observation suggested a potential issue with the quality of the source material as the palms were already severely infested with *Rhynchophorus ferrugineus*, which may have adversely affected the normal development of fruits and seed viability. Similarly this occurred with *P. canariensis* var. *canariensis* from Colombres and among the Indian samples with *P. pusilla* (Table 1).

On the contrary, several samples with good gemination rates (Figure 4) failed previously to germinate: *P. dactylifera* var. *dactylifera* "Negres" (E160-1489) and two *P. dactylifera* var. *dactylifera* (samples E160-1476 and E160-1477) from Fortuna and *P. dactylifera* var. *dactylifera* from Orihuela (E160-1485).

The sample of *P. theophrasti* from Datça (Turkey), possibly a hybrid between *P. theophrasti* and *P. dactylifera* according to Carreño [3], displayed a germination rate close to 50%. Notably, none of the tropical species from the Indian subcontinent and Southeast Asia (*Phoenix sylvestris*, *P. loureiroi*, *P. roebelenii*, *P. rupicola*, and *P. pusilla*) germinated, except for two originally labeled *P. acaulis* samples. The unexpectedly high germination in samples initially attributed to *P. acaulis*, considering that this species belonged to the

Indian group, confirmed previous questions regarding their identity as thousands of seeds labeled *P. acaulis* circulated in Europe in the early 2000s, producing plants resembling the so-called *P.* × "Palmeri", which Carreño [3] identified as a complex hybrid involving multiple parental species. However, the samples of *Phoenix sylvestris*, *P. loureiroi*, and *P. rupicola* had successfully germinated in 2011 or 2013 (Table 1).

The data presented in Figure 4 demonstrated that the year of seed collection, and consequently the seed age, exerted a substantially lesser effect on germination capacity than other examined factors. Specifically, the botanical origin of the seeds, defined by species, and the geographic region of origin were shown to be significantly more influential determinants of germination success.

There did not appear to be a clear relationship between the average days to germination and the final germination percentage of the samples. However, all the most successful samples with high germination percentages had an average germination time of 35–43 days (Figure 5).

### 3.3. *Phoenix Seedling Development*

#### 3.3.1. Leaf Number

The analysis of total leaf emergence patterns in *Phoenix* spp. seedlings revealed distinct species-level and variety-level trends, offering valuable insights for palm cultivation and breeding strategies.

Species-Level Patterns (Figure 6A):

All species exhibited synchronized major leaf emergence periods in September 2017 and March–July 2018. *P. dactylifera* demonstrated the most pronounced step pattern, with dramatic increases during these windows. The universal September emergence event suggested a strong seasonal trigger across species.

Growth rate differentiation was evident among species. *P. dactylifera* displayed three distinct growth phases: rapid growth from July to September 2017, a plateau from October 2017 to February 2018, and accelerated growth from March to July 2018. In contrast, *P. canariensis* showed a more constant, gradual increase throughout the year. The hybrid *P.* × "Palmeri" exhibited minimal winter growth but notable spring emergence.

Dormancy periods were observed in all species, with reduced or halted leaf production between November and February. *P. dactylifera* maintained the most stable plateau during this period, while *P. theophrasti* and the hybrid (*P.* × *hybrida*) showed nearly complete cessation of new leaf emergence.

Variety-Level Patterns (Figure 6B):

Significant intraspecific variation was noted within *P. dactylifera*. The standard *P. dactylifera* s.l. dramatically outperformed, in total leaf numbers, specialized varieties such as "Negres", "Candios de mala clase", and "Rojos", which showed substantially lower production rates. Similarly, *P. canariensis* varieties differed significantly, with var. *porphyrococca* performing better than var. *macrocarpa* or var. *canariensis*.

The response to seasonal triggers varied among varieties. The September leaf flush was strongest in *Phoenix dactylifera* L. s.l. and the hybrid *P dactylifera* × *P. canariensis*. Spring emergence timing varied slightly, with some varieties activating earlier in March compared with others in April–May. *P. canariensis* varieties showed a less pronounced seasonal response compared with *P. dactylifera* varieties.

The hybrid *Phoenix dactylifera* × *P. canariensis* exhibited intermediate behavior between its parent species, following the *P. dactylifera* emergence pattern but at a reduced magnitude. On a per-plant basis, considering only three individuals, this hybrid appeared quite productive. This hybrid exhibited growth and developmental rates analogous to

*P. dactylifera*; however, its leaf emergence frequency pattern more closely aligned with that of *P. canariensis*.

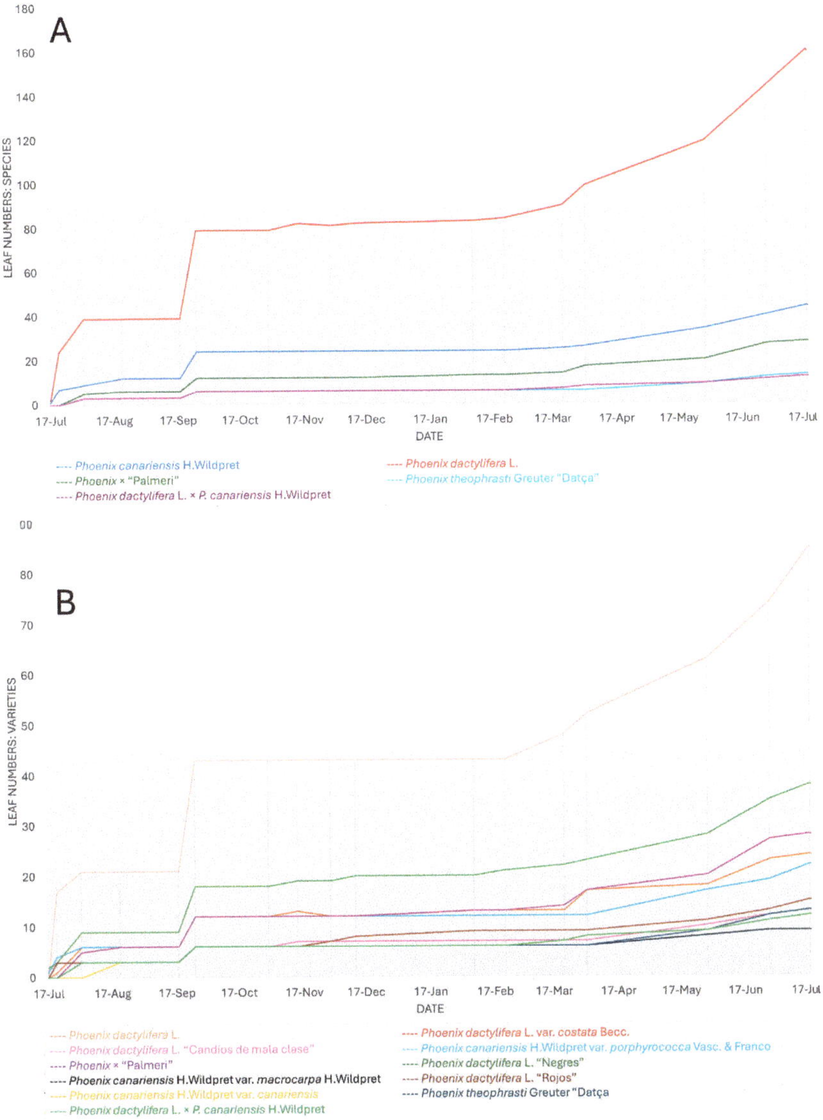


**Figure 6.** Leaf production dynamics in *Phoenix* spp. seedlings during the first 12 months post-germination, represented as total leaf count. (**A**) Total leaf production per species. The figure includes 42 of the 149 individuals germinated of *P. dactylifera*, 15 of 46 of *P. canariensis*, 6 of 16 of *P.* "Palmeri", 3 of 11 of *P. hybrida*, and 3 of 7 of *P. theophrasti*. (**B**) Total leaf production per variety. The figure includes 21 of the 72 individuals that germinated *P. dactylifera*. s.l., 6 of 31 of *P. dactylifera* "Negres", 9 of 28 of *P. canariensis* var. *porphyrococca*, 6 of 25 of *P dactylifera* var. *costata*, 6 of 16 of *P.* "Palmeri", 3 of 14 of *P. canariensis* var. *macrocarpa*, 3 of 13 of *P. dactylifera* "Candios de mala clase", 3 of 11 of *P. dactylifera* × *P. canariensis*, 6 of 8 of *P. dactylifera* "Rojos", 3 of 7 of *P. theophrasti* "Datça", and 3 of 4 of *P. canariensis* var. *canariensis*.

Species vs. Varieties—Key Insights:

The basic leaf emergence pattern, characterized by a September flush, winter dormancy, and spring flush, appeared genetically conserved across the *Phoenix* genus. However, the magnitude and duration of these flushes varied substantially between species and varieties. *P. dactylifera* genetics likely conferred greater leaf production potential regardless of variety.

All *Phoenix* palms responded to the same seasonal cues, albeit with varying sensitivity. The universal September flush suggested a response to late summer conditions, while the consistent winter dormancy indicated temperature or photoperiod sensitivity across the genus.

Hybrid vigor was not strongly evident in leaf production metrics in the case of *P.* × *hybrida*, suggesting that breeding for increased leaf production should focus on *P. dactylifera* genetics (Figure 7).

**Figure 7.** Leaf production dynamics in *Phoenix* spp. seedlings over the first 12 months. (**A–C**) E160-1486 *P. dactylifera* L., (**D–F**) E160-1494 *P.* × *hybrida* André (*P. dactylifera* L. × *P. canariensis* H. Wildpret), and (**G–I**) E160-1492 *P. theophrasti* Greuter "Datça". (**A**,**D**,**G**) 7-August-2017, (**B**,**E**,**H**) 7-February-2018, and (**C**,**F**,**I**) 30-May-2018. The black, square pots (**A**,**B**,**D**,**E**,**G**,**H**) were used in the first phase, germination and growth, until March 2018. In April 2018, they were transplanted into larger, round, brown pots (**C**,**F**,**I**).

Growth strategy differences were notable between species. *Phoenix dactylifera* demonstrated an "opportunistic" strategy with intense periodic leaf flushes, while *P. canariensis* showed a more "conservative" approach with steady, continuous leaf production. These strategies likely reflected adaptations to their native habitats.

Considering the mean number of leaves per individual, the analysis of key patterns and observations in *Phoenix* spp. seedlings also revealed the distinct growth phases and species-specific variations, offering insights into developmental biology and taxonomic relationships within the genus.

Growth Phases: Most taxa exhibited three distinct phases: initial emergence (Figures 6 and 7), a stable plateau period, and a final acceleration phase. The timing of these phases was relatively consistent across taxa, suggesting a conserved developmental program.

Species-Specific Patterns: *Phoenix dactylifera* varieties showed considerable variation. The "Negres" variety exhibited the most vigorous growth, reaching 6.33 leaves on average by the end, while the standard variety reached 4.05 leaves. In contrast, the "Rojos" variety showed remarkably slow development, producing only on average 2.50 leaves per individual. This suggested high intraspecific variability within *P. dactylifera*.

*Phoenix canariensis* varieties generally demonstrated slower leaf emergence, with all varieties having fewer leaves by the end compared with most *P. dactylifera* varieties. The variety *porphyrococca* showed particularly slow development, reaching only 2.44 leaves on average. The species average was 2.93 leaves by the final time point.

Hybrid Performance: *Phoenix* × "Palmeri" exhibited vigorous growth, reaching on average 4.67 leaves, suggesting possible heterosis or hybrid vigor. The *Phoenix dactylifera* × *P. canariensis* hybrid showed intermediate performance, with 4.00 leaves, falling between the two parent species, but it did not reach 5 leaves like its parents and was somewhat slower. The speed of its growth was more similar to *P. canariensis* since it was March when its third leaf developed, and in June its fourth final leaf appeared.

Initial Emergence Timing: *P. dactylifera* "Negres" uniquely started with a mean of 0.33 leaves at the first point, indicating earlier germination. In contrast, *P. canariensis* var. *canariensis* and *P. theophrasti* "Datça" showed delayed initial emergence.

Stability Periods: Most taxa demonstrated extended plateaus where the leaf number remained constant. The *Phoenix dactylifera* "Negres" showed the most continuous growth with fewer plateaus.

Final Growth Acceleration: All taxa exhibited accelerated leaf production in the final three to four time points, with the magnitude of this acceleration varying dramatically between taxa (Figure 7).

Taxonomic Significance: The analysis highlighted remarkable intraspecific variation within *Phoenix dactylifera*, with the final mean leaf counts ranging from 2.50 to 6.33 among its varieties, suggesting high genetic diversity or plasticity. Despite differences in growth rate, the overall pattern of initial emergence followed by a plateau and final acceleration was conserved across taxa, indicating a shared developmental program.

Hybrid performance suggested that leaf emergence traits may be inherited differently, with some hybrids showing intermediate inheritance and others possibly exhibiting heterosis. The patterns supported current taxonomic relationships, with varieties within species showing more similar patterns than between species, although notable exceptions existed, such as *P. dactylifera* "Rojos," which behaved more like a *P. canariensis* variety.

This analysis provides valuable insights into both developmental biology and taxonomic relationships within the *Phoenix* genus, highlighting the importance of seedling development traits in understanding palm evolution and classification.

### 3.3.2. Leaf Length

The growth patterns of *Phoenix* spp. seedlings were monitored over a 12-month period, revealing distinct variations among species and seasonal influences (Figure 8). *Phoenix dactylifera* varieties, notably "Rojos", exhibited the most vigorous growth, while *Phoenix theophrasti* "Datça" achieved the highest leaf length of 48 cm in April 2018. Conversely, *Phoenix canariensis* varieties showed slower but steady growth.

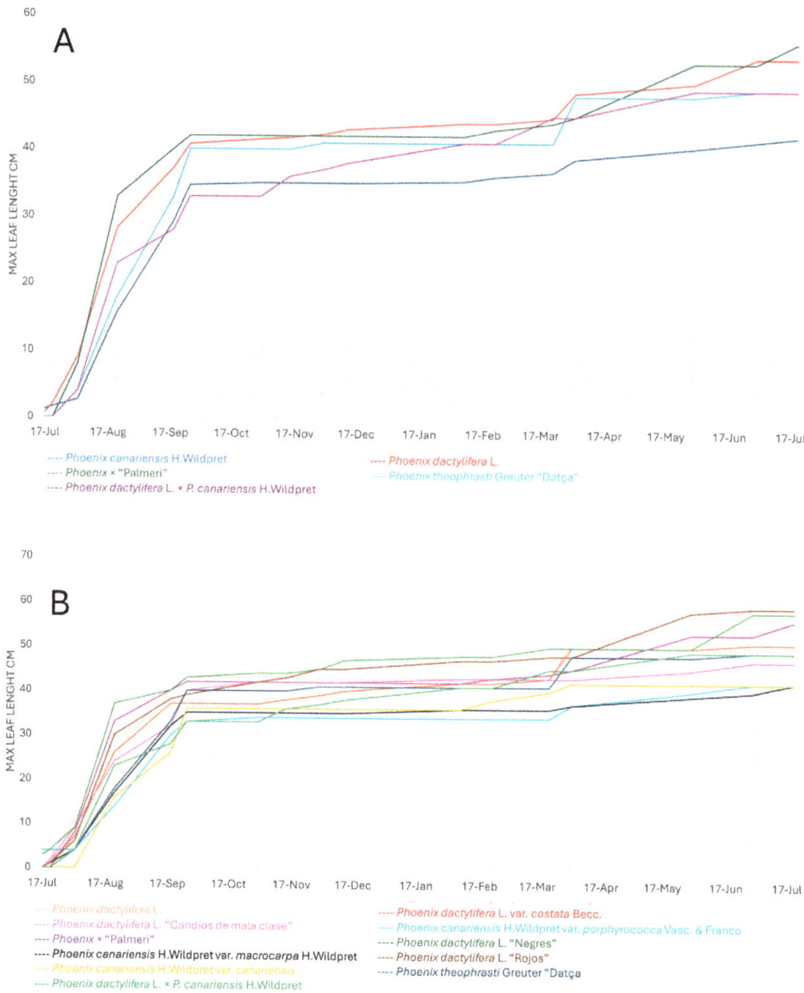


**Figure 8.** Temporal patterns of leaf development in *Phoenix* spp. seedlings during the initial 12-month post-germination period: maximum leaf length. (**A**) Maximum length per species. The figure includes 42 of the 149 individuals germinated of *P. dactylifera*, 15 of 46 of *P. canariensis*, 6 of 16 of *P. "Palmeri"*, 3 of 11 of *P. hybrida*, and 3 of 7 of *P. theophrasti*. (**B**) Maximum length per variety. The figure includes 21 of the 72 individuals that germinated *P. dactylifera*. s.l., 6 of 31 of *P. dactylifera* "Negres", 9 of 28 of *P. canariensis* var. *porphyrococca*, 6 of 25 of *P dactylifera* var. *costata*, 6 of 16 of *P.* "Palmeri", 3 of 14 of *P. canariensis* var. *macrocarpa*, 3 of 13 of *P. dactylifera* "Candios de mala clase", 3 of 11 of *P. dactylifera* × *P. canariensis*, 6 of 8 of *P. dactylifera* "Rojos", 3 of 7 of *P. theophrasti* "Datça", and 3 of 4 of *P. canariensis* var. *canariensis*.

Seasonal effects significantly influenced the growth patterns, with a noticeable slowdown during winter (December–February) and a growth surge in spring (March–April) (Figure 8). Some species exhibited slight decreases in average leaf length due to novel leaf emergence. This led us to instead represent the maximum length of the leaves in Figure 8.

By the final measurement on July 19th of the second year, most species reached mean leaf lengths between 28 and 38 cm, with *Phoenix dactylifera* "Rojos" and *Phoenix* "Palmeri" achieving the greatest average lengths. But this peak of leaf length was shared also with other *P. dactylifera* s.l. samples and the *P. dactylifera* "Negres" cultivar (Figure 8B). On average, the fastest-growing varieties were *P. dactylifera* "Rojos", *P.* "Palmeri," and *P. theophrasti* "Datça," while the slowest were *P. canariensis* var. *porphyrococca* and the hybrid *P. dactylifera* × *P. canariensis* (*P.* × *hybrida*).

The data suggested that different *Phoenix* species had distinctive growth patterns, which could inform selection for landscaping or agriculture. The hybrid did not exhibit hybrid vigor for leaf length, and there may be optimal times for interventions based on observed growth phases.

Key growth patterns included a rapid establishment phase from July to September, a winter plateau from November to February, a spring growth surge from March to April, and a maturation phase from May to July of the second year. The top performers in final leaf length were *Phoenix dactylifera* "Rojos" (mean 37.67 cm and maximum 59 cm), *P.* × "Palmeri" (mean 36.86 cm and maximum 56 cm), and *P. dactylifera* "Negres" (mean 33.63 cm and maximum 58 cm), while slower growers included *P. canariensis* var. *porphyrococca* (mean 26.82 cm and maximum 42 cm) and the hybrid (mean 28.08 cm and maximum 49 cm). Overall, *P. dactylifera* varieties achieved greater leaf lengths than *P. canariensis* varieties, with the hybrid showing intermediate performance.

## 4. Discussion

The diaspores of most palm species have morphophysiological dormancy. Date palm seedlings exhibit a "remote-tubular" germination pattern. The embryonic axis within the cotyledonary sheath extends downward via positive geotropism, allowing the seedling to develop at a distance from the seed. The persistent radicle breaks through the cotyledonary base, while the plumule emerges from the sheath's cleft or ligular extension. This remote germination strategy relocates the seedling from the original seed location [40–42].

Seed storage behavior can be orthodox or non-orthodox (intermediate or recalcitrant) [5–10,28]. We identified two distinct behavioral patterns regarding the influence of storage conditions involving seed desiccation and preservation at low temperatures, although above freezing, in the successful germination of seeds. The "tropical" group of species from the Indian subcontinent and Southeast Asia appeared unable to tolerate these storage conditions, rapidly losing germination capacity, likely due to their recalcitrant nature. This coincides with the generally accepted view that recalcitrance is more common in tropical and subtropical regions, particularly in humid broadleaved forests [14,15]. "Recalcitrant" seeds are desiccation intolerant. They are highly hydrated at shedding and do not survive drying below a moisture content between 30 and 65% depending on the species [7,43].

This finding is particularly relevant for the international seed trade as these species should ideally be sown from freshly harvested fruits. In contrast, the "Mediterranean-Saharan" group, broadly defined, includes species such as *Phoenix dactylifera* L., *P. canariensis* H. Wildpret, and *P. theophrasti* W. Greuter, whose seeds exhibit an apparently orthodox behavior. This, again, coincides with the generally accepted view that orthodox seeds predominate in temperate regions, arid environments, and dry forests [44,45]. Seeds that have better protection against damage that occurs under dry conditions survive longer;

hence, seed longevity is a manifestation of desiccation tolerance and is conventionally treated as a quantitative trait [28,46]. Obviously, other factors such as the degree of maturity of the fruit must be considered. In the case of *P. canariensis* it has been found that the best percentages of gemination are obtained from fully ripe fruits and under temperatures ranging between 20 and 30 °C and above [47]. Pre-treatments involving immersion in water at different temperatures can also improve germination rates [48]. In our experiment we worked with seeds of different species and varieties under uniform conditions without pre-treatment.

Archival botanical repositories, historical herbaria, and archaeological assemblages have yielded viable propagules with remarkable longevity, facilitating the regeneration of extinct germplasm and the restoration of extirpated taxa [49]. Documentary evidence provides the most robust validation of seed longevity, as exemplified by three notable cases: *Vachellia farnesiana* (L.) Wight & Arn. (*Acacia farnesiana* L.) and *Acacia melanoxylon* R.Br. collected in Egypt in 1856 and deposited in the Swedish Museum of Cultural History, which germinated after 151 years of storage; *Albizia julibrissin* Durazz. specimens, collected in 1793, which exhibited germination following inadvertent hydration during the British Museum's aerial bombardment of 1940; and *Nelumbium* specimens, originally accessioned in 1705 within the Hans Sloane collection at the British Museum, which demonstrated viability upon testing in 1942. However, seed age determination becomes substantially more problematic when derived from archaeological or geological contexts as both the absolute chronology of cultural or geological events may be imprecise and the temporal relationship between seeds and their encompassing matrix cannot always be definitively established [49–54].

However, from archaeological excavations in the Judean desert (1963–1991), 32 preserved date seeds from Masada, Qumran, Wadi Makukh, and Wadi Kelt sites were cultivated at Kibbutz Ketura. Six seeds were successfully germinated and named "Adam" (Masada), "Jonah", "Uriel", "Boaz", "Judith" (Qumran), and "Hannah" (Wadi Makukh) (Table 2). Radiocarbon dating was performed on seed shell fragments adhering to the seedlings' rootlets, including the previously germinated "Methuselah". After adjusting for modern carbon contamination (2–3%), the chronological distribution showed Methuselah, Hannah, and Adam (first to fourth centuries BCE), Judith and Boaz (mid-second century BCE to mid-first century CE), and Uriel and Jonah (first to second centuries CE) [34,35].

Table 2 provides a comprehensive overview of germination data for seeds of various date palm (*Phoenix*) species under different storage conditions and ages. The analysis reveals several key insights.

Ancient date palm seeds (*Phoenix dactylifera*) were discovered in archaeological sites in Israel, specifically Masada, Wadi Makukh, and Qumran. These seeds were approximately 1900–2200 years old. Remarkably, these ancient seeds exhibited a germination rate of around 19%, demonstrating their ability to retain viability over millennia under natural dry conditions.

In contrast, 144-year-old Kew Garden seeds, which included multiple varieties of *Phoenix dactylifera* collected from Baghdad in 1873, showed a 0% germination rate. These seeds were stored in sealed glass jars and later in acid-free boxes at approximately 16 °C at Kew Gardens, indicating a complete loss of viability under these conditions (Table 2).

Modern storage methods have yielded more promising results. Recent *Phoenix dactylifera* seeds, aged 9–11 years, exhibited an 88% germination rate when dried at 15% relative humidity and stored at −20 °C. Desiccated seeds stored at 5 °C showed a germination rate ranging from 47% to 87%. Similarly, *Phoenix canariensis* varieties stored for 2–17 years demonstrated variable germination rates, ranging from 52% to 93%. The best results were achieved through desiccated storage at 5 °C and low-humidity freezer storage.

**Table 2.** Longevity of viable *Phoenix* sp. seeds: Documented germination success following extended preservation.

| Species | Varieties | Storage Conditions | Germination Percentage | Age of the Seeds (Years) | Refs |
|---|---|---|---|---|---|
| *Phoenix dactylifera* | "Methuselah" and "Hannah" | Dry archaeological, Masada | c. 19% | c. 2200 | 1,2 |
| *Phoenix dactylifera* | "Adam" | Dry archaeological, Wadi Makukh | c. 19% | c. 2200 | 1 |
| *Phoenix dactylifera* | "Judith" and "Boaz" | Dry archaeological, Qumran | c. 19% | c. 2000 | 1 |
| *Phoenix dactylifera* | "Uriel" and "Jonah" | Dry archaeological, Qumran | c. 19% | c. 1900 | 1 |
| *Phoenix dactylifera* | Baghdad, 1873, 35,988 ("Drah Subbah") | Sealed glass jars and later acid-free card boxes, at c. 16 °C, at Kew | 0% | 144 | 4,5 |
| *Phoenix dactylifera* | Baghdad, 1873, 35,990 ("Zadie") | Sealed glass jars and later acid-free card boxes, at c. 16 °C, at Kew | 0% | 144 | 4,5 |
| *Phoenix dactylifera* | Baghdad, 1873, 35,991 ("Brem") | Sealed glass jars and later acid-free card boxes, at c. 16 °C, at Kew | 0% | 144 | 4,5 |
| *Phoenix dactylifera* | Baghdad, 1873, 35,992 ("Makawieh Eshgar") | Sealed glass jars and later acid-free card boxes, at c. 16 °C, at Kew | 0% | 144 | 4,5 |
| *Phoenix dactylifera* | Baghdad, 1873, 35,993 ("Etchrisieh") | Sealed glass jars and later acid-free card boxes, at c. 16 °C, at Kew | 0% | 144 | 4,5 |
| *Phoenix dactylifera* | Baghdad, 1873, 36,006 ("Tiburzel") | Sealed glass jars and later acid-free card boxes, at c. 16 °C, at Kew | 0% | 144 | 4,5 |
| *Phoenix dactylifera* | Baghdad, 1873, 36,007 ("Sin Mufta") | Sealed glass jars and later acid-free card boxes, at c. 16 °C, at Kew | 0% | 144 | 4,5 |
| *Phoenix dactylifera* | Baghdad, 1873, 36,008 ("Khaderawieh") | Sealed glass jars and later acid-free card boxes, at c. 16 °C, at Kew | 0% | 144 | 4,5 |
| *Phoenix dactylifera* | Baghdad, 1873, 36,011 ("Makawieh Ahmur") | Sealed glass jars and later acid-free card boxes, at c. 16 °C, at Kew | 0% | 144 | 4,5 |
| *Phoenix dactylifera* | Baghdad, 1873, 36,012 ("El Washa") | Sealed glass jars and later acid-free card boxes, at c. 16 °C, at Kew | 0% | 144 | 4,5 |
| *Phoenix dactylifera* | Baghdad, 1873, 36,021 ("Sukri") | Sealed glass jars and later acid-free card boxes, at c. 16 °C, at Kew | 0% | 144 | 4,5 |
| *Phoenix dactylifera* | Baghdad, 1873, 36,024 ("Bedraieh") | Sealed glass jars and later acid-free card boxes, at c. 16 °C, at Kew | 0% | 144 | 4,5 |
| *Phoenix dactylifera* | Baghdad, 1873, 36,025 ("Khustawieh") | Sealed glass jars and later acid-free card boxes, at c. 16 °C, at Kew | 0% | 144 | 4,5 |
| *Phoenix dactylifera* | Baghdad, 1873, 35,988 ("Drah Subbah") | Sealed glass jars and later acid-free card boxes, at c. 16 °C | 0% | 144 | 4,5 |
| *Phoenix dactylifera* | Balqa, Jordan, 2003 | Dried at 15% [RH], stored at −20 °C | 88% | 11 | 4 |
| *Phoenix dactylifera* | "Negres" | Desiccated, stored at 5 °C | 47–87% | 9 | 3 |
| *Phoenix canariensis* | var. *porphyrococca* Vasc. & Franco | Desiccated, stored at 5 °C | 93% | 7 | 3 |
| *Phoenix dactylifera* | var. *costata* Becc. | Desiccated, stored at 5 °C | 93% | 6 | 3 |
| *Phoenix canariensis* | var. *porphyrococca* Vasc. & Franco | Desiccated, stored at 5 °C | 93% | 4 | 3 |
| *Phoenix. canariensis* | | Dried at 15% [RH], stored at −20 °C | 72% | 17 | 6 |
| *Phoenix. canariensis* | | Dried at 15% [RH], stored at −20 °C | 52% | 6 | 6 |
| *Phoenix. canariensis* | | Dried at 15% [RH], stored at −20 °C | 72% | 2 | 6 |

[1] [35]; [2] [34]; [3] This research; [4] [49]; [5] [54]; [6] [55]. Abbreviation: RH—relative humidity.

The key findings from this analysis underscore the remarkable longevity potential of date palm seeds. Ancient seeds can retain viability for millennia under natural dry conditions, while modern cold storage methods with controlled humidity are effective for medium-term preservation. However, the 144-year-old seeds stored at moderate temperatures completely lost viability, highlighting the critical role of storage temperature and low humidity in maintaining seed viability. These data emphasize the importance of proper storage conditions for the preservation of date palm seeds.

We must clearly distinguish between the concept of seed orthodoxy as it pertains to long-term storage in germplasm banks following seed desiccation and orthodoxy as verified by experimental desiccation with immediate germination testing. These results may not always align, as seen with *P. roebelenii* seeds, which in our experience at the germplasm bank did not withstand storage, although Beltrame et al. [56] classify them as physiologically orthodox; unfortunately, we did not have samples with enough seeds to include this species in our study. Davis and Pritchard [57] report *P. dactylifera* seeds as orthodox and those of *P. reclinata* as doubtfully orthodox.

The preservation of viable seeds has been proposed as a potential mechanism for the restoration of extinct plant species through the utilization of historical specimens maintained in herbaria and natural history collections. The theoretical framework for plant species resurrection postulates that viable propagules—specifically seeds or spores—preserved within herbarium voucher specimens of extinct taxa could potentially facilitate de-extinction events. This paradigm suggests that botanical extinction is not ultimately determined by the demise of the final living specimen but rather by the loss of viability in the terminal propagule. Currently, the primary constraint in implementing plant de-extinction protocols lies in the methodological limitations of germination and in vitro cultivation techniques. This challenge is particularly acute given the irreplaceable nature of diaspores conserved in herbarium specimens of extinct taxa, which precludes experimental optimization through iterative attempts [58]. This potential restoration methodology holds particular significance for agricultural sciences, specifically in cases where the extinct taxa represent domesticated crops or their wild progenitors with agronomic importance.

Our findings, despite temporal and taxonomic constraints, indicate that *Phoenix* taxa indigenous to Mediterranean and Saharan regions, notably *P. dactylifera* and related taxa, demonstrate remarkable desiccation tolerance in their seeds. The data suggest that long-term seed viability can be maintained under relatively moderate storage conditions. Consequently, the potential exists for the recovery of near-extinct taxa, such as *P. arabica* and *P. caespitosa*, through the utilization of seeds from preserved herbarium specimens.

However, samples of *Phoenix sylvestris*, *P. loureiroi*, and *P. rupicola*, aged between 1644 and 2773 days (Table 1), which failed to germinate in this study, had successfully germinated four to six years earlier. This suggests a loss of viability under relatively moderate storage conditions and indicates that these species may exhibit recalcitrant behavior.

## 5. Conclusions

Our study revealed significant differences in seed behavior and seedling development across *Phoenix* species. *P. dactylifera*, *P.* × "Palmeri", and *P. canariensis* demonstrated orthodox seed storage behavior with the seed age having minimal impact on germination success, while *P. sylvestris*, *P. rupicola*, *P. pusilla*, and *P. loureiroi* exhibited potentially recalcitrant characteristics. Species identity and geographic origin proved substantially more influential than seed age in determining germination outcomes, with notable intraspecific variation observed. Seedling development followed a conserved seasonal pattern across all species (September flush, winter dormancy, and spring flush), although with marked differences in magnitude and duration, particularly among *P. dactylifera* varieties where "Negres" displayed superior vigor. These findings provide valuable guidance for ex situ germplasm conservation strategies, suggesting that orthodox-behaving *Phoenix* species can be effectively preserved through conventional seed banking, while potentially recalcitrant species require alternative conservation approaches. Additionally, the observed growth patterns inform cultivar selection for ornamental horticulture and breeding programs, particularly in regions with pronounced seasonal temperature fluctuations. Future research should explore the physiological and biochemical mechanisms underlying seed recalcitrance in certain *Phoenix* species, investigate potential dormancy in *P. dactylifera*, characterize metabolic changes during prolonged storage, and test germination and growth responses across broader genetic pools and environmental conditions.

**Supplementary Materials:** The following supporting information can be downloaded at https://www.mdpi.com/article/10.3390/horticulturae11050537/s1: Table S1: *Phoenix* seed samples for the germination and early development study.

**Author Contributions:** Conceptualization, C.O. and D.R.; methodology, D.R. and C.O.; software, D.R. and S.P.-P.; validation, D.J. and C.O.; formal analysis, D.R.; investigation, S.P.-P.; resources, C.O.; data curation, D.R. and S.P.-P.; writing—original draft preparation, D.R.; writing—review and editing, D.J. and C.O.; visualization, D.R. and S.P.-P.; supervision, C.O. and D.R.; project administration, C.O.; funding acquisition, C.O. All authors have read and agreed to the published version of the manuscript.

**Funding:** This research was funded by INIA projects RF2007-00010-C03 and RF2010-00006-C02 (European Regional Development Fund 2007–2013), Prospección y recogida de recursos fitogenéticos autóctonos de palmera datilera y especies silvestres emparentadas, RFP 2013-00004-00-00, RFP2017-00004-00-00, INIA/Agencia Estatal de Investigación within Plan Estatal de Investigación Científica y Técnica y de Innovación, 2013–2016, and from the Ministerio de Ciencia e Innovación of Spain, project CGL 2008-04635.

**Data Availability Statement:** The original contributions presented in the study are included in the article/Supplementary Materials; further inquiries can be directed to the corresponding author.

**Conflicts of Interest:** The authors declare no conflicts of interest.

## References

1. Beccari, O. Rivista monografica delle specie del genere *Phoenix* L. *Malesia* **1890**, *3*, 345–416.
2. Barrow, S. A Monograph of *Phoenix* L. (Palmae: *Coryphoideae*). *Kew Bull.* **1998**, *53*, 513–575. [CrossRef]
3. Carreño, E. Diversidad Genética en Especies del Género *Phoenix* L. Ph.D. Thesis, Universidad Miguel Hernández de Elche, Orihuela, Spain, 2017.
4. Johnson, D.V.; Al-Khayri, J.M.; Jain, S.M. Seedling date palms (*Phoenix dactylifera* L.) as genetic resources. *Emir. J. Food Agric.* **2013**, *25*, 809–830. [CrossRef]
5. Hong, T.D.; Ellis, R.H. *A Protocol to Determine Seed Storage Behaviour*; IPGRI Technical Bulletin No. 1; International Plant Genetic Resources Institute: Rome, Italy, 1996; Available online: https://cropgenebank.sgrp.cgiar.org/images/file/learning_space/technicalbulletin1.pdf (accessed on 1 May 2025).
6. Pammenter, N.W.; Berjak, P. Physiology of desiccation-sensitive (recalcitrant) seeds and the implications for cryopreservation. *Int. J. Plant Sci.* **2014**, *175*, 21–28. [CrossRef]
7. Berjak, P.; Pammenter, N.W. Implications of the lack of desiccation tolerance in recalcitrant seeds. *Front. Plant Sci.* **2013**, *4*, 478. [CrossRef]
8. Walters, C.; Berjak, P.; Pammenter, N.; Kennedy, K.; Raven, P. Preservation of recalcitrant seeds. *Science* **2013**, *339*, 915–916. [CrossRef]
9. Walters, C. Orthodoxy, recalcitrance and in-between: Describing variation in seed storage characteristics using threshold responses to water loss. *Planta* **2015**, *242*, 397–406. [CrossRef] [PubMed]
10. Barbedo, C.J.; Centeno, D.C.; Ribeiro, R.C.F. Do recalcitrant seeds really exist? *Hoehnea* **2013**, *40*, 583–593. [CrossRef]
11. Rajjou, L.; Debeaujon, I. Seed longevity: Survival and maintenance of high germination ability of dry seeds. *Comptes Rendus Biol.* **2008**, *331*, 796–805. [CrossRef]
12. Matilla, A.J. The Orthodox Dry Seeds Are Alive: A Clear Example of Desiccation Tolerance. *Plants* **2022**, *11*, 20. [CrossRef]
13. FAO. *Genebank Standards for Plant Genetic Resources for Food and Agriculture*; FAO: Rome, Italy, 2014; pp. 18–64.
14. Tweddle, J.C.; Dickie, J.B.; Baskin, C.C.; Baskin, J.M. Ecological aspects of seed desiccation sensitivity. *J. Ecol.* **2003**, *91*, 294–304. [CrossRef]

15. Wyse, S.V.; Dickie, J.B. Predicting the global incidence of seed desiccation sensitivity. *J. Ecol.* **2017**, *105*, 1082–1093. [CrossRef]
16. Obroucheva, N.V.; Lityagina, S.V.; Novikova, G.V.; Sin'kevich, I.A. Vacuolar status and water relations in embryonic axes of recalcitrant *Aesculus hippocastanum* seeds during stratification and early germination. *AoB Plants* **2012**, *2012*, pls008. [CrossRef] [PubMed]
17. Farnsworth, E. The ecology and physiology of viviparous and recalcitrant seeds. *Annu. Rev. Ecol. Syst.* **2000**, *31*, 107–138 [CrossRef]
18. Hamilton, K.N.; Offord, C.A.; Cuneo, P.; Deseo, M.A. A comparative study of seed morphology in relation to desiccation tolerance and other physiological responses in 71 Eastern Australian rainforest species. *Plant Species Biol.* **2013**, *28*, 51–62. [CrossRef]
19. Daws, M.I.; Garwood, N.C.; Pritchard, H.W. Prediction of desiccation sensitivity in seeds of woody species: A probabilistic model based on two seed traits and 104 species. *Ann. Bot.* **2006**, *97*, 667–674. [CrossRef] [PubMed]
20. Hay, F.R.; Probert, R.J. Advances in seed conservation of wild plant species: A review of recent research. *Conserv. Physiol.* **2013**, *1*, cot030. [CrossRef]
21. Sershen; Varghese, B.; Pammenter, N.W.; Berjak, P. Cryo-tolerance of zygotic embryos from recalcitrant seeds in relation to oxidative stress—A case study on two amaryllid species. *J. Plant Physiol.* **2012**, *169*, 999–1011. [CrossRef]
22. Ballesteros, D.; Pritchard, H.W. The cryobiotechnology of oaks: An integration of approaches for the long-term ex situ conservation of *Quercus* species. *Forests* **2020**, *11*, 1281. [CrossRef]
23. Hong, T.D.; Linington, S.; Ellis, R.H. *Seed Storage Behaviour: A Compendium*; Handbooks for Genebanks No. 4; International Plant Genetic Resources Institute: Rome, Italy, 1996; Available online: https://cgspace.cgiar.org/server/api/core/bitstreams/7e86d173-da38-406e-bd8f-7c971ce9d4b9/content (accessed on 1 May 2025).
24. Nagel, M.; Rehman Arif, M.; Rosenhauer, M.; Börner, A. Longevity of seeds—Intraspecific differences in the Gatersleben genebank collections. In *60. Tagung der Vereinigung der Pfl anzenzüchter und Saatgutkaufleute Österreichs Lehr- und Forschungszentrum für Landwirtschaft, Raumberg-Gumpenstein 2009*; Vereinigung der Pflanzenzüchter und Saatgutkaufleute Österreichs: St. Pölten, Austria, 2010; pp. 179–182.
25. Solberg, S.; Yndgaard, F.; Andreasen, C.; von Bot, R.; Loskutov, I.G.; Asdal, Å. Long-Term Storage and Longevity of Orthodox Seeds: A Systematic Review. *Front. Plant Sci.* **2020**, *11*, 1007. [CrossRef]
26. Nadarajan, J.; Walters, C.; Pritchard, H.W.; Ballesteros, D.; Colville, L. Seed Longevity—The Evolution of Knowledge and a Conceptual Framework. *Plants* **2023**, *12*, 471. [CrossRef]
27. Yashina, S.; Gubin, S.; Maksimovich, S.; Yashina, A.; Gakhova, E.; Gilichinsky, D. Regeneration of whole fertile plants from 30,000-y-old fruit tissue buried in Siberian permafrost. *Proc. Natl. Acad. Sci. USA* **2012**, *109*, 4008–4013. [CrossRef] [PubMed]
28. Farooq, M.A.; Zhang, X.; Zafar, M.M.; Ma, W.; Zhao, J. Roles of reactive oxygen species and mitochondria in seed germination. *Front. Plant Sci.* **2021**, *12*, 781734. [CrossRef]
29. Crowe, J.H. Anhydrobiosis: An unsolved problem. *Am. Nat.* **1971**, *105*, 563–573. [CrossRef]
30. Lloyd, F.E. Development and Nutrition of the Embryo, Seed and Carpel in the Date, *Phoenix dactylifera* L. *Mo. Bot. Gard. Annu. Rep.* **1910**, *1910*, 103–164. [CrossRef]
31. von Fintel, G.T.; Berjak, P.; Pammenter, N.W. Seed behaviour in *Phoenix reclinata* Jacquin, the wild date palm. *Seed Sci. Res.* **2004**, *14*, 197–204. [CrossRef]
32. Muhammad, M.; Ringim, A.S.; Dangora, I.I. Effects of different methods of breaking dormancy and seed germination rate in date palm (*Phoenix dactylifera* L.). *J. Res. For. Wildl. Environ.* **2017**, *9*, 28–35.
33. Mohammed, N.M.I.; Said, A.G.E. Date palm (*Phoenix dactylifera*) seeds germination. *Cell Biol. Dev.* **2018**, *2*, 63–68. [CrossRef]
34. Sallon, S.; Solowey, E.; Cohen, Y.; Korchinsky, R.; Egli, M.; Woodhatch, I.; Simchoni, O.; Kislev, M. Germination, genetics, and growth of an ancient date seed. *Science* **2008**, *320*, 1464. [CrossRef]
35. Sallon, S.; Cherif, E.; Chabrillange, N.; Solowey, E.; Gros-Balthazard, M.; Ivorra, S.; Terral, J.F.; Egli, M.; Aberlenc, F. Origins and insights into the historic Judean date palm based on genetic analysis of germinated ancient seeds and morphometric studies. *Sci. Adv.* **2020**, *6*, eaax0384. [CrossRef]
36. Obón, C.; Rivera, D.; Amorós, A.; Alcaraz, F.; Díaz, G.; Carreño, E.; Martínez-Rico, M.; Larrosa, E.; Laguna, E. El Banco de Germoplasma Español de Palmera Datilera y Especies Próximas. In *Huellas inéditas del VI Congreso Internacional de Etnobotánica/Recovered Tracks from the VIth International Congress of Ethnobotany, ICEB 2014*; Herrera, F., Hurrel, J., Tarifa, F., Hernández-Bermejo, J., Eds.; UCO Press: Córdoba, Spain, 2017; pp. 235–254.
37. AEMET. Agencia Estatal de Meteorología. 2025. Available online: https://www.aemet.es/es/portada (accessed on 20 July 2018).
38. Python. Python [software]. Version 3.11.5. Python Software Foundation. Available online: https://www.python.org/ (accessed on 8 March 2025).
39. Anthropic. Claude Analysis [Internet]. Anthropic—Cited 10 March 2025. Available online: https://claude.ai (accessed on 8 March 2025).
40. Baskin, J.M.; Baskin, C.C. What Kind of Seed Dormancy Might Palms Have? *Seed Sci. Res.* **2013**, *24*, 17–22. [CrossRef]

41. Biradar, N.; Mahabale, T. Studies on palms: Fruits, seeds and seed germination in the genus *Phoenix* L. *Proc. Indian Acad. Sci. Sect. B* **1969**, *70*, 55–65. [CrossRef]
42. Iossi, E.; Moro, F.V.; Sader, R. Seed anatomy and germination of *Phoenix roebelenii* O'Brien (Arecaceae). *Rev. Bras. De Sementes* **2006**, *28*, 121–128. [CrossRef]
43. Corbineau, F. The effects of storage conditions on seed deterioration and ageing: How to improve seed longevity. *Seeds* **2024**, *3*, 56–75. [CrossRef]
44. Duncan, C.; Schultz, N.; Lewandrowski, W.; Good, M.K.; Cook, S. Lower dormancy with rapid germination is an important strategy for seeds in an arid zone with unpredictable rainfall. *PLoS ONE* **2019**, *14*, e0218421. [CrossRef]
45. Khurana, E.; Singh, J.S. Ecology of seed and seedling growth for conservation and restoration of tropical dry forest: A review. *Environ. Conserv.* **2001**, *28*, 39–52. [CrossRef]
46. Ellis, R.H. The longevity of seeds. *HortScience* **1991**, *26*, 1119–1125. [CrossRef]
47. Pimenta, R.S.; Luz, P.B.; Pivetta, K.F.; Castro, A.D.; Pizetta, P.U. Efeito da maturação e temperatura na germinação de sementes de *Phoenix canariensis* hort. ex Chabaud-Arecaceae. *Rev. Árvore* **2010**, *34*, 31–38. [CrossRef]
48. Azad, M.S.; Rahman, M.T.; Matin, M.A. Seed germination techniques of *Phoenix dactylifera*: A new experience from Bangladesh. *Front. Agric. China* **2011**, *5*, 241–246. [CrossRef]
49. Porteous, G.; Nesbitt, M.; Kendon, J.P.; Prychid, C.J.; Stuppy, W.; Conejero, M.; Ballesteros, D. Assessing extreme seed longevity: The value of historic botanical collections to modern research. *Front. Plant Sci.* **2019**, *10*, 1181. [CrossRef]
50. Godwin, H. Evidence for longevity of seeds. *Nature* **1968**, *220*, 708–709. [CrossRef]
51. Toole, V.K. Ancient seeds; seed longevity. *J. Seed Technol.* **1986**, *10*, 1–23.
52. Zazula, G.; Harington, R.; Telka, A.; Brock, F. Radiocarbon dates reveal that Lupinus arcticus plants were grown from modern not Pleistocene seed. *New Phytol.* **2009**, *182*, 788–792. [CrossRef]
53. Leino, M.; Edqvis, J. Germination of 151-year old *Acacia* spp. seeds. *Genet. Resour. Crop Evol.* **2010**, *57*, 741–746. [CrossRef]
54. Colvill, W.H. Some observations on the vegetable productions and the rural economy of the province of Baghdad. *Bot. J. Linn. Soc.* **1875**, *14*, 399–410. [CrossRef]
55. Perez, M.A.G.; Cabrera-García, N.; Curbelo, L.; Sosa, P.A. Is the endemic *Phoenix canariensis* H. Wildpret an orthodox species? Implications for its conservation. *N. Z. J. Bot.* **2025**, 1–9. [CrossRef]
56. Beltrame, R.A.; Mendes Jasmim, J.; Duarte Vieira, H.; Acha, A.J. Desecación, almacenamiento y calidad fisiológica de las semillas de *Phoenix roebelenii* O'Brien (Arecaceae). *Rev. Fac. Cienc. Agrarias. Univ. Nac. Cuyo* **2022**, *54*, 25–34.
57. Davies, R.; Pritchard, H. Seed Conservation of Dryland Palms of Africa and Madagascar: Needs and Prospects. *For. Genet. Resour.* **1998**, *26*, 36–44.
58. Albani-Rocchetti, G.; Carta, A.; Mondoni, A.; Godefroid, S.; Davis, C.C.; Caneva, G.; Albrecht, M.A.; Alvarado, K.; Bijmoer, R.; Borosova, R.; et al. Selecting the best candidates for resurrecting extinct-in-the-wild plants from herbaria. *Nat. Plants* **2022**, *8*, 1385–1393. [CrossRef]

horticulturae

MDPI

Article

# Deficit Irrigation and Nitrogen Application Rate Influence Growth and Yield of Four Potato Cultivars (*Solanum tuberosum* L.)

**Abdulssamad M. H. Barka [1], Samuel Y. C. Essah [2] and Jessica G. Davis [3,*]**

[1] Department of Horticulture and Landscape Architecture, Colorado State University, Fort Collins, CO 80523, USA; abarka@rams.colostate.edu
[2] San Luis Valley Research Center, Department of Horticulture and Landscape Architecture, Colorado State University, Center, CO 81125, USA; samuel.essah@colostate.edu
[3] Agricultural Experiment Station, Colorado State University, Fort Collins, CO 80523, USA
* Correspondence: jessica.davis@colostate.edu

**Abstract**

Potatoes have high nitrogen (N) and irrigation requirements. Increasing water scarcity and environmental concerns highlight the need for efficient resource management. This study evaluated the effects of deficit irrigation and reduced N on yield and growth parameters in four potato cultivars (Canela Russet, Mesa Russet, Russet Norkotah3, and Yukon Gold) at Colorado State University's San Luis Valley Research Center over two growing seasons. Three irrigation levels (~70%, ~80%, and 100% ET replacement) and two N rates (165 and 131 kg/ha) were evaluated. Measurements included total and marketable yield, tuber size distribution, tuber bulking (TB), leaf area index (LAI), and stem and tuber numbers. Yield losses were absent with $\leq$18% irrigation reduction in Canela Russet, Mesa Russet, or Yukon Gold but occurred with larger deficits. Russet Norkotah3 experienced yield decline with 16–23% reductions in irrigation. A twenty percent reduction in N application had no effect on Mesa Russet or Russet Norkotah3 yields, while the other varieties experienced a yield decline in one out of two years. Early-season LAI and late-season TB were positively correlated with yield, particularly for Canela Russet and Russet Norkotah3. These findings suggest irrigation and N inputs can be reduced without compromising productivity, but reductions must be determined on a cultivar-by-cultivar basis.

**Keywords:** water conservation; leaf area index; tuber bulking

## 1. Introduction

Potato (*Solanum tuberosum* L.) is an important food crop in many countries and is the fourth most important staple crop in the world after rice, wheat, and maize due to its short growing cycle and large production per unit area [1,2]. The growing population's demand for food puts pressure on land and irrigation water supplies, and this contributes to increased pressure on water resource availability in the future. Therefore, it is crucial to improve water and nutrient use efficiency in the field by using efficient and effective irrigation and nutrient management [3]. Water and nitrogen (N) supplies are key challenges in the long-term production of potatoes [4], since increased or decreased irrigation and N application during the growing season can have adverse effects on the quantity and/or quality of potatoes [5].

Adequate N is essential for growth, development, productivity, and tuber quality [6]. Nitrogen is the most important limiting nutrient, influencing the productivity and quality of the crop; however, excessive N can result in nitrate leaching to the groundwater [7].

Likewise, excessive N fertilizer use may reduce tuber yield and quality [8]. Moreover, high N application may increase the degree of branching and the number of leaves per plant, delay maturity, and reduce tuber quality and dry matter concentration [7,9]. Management of N fertilization rate, timing, and method of application and water scheduling are important to improve N uptake and maintain high yield and quality while reducing N leaching [10]. Therefore, N use efficiency (NUE) must be improved to increase yields, reduce production costs, and decrease environmental pollution.

The availability of agricultural irrigation water has decreased due to an increased demand for water from non-agricultural sectors. Arid and semi-arid regions, including the San Luis Valley of southern Colorado (USA), are the areas most negatively impacted by irrigation water shortages [4]. For crop production in arid and semi-arid areas where rainfall is low and evapotranspiration is very high, declining irrigation water availability has become the main concern for maintaining crop productivity [3]. Therefore, the increasing demand for water and decreasing availability of agricultural irrigation water resources require the optimization of irrigation management practices to increase crop productivity and improve irrigation water use efficiency (WUE) [4,11,12].

Excessive irrigation can promote vegetative growth and reduce potato crop yield while increasing groundwater contamination [13]. On the other hand, deficit irrigation can increase productivity while saving irrigation water, especially in arid and semi-arid regions [4]. Scientists have shown that it is possible to increase WUE by 15 to 40% through improved nutrient and soil management practices, which can positively affect crop yield while conserving water [14,15].

The most sensitive potato growth stages to deficit irrigation are tuber bulking and tuber ripening [16]. Many studies report that potatoes are very sensitive to water stress due to their sparse and shallow root system [12,16]. Deficit irrigation has achieved success with several crops; however, it is difficult to manage potatoes because irrigation is essential for higher production of potatoes [12]. Sufficient irrigation supply before and during tuber initiation increases the number of tubers per plant, while after tuber initiation, tuber size is promoted by irrigation [16]. In addition, leaf area index (LAI) and biomass under deficit irrigation treatment are significantly less than under full irrigation, limiting photosynthate production [12].

Understanding the interactions between water and N over time is very important for improving the sustainability of agriculture in arid and semi-arid areas [17]. Interactions between water and N regulate the geochemical cycling of N and influence crop yield, grain size and protein concentration, photosynthesis, senescence and root architecture, root-to-shoot translocation, and soil microbial enzyme activity [17].

Soil water content directly affects the availability of N for crop uptake, and N affects plant canopy size and crop water use, which controls evapotranspiration [17]. The ability of plants to take up N is influenced by the irrigation regime and amount of water in the soil [18]. Therefore, increasing the availability of N and water to crops are management practices that often result in increased yield and profit [17]. The issues of water and N supply are interconnected and should not be considered separately [7]. Water use between 80% and 100% of ET was optimal to achieve maximum N uptake, high yield, biomass, and quality of an irrigated potato crop in Italy [18]. Although their study, conducted under different environmental and soil conditions in Italy, offers a useful reference, it does not fully account for the unique characteristics of the San Luis Valley, such as its high elevation, cool climate, alkaline soils, and potato varieties. Therefore, while informative, existing research provides limited insight into how these practices affect potato cultivars grown in the San Luis Valley.

Since irrigation water and N management have been high-priority issues for global agricultural production and the global economy, there is an urgent need to use irrigation water and N fertilizer efficiently in potato production systems while sustaining productivity. The objective of this study was to evaluate the effect of different levels of irrigation and N application rates on tuber yield and growth parameters of four potato cultivars grown in the San Luis Valley of Colorado (USA). This research will provide guidance for farmers striving to conserve water and N while maintaining productivity.

## 2. Materials and Methods

### 2.1. Experimental Site

Field experiments were laid out at Colorado State University's San Luis Valley Research Center (latitude 37°43′ N, longitude 106°9′ W, and 2310 m altitude) during the 2016 and 2017 potato cropping seasons. The soil is classified as a Norte gravelly sandy loam [loamy-skeletal, mixed (calcareous), frigid Aquic Ustorthents]. Typically, the area receives annual rainfall of 10 to 18 cm. Prior to the potatoes, the preceding crop was dry beans. Soil pH was 8.0–8.1, and residual soil $NO_3$-N was 1.86 mg $kg^{-1}$ and 1.98 mg $kg^{-1}$ in 2016 and 2017, respectively.

Four potato cultivars (certified, disease-free seed) were evaluated in the experiment: Canela Russet (medium maturity with high specific gravity), Mesa Russet (medium maturity with medium specific gravity), Russet Norkotah3 (early maturity, fresh market, with medium to high specific gravity), and Yukon Gold (early to medium maturity, with yellow flesh, and high specific gravity).

### 2.2. Experimental Design and Field Layout

Experiments were laid out with six treatments (3 irrigation levels × 2 fertilizer treatments) and four replications. Each plot consisted of four rows, 8 m long, and 0.9 m between rows. Cut or whole seeds were planted 0.3 m apart within rows. Cut seed (~80 to 85 g) of each cultivar was planted to a depth of 10 to 13 cm using a potato planter on 19 May 2016 and 3 June 2017. The four cultivars were planted in separate blocks as independent experiments.

### 2.3. Irrigation Water Treatments

Daily reference evapotranspiration (ET) was estimated using the Hargreaves method [19], with data obtained from a weather station located at the San Luis Valley Research Center (https://coagmet.colostate.edu/ (accessed on 2 April 2025)). There were three irrigation treatments: the control, which provided 100% ET replacement for the crop, and two deficit irrigation levels targeting 80% and 70% of ET replacement (Table 1). All experimental plots were provided with 100% ET replacement until initiation of tuber bulking and were then started on different levels of deficit irrigation about 60 days after planting (DAP). A single-line source irrigation technique was used to achieve deficit irrigation levels (Figure 1); therefore, irrigation treatments could not be randomized. However, soil properties were uniform within the study area, thus limiting potential bias due to the experimental design.

Unfortunately, due to variability in the irrigation system, Mesa Russet received a higher ET percentage for the deficit irrigation treatments than the other varieties; this was particularly true for the moderate irrigation treatment, where Mesa Russet received 90–94% ET compared to the other varieties, which received 77–85% ET. Although the actual amount of water applied under the IRRG2 treatment was similar across cultivars in both years, the percentage of ET replacement for Mesa Russet appears higher because the full irrigation (IRRG3) treatment for this cultivar received less total water (49 and 50 cm) for 2016 and 2017, respectively, compared to most of the other cultivars (52–56 cm). As a result, the same volume of water under IRRG2 represents a greater proportion of full ET for Mesa Russet, artificially inflating its percentage value relative to the other cultivars.

**Table 1.** Irrigation treatments for four potato cultivars grown in 2016 and 2017 at the Colorado State University San Luis Valley Research Center. Irrigation amounts are given as a percentage of evapotranspiration replacement and in parentheses as a depth measurement in cm. Deficit irrigation treatments began about 60 days after planting. IRRG1 = lowest irrigation treatment, IRRG2 = moderate irrigation treatment, and IRRG3 = full irrigation treatment.

| Cultivar and Year | Irrigation Treatment | | |
|---|---|---|---|
| | IRRG1 | IRRG2 | IRRG3 |
| Canela Russet | | | |
| 2016 | 67% (36 cm) | 84% (46 cm) | 100% (54 cm) |
| 2017 | 55% (31 cm) | 82% (46 cm) | 100% (56 cm) |
| Mesa Russet | | | |
| 2016 | 75% (37 cm) | 94% (46 cm) | 100% (49 cm) |
| 2017 | 66% (33 cm) | 90% (45 cm) | 100% (50 cm) |
| Russet Norkotah3 | | | |
| 2016 | 67% (36 cm) | 84% (45 cm) | 100% (54 cm) |
| 2017 | 61% (34 cm) | 77% (43 cm) | 100% (56 cm) |
| Yukon Gold | | | |
| 2016 | 69% (36 cm) | 85% (44 cm) | 100% (52 cm) |
| 2017 | 61% (28 cm) | 83% (38 cm) | 100% (46 cm) |

**Figure 1.** Photograph illustrating the line-source irrigation system and irrigation treatments. Two lines of irrigation sprinklers border the plots (seen on the left and right sides of the image). All experimental plots were provided with 100% ET replacement until initiation of tuber bulking. At ~60 days after planting, the line source on the right side of the image was turned off to achieve the target deficit irrigation levels.

### 2.4. Fertilizer Application Treatments

Before planting potatoes, 68 kg N ha$^{-1}$ [urea ammonium nitrate (32-0-0)] was applied to all plots, banded 5 cm below and 5 cm to the side of each seed piece. In addition, 67 kg ha$^{-1}$ of ammonium phosphate (10-34-0) and 45 kg ha$^{-1}$ of potash [KCl (0-0-60)] were applied to all plots as pre-plant fertilizers based on Colorado State University (CSU) fertilizer recommendations for potatoes [20].

Two N application rates were applied to separate plots within each irrigation treatment. After tuberization, in-season N fertilizer was applied in three split applications at weekly intervals, with the first in-season N application at 64 and 50 days after planting (DAP) in 2016 and 2017, respectively, to achieve total N rates of either 131 or 165 kg N ha$^{-1}$ (Table 2). In-season N fertilizer was applied using a boom sprayer and was immediately watered in after spraying.

**Table 2.** Nitrogen fertilizer application treatments used in 2016–2017 potato experiments at the Colorado State University San Luis Valley Research Center.

| Treatment | Low Nitrogen Application (N1) | Recommended Nitrogen Application (N2) |
|---|---|---|
| | ------------------------kg N ha$^{-1}$------------------------ | |
| Pre-plant | 75 | 75 |
| In-season | 56<br>(split application 22/22/12) | 90<br>(split application 34/34/22) |
| Total nitrogen applied | 131 | 165 |

### 2.5. Data Collection

#### 2.5.1. Irrigation Water Measurement

Rain gauges were installed in the middle of each line of plots above the crop canopy to catch and measure the rainfall and the amount of water applied with each irrigation (Table 1). Reading was performed immediately after water application to prevent evaporation from the rain gauges.

#### 2.5.2. Plant Sampling

Two weeks after deficit irrigation and N application rate treatments were initiated, two plants were sampled from each plot for a total of eight sampled plants per treatment. One of the two middle rows was maintained for sampling, and the other was for harvesting. Samples were collected 82, 103, and 113 DAP in 2016 and 104, 111, and 118 DAP in 2017 to measure the number of stems, number of tubers, leaf area index (LAI), and tuber bulking. Sampled plants were separated into aboveground vegetative growth (stems and leaves) and tubers. Tubers were washed and weighed to quantify tuber bulking at each sampling date. A leaf area meter (LI-COR, Inc., LI-3100 Area Meter, Lincoln, NE, USA) was used to measure total leaf area. The leaf area index was calculated as green leaf area/unit area of land. The number of stems and tubers per plant was counted in each plot.

#### 2.5.3. Tuber Yield and Tuber Size Distribution

Potatoes were harvested at 115 DAP in 2016 (27 September) and 120 DAP (10 October) in 2017. Vines were killed by mechanical flailing, and a one-row tractor-mounted potato digger was used to dig one of two middle rows that had been reserved for end-of-season tuber yield and tuber size distribution. The harvested tubers from each plot were weighed for total tuber yield. Tubers were separated into various size distribution groups based on weight (>114 g, 170 g, 285 g, 114–285 g, and 170–454 g). Premium potatoes are those greater than 170 g, and marketable yield is the yield of potatoes >114 g.

2.5.4. Statistical Analysis

Statistical analysis was conducted on all data collected, using analysis of variance (ANOVA). Data analysis was conducted using the PROC MIXED procedure in SAS (Statistical Analysis System, version 9.4, Cary, NC, USA) with water and nitrogen treatments as main effects. Cultivars were analyzed separately. Fisher's protected least significant difference (LSD) test was used for mean separation with a *p*-value of 0.05. If significant interaction ($p < 0.05$) between irrigation and N treatments was detected, pairwise comparisons were performed on the interaction using the SLICE statement to compare water levels within each N level using the Tukey adjustment. The Pearson Correlation Coefficient Analysis was used to evaluate the relationship between total yield vs. leaf area index (LAI) and tuber bulking (TB) for each of the cultivar by year combinations.

## 3. Results

### *3.1. Climate*

The San Luis Valley Research Center received precipitation during the 2016 and 2017 growing seasons, as shown in Figure 2 (https://coagmet.colostate.edu/ accessed on 2 April 2025). In 2016, precipitation was fairly evenly distributed, except for a major peak in August, aligning with the tuber bulking stage, and total precipitation was 86 mm. In 2017, precipitation was highly variable, with dry early-season conditions and peaks occurring in July and September, with total precipitation of 108 mm. Although total rainfall was higher in 2017 compared to 2016, the poor distribution and timing of precipitation may have limited its benefit for potato growth and yield. Research indicates that not only the amount but also the distribution and timing of precipitation significantly influences potato yields [21].

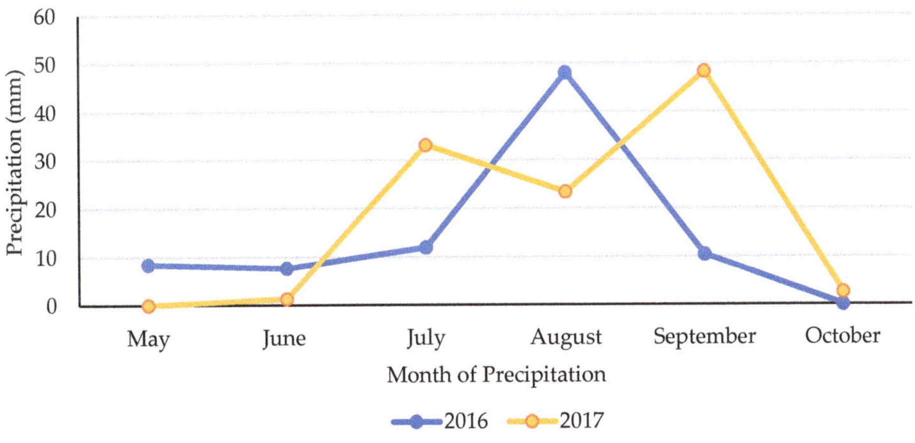


**Figure 2.** Monthly precipitation at the San Luis Valley Research Center in Colorado from potato planting until harvest in 2016 and 2017.

Growing Degree Days (GDDs) are beneficial for tracking growth parameters and can be used as an indicator of yield [22]. The results from two years of GDD indicate that cumulative GDD was slightly higher in 2017 than in 2016, particularly during the early and mid-growth stages (Figure 3) (https://coagmet.colostate.edu/ accessed on 2 April 2025). However, the difference between the two years was minimal and diminished by the end of the growing season.

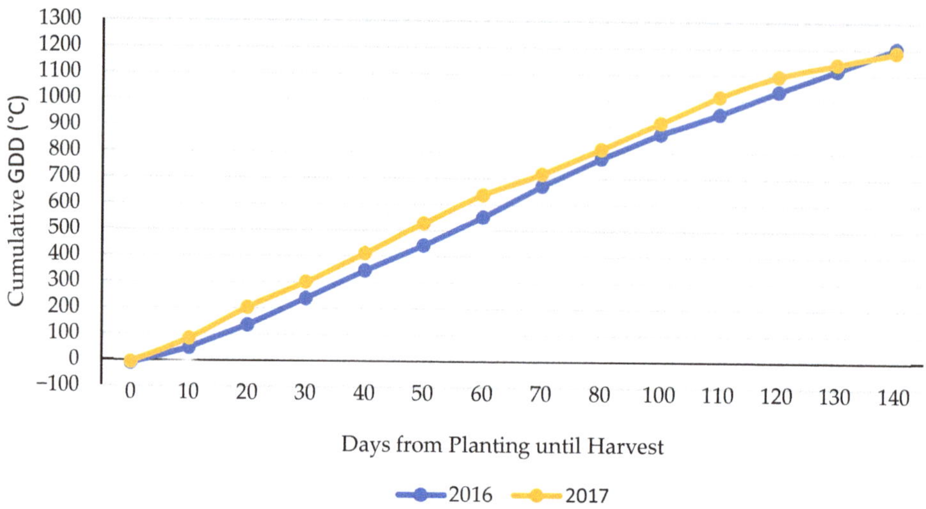


**Figure 3.** Cumulative Growing Degree Days (GDDs) at the San Luis Valley Research Center in Colorado from potato planting until harvest in 2016 and 2017.

### 3.2. Total Yield and Tuber Size Distribution

Each potato variety was grown in a separate experimental block; therefore, the varieties cannot be directly compared and will be discussed individually.

#### 3.2.1. Canela Russet

In 2016 and 2017, reducing irrigation by 16% and 18% (IRRG2), respectively, resulted in total yields of Canela Russet that were not significantly different from full irrigation (IRRG3); however, IRRG2 did reduce the yield of some size fractions in 2017 (Table 3). In addition, a reduction of 33% and 45% for 2016 and 2017, respectively (IRRG1), significantly decreased total yield and marketable yield (114–285 g or 114–454 g). In 2016, nitrogen treatment had no effect on yield. However, in 2017, the recommended nitrogen level (N2) significantly increased total yield and marketable tubers (114–285 g and 114–454 g) compared to the lower nitrogen level (N1). No significant interactions were observed in either year.

**Table 3.** Effect of deficit irrigation and nitrogen fertilizer application rate on potato yield and tuber size distribution of Canela Russet in 2016 and 2017.

| Treatments | 2016 | | | | | | |
|---|---|---|---|---|---|---|---|
| | **Total Y** | **>114 g** | **>170 g** | **>285 g** | **114–285 g** | **114–454 g** | **170–454 g** |
| Water | Yield (Mg ha$^{-1}$) | | | | | | |
| IRRG3 | 45 a | 39 a | 31 a | 10 b | 29 a | 37 a | 29 a |
| IRRG2 | 47 a | 41 a | 33 a | 13 a | 28 a | 39 a | 31 a |
| IRRG1 | 36 b | 28 b | 20 b | 6 c | 23 b | 28 b | 19 b |
| Nitrogen | | | | | | | |
| N1 | 42 a | 36 a | 27 a | 10 a | 27 a | 35 a | 26 a |
| N2 | 42 a | 36 a | 29 a | 10 a | 26 a | 35 a | 27 a |

**Table 3.** *Cont.*

| **Treatments** | | **2016** | | | | | | |
|---|---|---|---|---|---|---|---|---|
| | | **Total Y** | **>114 g** | **>170 g** | **>285 g** | **114–285 g** | **114–454 g** | **170–454 g** |
| Source of Variation | DF | *p*-Value | | | | | | |
| Water | 2 | 0.0004 | <0.0001 | <0.0001 | 0.0006 | 0.0036 | <0.0001 | 0.0001 |
| Nitrogen | 1 | 0.9468 | 1.0000 | 0.3563 | 0.9173 | 0.5906 | 0.6904 | 0.5394 |
| W × N | 2 | 0.9600 | 0.9941 | 0.8846 | 0.1833 | 0.5170 | 0.5166 | 0.7572 |
| | | **2017** | | | | | | |
| Water | | | | | | | | |
| IRRG3 | | 40 a | 35 a | 27 a | 9 a | 26 a | 34 a | 26 a |
| IRRG2 | | 38 a | 30 b | 20 b | 6 ab | 24 a | 29 b | 19 b |
| IRRG1 | | 27 b | 19 c | 10 c | 2 b | 17 b | 19 c | 10 c |
| Nitrogen | | | | | | | | |
| N1 | | 32 b | 24 b | 15 a | 3 b | 20 b | 24 b | 15 b |
| N2 | | 38 a | 32 a | 23 b | 7 a | 25 a | 31 a | 22 a |
| Source of Variation | DF | *p*-Value | | | | | | |
| Water | 2 | <0.0001 | <0.0001 | <0.0001 | 0.0069 | 0.0010 | <0.0001 | <0.0001 |
| Nitrogen | 1 | 0.0003 | <0.0001 | 0.0001 | 0.0212 | 0.0214 | <0.0001 | <0.0001 |
| W × N | 2 | 0.1269 | 0.3913 | 0.6799 | 0.1285 | 0.7071 | 0.2117 | 0.7169 |

Means within the same columns followed by the same letters are not significantly different (ns) at the 0.05 level of probability. IRRG1 = lowest irrigation treatment, IRRG2 = moderate irrigation treatment, and IRRG3 = full irrigation treatment. N1: Nitrogen reduced level (131 kg N ha$^{-1}$). N2: Nitrogen recommended level (165 kg N ha$^{-1}$). DF = degrees of freedom. W × N = The interaction between water (W) and nitrogen (N) treatments.

3.2.2. Mesa Russet

In 2016 and 2017, reducing irrigation by 6% or 10% (IRRG2) had no significant effect on total yield or marketable tubers of Mesa Russet (114–285 g and 114–454 g) compared to full irrigation (IRRG3) (Table 4). However, a reduction of 25% or 34% (IRRG1) resulted in significantly lower total yield and marketable tubers in both years. No significant differences were observed between nitrogen levels in either year, and no significant interactions were observed across all tuber size categories.

**Table 4.** Effect of deficit irrigation and nitrogen fertilizer application rate on potato yield and tuber size distribution of Mesa Russet in 2016 and 2017.

| | **2016** | | | | | | |
|---|---|---|---|---|---|---|---|
| **Treatments** | **Total Y** | **>114 g** | **>170 g** | **>285 g** | **114–285 g** | **114–454 g** | **170–454 g** |
| Water | Yield (Mg ha$^{-1}$) | | | | | | |
| IRRG3 | 43 a | 28 a | 17 a | 3 a | 26 a | 28 a | 17 a |
| IRRG2 | 48 a | 34 a | 19 a | 4 a | 29 a | 34 a | 19 a |
| IRRG1 | 31 b | 14 b | 5 b | 0 b | 14 b | 14 b | 5 b |
| Nitrogen | | | | | | | |
| N1 | 41 a | 27 a | 14 a | 2 a | 24 a | 27 a | 14 a |
| N2 | 40 a | 24 a | 13 a | 3 a | 22 a | 24 a | 13 a |

**Table 4.** *Cont.*

| Treatments | | Total Y | >114 g | >170 g | >285 g | 114–285 g | 114–454 g | 170–454 g |
|---|---|---|---|---|---|---|---|---|
| | | | | **2016** | | | | |
| Source of Variation | DF | | | | *p*-Value | | | |
| Water | 2 | 0.0001 | <0.0001 | 0.0001 | 0.0018 | 0.0001 | <0.0001 | 0.0001 |
| Nitrogen | 1 | 0.7835 | 0.4327 | 0.8909 | 0.4006 | 0.2299 | 0.4327 | 0.8909 |
| W × N | 2 | 0.2372 | 0.5445 | 0.5000 | 0.2428 | 0.6346 | 0.5445 | 0.5000 |
| | | | | **2017** | | | | |
| Water | | | | | | | | |
| IRRG3 | | 38 a | 28 a | 16 a | 6 a | 22 b | 28 a | 16 a |
| IRRG2 | | 36 a | 30 a | 15 a | 4 a | 27 a | 30 a | 15 a |
| IRRG1 | | 26 b | 14 b | 4 b | 0 b | 14 c | 14 b | 4 b |
| Nitrogen | | | | | | | | |
| N1 | | 34 a | 25 a | 12 a | 3 a | 22 a | 25 a | 12 a |
| N2 | | 33 a | 23 a | 12 a | 3 a | 20 a | 23 a | 12 a |
| Source of Variation | DF | | | | *p*-Value | | | |
| Water | 2 | <0.0001 | <0.0001 | <0.0001 | 0.0008 | <0.0001 | <0.0001 | <0.0001 |
| Nitrogen | 1 | 0.5664 | 0.3012 | 0.9165 | 0.7855 | 0.1822 | 0.2980 | 0.9165 |
| W × N | 2 | 0.7432 | 0.9026 | 0.7312 | 0.9546 | 0.7778 | 0.9051 | 0.7312 |

Means within the same columns followed by the same letters are not significantly different (ns) at the 0.05 level of probability. IRRG1 = lowest irrigation treatment, IRRG2 = moderate irrigation treatment, and IRRG3 = full irrigation treatment. N1: Nitrogen reduced level (131 kg N ha$^{-1}$). N2: Nitrogen recommended level (165 kg N ha$^{-1}$). DF = degrees of freedom. W × N = The interaction between water (W) and nitrogen (N) treatments.

### 3.2.3. Russet Norkotah3

A significant difference was observed in Russet Norkotah3 yields among irrigation treatments in both 2016 and 2017 (Table 5). In both years, full irrigation (IRRG3) produced the highest total yield and marketable tuber yield (114–285 g and 114–454 g). Reducing irrigation by 16% (IRRG2) reduced yield in most tuber size categories compared to full irrigation, and reducing irrigation by 33% or 39% (IRRG1) for 2016 and 2017, respectively, led to significant yield losses in every category. Nitrogen treatments had no effect on total yield or size fractions in either year. No significant interactions were found except for the category >285 g (large tuber), which showed a significant interaction of water within the N level in 2016. These findings indicate that reducing irrigation by more than 16% ET negatively affected Russet Norkotah3 tuber yields.

**Table 5.** Effect of deficit irrigation and nitrogen fertilizer application rate on potato yield and tuber size distribution of Russet Norkotah3 in 2016 and 2017.

| Treatments | Total Y | >114 g | >170 g | >285 g | 114–285 g | 114–454 g | 170–454 g |
|---|---|---|---|---|---|---|---|
| | | | **2016** | | | | |
| Water | | | | Yield (Mg ha$^{-1}$) | | | |
| IRRG3 | 49 a | 41 a | 31 a | 10 a | 32 a | 41 a | 31 a |
| IRRG2 | 44 b | 36 b | 25 b | 8 ab | 28 ab | 36 b | 25 b |
| IRRG1 | 40 b | 30 c | 18 c | 5 b | 25 b | 29 c | 18 c |

**Table 5.** *Cont.*

| | | | | 2016 | | | | |
|---|---|---|---|---|---|---|---|---|
| **Treatments** | **Total Y** | **>114 g** | **>170 g** | **>285 g** | **114–285 g** | **114–454 g** | | **170–454 g** |
| Nitrogen | | | | | | | | |
| N1 | | 46 a | 36 a | 25 a | 7 a | 29 a | 36 a | 25 a |
| N2 | | 43 a | 35 a | 24 a | 8 a | 27 a | 35 a | 24 a |
| Interaction | | | | | | | | |
| Nitrogen | Water | | | | | | | |
| N1 | IRRG3 | 52 a | 42 a | 32 a | 7 a | 35 a | 42 a | 32 s |
| | IRRG2 | 45 a | 35 a | 24 a | 8 a | 28 a | 35 a | 24 s |
| | IRRG1 | 41 a | 31 a | 19 a | 7 a | 24 a | 30 a | 19 s |
| N2 | IRRG3 | 46 a | 40 a | 30 a | 12 a | 28 a | 40 s | 30 s |
| | IRRG2 | 44 a | 36 a | 26 a | 8 ab | 28 a | 36 s | 26 a |
| | IRRG1 | 39 a | 28 a | 17 a | 3 b | 25 a | 28 s | 17 a |
| Source of Variation | DF | | | | *p*-Value | | | |
| Water | 2 | 0.0024 | 0.0002 | <0.0001 | 0.0210 | 0.0157 | <0.0001 | <0.0001 |
| Nitrogen | 1 | 0.1868 | 0.3368 | 0.4875 | 0.6701 | 0.2077 | 0.3442 | 0.5861 |
| W × N | 2 | 0.5043 | 0.7196 | 0.5592 | 0.0484 | 0.1032 | 0.6447 | 0.6210 |
| | | | | 2017 | | | | |
| **Treatments** | | **Total Y** | **>114 g** | **>170 g** | **>285 g** | **114–285 g** | **114–454 g** | **170–454 g** |
| Water | | | | | Yield (Mg ha$^{-1}$) | | | |
| IRRG3 | | 43 a | 31 a | 24 a | 13 a | 18 a | 26 a | 19 a |
| IRRG2 | | 27 b | 15 b | 9 b | 3 b | 11 b | 14 b | 9 b |
| IRRG1 | | 24 b | 12 b | 8 b | 2 b | 10 b | 12 b | 8 b |
| Nitrogen | | | | | | | | |
| N1 | | 31 a | 19 a | 14 a | 7 a | 12 a | 18 a | 12 a |
| N2 | | 32 a | 19 a | 14 a | 6 a | 14 a | 17 a | 11 a |
| Source of Variation | DF | | | | *p*-Value | | | |
| Water | 2 | <0.0001 | <0.0001 | <0.0001 | <0.0001 | 0.0056 | <0.0001 | 0.0007 |
| Nitrogen | 1 | 0.4777 | 0.8019 | 0.9279 | 0.4651 | 0.3782 | 0.7336 | 0.5909 |
| W × N | 2 | 0.6271 | 0.7275 | 0.4077 | 0.4388 | 0.9980 | 0.6074 | 0.5108 |

Means within the same columns followed by the same letters are not significantly different (ns) at 0.05 level of probability. IRRG1 = lowest irrigation treatment, IRRG2 = moderate irrigation treatment, and IRRG3 = full irrigation treatment. N1: Nitrogen reduced level (131 kg N ha$^{-1}$). N2: Nitrogen recommended level (165 kg N ha$^{-1}$). DF = degrees of freedom. W × N = The interaction between water (W) and nitrogen (N) treatments.

#### 3.2.4. Yukon Gold

In 2016 and 2017, reducing irrigation by 15% or 17% (IRRG2) had no significant effect on total yield or marketable tubers (114–285 g and 114–454 g) of Yukon Gold compared to full irrigation (IRRG3) (Table 6). However, a reduction of 31% or 39% (IRRG1) significantly decreased total yield and marketable tuber yield in both years. The effects of N varied; lower N levels produced higher total yields but were not significant for marketable yield (114–285 g and 114–454 g) in 2016, while higher N levels increased total and marketable yields in 2017. No significant interactions were observed, except for total and marketable yields in 2016, when full irrigation (IRRG3) resulted in the highest yield under low N (N1), while the recommended nitrogen level (N2) increased tolerance of moderate water stress (IRRG2). However, severe water deficit (IRRG1) reduced yield regardless of the N rate.

**Table 6.** Effect of deficit irrigation and nitrogen fertilizer application rate on potato yield and tuber size distribution of Yukon Gold in 2016 and 2017.

| | | | | 2016 | | | | |
|---|---|---|---|---|---|---|---|---|
| **Treatments** | | **Total Y** | **>114 g** | **>170 g** | **>285 g** | **114–285 g** | **114–454 g** | **170–454 g** |
| Water | | | | | Yield (Mg ha$^{-1}$) | | | |
| IRRG3 | | 41 a | 33 a | 23 a | 7 a | 28 a | 34 a | 23 a |
| IRRG2 | | 40 a | 31 a | 18 b | 3 b | 28 a | 31 a | 18 b |
| IRRG1 | | 34 b | 21 b | 11 c | 1 b | 20 b | 21 b | 11 c |
| Nitrogen | | | | | | | | |
| N1 | | 40 a | 30 a | 19 a | 4 a | 26 a | 30 a | 19 a |
| N2 | | 37 b | 26 b | 16 b | 2 a | 25 a | 27 a | 16 b |
| Interaction | | | | | | | | |
| Nitrogen | Water | | | | | | | |
| N1 | IRRG3 | 45 a | 37 a | 27 a | 9 a | 27 a | 37 a | 27 a |
| | IRRG2 | 38 b | 30 b | 18 a | 3 a | 27 a | 30 a | 18 a |
| | IRRG1 | 38 b | 23 c | 12 a | 0 a | 23 a | 23 a | 12 a |
| N2 | IRRG3 | 38 a | 29 a | 19 a | 4 a | 28 a | 32 | 19 a |
| | IRRG2 | 41 a | 31 a | 18 a | 2 a | 29 a | 31 a | 18 a |
| | IRRG1 | 31 b | 19 b | 10 a | 1 a | 18 a | 19 a | 10 a |
| Source of Variation | DF | | | | *p*-Value | | | |
| Water | 2 | 0.0031 | <0.0001 | <0.0001 | 0.0002 | 0.0012 | <0.0001 | <0.0001 |
| Nitrogen | 1 | 0.0245 | 0.0116 | 0.0253 | 0.0741 | 0.5888 | 0.1247 | 0.0370 |
| W × N | 2 | 0.0262 | 0.0316 | 0.0712 | 0.0537 | 0.1716 | 0.2545 | 0.1004 |
| | | | | **2017** | | | | |
| **Treatments** | | **Total Y** | **>114 g** | **>170 g** | **>285 g** | **114–285 g** | **114–454 g** | **170–454 g** |
| Water | | | | | Yield (Mg ha$^{-1}$) | | | |
| IRRG3 | | 37 a | 28 a | 18 a | 6 a | 22 ab | 28 a | 18 a |
| IRRG2 | | 36 a | 27 a | 15 b | 2 b | 25 a | 27 a | 15 b |
| IRRG1 | | 29 b | 20 b | 10 c | 1 b | 19 b | 20 b | 10 c |
| Nitrogen | | | | | | | | |
| N1 | | 32 b | 23 b | 13 b | 3 a | 20 b | 23 b | 13 b |
| N2 | | 36 a | 27 a | 16 a | 3 a | 24 a | 27 a | 16 a |
| Source of Variation | DF | | | | *p*-Value | | | |
| Water | 2 | 0.0021 | 0.0009 | 0.0003 | <0.0001 | 0.0329 | 0.0009 | 0.0003 |
| Nitrogen | 1 | 0.0142 | 0.0143 | 0.0207 | 0.5377 | 0.0361 | 0.0143 | 0.0207 |
| W × N | 2 | 0.6485 | 0.1862 | 0.2013 | 0.1641 | 0.1934 | 0.1862 | 0.2013 |

Means within the same columns followed by the same letters are not significantly different (ns) at the 0.05 level of probability. IRRG1 = lowest irrigation treatment, IRRG2 = moderate irrigation treatment, and IRRG3 = full irrigation treatment. N1: Nitrogen reduced level (131 kg N ha$^{-1}$). N2: Nitrogen recommended level (165 kg N ha$^{-1}$). DF = degrees of freedom. W × N = The interaction between water (W) and nitrogen (N) treatments.

### 3.3. Tuber Bulking (TB)

#### 3.3.1. Canela Russet

Water treatment significantly affected TB of Canela Russet at all measured stages ($p < 0.0001$) in 2016 and 2017 (Table 7). IRRG3 (full irrigation) had the highest TB, followed by IRRG2 (moderate deficit irrigation), while IRRG1 (lowest deficit irrigation) had the lowest values in both years. In 2016, N showed a significant effect on TB at the end of the season at 113 DAP ($p < 0.0001$), with N1 (low N) producing a higher TB than N2 (recommended N). However, in 2017, N treatment showed a significant effect on TB at all three sampling dates; the recommended level of N (N2) had greater tuber bulking than the lower N rate (N1). The interaction between the water and N treatments was significant at all stages in both years, and full irrigation consistently produced the highest tuber bulking across both N rates, except at the low N (N1) rate in 2016.

**Table 7.** Effect of deficit irrigation and nitrogen fertilizer application rate on potato tuber bulking of Canela Russet in 2016 and 2017.

| | | **Years and Days After Planting (DAP)** | | | | | |
|---|---|---|---|---|---|---|---|
| | | **2016** | | | **2017** | | |
| Treatments | | 82 DAP | 103 DAP | 113 DAP | 104 DAP | 111 DAP | 118 DAP |
| Water | | | | | | | |
| IRRG3 | | 765.69 a | 1120.31 a | 1211.48 a | 979.41 a | 1145.62 a | 1144.53 a |
| IRRG2 | | 726.94 b | 909.84 b | 1140.61 b | 960.72 a | 892.17 b | 928.29 b |
| IRRG1 | | 460.53 c | 596.02 c | 678.53 c | 565.40 b | 718.50 c | 664.66 c |
| Nitrogen | | | | | | | |
| N1 | | 660.82 a | 884.19 a | 1054.08 a | 808.79 b | 886.43 b | 821.32 b |
| N2 | | 641.28 a | 866.60 a | 966.34 b | 861.56 a | 951.09 a | 1003.67 a |
| Interaction | | | | | | | |
| Nitrogen | Water | | | | | | |
| N1 | IRRG3 | 695.71 b | 1236.06 a | 1158.03 b | 1017.41 a | 1023.76 a | 1038.91 a |
| | IRRG2 | 787.31 a | 824.14 b | 1263.39 a | 917.41 b | 947.50 b | 754.55 b |
| | IRRG1 | 499.44 c | 539.59 c | 740.81 c | 491.55 c | 688.04 c | 670.49 c |
| N2 | IRRG3 | 835.66 a | 1004.56 a | 1123.20 a | 941.40 a | 1267.48 a | 1250.15 a |
| | IRRG2 | 666.58 b | 995.55 a | 1159.56 a | 1004.03 a | 836.84 b | 1102.04 b |
| | IRRG1 | 421.61 c | 652.45 b | 616.25 b | 639.25 b | 748.96 c | 658.83 c |
| Source of Variation | DF | *p*-Value | | | | | |
| Water | 2 | <0.0001 | <0.0001 | <0.0001 | <0.0001 | <0.0001 | <0.0001 |
| Nitrogen | 1 | 0.1758 | 0.1086 | <0.0001 | 0.0021 | <0.0001 | <0.0001 |
| W × N | 2 | <0.0001 | <0.0001 | 0.0051 | <0.0001 | <0.0001 | <0.0001 |

Means within the same columns followed by the same letters are not significantly different (ns) at the 0.05 level of probability. IRRG1 = lowest irrigation treatment, IRRG2 = moderate irrigation treatment, and IRRG3 = full irrigation treatment. N1: Nitrogen reduced level (131 kg N $ha^{-1}$). N2: Nitrogen recommended level (165 kg N $ha^{-1}$). DF = degrees of freedom. W × N = The interaction between water (W) and nitrogen (N) treatments.

### 3.3.2. Mesa Russet

Water levels significantly affected TB at all sampling dates in both years (Table 8). In 2017, IRRG3 (full irrigation) produced the highest TB on every sampling date, but in 2016, results were more variable. Nitrogen had a strong effect on TB at most stages in 2016 and 2017, where N1 (low nitrogen) led to higher TB than N2 in both years. Water × Nitrogen interactions were significant on every sampling date.

**Table 8.** Effect of deficit irrigation and nitrogen fertilizer application rate on potato tuber bulking of Mesa Russet in 2016 and 2017.

| | | Years and Days After Planting (DAP) | | | | | |
|---|---|---|---|---|---|---|---|
| | | 2016 | | | 2017 | | |
| Treatments | | 82 DAP | 103 DAP | 113 DAP | 104 DAP | 111 DAP | 118 DAP |
| Water | | | | | | | |
| IRRG3 | | 947.42 a | 908.46 c | 1094.40 b | 801.14 a | 845.79 a | 969.53 a |
| IRRG2 | | 871.80 b | 1104.71 a | 1050.23 c | 670.01 b | 800.14 b | 847.13 b |
| IRRG1 | | 798.83 c | 1054.44 b | 1198.81 a | 643.51 b | 761.83 c | 776.47 c |
| Nitrogen | | | | | | | |
| N1 | | 943.50 a | 1236.05 a | 1187.33 a | 740.73 a | 827.55 a | 867.60 a |
| N2 | | 801.86 b | 809.03 b | 1041.63 b | 669.04 b | 777.62 b | 861.15 a |
| Interaction | | | | | | | |
| Nitrogen | Water | | | | | | |
| N1 | IRRG3 | 985.50 a | 1187.99 b | 1164.03 b | 880.99 a | 959.15 a | 995.31 a |
| | IRRG2 | 910.81 b | 1289.80 a | 1046.50 c | 764.04 b | 775.68 b | 790.69 b |
| | IRRG1 | 934.19 ab | 1230.36 ab | 1351.45 a | 577.15 c | 747.83 b | 816.79 b |
| N2 | IRRG3 | 909.34 a | 628.94 b | 1024.78 a | 721.29 a | 732.43 b | 943.74 a |
| | IRRG2 | 832.79 b | 919.61 a | 1053.95 a | 575.98 b | 824.61 a | 903.56 a |
| | IRRG1 | 663.46 c | 878.53 a | 1046.18 a | 709.86 ab | 775.83 ab | 736.15 b |
| Source of Variation | DF | *p*-Value | | | | | |
| Water | 2 | <0.0001 | <0.0001 | <0.0001 | <0.0001 | 0.0007 | <0.0001 |
| Nitrogen | 1 | <0.0001 | <0.0001 | <0.0001 | 0.0017 | 0.0028 | 0.6074 |
| W × N | 2 | <0.0001 | <0.0001 | <0.0001 | <0.0001 | <0.0001 | <0.0001 |

Means within the same columns followed by the same letters are not significantly different (ns) at the 0.05 level of probability. IRRG1 = lowest irrigation treatment, IRRG2 = moderate irrigation treatment, and IRRG3 = full irrigation treatment. N1: Nitrogen reduced level (131 kg N $ha^{-1}$). N2: Nitrogen recommended level (165 kg N $ha^{-1}$). DF = degrees of freedom. W × N = The interaction between water (W) and nitrogen (N) treatments.

### 3.3.3. Russet Norkotah3

Water treatments significantly affected TB of Russet Norkotah3 at all sampling dates in both years (Table 9). IRRG3 (full irrigation) had the highest TB in 2016, but in 2017, IRRG2 had higher TB in two out of three sampling times. IRRG1 had the lowest TB in all years and sampling times. Nitrogen treatment sometimes had a significant effect on TB, but the effect was inconsistent. The interaction between water and N treatments was significant on all dates; full irrigation (IRRG3) usually produced the highest TB at both nitrogen rates.

**Table 9.** Effect of deficit irrigation and nitrogen fertilizer application rate on potato tuber bulking of Russet Norkotah3 in 2016 and 2017.

| | Years and Days After Planting (DAP) | | | | | |
|---|---|---|---|---|---|---|
| | 2016 | | | 2017 | | |
| Treatments | 82 DAP | 103 DAP | 113 DAP | 104 DAP | 111 DAP | 118 DAP |
| Water | | | | | | |
| IRRG3 | 937.59 a | 1251.69 a | 1433.91 a | 956.39 a | 903.32 b | 1020.69 b |
| IRRG2 | 928.16 a | 1140.33 b | 1223.40 b | 735.94 b | 997.32 a | 1094.78 a |
| IRRG1 | 585.36 b | 796.36 c | 865.77 c | 631.38 c | 742.75 c | 706.66 c |
| Nitrogen | | | | | | |
| N1 | 736.40 b | 1091.75 a | 1179.41 a | 787.25 a | 831.48 b | 955.67 a |
| N2 | 897.68 a | 1033.84 b | 1169.30 a | 761.89 a | 930.78 a | 925.75 a |

**Table 9.** *Cont.*

| | | Years and Days After Planting (DAP) | | | | | |
|---|---|---|---|---|---|---|---|
| | | 2016 | | | 2017 | | |
| Interaction | | | | | | | |
| Nitrogen | Water | | | | | | |
| N1 | IRRG3 | 822.58 a | 1180.60 a | 1299.46 b | 1018.80 a | 977.31 a | 899.34 b |
| | IRRG2 | 844.10 a | 1203.44 a | 1355.71 a | 710.18 b | 838.93 b | 1098.68 a |
| | IRRG1 | 542.51 b | 717.48 b | 883.06 c | 632.79 c | 678.21 c | 869.00 b |
| N2 | IRRG3 | 1052.61 a | 132,278 a | 1568.35 a | 893.98 a | 829.33 b | 1142.05 a |
| | IRRG2 | 1012.21 a | 1091.09 b | 1091.09 b | 761.71 b | 1155.71 a | 1090.89 a |
| | IRRG1 | 628.21 b | 848.48 c | 848.48 c | 629.98 c | 807.29 b | 544.31 b |
| Source of Variation | DF | *p*-Value | | | | | |
| Water | 2 | <0.0001 | <0.0001 | <0.0001 | <0.0001 | <0.0001 | <0.0001 |
| Nitrogen | 1 | <0.0001 | 0.0032 | 0.4210 | 0.0546 | <0.0001 | 0.0496 |
| W × N | 2 | 0.0014 | <0.0001 | <0.0001 | <0.0001 | <0.0001 | <0.0001 |

Means within the same columns followed by the same letters are not significantly different (ns) at the 0.05 level of probability. IRRG1 = lowest irrigation treatment, IRRG2 = moderate irrigation treatment, and IRRG3 = full irrigation treatment. N1: Nitrogen reduced level (131 kg N ha$^{-1}$). N2: Nitrogen recommended level (165 kg N ha$^{-1}$). DF = degrees of freedom. W × N = The interaction between water (W) and nitrogen (N) treatments.

#### 3.3.4. Yukon Gold

Water treatments significantly affected the TB of Yukon Gold at all stages in both years (Table 10). In 2016, IRRG3 had the highest TB except at 113 DAP, when the moderate irrigation IRRG2 was highest, and the low irrigation water treatment (IRRG1) had the lowest TB on every sampling date. In 2017, TB was highest at 104 DAP with moderate irrigation and at 111 DAP with full or moderate irrigation levels. Nitrogen significantly influenced tuber bulking at 82 and 113 DAP in 2016 and 104 and 118 DAP in 2017; N1 (low nitrogen) produced higher TB than N2 in most cases. The interaction was significant in most sampling times in both years, and full or moderate irrigation generally produced the highest TB.

**Table 10.** Effect of deficit irrigation and nitrogen fertilizer application rate on potato tuber bulking of Yukon Gold in 2016 and 2017.

| | | Years and Days After Planting (DAP) | | | | | |
|---|---|---|---|---|---|---|---|
| | | 2016 | | | 2017 | | |
| Treatments | | 82 DAP | 103 DAP | 113 DAP | 104 DAP | 111 DAP | 118 DAP |
| Water | | | | | | | |
| IRRG3 | | 990.53 a | 1215.50 a | 1274.23 b | 770.75 b | 820.34 a | 913.71 b |
| IRRG2 | | 946.89 b | 1143.19 b | 1431.72 a | 815.88 a | 808.19 a | 853.22 c |
| IRRG1 | | 617.36 c | 904.73 c | 821.44 c | 805.83 ab | 732.84 b | 982.03 a |
| Nitrogen | | | | | | | |
| N1 | | 892.00 a | 1077.92 a | 1262.24 a | 750.03 b | 776.71 a | 994.53 a |
| N2 | | 811.19 b | 1097.69 a | 1089.35 b | 844.94 a | 797.54 a | 838.10 b |
| Interaction | | | | | | | |
| Nitrogen | Water | | | | | | |
| N1 | IRRG3 | 1105.59 a | 1176.53 a | 1538.83 a | 715.23 b | 736.74 b | 855.21 a |
| | IRRG2 | 928.78 b | 1144.68 a | 1410.06 b | 865.68 a | 899.35 a | 770.19 a |
| | IRRG1 | 641.63 c | 912.55 b | 837.83 c | 669.20 b | 694.05 b | 888.91 a |

**Table 10.** *Cont.*

| | | Years and Days After Planting (DAP) | | | | | |
|---|---|---|---|---|---|---|---|
| | | 2016 | | | 2017 | | |
| N2 | IRRG3 | 875.48 b | 1254.48 a | 1009.63 b | 826.28 b | 903.94 a | 972.20 a |
| | IRRG2 | 965.00 a | 1141.70 a | 1453.38 a | 766.08 b | 717.04 c | 936.25 a |
| | IRRG1 | 593.09 c | 896.90 b | 805.05 c | 942.46 a | 771.64 b | 1075.15 a |
| Source of Variation | DF | *p*-Value | | | | | |
| Water | 2 | <0.0001 | <0.0001 | <0.0001 | 0.0473 | <0.0001 | <0.0001 |
| Nitrogen | 1 | <0.0001 | 0.1921 | <0.0001 | <0.0001 | 0.0939 | <0.0001 |
| W × N | 2 | <0.0001 | 0.0356 | <0.0001 | <0.0001 | <0.0001 | 0.2324 |

Means within the same columns followed by the same letters are not significantly different (ns) at the 0.05 level of probability. IRRG1 = lowest irrigation treatment, IRRG2 = moderate irrigation treatment, and IRRG3 = full irrigation treatment. N1: Nitrogen reduced level (131 kg N $ha^{-1}$). N2: Nitrogen recommended level (165 kg N $ha^{-1}$). DF = degrees of freedom. W × N = The interaction between water (W) and nitrogen (N) treatments.

### 3.4. Leaf Area Index (LAI)

#### 3.4.1. Canela Russet

Water treatment significantly affected the LAI of Canela Russet at 82 and 103 DAP in 2016 and 104 DAP in 2017 (Table 11). The highest LAI was consistently observed under IRRG3 (full irrigation) in both years, while IRRG1 had the lowest LAI. Nitrogen levels generally had no significant effect on LAI at any stage in 2016 or 2017. In general, there was no significant interaction effect on LAI in 2016, but in 2017 at 104 DAP, there was a significant interaction of levels of irrigation within N treatment.

**Table 11.** Effect of deficit irrigation and nitrogen fertilizer application rate on leaf area index (LAI) of Canela Russet in 2016 and 2017.

| | | Years and Days After Planting (DAP) | | | | | |
|---|---|---|---|---|---|---|---|
| | | 2016 | | | 2017 | | |
| Treatments | | 82 DAP | 103 DAP | 113 DAP | 104 DAP | 111 DAP | 118 DAP |
| Water | | | | | | | |
| IRRG3 | | 2.9625 a | 2.550 a | 0.375 a | 1.4875 a | 1.100 a | - |
| IRRG2 | | 2.7375 a | 2.1875 a | 0.225 a | 1.0625 b | 0.775 a | - |
| IRRG1 | | 1.4375 b | 1.0625 b | 0.2125 a | 0.500 c | 0.5875 a | - |
| Nitrogen | | | | | | | |
| N1 | | 2.425 a | 1.9334 a | 0.29167 a | 1.0667 a | 1.6560 a | - |
| N2 | | 2.333 a | 1.9334 a | 0.2500 a | 0.9667 a | 0.5917 b | - |
| Interaction | | | | | | | |
| Nitrogen | Water | | | | | | |
| N1 | IRRG3 | 2.600 a | 2.6250 a | 0.350 a | 1.825 a | 0.875 a | - |
| | IRRG2 | 3.000 a | 2.175 a | 0.275 a | 0.75 b | 0.500 a | - |
| | IRRG1 | 1.675 a | 1.00 a | 0.250 a | 0.325 b | 0.400 a | - |
| N2 | IRRG3 | 3.325 a | 2.475 a | 0.400 a | 1.150 a | 1.325 a | - |
| | IRRG2 | 2.475 a | 2.200 a | 0.175 a | 1.375 a | 1.05 a | - |
| | IRRG1 | 1.200 a | 1.125 a | 0.175 a | 0.675 a | 0.775 a | - |

**Table 11.** *Cont.*

| | | Years and Days After Planting (DAP) | | | | | |
|---|---|---|---|---|---|---|---|
| | | 2016 | | | 2017 | | |
| Source of Variation | DF | p-Value | | | | | |
| Water | 2 | 0.0002 | <0.0001 | 0.2236 | 0.0007 | 0.1291 | - |
| Nitrogen | 1 | 0.7230 | 1.000 | 0.6166 | 0.5679 | 0.00322 | - |
| W × N | 2 | 0.1039 | 0.8130 | 0.7289 | 0.0155 | 0.9365 | - |

Means within the same columns followed by the same letters are not significantly different (ns) at the 0.05 level of probability. IRRG1 = lowest irrigation treatment, IRRG2 = moderate irrigation treatment, and IRRG3 = full irrigation treatment. N1: Nitrogen reduced level (131 kg N $ha^{-1}$). N2: Nitrogen recommended level (165 kg N $ha^{-1}$). DF = degrees of freedom. W × N = The interaction between water (W) and nitrogen (N) treatments.

### 3.4.2. Mesa Russet

Water had a significant effect on the LAI of Mesa Russet at 82 and 113 DAP in 2016 and 104 and 111 DAP in 2017 (Table 12). In most cases, IRRG3 had the highest LAI, while IRRG1 had the lowest. Nitrogen had no significant effect on LAI in either year. There was no significant interaction observed in 2017, but in 2016, the interaction between water and N levels was significant at 113 DAP.

**Table 12.** Effect of deficit irrigation and nitrogen fertilizer application rate on leaf area index (LAI) of Mesa Russet in 2016 and 2017.

| | | Years and Days After Planting (DAP) | | | | | |
|---|---|---|---|---|---|---|---|
| | | 2016 | | | 2017 | | |
| Treatments | | 82 DAP | 103 DAP | 113 DAP | 104 DAP | 111 DAP | 118 DAP |
| Water | | | | | | | |
| IRRG3 | | 3.5375 a | 3.000 a | 1.4375 b | 1.3250 a | 1.1375 a | - |
| IRRG2 | | 2.9750 ab | 2.4625 a | 0.9625 c | 0.7625 b | 0.6875 b | - |
| IRRG1 | | 2.400 b | 2.350 a | 1.700 a | 0.600 b | 0.5875 b | - |
| Nitrogen | | | | | | | |
| N1 | | 2.9417 a | 2.7250 a | 1.2750 a | 0.9833 a | 0.8833 a | - |
| N2 | | 3.000 a | 2.4833 a | 1.4583 a | 0.8083 a | 0.7250 a | - |
| Interaction | | | | | | | |
| Nitrogen | Water | | | | | | |
| N1 | IRRG3 | 3.325 a | 3.200 a | 1.275 b | 1.575 a | 0.925 a | - |
| | IRRG2 | 2.900 a | 2.175 a | 0.75 c | 0.800 a | 0.625 a | - |
| | IRRG1 | 2.600 a | 2.800 a | 1.800 a | 0.575 a | 0.625 a | - |
| N2 | IRRG3 | 3.750 a | 2.800 a | 1.600 a | 1.075 a | 1.350 a | - |
| | IRRG2 | 3.050 a | 2.525 a | 1.175 a | 0.725 a | 0.750 a | - |
| | IRRG1 | 2.200 a | 2.125 a | 1.600 a | 0.625 a | 0.550 a | - |
| Source of Variation | DF | p-Value | | | | | |
| Water | 2 | 0.0336 | 0.3600 | <0.001 | 0.0004 | <0.0001 | - |
| Nitrogen | 1 | 0.859 | 0.5388 | 0.0761 | 0.1748 | 0.0601 | - |
| W × N | 2 | 0.5799 | 0.5435 | 0.0376 | 0.1929 | 0.0563 | - |

Means within the same columns followed by the same letters are not significantly different (ns) at 0.05 level of probability. IRRG1 = lowest irrigation treatment, IRRG2 = moderate irrigation treatment, and IRRG3 = full irrigation treatment. N1: Nitrogen reduced level (131 kg N $ha^{-1}$). N2: Nitrogen recommended level (165 kg N $ha^{-1}$). DF = degrees of freedom. W × N = The interaction between water (W) and nitrogen (N) treatments.

3.4.3. Russet Norkotah3

Water treatment significantly affected the LAI of Russet Norkotah3 on most sampling dates ($p < 0.05$) in both years; IRRG3 had the highest LAI, and IRRG1 was the lowest (LAI decreased with increased water deficit) (Table 13). The N treatment was only significant on 111 DAP in 2017, and the interaction never had a significant effect on LAI.

**Table 13.** Effect of deficit irrigation and nitrogen fertilizer application rate on leaf area index (LAI) of Russet Norkotah3 in 2016 and 2017.

| | | Years and Days After Planting (DAP) | | | | | |
|---|---|---|---|---|---|---|---|
| | | 2016 | | | 2017 | | |
| Treatments | | 82 DAP | 103 DAP | 113 DAP | 104 DAP | 111 DAP | 118 DAP |
| Water | | | | | | | |
| IRRG3 | | 2.9875 a | 2.5625 a | 0.6250 a | 1.9875 a | 1.6375 a | - |
| IRRG2 | | 2.4875 a | 1.9250 ab | 0.4250 ab | 1.3875 ab | 1.1125 b | - |
| IRRG1 | | 1.4250 b | 1.2000 b | 0.1500 b | 0.9250 b | 0.5000 c | - |
| Nitrogen | | | | | | | |
| N1 | | 2.5500 a | 2.0583 a | 0.4833 a | 1.5083 a | 1.3500 a | - |
| N2 | | 2.0500 a | 1.7333 a | 0.3167 a | 1.3583 a | 0.8167 b | - |
| Source of Variation | DF | *p*-Value | | | | | |
| Water | 2 | 0.0035 | 0.0261 | 0.0636 | 0.0132 | 0.0002 | - |
| Nitrogen | 1 | 0.1450 | 0.3929 | 0.2915 | 0.5724 | 0.0072 | - |
| W × N | 2 | 0.9981 | 0.6490 | 0.8588 | 0.0825 | 0.1770 | - |

Means within the same columns followed by the same letters are not significantly different (ns) at the 0.05 level of probability. IRRG1 = lowest irrigation treatment, IRRG2 = moderate irrigation treatment, and IRRG3 = full irrigation treatment. N1: Nitrogen reduced level (131 kg N $ha^{-1}$). N2: Nitrogen recommended level (165 kg N $ha^{-1}$). DF = degrees of freedom. W × N = The interaction between water (W) and nitrogen (N) treatments.

3.4.4. Yukon Gold

Water treatment significantly affected the LAI of Yukon Gold at 82 DAP in 2016 and 104 and 111 DAP in 2017 (Table 14). IRRG3 generally had the highest LAI, while IRRG1 had the lowest. Nitrogen had no significant effect at any stage, and the interaction was not significant on any sampling date.

**Table 14.** Effect of deficit irrigation and nitrogen fertilizer application rate on leaf area index (LAI) of Yukon Gold in 2016 and 2017.

| | Years and Days After Planting (DAP) | | | | | |
|---|---|---|---|---|---|---|
| | 2016 | | | 2017 | | |
| Treatments | 82 DAP | 103 DAP | 113 DAP | 104 DAP | 111 DAP | 118 DAP |
| Water | | | | | | |
| IRRG3 | 1.5863 a | 0.3375 a | 0.02500 a | 0.6125 a | 0.400 a | 0.0000 a |
| IRRG2 | 1.6863 a | 0.2188 a | 0.01250 a | 0.4375 ab | 0.11250 b | 0.1250 a |
| IRRG1 | 1.0625 b | 0.1825 a | 0.00000 a | 0.2250 b | 0.06250 b | 0.0000 a |
| Nitrogen | | | | | | |
| N1 | 1.5000 a | 0.2992 a | 0.01667 a | 0.4667 a | 0.21667 a | 0.08333 a |
| N2 | 1.3900 a | 0.1933 a | 0.00833 a | 0.3833 a | 0.16667 a | 0.0000 a |

**Table 14.** *Cont.*

| | | Years and Days After Planting (DAP) | | | | | |
|---|---|---|---|---|---|---|---|
| | | 2016 | | | 2017 | | |
| Source of Variation | DF | $p$-Value | | | | | |
| Water | 2 | 0.0226 | 0.4492 | 0.3874 | 0.0171 | 0.0036 | 0.3874 |
| Nitrogen | 1 | 0.5448 | 0.3147 | 0.5709 | 0.4102 | 0.5143 | 0.3306 |
| W × N | 2 | 0.6797 | 0.4081 | 0.7209 | 0.9516 | 0.1512 | 0.3874 |

Means within the same columns followed by the same letters are not significantly different (ns) at the 0.05 level of probability. IRRG1 = lowest irrigation treatment, IRRG2 = moderate irrigation treatment, and IRRG3 = full irrigation treatment. N1: Nitrogen reduced level (131 kg N ha$^{-1}$). N2: Nitrogen recommended level (165 kg N ha$^{-1}$). DF = degrees of freedom. W × N = The interaction between water (W) and nitrogen (N) treatments.

### *3.5. Stem and Tuber Number*

The numbers of stems and tubers per plant were rarely significant ($p > 0.05$) across water, N treatments, or their interaction for all four potato cultivars (Canela Russet, Mesa Russet, Russet Norkotah3 and Yukon Gold) in either 2016 (Table 15) or 2017 (Table 16).

**Table 15.** Effect of deficit irrigation and nitrogen fertilizer application rate on number of stems and tubers in four potato cultivars (Canela Russet, Mesa Russet, Russet Norkotah3, and Yukon Gold) 103 days after planting (DAP) in 2016.

| Cultivars | | Canela Russet | | Mesa Russet | | Russet Norkotah3 | | Yukon Gold | |
|---|---|---|---|---|---|---|---|---|---|
| Treatments | | No. of Stems Plant$^{-1}$ | No. of Tubers Plant$^{-1}$ | No. of Stems Plant$^{-1}$ | No. of Tubers Plant$^{-1}$ | No. of Stems Plant$^{-1}$ | No. of Tubers Plant$^{-1}$ | No. of Stems Plant$^{-1}$ | No. of Tubers Plant$^{-1}$ |
| Water | | | | | | | | | |
| IRRG3 | | 3.12 ab | 7.50 a | 3.62 a | 8.62 a | 4.25 a | 9.12 a | 3.00 a | 8.75 a |
| IRRG2 | | 2.81 b | 6.62 a | 3.94 a | 10.56 a | 4.56 a | 9.62 a | 3.18 a | 9.12 a |
| IRRG3 | | 3.75 a | 6.56 a | 4.06 a | 11.12 a | 4.56 a | 7.75 a | 3.25 a | 9.12 a |
| Nitrogen | | | | | | | | | |
| N1 | | 3.17 a | 7.25 a | 4.08 a | 11.46 a | 3.96 a | 9.08 a | 3.21 a | 9.33 a |
| N2 | | 3.29 a | 6.54 a | 3.67 a | 8.75 a | 4.96 a | 8.58 a | 3.08 a | 8.67 a |
| Interaction | | | | | | | | | |
| | IRRG3 | 3.12 a | 7.62 a | 4.00 a | 10.25 a | 4.25 a | 9.38 a | 3.12 a | 9.12 a |
| N1 | IRRG2 | 3.38 a | 6.38 a | 4.25 a | 11.88 a | 3.75 a | 10.88 a | 3.38 a | 9.38 a |
| | IRRG1 | 3.00 a | 5.62 a | 4.00 a | 12.25 a | 3.88 a | 7.00 a | 3.12 a | 9.50 a |
| | IRRG3 | 3.12 b | 7.38 a | 3.25 a | 7.00 a | 4.25 a | 8.88 a | 2.88 a | 8.38 a |
| N2 | IRRG2 | 2.25 b | 6.88 a | 3.62 a | 9.25 a | 5.38 a | 8.38 a | 3.00 a | 8.88 a |
| | IRRG1 | 4.50 a | 7.50 a | 4.12 a | 10.00 a | 5.25 a | 8.50 a | 3.38 a | 8.75 a |
| Source of Variation | DF | $p$-Value | | | | | | | |
| Water | 2 | 0.0443 | 0.4680 | 0.7823 | 0.3029 | 0.8440 | 0.0980 | 0.8856 | 0.9257 |
| Nitrogen | 1 | 0.6669 | 0.3113 | 0.4347 | 0.0584 | 0.0624 | 0.4773 | 0.7744 | 0.4672 |
| W × N | 2 | 0.0054 | 0.4491 | 0.7631 | 0.9538 | 0.3850 | 0.0867 | 0.8225 | 0.9914 |

Means within the same columns followed by the same letters are not significantly different (ns) at the 0.05 level of probability according to the LSD test. IRRG1 = lowest irrigation treatment, IRRG2 = moderate irrigation treatment, and IRRG3 = full irrigation treatment. N1: Nitrogen reduced level (131 kg N ha$^{-1}$). N2: Nitrogen recommended level (165 kg N ha$^{-1}$). DF = degrees of freedom. W × N = The interaction between water (W) and nitrogen (N) treatments.

**Table 16.** Effect of deficit irrigation and nitrogen fertilizer application rate on number of stems and tubers in four potato cultivars (Canela Russet, Mesa Russet, Russet Norkotah3, and Yukon Gold) 110 days after planting (DAP) in 2017.

| Cultivars | | Canela Russet | | Mesa Russet | | Russet Norkotah3 | | Yukon Gold | |
|---|---|---|---|---|---|---|---|---|---|
| Treatments | | No. of Stems Plant$^{-1}$ | No. of Tubers Plant$^{-1}$ | No. of Stems Plant$^{-1}$ | No. of Tubers Plant$^{-1}$ | No. of Stems Plant$^{-1}$ | No. of Tubers Plant$^{-1}$ | No. of Stems Plant$^{-1}$ | No. of Tubers Plant$^{-1}$ |
| Water | | | | | | | | | |
| IRRG3 | | 2.69 a | 6.06 a | 3.38 a | 6.81 a | 4.62 a | 6.25 a | 3.56 a | 6.31 a |
| IRRG2 | | 2.94 a | 6.50 a | 4.44 a | 7.12 a | 4.75 a | 6.00 a | 3.94 a | 6.75 a |
| IRRG3 | | 2.87 a | 6.44 a | 3.69 a | 7.19 a | 4.25 a | 5.94 a | 3.25 a | 6.12 a |
| Nitrogen | | | | | | | | | |
| N1 | | 2.62 a | 6.58 a | 4.42 a | 7.29 a | 4.62 a | 5.83 a | 3.71 a | 6.38 a |
| N2 | | 3.04 a | 6.08 a | 3.25 b | 6.79 a | 4.46 a | 6.29 a | 3.46 a | 6.42 a |
| Interaction | | | | | | | | | |
| | IRRG3 | 2.00 a | 5.00 b | 4.00 a | 6.75 a | 5.00 a | 7.00 a | 3.75 a | 6.38 a |
| N1 | IRRG2 | 3.12 a | 8.00 a | 5.38 a | 7.88 a | 4.25 a | 5.00 a | 4.38 a | 7.00 a |
| | IRRG1 | 2.75 a | 6.75 a | 3.88 a | 7.25 a | 4.62 a | 5.50 a | 3.00 a | 5.75 a |
| | IRRG3 | 3.38 a | 7.12 a | 2.75 a | 6.88 a | 4.25 a | 5.50 a | 3.38 a | 6.25 a |
| N2 | IRRG2 | 2.75 a | 5.00 a | 3.50 a | 6.38 a | 5.25 a | 7.00 a | 3.50 a | 6.50 a |
| | IRRG1 | 3.00 a | 6.12 a | 3.50 a | 7.12 a | 3.88 a | 6.38 a | 3.50 a | 6.50 a |
| Source of Variation | DF | | | | *p*-Value | | | | |
| Water | 2 | 0.8014 | 0.8744 | 0.2092 | 0.8850 | 0.7313 | 0.8961 | 0.2134 | 0.7875 |
| Nitrogen | 1 | 0.2057 | 0.5095 | 0.0264 | 0.4595 | 0.7579 | 0.4357 | 0.4249 | 0.9565 |
| W × N | 2 | 0.1016 | 0.0373 | 0.4590 | 0.5682 | 0.3242 | 0.0637 | 0.2068 | 0.7875 |

Means within the same columns followed by the same letters are not significantly different (ns) at the 0.05 level of probability according to the LSD test. IRRG1 = lowest irrigation treatment, IRRG2 = moderate irrigation treatment, and IRRG3 = full irrigation treatment. N1: Nitrogen reduced level (131 kg N ha$^{-1}$). N2: Nitrogen recommended level (165 kg N ha$^{-1}$). DF = degrees of freedom. W × N = The interaction between water (W) and nitrogen (N) treatments.

### 3.6. Correlation of Tuber Bulking and Leaf Area Index with Yield

In general, LAI was significantly positively correlated with total yield, especially for Russet Norkotah3 and Canela Russet at the early sampling dates (Table 17). These correlations may be due to increased leaf area enhancing the plant's photosynthetic capacity, which in turn, supports greater tuber development (tuber bulking). In general, TB presented a high positive correlation with total yield, particularly for Canela Russet, which was significant for every sampling date in both years; TB was generally a better yield predictor than LAI for Canela Russet, as shown by their higher correlation coefficients at every sampling date in both years. Yukon Gold and Russet Norkotah3 also showed significant correlation between TB and yield, but only about half the time (Table 17). In addition, the results showed that early-stage LAI (82 DAP) and late-stage TB (113 DAP) were generally most highly correlated with yield.

**Table 17.** Pearson correlation coefficients for total yield vs. leaf area index (LAI) and tuber bulking (TB) for four potato cultivars for 2016 and 2017.

| | | LAI | | | TB | | |
|---|---|---|---|---|---|---|---|
| Year | Cultivar | 82 DAP | 103 DAP | 113 DAP | 82 DAP | 103 DAP | 113 DAP |
| 2016 | Canela Russet | 0.640 ** | 0.560 ** | 0.231 ns | 0.704 *** | 0.589 ** | 0.752 *** |
| | Mesa Russet | 0.435 * | −0.070 ns | −0.612 ** | 0.253 ns | 0.097 ns | −0.383 ns |
| | Russet Norkotah3 | 0.427 * | 0.591 ** | 0.503 * | 0.201 ns | 0.583 ** | 0.496 * |
| | Yukon Gold | 0.339 ns | 0.360 ns | 0.107 ns | 0.728 *** | 0.472 * | 0.660 *** |

**Table 17.** *Cont.*

| | | LAI | | | TB | | |
|---|---|---|---|---|---|---|---|
| | | **104 DAP** | **111 DAP** | **118 DAP** | **104 DAP** | **111 DAP** | **118 DAP** |
| **2017** | Canela Russet | 0.515 * | 0.500 * | — | 0.777 *** | 0.773 *** | 0.858 *** |
| | Mesa Russet | 0.225 ns | 0.276 ns | — | 0.399 ns | 0.265 ns | 0.420 * |
| | Russet Norkotah3 | 0.555 ** | 0.536 ** | — | 0.799 *** | 0.124 ns | 0.369 ns |
| | Yukon Gold | 0.406 * | 0.298 ns | −0.088 ns | 0.222 ns | 0.366 | 0.03 ns |

ns = not significant. * Significant at $p < 0.05$. ** Significant at $p < 0.01$. *** Significant at $p < 0.001$.

## 4. Discussion

### *4.1. Total Yield and Tuber Size Distribution*

Canela Russet maintained high total and marketable yields with a 16–18% reduction in irrigation (Table 3). However, in 2017, this cultivar exhibited sensitivity to reduced irrigation and N inputs, as the large-size marketable yield was significantly lower under deficit irrigation, and most tuber sizes were smaller under reduced N application. The total yield of 38–47 Mg ha$^{-1}$ was achieved under full and moderate deficit irrigation, and this aligns with findings by Colorado State University (CSU), where yield under full irrigation ranged from 24–52 Mg ha$^{-1}$ [23]. Our results demonstrate Canela Russet's ability to maintain productivity under moderate resource constraints.

Mesa Russet also showed results (Table 4) that confirmed that moderate reduction in water and N had no effect on total or marketable yield. Research by Colorado State University (CSU) showed that the average yield for Mesa Russet is between 39 and 50 Mg ha$^{-1}$ [24]; our results showed the average yields under full and moderate deficit irrigation to be 38–48 Mg/ha. These results align with previous research showing that moderate irrigation deficits did not compromise yield, due to reduced leaching and improved nutrient availability [25,26]. It is important to note that Mesa Russet received higher irrigation rates under deficit on a % ET basis than those of the other cultivars in both years, although the actual irrigation amounts were similar, thus confounding cultivar comparisons (Table 1).

On the other hand, Russet Norkotah3 exhibited significant yield reductions under moderate deficit irrigation (Table 5), confirming its high sensitivity to water stress. Russet Norkotah3 is the earliest maturing variety evaluated in this study and is, therefore, less drought tolerant. Similar yield declines have previously been reported under drought stress [27]. Due to its earliness, it needs a large quantity of water to build its vegetative structure (photosynthetic apparatus) early in the season to support early TB and early tuber maturity. In addition, Russet Norkotah3 has earlier canopy senescence and a shallower root system, leaving it more vulnerable to water stress during late-season bulking [28]. The yield potential for Russet Norkotah has been reported to be 45 Mg ha$^{-1}$ [29]. Our plots yielded 43–46 Mg ha$^{-1}$ under full irrigation; however, any reduction in irrigation water resulted in a yield decrease, indicating sensitivity to even slight reductions in irrigation. These findings are supported by others who demonstrated that limited soil water reduces potato productivity, restricting growth, tuber size, and overall yield [12,16,30]. However, Russet Norkotah3 maintained yields with reduced N application in both years.

The moderate deficit irrigation treatment had no significant effect on total or marketable yield of Yukon Gold in either year (Table 6). This supports previous research that found that deficit irrigation strategies, when applied either during specific growth periods or throughout the season, do not necessarily reduce yields [31]. The most severe negative effects of deficit irrigation occur during mid-to-late-season TB [32]. Average total yield of Yukon Gold has been reported to range from 26 to 57 Mg ha$^{-1}$ under full irrigation [33]. In contrast, our results with full or moderate deficit irrigation indicate yields of 37–41 Mg ha$^{-1}$, within the average range.

Our findings indicate that cultivars responded differently to reductions in irrigation and N application rates. While Canela Russet, Yukon Gold, and Mesa Russet tolerated moderate irrigation deficits without yield loss, Russet Norkotah3 required full irrigation for optimal productivity. A twenty percent reduction in N had no effect on Mesa Russet or Russet Norkotah3 yields but reduced the yields of the other varieties in one out of two years. These results highlight the need for each cultivar to be managed with specific water and nutrient management strategies to optimize yield, improve resource efficiency, and ensure sustainable potato production. Overall, a moderate reduction in both irrigation and N can be implemented without compromising the productivity of most potato cultivars, and moderate irrigation reduction also reduces the potential for nitrate leaching [34].

The San Luis Valley produces about 20,234 ha of potatoes annually. If Canela Russet or Yukon Gold was grown on all of this land, and irrigation was reduced by 8.5 cm annually without affecting yield, this would result in a savings of 17.2 billion L (17.2 million m$^3$) of water annually in the San Luis Valley alone. This water savings is equivalent to 850 m$^3$ ha$^{-1}$, within the range reported by others for a 20% reduction in irrigation [35]. Alternatively, if Russet Norkotah3 or Mesa Russet were grown, then the N application rate could be reduced by 34 kg N ha$^{-1}$ or 688 Mg N yr$^{-1}$.

### 4.2. Tuber Bulking (TB)

Tuber bulking demonstrated a strong and consistent positive correlation with total yield for Canela Russet (Table 17). Therefore, factors such as water and N availability, which influence physiological processes during TB, can alter the duration and effectiveness of this critical period, ultimately affecting total tuber yield. Our findings are in agreement with previously reported results showing that a reduction in irrigation water by 10% to 21% did not influence TB of Rio Grande Russet potatoes [13]. The TB rate and duration are important because they directly influence yield [36]. A faster TB rate is associated with higher yield and depends upon the amount of total available N and soil moisture. Insufficient moisture levels throughout the various stages of potato growth, particularly during tuber initiation and bulking, significantly affect the yield of potato crops [37], and excessive N may delay TB [38]. Reducing water gradually during TB provides a more favorable opportunity for the crop to adapt to stressful water conditions compared to exposing the crop to continuous water stress throughout the whole growing season [39]. The timing and rate of N fertilizer application also have a significant impact on tuber bulking; application of N at planting, followed by additional applications during TB, resulted in increased tuber size and yield [40]. Potatoes are sensitive to water stress during TB [41]. Specifically, deficit irrigation during the initial TB stage has led to reductions in tuber grade from US Number One tubers to US Number Two tubers [42]. The ability of plants to obtain N is influenced not only by the presence of N in the soil but also by the availability of water stored in the soil [43]. There is an optimal range of water and N requirements for each stage during the potato growing season, and this varies by cultivar. Providing inadequate water and N can have negative effects on TB as well as yield. Likewise, it is important to maintain a balanced and appropriate supply of water and N throughout the growing season to maximize potato production.

### 4.3. Leaf Area Index (LAI)

LAI was significantly positively correlated with total yield for Canela Russet and Russet Norkotah3; however, this pattern was not consistent for Mesa Russet or Yukon Gold (Table 17). Mesa Russet and Yukon Gold have broader leaves compared to the other two cultivars. Together with the geometric orientation of the leaves of Mesa Russet and Yukon Gold, there is a possibility of mutual shading among the leaves, thus preventing many of the middle and lower leaves from photosynthesizing optimally. These middle and lower leaves draw photosynthates from the plant at the expense of tuber growth, without contributing much photosynthate production [44]. Therefore, early and mid-season LAI measurements could be used as a yield predictor for Canela Russet and Russet Norkotah3, but not for Mesa Russet or Yukon Gold. In addition, most of the cultivars showed either a significant positive correlation between LAI and yield at 82 DAP, which corresponds to peak leaf growth. After 100 DAP, leaf senescence begins, and the correlation generally weakens as tubers mature [45].

Previous research has reported that LAI is an indicator of a potato plant's ability to maximize light interception and photosynthetic capacity, which are positively associated with growth, development, and yield [46,47]. In addition, others have found that the leaf area of the potato plant at the time of tuber initiation and its capacity to capture solar radiation during the TB stage have a major impact on final tuber yield [48]. This relationship aligns with previous findings that show that prolonged maintenance of the green canopy extends the TB period, thereby increasing the crop yield [49]. In addition, yield has been reported to be correlated with the net photosynthetic rate during the TB stage [25].

There were significant effects of reduced water and N levels on LAI. The IRRG1 treatment significantly reduced LAI of Russet Norkotah3 on all sampling dates in both years (Table 13). In addition, N rate had no effect on LAI of Mesa Russet or Yukon Gold in either year and rarely affected LAI of the other two varieties. Our findings are supported by several studies showing that LAI increased significantly with increased irrigation water level [50,51]. Our findings indicate that higher irrigation levels initially promote greater LAI, but the effect diminishes over time. Water stress has been shown to decrease the photosynthetic rate and leaf area [52], and net photosynthetic rate, total leaf area, and aboveground biomass are optimized when plants receive sufficient N [32]. In addition, different amounts of N affect LAI differently depending on the phenological stage of potato [53]. LAI is significantly affected by water and N supply, resulting in lower LAI under less irrigation or lower N rates [34]. In previous research in the San Luis Valley, reducing irrigation by more than 15–20% significantly reduced LAI [13]. Moreover, it is possible to reduce the amount of water and N applied without impacting the LAI, and this relationship varies with cultivar.

## 5. Conclusions

Reducing irrigation by 15–18% did not significantly affect total yield, LAI, stem number, or tuber number in Canela Russet or Yukon Gold under most conditions. However, both varieties exhibited a significant reduction in yield under 20% reduced N application in one out of two years. Mesa Russet also exhibited no yield reduction in IRRG2 or reduced N treatments; however, due to the variability of the irrigation system, IRRG2 only represented a 6–10% reduction in water application. In contrast, Russet Norkotah3 showed significant sensitivity to reductions in irrigation, resulting in yield decline; however, the IRRG2 treatment represented a 16–23% reduction in water application, more than the other varieties. A twenty percent N reduction had no effect on Russet Norkotah3 or Mesa Russet yield.

There was significant positive correlation between yield, LAI, and tuber bulking for Canela Russet and Russet Norkotah3. Although the cultivars responded differently to N application under various irrigation levels, N rate alone had little to no significant effect across treatments. Clearly, the optimal balance of irrigation and N input is cultivar dependent. However, further investigation should explore genotype-by-environment interactions and physiological traits such as root architecture, stomatal conductance, and chlorophyll content to better understand cultivar-specific drought tolerance. Additionally, studies should examine N concentrations in the soil and groundwater throughout the entire growing season to assess the environmental benefits of reduced N application.

**Author Contributions:** Conceptualization, A.M.H.B. and S.Y.C.E.; methodology, A.M.H.B. and S.Y.C.E.; formal analysis, A.M.H.B.; investigation, A.M.H.B.; resources, S.Y.C.E.; writing—original draft, A.M.H.B.; writing—review and editing, S.Y.C.E. and J.G.D.; supervision, S.Y.C.E. All authors have read and agreed to the published version of the manuscript.

**Funding:** This research was funded by the Colorado Agricultural Experiment Station, the Colorado Potato Administrative Committee, and the Ministry of Higher Education and Scientific Research of Libya.

**Data Availability Statement:** Data are contained within the article.

**Acknowledgments:** We gratefully acknowledge the technical support provided by Mercy Essah in the field and in the lab.

**Conflicts of Interest:** The authors declare no conflicts of interest. The funders had no role in the design of the study; in the collection, analyses, or interpretation of data; in the writing of the manuscript; or in the decision to publish the results.

## References

1. Gültekin, R.; Ertek, A. Effects of deficit irrigation on the potato tuber development and quality. *Int. J. Agric. Environ. Food Sci.* **2018**, *2*, 93–98. [CrossRef]
2. Low, J.; Nyongesa, M.; Quinn, S.; Parker, M. (Eds.) *Potato and Sweet Potato in Africa: Transforming the Value Chains for Food and Nutrition Security*; CABI International: Wallingford, UK, 2015.
3. Kassaye, K.T.; Yilma, W.A.; Fisha, M.H.; Haile, D.H. Yield and water use efficiency of potato under alternate furrows and deficit irrigation. *Int. J. Agron.* **2020**, *2020*, 8869098. [CrossRef]
4. Rathore, R.S.; Singh, A.K.; Chaudhary, A.; Kumar, R.; Chauhan, P.S.; Rathore, D.S.; Singh, D. Effect of different irrigation methods and fertigation levels on yield of potato (*Solanum tuberosum* L.) under Western Rajasthan condition. *Int. J. Curr. Microbiol. Appl. Sci.* **2021**, *10*, 3146–3150. [CrossRef]
5. Crosby, T.W.; Wang, Y. Effects of irrigation management on chipping potato (*Solanum tuberosum* L.) production in the Upper Midwest of the U.S. *Agronomy* **2021**, *11*, 768. [CrossRef]
6. Maheswari, M.; Murthy, A.N.G.; Shanker, A.K. Nitrogen nutrition in crops and its importance in crop quality. In *The Indian Nitrogen Assessment*; Elsevier: Amsterdam, The Netherlands, 2017; pp. 175–186.
7. Haverkort, A.J.; MacKerron, D.K.L. (Eds.) *Management of Nitrogen and Water in Potato Production*; Wageningen Pers. Book: Wageningen, The Netherlands, 2000; p. 353.
8. Öztürk, E.; Kavurmacı, Z.; Kara, K.; Polat, T. The effects of different nitrogen and phosphorus rates on some quality traits of potato. *Potato Res.* **2010**, *53*, 309–312. [CrossRef]
9. Fageria, N.K.; Baligar, V.C. Enhancing nitrogen use efficiency in crop plants. *Adv. Agron.* **2005**, *88*, 97–185.
10. Badr, M.A.; El-Tohamy, W.A.; Zaghloul, A.M. Yield and water use efficiency of potato grown under different irrigation and nitrogen levels in an arid region. *Agric. Water Manag.* **2012**, *110*, 9–15. [CrossRef]
11. Zin El-Abedin, T.K.; Mattar, M.A.; Alazba, A.A.; Al-Ghobari, H.M. Comparative effects of two water-saving irrigation techniques on soil water status, yield, and water use efficiency in potato. *Sci. Hortic.* **2017**, *225*, 525–532. [CrossRef]
12. Liu, F.; Shahnazari, A.; Andersen, M.N.; Jacobsen, S.; Jensen, C.R. Effects of deficit irrigation (DI) and partial root drying (PRD) on gas exchange, biomass partitioning, and water use efficiency in potato. *Sci. Hortic.* **2006**, *109*, 113–117. [CrossRef]
13. Essah, S.Y.; Andales, A.A.; Bauder, T.A.; Holm, D.G. Response of two Colorado Russet potato cultivars to reduced irrigation water use. *Am. J. Potato Res.* **2020**, *97*, 221–233. [CrossRef]
14. Hatfield, J.L.; Sauer, T.J.; Prueger, J.H. Managing soils to achieve greater water use efficiency: A review. *Agron. J.* **2001**, *93*, 271–280. [CrossRef]
15. Zhang, H.; Wang, X.; You, M.; Liu, C. Water-yield relations and water-use efficiency of winter wheat in the North China Plain. *Irrig. Sci.* **1999**, *19*, 37–45. [CrossRef]
16. Karam, F.; Amacha, N.; Fahed, S.; Asmar, T.E.; Domínguez, A. Response of potato to full and deficit irrigation under semiarid climate: Agronomic and economic implications. *Agric. Water Manag.* **2014**, *142*, 144–151. [CrossRef]

17. Sadras, V.O.; Hayman, P.T.; Rodriguez, D.; Monjardino, M.; Bielich, M.; Unkovich, M.; Wang, E. Interactions between water and nitrogen in Australian cropping systems: Physiological, agronomic, economic, breeding, and modelling perspectives. *Crop Pasture Sci.* **2016**, *67*, 1019–1053. [CrossRef]
18. Dalla Costa, L.; Delle Vedove, G.; Gianquinto, G.; Giovanardi, R.; Peressotti, A. Yield, water use efficiency, and nitrogen uptake in potato: Influence of drought stress. *Potato Res.* **1997**, *40*, 19–34. [CrossRef]
19. Hargreaves, G.H.; Allen, R.G. History and evaluation of Hargreaves evapotranspiration equation. *J. Irrig. Drain. Eng.* **2003**, *129*, 53–63. [CrossRef]
20. Essah, S.Y.C.; Davis, J.G. *Fertilizing Potatoes in Colorado*; Colorado State University Extension: Fort Collins, CO, USA, 2019; Available online: https://extension.colostate.edu/docs/pubs/crops/00541.pdf (accessed on 1 May 2016).
21. Mulders, P.J.; van den Heuvel, E.R.; van de Molengraft, M.J.G.; Heemels, W.P.M.H.; Reidsma, P. Extreme drought and rainfall had a large impact on potato production in the Netherlands between 2015 and 2020. *Commun. Earth Environ.* **2024**, *5*, 496. [CrossRef]
22. Barman, B.; Goswami, B.; Hussain, R.; Khanikar, P.G.; Deka, R.L.; Singh, K.S.G.; Singh, A. Influence of GDD in different growth parameters of potato at agroclimatic condition of Jorhat. *Pharma Innov. J.* **2019**, *8*, 247–248.
23. Canela Russet. Available online: https://www.potatoslvrc.com/wp-content/uploads/2014/02/Canela-Cultivar-sheet.pdf (accessed on 15 July 2025).
24. Mesa Russet. Available online: https://www.potatoslvrc.com/wp-content/uploads/2014/02/Mesa-Russet.pdf (accessed on 15 July 2025).
25. Li, W.; Xiong, B.; Wang, S.; Deng, X.; Yin, L.; Li, H. Regulation effects of water and nitrogen on the source-sink relationship in potato during the tuber bulking stage. *PLoS ONE* **2016**, *11*, e0146877. [CrossRef] [PubMed]
26. Fouda, S.; Rady, M.M.; Moursy, A.A. Impact of deficit irrigation and nitrogen fertilizers on potato yield and nutrient status under sandy soil conditions. *J. Appl. Sci. Res.* **2012**, *8*, 3045–3055.
27. Gervais, T.; Creelman, A.; Li, X.Q.; Bizimungu, B.; De Koeyer, D.; Dahal, K. Potato response to drought stress: Physiological and growth basis. *Front. Plant Sci.* **2021**, *12*, 698060. [CrossRef] [PubMed]
28. Hill, D.; Nelson, D.; Hammond, J.; Bell, L. Morphophysiology of potato (*Solanum tuberosum*) in response to drought stress: Paving the way forward. *Front. Plant Sci.* **2021**, *11*, 597554. [CrossRef] [PubMed]
29. Miller, J.C.; Tai, G.C.; Ouellette, B.; Miller, J.P. Discriminating Russet Norkotah intraclonal selections using canonical and cluster analysis. *Am. J. Potato Res.* **2004**, *81*, 203–207. [CrossRef]
30. Yuan, B.Z.; Nishiyama, S.; Kang, Y. Effects of different irrigation regimes on the growth and yield of drip-irrigated potato. *Agric. Water Manag.* **2003**, *63*, 153–167. [CrossRef]
31. Kirda, C. Deficit irrigation scheduling based on plant growth stages showing water stress tolerance. *Food Agric. Organ. United Nations Deficit Irrig. Pract. Water Rep.* **2002**, *22*, 3–10.
32. Alva, A.K.; Ren, H.; Moore, A.D. Water and nitrogen management effects on biomass accumulation and partitioning in two potato cultivars. *Am. J. Plant Sci.* **2012**, *3*, 164–170. [CrossRef]
33. Holm, D.G.; Gray, C.; Niebaum, B. *Research Progress Report: Potato Breeding and Selection*; San Luis Valley Research Center: Center, CO, USA, 2020. Available online: https://www.potatoslvrc.com/wp-content/uploads/2021/08/2020-Research-Progress-Report.pdf (accessed on 22 May 2025).
34. Shrestha, B.; Darapuneni, M.; Stringam, B.L.; Lombard, K.; Djaman, K. Irrigation water and nitrogen fertilizer management in potato (*Solanum tuberosum* L.): A review. *Agronomy* **2023**, *13*, 2566. [CrossRef]
35. Shrestha, B.; Stringam, B.L.; Darapuneni, M.K.; Lombard, K.A.; Sanogo, S.; Higgins, C.; Djaman, K. Effect of irrigation and nitrogen management on potato growth, yield, and water and nitrogen use efficiencies. *Agronomy* **2024**, *14*, 560. [CrossRef]
36. Ojala, J.C.; Stark, J.C.; Kleinkopf, G.E. Influence of irrigation and nitrogen management on potato yield and quality. *Am. Potato J.* **1990**, *67*, 29–43. [CrossRef]
37. Ierna, A.; Mauromicale, G. Potato growth, yield, and water productivity response to different irrigation and fertilization regimes. *Agric. Water Manag.* **2018**, *201*, 21–26. [CrossRef]
38. Westermann, D.T.; Tindall, T.A.; James, D.W.; Hurst, R.L. Nitrogen and potassium fertilization of potatoes: Yield and specific gravity. *Am. Potato J.* **1994**, *71*, 417–431. [CrossRef]
39. Da Ros, L.; Elferjani, R.; Soolanayakanahally, R.; Kagale, S.; Pahari, S.; Kulkarni, M.; Wahab, J.; Bizimungu, B. Drought-induced regulatory cascades and their effects on the nutritional quality of developing potato tubers. *Genes* **2020**, *11*, 864. [CrossRef] [PubMed]
40. Bhattarai, S.P.; Midmore, D.J.; Pendergast, L. The importance of timing and rate of nitrogen fertilizer application on potato yield and quality. *Field Crops Res.* **2006**, *96*, 185–195.
41. van Loon, C.D. The effect of water stress on potato growth, development, and yield. *Am. Potato J.* **1981**, *58*, 51–69. [CrossRef]
42. Eldredge, E.P.; Holmes, Z.A.; Mosley, A.R.; Shock, C.C.; Stieber, T.D. Effects of transitory water stress on potato tuber stem-end reducing sugar and fry color. *Am. Potato J.* **1996**, *73*, 517–530. [CrossRef]

43. Nguyen, T.T.N.; Xu, C.Y.; Tahmasbian, I.; Che, R.; Xu, Z.; Zhou, X.; Bai, S.H. Effects of biochar on soil available inorganic nitrogen: A review and meta-analysis. *Geoderma* **2017**, *288*, 79–96. [CrossRef]
44. Vos, J.; van der Putten, P.E.L. Effects of partial shading of the potato plant on photosynthesis of treated leaves, leaf area expansion and allocation of nitrogen and dry matter in component plant parts. *Eur. J. Agron.* **2001**, *14*, 209–220. [CrossRef]
45. Luo, S.; He, Y.; Li, Q.; Jiao, W.; Zhu, Y.; Zhao, X. Nondestructive estimation of potato yield using relative variables derived from multi-period LAI and hyperspectral data based on weighted growth stage. *Plant Methods* **2020**, *16*, 150. [CrossRef] [PubMed]
46. Boyd, N.S.; Gordon, R.; Martin, R.C. Relationship between leaf area index and ground cover in potato under different management conditions. *Potato Res.* **2002**, *45*, 117–129. [CrossRef]
47. Cheng, M.; Wang, H.; Zhang, F.; Wang, X.; Liao, Z.; Zhang, S.; Yang, Q.; Fan, J. Effects of irrigation and fertilization regimes on tuber yield, water-nutrient uptake and productivity of potato under drip fertigation in sandy regions of northern China. *Agric. Water Manag.* **2023**, *287*, 108459. [CrossRef]
48. Deshi, K.E.; Obasi, M.O.; Odiaka, N.I.; Kalu, B.A.; Ifenkwe, O.P. Leaf area index values of potato (*Solanum tuberosum* L.) stored for different periods in different kinds of stores. *IOSR J. Agric. Vet. Sci.* **2015**, *8*, 9–19.
49. Horsfield, A.; Wicks, T.; Davies, K.; Wilson, D.; Paton, S. Effect of fungicide use strategies on the control of early blight (*Alternaria solani*) and potato yield. *Australas. Plant Pathol.* **2010**, *39*, 368–375. [CrossRef]
50. Liu, F.; Song, Q.; Zhao, J.; Mao, L.; Bu, H.; Hu, Y.; Zhu, X.G. Canopy occupation volume as an indicator of canopy photosynthetic capacity. *New Phytol.* **2021**, *232*, 941–956. [CrossRef] [PubMed]
51. Kashyap, P.S.; Panda, R.K. Effect of irrigation scheduling on potato crop parameters under water stressed conditions. *Agric. Water Manag.* **2003**, *59*, 49–66. [CrossRef]
52. Deblonde, P.M.K.; Ledent, J.F. Effects of moderate drought conditions on green leaf number, stem height, leaf length, and tuber yield of potato cultivars. *Eur. J. Agron.* **2001**, *14*, 31–41. [CrossRef]
53. Villa, C.M.; Boschetti, L.; Confalonieri, R.; Acutis, M. Impact of nitrogen supply on canopy development and LAI in the Granola potato variety. *Field Crops Res.* **2017**, *209*, 49–57. [CrossRef]

horticulturae

MDPI

*Article*

# Proximal and Remote Sensing Monitoring of the 'Spinoso sardo' Artichoke Cultivar on Organic and Conventional Management

**Alessandro Deidda [1], Alberto Sassu [1], Luca Ghiani [2], Maria Teresa Tiloca [1] , Luigi Ledda [3], Marco Cossu [1], Paola A. Deligios [3] and Filippo Gambella [1,4,*]**

1 Department of Agricultural Sciences, University of Sassari, Viale Italia 39 a, 07100 Sassari, Italy; adeidda1@uniss.it (A.D.); asassu@uniss.it (A.S.); mtiloca@uniss.it (M.T.T.); mcossu2@uniss.it (M.C.)
2 Department of Biomedical Sciences, University of Sassari, Viale San Pietro 43/B, 07100 Sassari, Italy; lghiani@uniss.it
3 Department of Agricultural, Food and Environmental Sciences, Marche Polytechnic University, Via Brecce Bianche 11, 60131 Ancona, Italy; l.ledda@staff.univpm.it (L.L.); p.a.deligios@staff.univpm.it (P.A.D.)
4 Interdepartmental Center IA—Innovative Agriculture Loc. Surigheddu, 127 bis, Km 28,500, 07041 Alghero, Italy
* Correspondence: gambella@uniss.it

**Abstract**

The development of new techniques to improve crop management, especially through precision agriculture methods and innovations, is crucial for increasing crop yield and ensuring high-quality production. The horticultural sector is particularly vulnerable to inefficiencies in crop management due to the complex and costly processes required for producing marketable products. Optimal nutritional inputs and effective disease management are crucial for maintaining commercial standards. This two-year study investigated the physiological differences between organic and conventional crop management of the Sardinian 'Spinoso sardo' artichoke ecotype (*Cynara cardunculus* var. *scolymus* L.) by integrating a multiplex force-A (MFA) fluorometer and unmanned aerial systems (UASs) equipped with a multispectral camera capable of analysing the NDVI vegetation index. Using both proximal and remote sensing instruments, physiological and nutritional variations in the growth cycle of artichokes were identified, distinguishing between traditional and two organic management practices. The two-year MFA experiment revealed physiological variability and different trends among the three management practices, indicating that MFA proximal sensing is a valuable tool for detecting physiological differences, particularly in chlorophyll activity and nitrogen content. In contrast, the UAS survey was less effective at distinguishing between management types, likely due to its limited use during the second year and the constrained timeframe of the multitemporal analysis. The analysis of the MFA fluorimetric indices suggested significant differences among the plots monitored due to the ANOVA statistical analysis and Tukey test, showing greater adaptability of the conventional system in managing production inputs, unlike the organic systems, which showed higher variability within the plots and across the survey years, indicating aleatory trends due to differences in crop management.

**Keywords:** precision agriculture; horticulture; fluorometer; UAS; organic management

## 1. Introduction

In recent decades, innovations in agricultural practices have led to significantly higher crop yields per unit area compared to the previous century [1]. This evolution is related to the gradual rise in the global population [2] and, consequently, increased demand for essential goods, all aimed at maximising production efficiency and ensuring food security [3]. In the agricultural scenario, horticultural crops are crucial, accounting for more than a billion tonnes in 2022 [4]. Traditionally, conventional horticultural crop management involves numerous interventions in soil management, plant nutrition, and disease control [5], with massive treatments to achieve production goals and maintain quality standards [6]. Recently, the emergence of organic management for horticultural crops has led to a significant reduction in production inputs, tillage, and the total elimination of chemical plant protection products [7]. Artichoke cultivation (*Cynara cardunculus* var. *scolymus* L.) represents a typical species of the Mediterranean area in the Asteraceae family, derived from domesticated forms of the wild cardoon (*Cynara cardunculus* var. *sylvestris* Lam.); it occupies a relevant place among vegetable crops [8], counting several varieties and management methods found in the regions where it is cultivated [9]. Artichoke cultivation is considered a high-value agricultural activity, attributable to its substantial yield and economic return per hectare [10], and, depending on the region, it exhibits a wide range of differentiated ecotypes. With regard to its commercial diffusion in Sardinia, one of the principal ecotypes in Sardinia is the "Spinoso sardo", due to its wide diffusion on the island. Among artichoke ecotypes, the "Spinoso sardo" variety is an autumn–winter re-flowering variety [11]. Traditionally, it is planted with semi-dormant offshoots during the summer and is subject to a forcing technique. This method uses early awakening through irrigation to start the production cycle sooner, allowing the plants to produce first-order flower heads by October or November. Therefore, the forcing technique anticipates the plant's cycle, allowing for an earlier commercial output. It highlights the significance of cultivation techniques and how different management methods can impact the plant's physiological development [12].

Conventional artichoke management involves supplementing the soil with mineral fertilisers, weed control, phytoiatric treatments, and, in some cases, the removal of senescent crop residues for potential energy use at the end of the season, despite their low energy yield [13]. This cultivation system is typically based on continuous artichoke monoculture. Traditionally, artichoke cultivation depends on conventional soil management and uniform fertilisation practices throughout the growing season. As a nutrient-absorbing species, artichoke requires substantial inputs of nitrogen and phosphorus–potassium fertilisers, which are typically supplied through mineral fertilisers applied during soil preparation and throughout the growing season [14]. These practices influence both the physiological crop activity and the chemical–structural composition of the soil. In recent years, organic management systems have emerged as a sustainable alternative to conventional cultivation [15]. This approach enhances crop nutrition through intercrops cultivation, which is subsequently incorporated into the soil, promoting natural mineralisation processes and reducing dependency on synthetic fertilisers [16–19]. While the organic system offers more sustainable management and less environmental impact than traditional systems, organic cultivation presents the risk of unpredictable production from both a nutritional and phytosanitary point of view, as the control of production inputs is more limited and potentially less efficient. Nonetheless, traditional cultivation methods, as with many other cropping systems, often lack strategies for the selective application of inputs based on the heterogeneous characteristics within a field. Uniform field management—encompassing tillage and nutrient application—frequently fails to account for spatial variability in soil fertility and crop needs [12,20]. The gradual and site-specific management of agricultural inputs is one of the key innovations introduced by precision agriculture (PA). PA seeks to address intra-field variability by employing advanced technologies and analytical methods for crop monitoring, providing the basis for decision support systems (DSSs) that guide more efficient and sustainable management strategies. Crop monitoring is a critical component of DSS development [21], enabling real-time data acquisition through dedicated sensors. This data is then processed into actionable information, allowing operators to make informed decisions. A central objective of such monitoring is to generate spatial information that distinguishes local field conditions, thereby facilitating the optimisation of input use. This helps to prevent the over- or under-application of resources in areas where conventional, uniform management might lead to inefficiencies. Proximal sensing systems have emerged as a highly effective tool for localised crop monitoring [22]. By providing high-resolution, site-specific data, these systems capture the physiological variability within the field and enable timely, targeted interventions [23]. The proximal sensing application strategy could integrate and enhance the evaluation of site-specific input applications, such as localised nitrogen fertilisation, water management, and plant protection product (PPP) application [24]. These tools, equipped with the latest technologies, offer rapid analysis and immediate access to critical information, enhancing decision-making processes at the field level. Proximal sensing can provide information ranging from physiological to quantitative variables. Among the various existing instruments, fluorimetric sensors deserve particular mention, as they can rapidly obtain non-destructive information on the crop's physiological variables. This application is applied in numerous cropping systems, from herbaceous to tree crops. Fluorimetry has multiple applications in horticulture as well, enabling the analysis of different physiological characteristics of crops both in the field and in the laboratory [24,25]. In the field of artichoke cultivation, fluorimetry has been used to analyse chlorophyll, chlorosis disease [26], and biomass composition [27,28]. In parallel, remote sensing technologies, using satellite platforms, manned aircraft, and unmanned aerial systems (UASs), have expanded the scale and efficiency of monitoring practices through multiple sensors and vegetational indices [29]. In fact, the application of these techniques has contributed to more rapid crop analysis, reducing the operational costs of acquiring physiological data in the field. Traditionally, analyses conducted for product evaluation involved destructive sampling, which negatively affects production. Over the past decade, these tools have become integral to agricultural surveillance, enabling the collection of data across large areas with increasingly reliable results [30], regardless of variations in ground resolution associated with these different systems [31].

Understanding field variability, achieved through site-specific physiological evaluation of the crop, could enable more efficient management of production resources. This process would allow for a more targeted and potentially localised assessment of individual plants, considering their physiological needs to optimise crop management.

The present study involved non-destructive sampling of 'Spinoso sardo' Sardinian ecotype artichoke leaves during the 2018/2019 and 2019/2020 seasons, covering the vegetative growth and reproductive phases of the plants, monitoring three different cultivation management systems, classified as conventional or organic, using a proximal fluorimetric sensor. The primary purpose is to evaluate the ability and efficiency of fluorimetric indices in detecting any physiological differences in leaves based on the different cultivation techniques adopted. Secondly, due to the high intensity of the work involved in MFA analysis on a wide range of samples, an aerial survey was performed using a UAS to acquire the NDVI vegetation index. The UAS analysis was performed to observe the dynamics of the remote sensing application and compare it with fluorimetric data, evaluating the potential for using a UAS as a substitute instead of MFA.

## 2. Materials and Methods

### 2.1. Site and Experimental Design

The experiment took place at the experimental fields of the Agriculture Department, University of Sassari, located in Ottava (SS, Italy, 40°46′31″ N; 8°29′12″ E, WGS84 Coordinate System, 81 m above ground level) during the 2018/2019 and 2019/2020 seasons. These two seasons were characterised by a similar temperature trend, with a 10% reduction in natural water supply from precipitation, settling at around 450 mm.

The monitoring operation was performed in an ongoing experiment plot [16] identified in Figure 1, covering an area of approximately 2500 m$^2$. The artichoke plants were distributed in rows oriented in a northwest–southeast direction, with a row spacing of 0.7 m × 1.4 m (along the row and between rows, respectively), resulting in a density of 9524 plants/ha. To minimise the canopy border effect, two rows were planted on the borders, and two additional rows were placed between the crop management sections to prevent contamination. Artichokes were planted in July using semi-dormant offshoots as propagation organs. The adopted experimental design considered three different types of agronomic management for the artichoke, as follows:

- Conventional (CON) management plot;
- Organic plot with annual presence of artichoke (ORG-I);
- Organic plot, alternated annually with cauliflower (ORG-II);

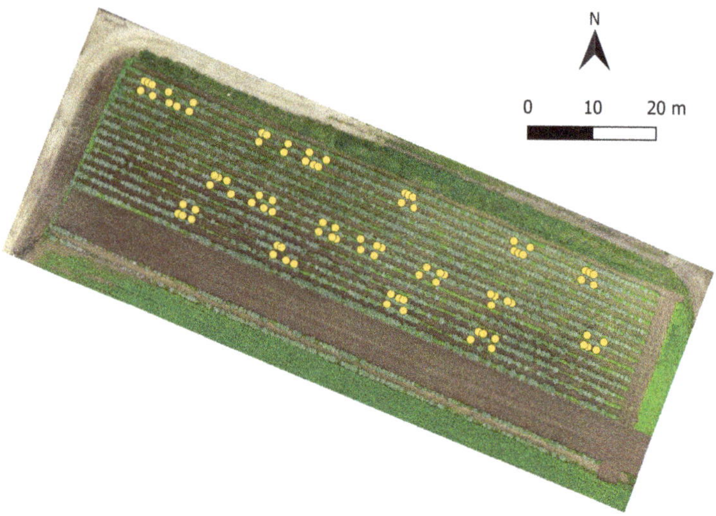


**Figure 1.** Experimental field under analysis. The plants monitored during the first survey year are marked in yellow. Sampling operations covered specific parts of the field, where the operator visually assessed the location of the measurements around some checking points placed along the field.

The experimental design compares conventional (CON) practices with two alternative organic management types, where artichoke succession occurs annually and biennially in the field (ORG-I and ORG-II, respectively). The management of CON involves monocropping in the same plot, with conventional agricultural practices including mineral fertilisation, weeding, phytoiatric treatments, and the soil incorporation of senescent dried crop residues toward the end of the crop cycle during the spring season.

The artichoke-growing cycle on the annual organic management ORG-I was interrupted early at the end of the marketable harvest period (mid-April), and the fresh residues were chopped and ploughed into the soil. To restock the soil's nitrogen, a short-cycle legume, French bean (*Phaseolus vulgaris* L. cv. Bronco) (Monsanto Agricoltura Italia SpA, Milano, Italy), was planted in the ORG-I plot. The French bean was interrupted at the reproductive stage, when the plants produced the first pods (end of flowering). The biomass was incorporated in a fresh state to increase the soil's nutrient content. Fresh residues from this bean crop were also incorporated into the soil at the end of June, before the new growing season for artichoke began.

The biennial rotation of artichokes was managed using cauliflower (*Brassica oleracea* L. var. *botrytis* cv. Nautilus). Artichoke and cauliflower species were cultivated alternately in the two plots in the two survey years. *Pisum sativum* L. cv. Attika (Limagrain Verneuil Holding, France), used as a legume cover crop, was sown in the inter-row spaces of artichoke and cauliflower in February. At the end of the primary crop's marketable harvest period, and when the peas were flowering (mid-April), artichoke, cauliflower, and fresh pea residues were incorporated into the soil.

According to organic farming principles, no phytosanitary treatments or chemical fertilisation were applied in the ORG-I and ORG-II management types, and the irrigation criteria followed the ongoing experiment [32,33]. The surveys were conducted during the crop's critical phenological phases, with monitoring operations scheduled according to the weather trend on a three-week basis.

Table 1 reports the details of the survey days during the experiment and their corresponding BBCH (Biologische Bundesanstalt, Bundessortenamt, and CHemical industry) code identification [34].

**Table 1.** Experimental survey days with the associated day of year (DOY) and BBCH scale.

| Date | DOY | BBCH |
|---|---|---|
| 21 December 2018 | 355 | 59 |
| 16 January 2019 | 16 | 59 |
| 13 February 2019 | 44 | 61 |
| 7 March 2019 | 66 | 67 |
| 22 March 2019 | 81 | 69 |
| 29 November 2019 | 333 | 55 |
| 27 December 2019 | 361 | 59 |
| 16 January 2020 | 16 | 59 |
| 24 February 2020 | 55 | 69 |

Fluorimetric data acquisition involved five representative plants of "Spinoso sardo" positioned around several checkpoints distributed along the field, as some crop failures were found along the rows due to the abortion of the offshoots. As shown in Figure 2a, the monitorable plants have large, well-expanded leaves, apparently not diseased and of normal development. The non-monitorable plants, as observed in Figure 2b, have poor development, with inadequately expanded leaves that cannot be observed with the monitoring instrument.

MFA field measurements were taken on clear days after the morning dew on the plants had evaporated. Figure 3 shows the thesis arrangement in the field; the ORG-II thesis appears on different plots in alternate years with the cauliflower thesis due to their biennial rotation. Figure 3 displays the plots involved in the survey, with experimental rows (Thesis, "T") spaced by intermediate border rows ("B") to isolate the different treatments.

**Figure 2.** Example of sample selection during the sampling preparation phase. On the right, well-developed plants are adequate for multi-temporal monitoring (**a**); on the left, plants are unsuitable for monitoring (**b**). The preliminary identification of the plants permitted their subsequent monitoring along the artichoke production cycle.

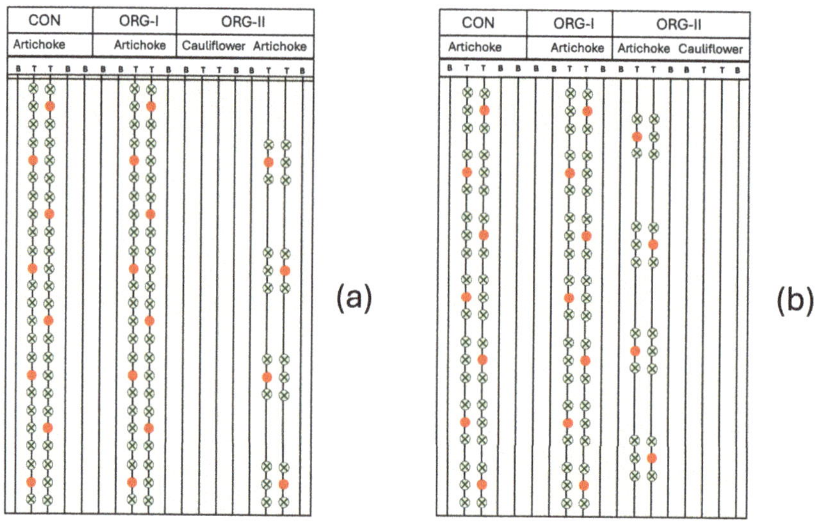


**Figure 3.** Experimental design adopted for 2018/2019 (**a**) and for 2019/2020 (**b**). From left to right, the monitored thesis (T) separated by the border (B) rows are as follows: Conventional (CON) plot; organic plot (ORG-I) in a monocropping and annual rotation with French bean; organic plot (ORG-II) in a biennial rotation with cauliflower, as observable in both designs. Checking points are marked with red circles, and artichoke plants are marked with green crossed-out circles.

### 2.2. Fluorimetric Analysis

The instrument adopted for the non-destructive monitoring of plants is a hand-held multi-parameter fluorescence device capable of observing the physiological evolution of the crop and extrapolating its main nutritional characteristics [25]. The instrument used was the ForceA Multiplex 3 portable fluorometer (MFA, Orsay, France). The detection system comprises three light-emitting diode (LED) channels that emit pulsed radiation at wavelengths in the visible spectrum (RGB): 470 nm (blue), 516 nm (green), and 635 nm (red), respectively. The instrument features six LEDs that operate at a wavelength of 375 nm in the ultraviolet spectrum [35]. All the detectors are enclosed in a plexiglass structure to protect and isolate input and output radiation. At the centre of the sensor array, there are three sensors responsible for fluorescence detection in three delimited ranges, as follows:

far-red fluorescence (FRF), near-red fluorescence (RF), and blue–green fluorescence (BG-F). The signals processed by MFA provide a total of 24 specific indices, which estimate the main physiological properties of the crop [36]. For this research, the fluorimetric analysis involved the SFR-G, SFR-R, NBI-G, NBI-R, FLAV, and BRR-FRF indices, evaluating chlorophyll, nitrogen, and flavonoid content, over the water stress and other abiotic stress indices. According to the literature, the choice of these indices is based on their ability to investigate the main physiological characteristics of the plant, providing a comprehensive view of the plant's state [37]. The measurements include the sample fluorescence ratio with excitation channels in the green and red wavelengths (SFR-G and SFR-R) for assessing chlorophyll content and yield [37], and the nitrogen balance index, which utilises excitation channels in red and green (NBI-G and NBI-R) to monitor total nitrogen levels [38]. The flavonoid index (FLAV), defined as the logarithm of the ratio of red to UV excitation of chlorophyll fluorescence in the RF, is widely used in leaf surveys for flavonoid estimation [39]. Furthermore, the blue-to-red emission ratio index (BRR-FRF) is a comprehensive indicator. Depending on the specific crop analysed, it can identify stress, nutritional deficiencies in field crops, the presence of pathogens, or grape ripening [40].

In the experimental design, the MFA instrument was used to monitor the artichoke vegetative architecture, dividing it into three different layers, as follows: pre-senescent adult leaves (first stage), adult leaves (second stage), and newly formed leaves (third stage), respectively. The MFA analysed each leaf at three different points—the apical, central and basal parts—for a total of nine leaves per plant, at a 10 cm distance between the sample and the sensors. The application of a three-layer analysis was necessary to obtain a homogeneous multitemporal analysis; during the growing period, and until its final stages, all the measurements focused on the same kind of leaves, guaranteeing the uniform analysis of the plant.

The indices used in the MFA monitoring are listed below.

$$SFR-R=\frac{FRF_{R}}{RF_{R}}, \tag{1}$$

$$SFR-G=\frac{FRF_{G}}{RF_{G}}, \tag{2}$$

$$FLAV=log(\frac{FRF_{R}}{FRF_{UV}}), \tag{3}$$

$$BRR-FRF=\frac{YF_{UV}}{FRF_{G}}, \tag{4}$$

$$NBI-R=\frac{FRF_{UV}}{RF_{R}}, \tag{5}$$

$$NBI-G=\frac{FRF_{UV}}{RF_{G}}, \tag{6}$$

### 2.3. UAS Survey

During the 2019–2020 season, MFA monitoring was combined with a UAS survey to observe the different management types through a vegetation index. To avoid environmental variability in remote and proximal sensing data acquisition, the surveys were conducted on the same day. The aerial system involved in the monitoring operations was a commercial UAS Phantom 4 Pro (DJI, Shenzhen, China) with an RGB CMOS 1-inch 20-megapixel camera, equipped with a multispectral red–green–near-infrared (RGN) Survey 3 (Mapir) sensor mounted on the UAS frame. The Mapir calibration target was positioned on the ground due to reflectance corrections for raw data acquisition. The flights were planned on sunny days, in order to maintain a constant meteorological situation and ensure homogeneous and adequate image acquisition conditions. All the surveys were performed at 40 m above ground level (AGL), with a ground sampling distance (GSD) of 1.096 cm. To acquire a high-resolution orthomosaic reconstruction, the overlap values were 75% and 85% for frontal and side directions, respectively. The RGB camera was necessary to reconstruct and segment the canopy into three dimensions, allowing for the digitalisation of individual artichoke plants. This operation facilitated plant-specific analysis using the structure-from-motion system Agisoft Metashape (St. Petersburg, Russia). Specifically, the RGB orthomosaic, combined with the height values obtained from the digital terrain model (DTM) and the digital surface model (DSM), enabled the identification of individual plants using the canopy height model (CHM) segmentation technique. The reconstructed plants allowed the analysis of vegetative crops through the normalised difference vegetation index (NDVI) derived from the bands collected by the RGN sensor, enabling the assessment of physiological variability between the three management methods adopted. Below is the equation for the extraction of the NDVI index.

$$NDVI = \frac{NIR_{(850\ \text{nm})} - Red_{(550\ \text{nm})}}{NIR_{(850\ \text{nm})} + Red_{(550\ \text{nm})}}, \quad (7)$$

### 2.4. Data Processing and Software Elaboration

The raw data from MFA and UAS were pre-processed prior to statistical analysis. Regarding the fluorimetric analysis, the three MFA measurements taken from each leaf were averaged. The result was then averaged across the three leaf layers to yield a single value for each plant, representing its overall trend. The CHM technique for the canopy reconstruction involved subtracting the DTM layer from the DSM layer to isolate the heights of the plants effectively [41]. The digital layer segmentation process was performed on QGIS using the following formula:

$$CHM = DSM - DTM \quad (8)$$

The reconstruction and extraction of individual plants through the CHM model enabled the calculation of the NDVI vegetation index for the examined plants, as the multispectral sensor was externally implemented in the UAS frame, and the NIR RAW data did not contain GNSS positioning information, which required the overlaying of the CHM model.

The preliminary dataset construction and processing allowed the association of the fluorimetric measurements with the corresponding vegetation indices for individual plants. The MFA data were processed using the open-access software R for statistical analysis and reconstruction of the relative graphs. After the preliminary statistical analysis check for normality distribution, the analysis of variance (ANOVA) was used to identify significant differences among the studied groups, and subsequently, the Tukey test [42] was performed to assess the subdivision between the CON, ORG-I, and ORG-II groups. The statistical results were integrated by incorporating the respective graphs for each variable, which illustrated the evolution of indices over the monitoring period.

The indices derived from the MFA were categorised into the following four groups: (a) date, (b) management, (c) checking points, and (d) index. The "date" refers to the measurement days, "management" pertains to the type of cultivation practice applied, and "checking points" indicate the relative positions of the plants to the sample points. The "index" represents the value of interest, averaged across different leaf stages and plants at the various checking points. Among these variables, "management" serves as the key discriminating factor, as it represents the primary source of variation in the fluorimetric indices. Additionally, potential interactions between "management" and "date" (date × management), as well as between "management" and "checking points" (management × checking points), were explored to assess whether the survey dates or the spatial arrangement of the checking points in the field could contribute to variability in the fluorimetric indices. The NDVI data from UAS followed the same MFA statistical analysis. They were then compared with the MFA indices using a correlation matrix in R through the ggcorrplot and emmeans libraries to observe any correlations between UAS and MFA data.

## 3. Results

### 3.1. Fluorimetric Analysis

The present subsection explores the statistical results of the MFA survey in the 2018/2019 and 2019/2020 research years.

#### 3.1.1. First Survey Year

The ANOVA analysis for the SFR indices displayed in Table 2 observed a highly significant statistical difference, evidencing a different canopy evolution among the management during the survey period. Table 2 also reports the Tukey test analysis to identify any affinities between ORGs and CON based on the chlorophyll detection indices. The comparison between ORG-I and ORG-II did not indicate a significant difference, but a difference emerged when they were compared with CON management.

**Table 2.** ANOVA and Tukey test results on SFR-G and SFR-R.

| Index | SFR-G | | SFR-R | |
|---|---|---|---|---|
| | F-Value | *p*-Value | F-Value | *p*-Value |
| Management | 23.609 | $1.69 \times 10^{-10}$ *** | 22.401 | $5.09 \times 10^{-10}$ *** |
| Date × Management | 5.314 | $2.12 \times 10^{-6}$ *** | 5.337 | $1.97 \times 10^{-6}$ *** |
| Tukey Test | *p*-value | | *p*-value | |
| CON-ORG-I | $<1.0 \times 10^{-16}$ *** | | $<1.0 \times 10^{-10}$ *** | |
| CON-ORG-II | $3.0 \times 10^{-4}$ *** | | 0.030 ** | |
| ORG-I-ORG-II | 0.205 ns | | 0.080 ns | |

The level of significance (*p*-value) is shown in the table, where ** indicates $p < 0.01$, and *** indicates $p < 0.001$. ns represents no significant difference.

The results from the three management types were plotted in a line graph to observe the seasonal trend across the survey days. The SFR-G and SFR-R graphs in Figure 4 show that, for the CON management, photosynthetic activity was initially slightly lower than in the ORG-I and ORG-II theses, but rose to higher values starting in January 2019. The ORG management types, however, did not provide a statistical discrimination between them, but the graphs in Figure 4 highlight the SFR index decrease for ORG-II management, which is 23% lower than the other plots.

The BRR-FRF index did not provide a statistical discrimination between the treatments during the first survey year, as observed in the ANOVA analysis and Tukey test results. Different from the stress index, the FLAV index showed a significant difference between the ORG-I and ORG-II treatments. The index increased steadily over the days of monitoring, increasing tenfold on the last day. Table 3 shows the results of the ANOVA and the Tukey analysis.

The graphical reconstruction of the stress and flavonol indices during the survey confirms the low level of significance among the management types due to the overlapping of the obtained values. In particular, in the graph in Figure 5a, the trends of the three management types run side by side during all the sampling dates, separating only at the end of the cycle (22 March 2019). In this survey year, the ORG-I and ORG-II treatments show a much higher stress level than the CON management, especially in the last monitoring survey, despite the FLAV index in Figure 5b, which indicates an increasing and constant trend in all three management types. The ORG-II plot, despite the lack of a significant response, showed higher values than the other management types for most of the time (+11%), a level that the ORG-I management type reached in the last survey.

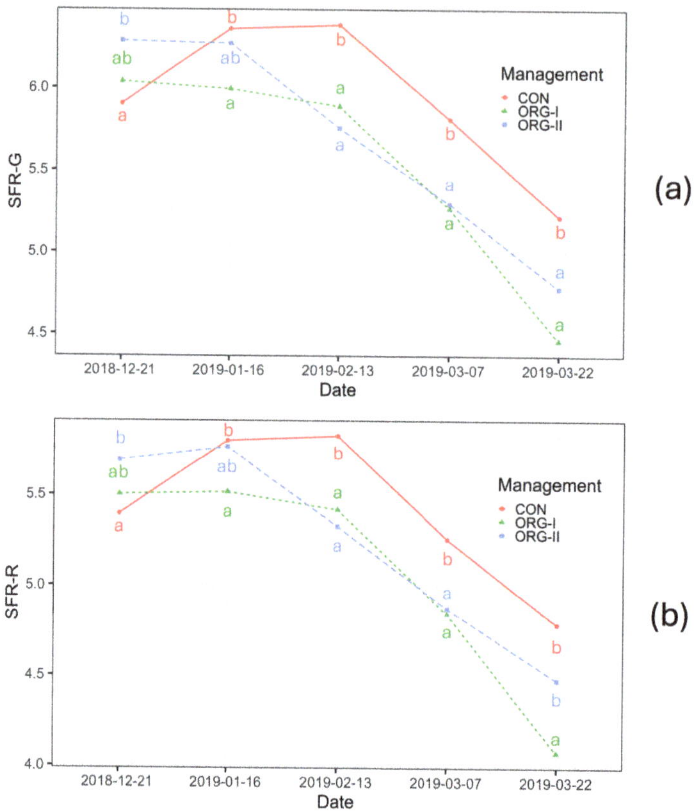


**Figure 4.** Evolution of SFR indices through the first survey year. On the x-axis are the survey dates, and on the y-axis are the index values. The two graphs show the SFR-G index (**a**) and the SFR-R (**b**). The letters in the graphs indicate the differences between the various management systems identified in the Tukey test.

**Table 3.** ANOVA results on BRR-FRF and FLAV.

| **Index** | **BRR-FRF** | | **FLAV** | |
|---|---|---|---|---|
| | **F-Value** | ***p*-Value** | **F-Value** | ***p*-Value** |
| Management | 1.165 | 0.313 ns | 2.945 | 0.054 ns |
| Date × Management | 1.140 | 0.335 ns | 1.546 | 0.139 ns |
| Tukey Test | *p*-value | | *p*-value | |
| CON-ORG-I | 0.449 ns | | 0.770 ns | |
| CON-ORG-II | 0.371 ns | | 0.156 ns | |
| ORG-I-ORG-II | 0.937 ns | | 0.044 * | |

The level of significance (*p*-value) is shown in the table, where * indicates $p < 0.05$.ns represents no significance.

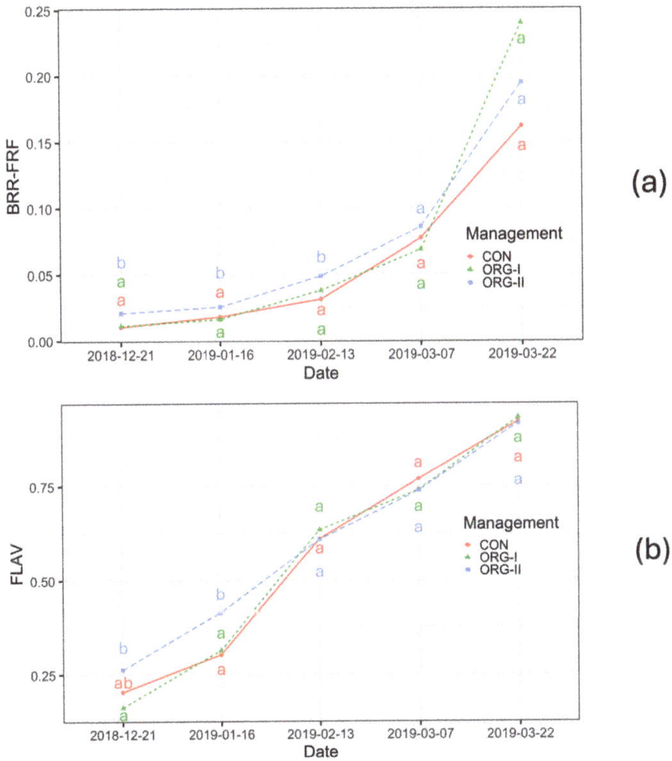


**Figure 5.** BRR-FRF (**a**) and FLAV (**b**) indices, indicating the evolution of stress and flavonol indices during the season. The letters in the graphs indicate the differences between the various management systems identified in the Tukey test.

The final aspect considered is the total nitrogen content, observed using the NBI-G and NBI-R indices. The statistical analysis, as with the previous ones, identified fluctuations in the nitrogen fluorimetric indices among the survey dates. Unlike the previous indices, especially those focusing on chlorophyll activity, the Tukey test revealed a significant difference between the ORG-I and ORG-II treatments, while showing no particular differences with the CON thesis. Table 4 shows the ANOVA and Tukey test statistical results.

**Table 4.** ANOVA results on NBI-G and NBI-R.

| **Index** | **NBI-G** | | **NBI-R** | |
|---|---|---|---|---|
| | **F-Value** | ***p*-Value** | **F-Value** | ***p*-Value** |
| Management | 6.468 | 0.002 ** | 5.662 | 0.004 ** |
| Date × Management | 3.460 | 0.001 *** | 2.786 | 0.005 ** |
| Tukey Test | *p*-value | | *p*-value | |
| CON-ORG-I | 0.132 ns | | 0.089 ns | |
| CON-ORG-II | 0.118 ns | | 0.277 ns | |
| ORG-I-ORG-II | 0.001 *** | | 0.003 ** | |

The level of significance (*p*-value) is shown in the table, where ** indicates $p < 0.01$, and *** indicates $p < 0.001$. ns represents no significance.

The fluorimetric data in Figure 6 indicate that the nitrogen indices show a constant inflexion throughout the survey dates, with continuous overlapping of values between the three management types, consolidating what was identified by the statistical analyses.

These graphs, despite the low test significance, reflect what is indicated by the SFR, BRR-FRF, and FLAV indices. For each NBI index, the decrease fell to more than 80% during the first monitoring year, according to the other fluorimetric indices observed above.

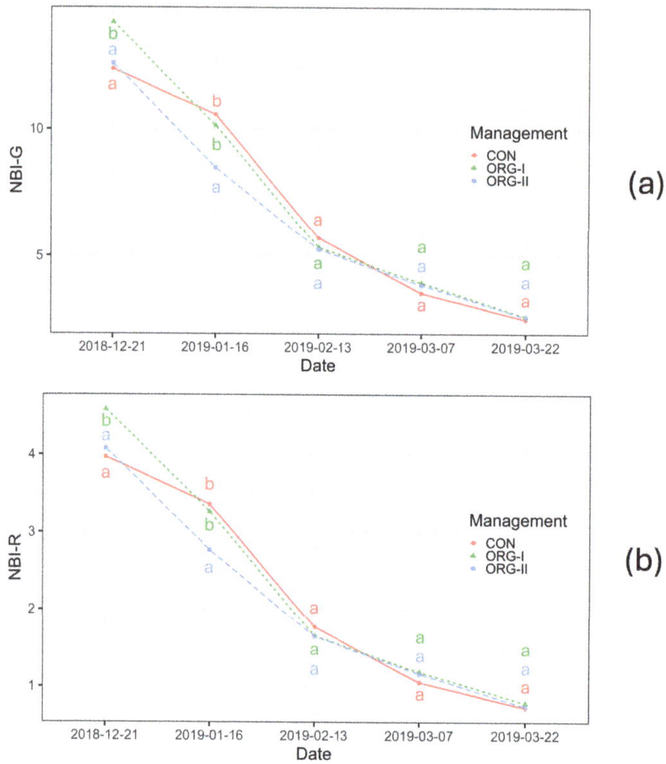


**Figure 6.** NBI-G (**a**) and NBI-R (**b**) on leaves during the first survey year. The letters in the graphs indicate the differences between the various management systems identified in the Tukey test.

### 3.1.2. Second Survey Year

The second monitoring year is characterised by different evolutionary dynamics compared to the previous one. Regarding the chlorophyll indices SFR-G and SFR-R, the ANOVA analysis in Table 5 reports a significant difference among the three management systems, highlighting statistical variability across the survey dates. According to the Tukey test, the primary differences are observed between the CON and ORG-I management systems due to the high significance values and, for the SFR-R index, between the CON and ORG-II management systems. For both fluorimetric indices, no significant differences are reported between ORG-I and ORG-II.

The graph construction of the SFR indices in Figure 7 shows the evolution of the three theses during the multi-temporal analysis. The ORG management types display high chlorophyll values in the early stages of the cycle, peaking in December and then undergoing a slow but continuous decrease until the end of the plant's physiological cycle. The CON thesis, on the other hand, shows the opposite trend to the ORG theses, starting with values consistently lower than those of the ORG types in the early growth stages, then experiencing a rapid escalation (+16.6% and 18.9% for SFR-G and SFR-R, respectively) until it surpasses both organic theses in January (+8.9% and 9.4% for SFR-G and SFR-R), remaining steady until the end of the cycle.

**Table 5.** ANOVA results on SFR-G and SFR-R.

| Index | SFR-G | | SFR-R | |
|---|---|---|---|---|
| | F-Value | *p*-Value | F-Value | *p*-Value |
| Management | 7.388 | <0.001 *** | 7.622 | <0.001 *** |
| Date × Management | 17.743 | $<2.000 \times 10^{-16}$ *** | 17.798 | $<2.000 \times 10^{-16}$ *** |
| **Tukey Test** | *p*-value | | *p*-value | |
| CON-ORG-I | <0.001 *** | | <0.001 *** | |
| CON-ORG-II | 0.342 ns | | 0.039 * | |
| ORG-I-ORG-II | 0.133 ns | | 0.677 ns | |

The level of significance (*p*-value) is shown in the table, where * indicates $p < 0.05$, and *** indicates $p < 0.001$. ns represents no significance.

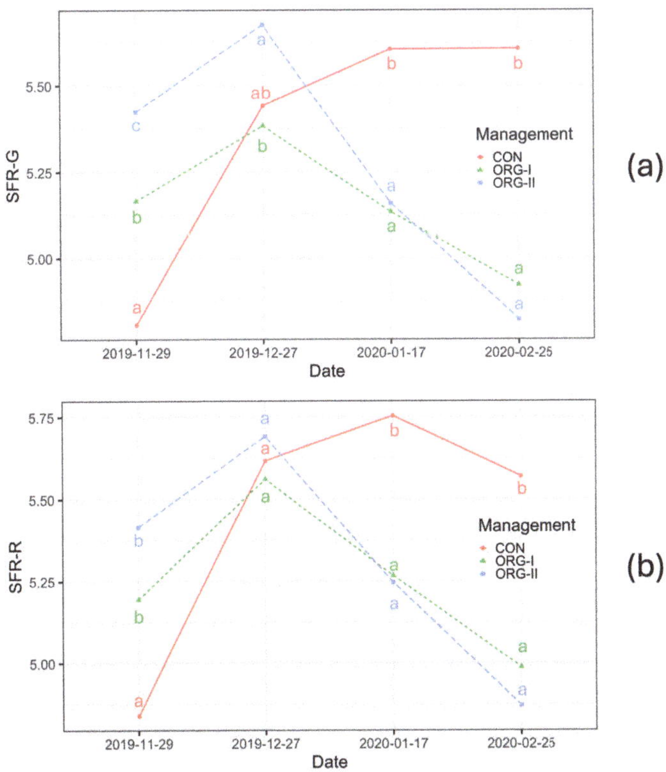


**Figure 7.** The SFR-G (**a**) and SFR-R (**b**) index trends during the 2019–2020 survey. The letters in the graphs indicate the differences between the various management systems identified in the Tukey test.

The spatial–temporal variability of the BRR-FRF-2 and FLAV-2 indices is presented in Table 6, revealing significant differences among the management systems, in contrast to the results from the previous year. ANOVA analysis of the management and date variables indicated a significant separation among the management systems, unlike the 2018/2019 season. All pairwise comparisons performed using the Tukey test showed significant differences between the CON and ORG management systems, while no statistical significance was found between ORG-I and ORG-II. Table 6 summarises the analysis of BRR-FRF and FLAV indices.

**Table 6.** ANOVA results on BRR-FRF and FLAV.

| Index | BRR-FRF | | FLAV | |
|---|---|---|---|---|
| | F-Value | $p$-Value | F-Value | $p$-Value |
| Management | 15.676 | $3.12 \times 10^{-7}$ *** | 32.590 | $1.19 \times 10^{-13}$ |
| Date × Management | 6.148 | $<3.96 \times 10^{-6}$ *** | 4.150 | $4.89 \times 10^{-4}$ *** |
| Tukey Test | $p$-Value | | $p$-Value | |
| CON-ORG-I | <0.001 *** | | <0.001 *** | |
| CON-ORG-II | 0.006 ** | | <0.001 *** | |
| ORG-II-ORG-I | 0.192 ns | | 0.207 ns | |

The level of significance ($p$-value) is shown in the table, where , ** indicates $p < 0.01$, and *** indicates $p < 0.001$. ns represents no significance.

The graphical results of the BRR-FRF and FLAV indices provide a valuable overview of the temporal evolution of the three management systems. The BRR-FRF index (Figure 8a) shows that the values of the three plots are similar during the early growth stage but begin to diverge from late December until the end of the monitoring period, with the ORG systems displaying significantly higher stress levels compared to the CON management (+49.3%). Similarly, the FLAV index (Figure 8b) exhibits the same trend, indicating that ORGs show higher values than the CON management. Throughout the production cycle, the CON system consistently maintains lower levels of stress and flavonoids than the ORG systems (−28.3%). These findings align with the SFR indices, where the lower stress and flavonol values observed in the CON management complement the higher chlorophyll content.

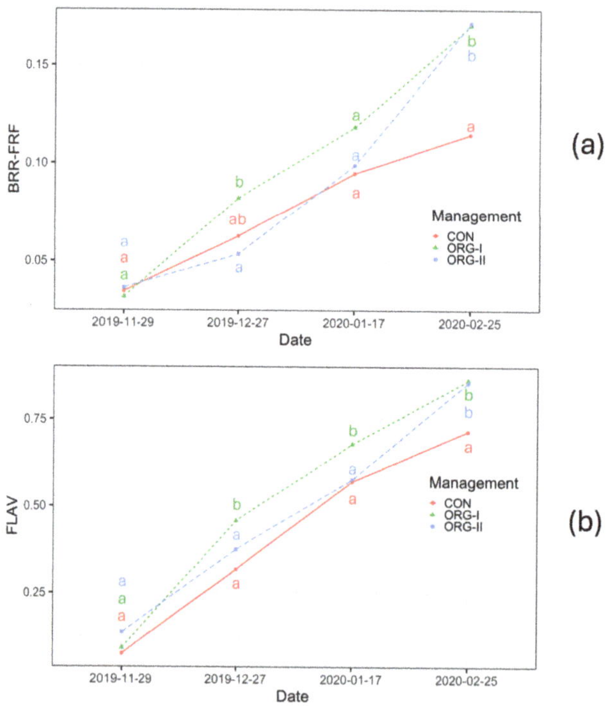


**Figure 8.** The stress index BRR-FRF (**a**) and the flavonol index FLAV (**b**) during the second survey year. The letters in the graphs indicate the differences between the various management systems identified in the Tukey test.

According to the previous indices, the statistical analysis of the NBI-G and NBI-R MFA indices demonstrated significant differentiation among the management systems. The ANOVA analysis revealed high variability in both the management and date variables, as indicated by the significance values. The Tukey test confirmed a distinction between the CON and ORG systems, similar to the BRR-FRF and FLAV indices, with the CON system being statistically different from the ORG system. However, no significant differences were observed between ORG-I and ORG-II. The ANOVA and Tukey statistical results are reported in Table 7.

**Table 7.** ANOVA results on NBI-G and NBI-R.

| **Index** | **NBI-G** | | **NBI-R** | |
|---|---|---|---|---|
| | **F-Value** | ***p*-Value** | **F-Value** | ***p*-Value** |
| Management | 15.02 | $5.69 \times 10^{-7}$ *** | 15.119 | $5.19 \times 10^{-7}$ *** |
| Date × Management | 5.42 | $2.31 \times 10^{-5}$ *** | 5.667 | $1.27 \times 10^{-5}$ *** |
| Tukey Test | *p*-value | | *p*-value | |
| CON-ORG-I | <0.001 *** | | <0.001 *** | |
| CON-ORG-II | 0.017 * | | 0.007 ** | |
| ORG-I-ORG-II | 0.115 ns | | 0.214 ns | |

The level of significance (*p*-value) is shown in the table, where * indicates $p < 0.05$, ** indicates $p < 0.01$, and *** indicates $p < 0.001$. ns represents no significance.

The NBI-G and NBI-R graphs in Figure 9 show a general decline in nitrogen content, which gradually decreases throughout the growth stages, averaging a 77% reduction across the management systems. As indicated by the Tukey test, although ORG-I and ORG-II exhibit relatively similar MFA values, both differ significantly from the CON system, consistent with previous observations from other indices. Notably, on the first monitoring day, the CON system displays lower NBI values compared to ORG-I but rises to higher values on all subsequent survey dates. Figure 9 resumes the MFA management evolution during the surveys.

This observation reflects what was highlighted in the SFR, BRR-FRF, and FLAV indices, where the CON thesis consistently maintains higher levels during the intermediate and final phases of the artichoke's biological cycle, suggesting a higher nitrogen content than that of the ORG types.

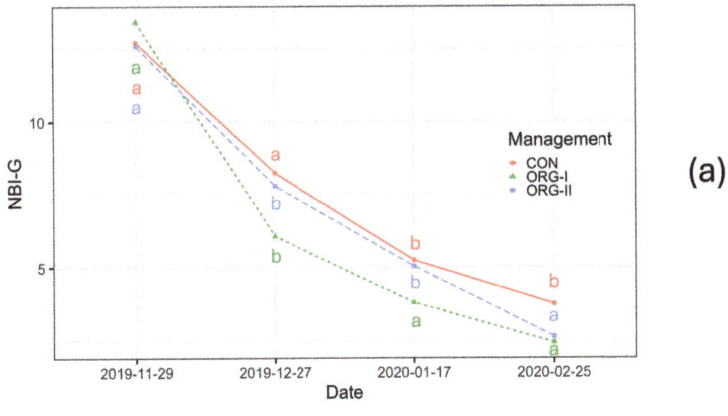


**Figure 9.** *Cont.*

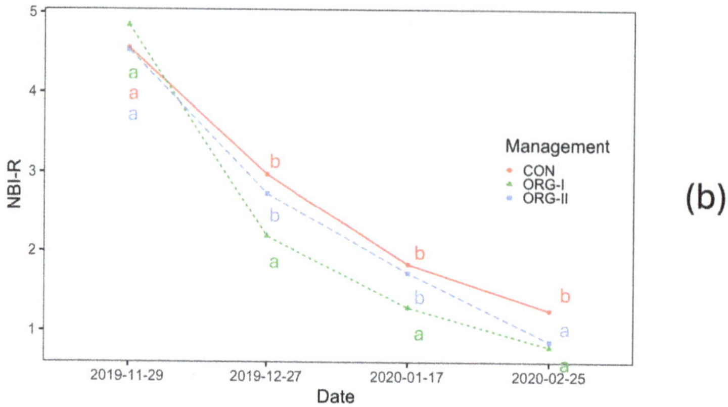


**Figure 9.** Trends of the NBI-G (**a**) and NBI-R (**b**) indices during the second survey year. The letters in the graphs indicate the differences between the various management systems identified in the Tukey test.

### 3.2. UAS Survey

The remote analysis performed by the UAS allowed the acquisition of vegetative vigour using the NDVI vegetation index. The plant segmentation performed in QGIS with the CHM technique enabled analysis exclusively on the plants, without the soil and the weeds within the field. The multi-temporal survey was conducted on the last three dates of the second year of monitoring, during the period of maximum vegetative–reproductive activity. The ANOVA analysis performed on the three management types did not observe a comparable trend to the fluorometric indices, as it did not discriminate particular differences between the plots. As observable in Table 8, no statistical difference was found between the treatments during the season, while instead, the ANOVA analysis evidenced a variability among the checking points (management×checking points), suggesting an internal heterogeneity on the three plots. Further analysis of singular surveys found that only on the last date (24 February 2020) was there significant variability between plots, with the ORG-II management differentiating itself from the others (+13.5%).

**Table 8.** ANOVA analysis on the NDVI index, where *p*-value < 0.05. The high *p*-value for the management variable suggests that the UAS did not evidence particular differences between the treatments, but rather indicated significant differences within the plots.

| ANOVA | NDVI | |
|---|---|---|
| | F-Value | *p*-Value |
| Management | 5.9524 | 0.002 ** |
| Date×Management | 1.9541 | 0.116 |
| Tukey Test | *p*-value | |
| ORG-II-ORG-I | 0.835 ns | |
| CON-ORG-I | 0.676 ns | |
| CON-ORG-II | 0.985 ns | |

The level of significance (*p*-value) is shown in the table, where ** indicates $p < 0.01$, and ns represents no significance.

The NDVI data development is summarised in Figure 10, comparing the three plots during the survey period. The data graph confirms what was observed in statistical analysis, where the CON and ORGs NDVI values follow a uniform development.

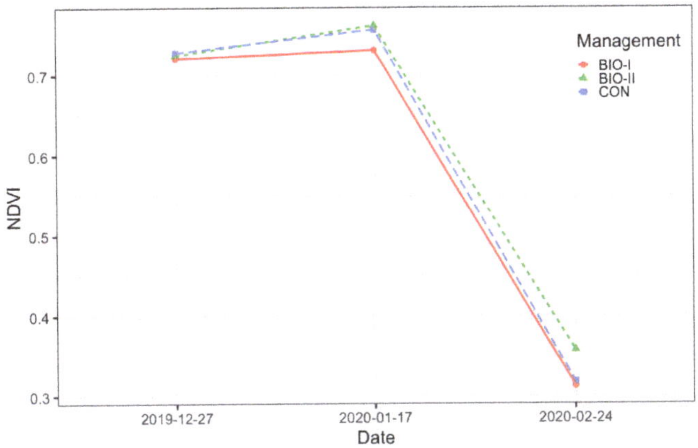


**Figure 10.** NDVI indices performance from the UAS survey. The field evolution across the three management plots did not identify any differences between treatments, pointing to a uniform tendency.

### 3.3. Correlation Matrix

Lastly, a correlation matrix was created between the NDVI and the MFA-generated index, as shown in Figure 11. The results show that the NDVI, SFR, and NBI indices perform unidirectionally, indicating a direct proportional correlation among them. As observed in the graphs in Figures 5 and 8, the BRR-FRF stress index and FLAV show an inverse correlation with the other indices. Specifically, high correlation values were observed between the NBI indices and the FLAV and BRR-FRF indices, suggesting that total leaf nitrogen content is closely related to stress and vegetative decay, with this result being associated with the more moderate relationship detected with the SFR data.

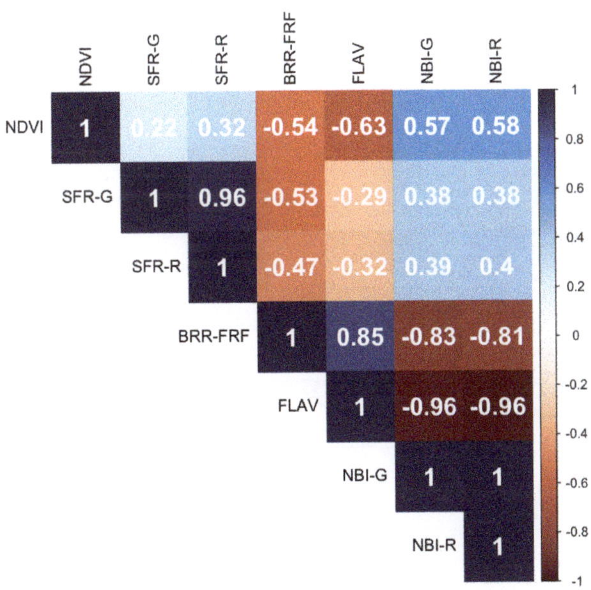


**Figure 11.** Matrix correlation of all variables considered in the experiment, converging MFA and UAS indices. The palette legend illustrates the proportional correlation (blue) to inverse correlation (red) between the indices. The white numbers in the matrix represent the correlation coefficients.

## 4. Discussion

The fluorimetric analysis conducted with the MFA in the three management areas provided a similar overview during the two seasons. Statistical analysis of the chlorophyll indices (SFR) showed a different evolution between the ORG and CON theses. The survey performed with the MFA revealed that the traditional CON management starts with lower SFR values. It reverses the trend toward December and maintains a higher level of photosynthetic activity until the senescence phase compared with the ORG management. This phenomenon can be explained by the effect of the agronomic strategy that was applied. The CON field relied on chemical fertilisation multiple times throughout the year, ensuring a continuous nutrient uptake. In contrast, the ORG-I and ORG-II plots only received fresh residues for nutritional supplements, as expected under organic conduction. In line with this assumption, leaf nitrogen analysis using the NBI indices led to similar conclusions, suggesting different developments between organic and conventional management. In fact, during the mid-periods of the two-year study, the nitrogen indices in the CON treatment were continuously higher than in the ORGs. This phenomenon suggests that the effect of fertilisation in the CON treatment extended the phenological phases, showing higher photosynthetic activity and total nitrogen content than in the ORGs. The NBI data did not identify a substantial and stable difference between the theses during the two monitoring years. This result may be explained by the plant's ability to manage nitrogen, as the uniform course of the three theses suggests that crop management did not affect leaf nitrogen uptake and management. It is important to note that the ORG-I and ORG-II management types can be compared with the CON during the first and second years of monitoring, respectively. This similarity is particularly evident during the winter period (December–January), i.e., at the crucial time of harvesting. It could suggest variable nitrogen management in the two ORG theses, depending on soil conditions and the availability of productive soil inputs, particularly due to the drier weather trend in the 2019–2020 season. This result was also observed for stress and flavonoid indices, which continued regularly in the three management types throughout the monitoring period, except peaking at higher values for the two organic management types than the CON one on the last monitoring day in 2020.

The BRR-FRF stress index trend significantly reflected the decline of the artichoke cycle, particularly in the ORG management, showing higher values in the cycle ending, while the chlorophyll and nitrogen indices showed decreasing values. The lower BRR-FRF and FLAV values in the CON thesis over the two seasons suggest that CON plants managed to extend their vegetative cycle, compared to the organic theses.

The MFA sensor and associated indices proved to be efficient tools for monitoring the physiological status of artichoke plants, revealing statistically significant differences among the three crop management systems. The non-destructive methods agree with existing literature, confirming that the assessment of physiological variables in fruits and vegetables is highly correlated with conventional destructive analysis methods [43]. This finding underscores the potential of non-destructive sensors in horticulture, enabling timely and targeted agronomic interventions through spatial and temporal management.

According to Kim et al. (2022) [44], integrating proximal and remote sensing techniques, such as using a fluorometer in combination with a UAS-mounted NIR sensor, can be a valuable approach for crop monitoring and management, provided proper calibration within the experimental design. However, in this experiment, the UAS-based aerial analysis did not reveal significant differences between the plots. This suggests a lower sensitivity of the UAS in detecting variations among the different management types, making its performance less comparable to that of the MFA. A first consideration is that the application of the UAS was conducted only in the second year of monitoring, reducing the possibility of detecting a significant difference between the three management types throughout the two-year experiment. As shown in Figure 9, NDVI values remained similar between treatments, following the same trend until the end of the analysis. Additionally, remote sensing measurements from UAS captured the entire plant, unlike the MFA survey method, which focused exclusively on leaves. By including all leaves and flower heads, UAS altered the NDVI values of representative leaves in monitored plants [45]. This highlights the broader function of the NDVI index, which serves as a general measure of vegetative vigour and should be considered differently depending on the research conducted. Another consideration for UAS monitoring is based on the multitemporal survey during the artichoke growth cycle, as the NDVI data were collected when the artichoke had already reached the vegetative and reproductive growth peak. Since important differences were observed in the early stages, one or more overflights in the early stages of the cycle would have helped to reinforce the UAS observations, possibly in support of additional vegetation indices to obtain a more complete view of the progress of the three treatments. Furthermore, the UAS was only used for a few months during the two-year experiment, so a more complete trend could not be followed as in the case of the MFA. Over the temporal limitation, another aspect has to be considered. The experiment, therefore, was conducted at a single site, where the pedological and climatic conditions play an important role in physiological development and the respective fluorimetric response.

Overall, the MFA system was able to distinguish the CON thesis from the ORGs during the survey period. However, proximal monitoring does not offer significant advantages in speed and convenience, as it requires the operator to spend several hours performing the monitoring. Additionally, to ensure reliable results, more than a thousand singular measurements with the MFA are provided to achieve a sufficient sample size. Furthermore, the MFA's interaction window should always be occupied by the sample to avoid interference from external elements, such as impurities, weeds, etc., which could lead to inaccurate readings of the physiological parameters. Due to these operational requirements, the MFA analysis can be employed as a high-performance sensor for conducting site-specific field investigations, following an initial crop assessment using an alternative wide-range system analysis.

## 5. Conclusions

The MFA showed a significant characterisation of plant physiology between the three crop management types analysed. The parameters of chlorophyll, stress, flavonoids, and, to a limited extent, nitrogen were adequate to observe the progression of the various theses throughout the entire crop season. The statistical analysis showed that the fluorimetric indices systematically correlated with the chlorophyll content over the two years between the ORG-I and ORG-II theses, suggesting a homogeneous trend, as indicated by the low significance value of the Tukey test. The different performance of the NDVI compared to the fluorimetric indices suggests that two high-performance instruments, such as the UAS and the MFA, do not always provide comparable results, as has been observed in other research contexts [46], as they perceive the vegetative characteristics of the crop differently. Consequently, the UAS cannot replace proximal sensing monitoring techniques for evaluating physiological differences in artichoke management. Future research activities will focus on characterising the fluorimetric indices provided by the MFA with the quantitative values of chlorophyll, nitrogen, and flavonoids to construct the respective agronomic calibration curves by correlating the indices obtained from the MFA with their equivalent quantitative values. This correlation, associated with crop control and the management of production inputs, could provide a valuable monitoring and decision-support tool for field operators to better control the main physiological characteristics of the artichoke, ultimately enhancing precision in cultivation management and input optimisation. The integration of advanced technologies, such as multi-sensor systems combining thermal imaging, hyperspectral data, and LiDAR, can capture a more comprehensive physiological and structural profile of the crop. These approaches could contribute to the development of more precise, repeatable, and immediate decision-support tools for the sustainable management of artichoke cultivation. Future research activities on artichoke management will also consider different fields and agroclimatic conditions to test and validate proximal and remote sensing applications through a longer multitemporal analysis, examining the responses from the two approaches during all the artichoke stages.

**Author Contributions:** Conceptualisation, F.G. and L.L.; methodology, A.D., M.T.T., F.G. and A.S.; software, A.D.; validation, A.D., A.S. and L.G.; formal analysis, A.D.; investigation, A.D. and A.S.; resources, F.G.; data curation, A.D. and A.S.; writing—original draft preparation, A.D.; writing—review and editing, A.S., F.G., P.A.D. and M.C.; visualization, A.D., A.S. and F.G.; supervision, F.G. and L.L.; project administration, F.G. and L.L.; funding acquisition, F.G. and L.L. All authors have read and agreed to the published version of the manuscript.

**Funding:** This research was funded by Advanced Technologies for LANds management and Tools for Innovative Development of an EcoSustainable agriculture, ATLANTIDE (CUP J88D20000070002). This project has received funding from the "Regione Autonoma della Sardegna" (RAS) by "Progetti di Ricerca e Sviluppo (R&S)", and this work is part of the activities of the project "ATLANTIDE", topic of Work Package 7. This research is also supported by Project RESTART FSC 2014-2020 (CUP D66C18000260002), and Project PEROLACRU PSR 16.1 (CUP H82E20000070006) funded by the Autonomous Region of Sardinia.

**Data Availability Statement:** The data that support the findings of this study are available from the corresponding author upon reasonable request.

**Conflicts of Interest:** The authors declare no conflicts of interest.

## Abbreviations

The following abbreviations are used in this manuscript:

| | |
|---|---|
| AGL | above ground level |
| ANOVA | analysis of variance |
| BBCH | Biologische Bundesanstalt, Bundessortenamt, and CHemical industry |
| BRR-FRF | blue-green to far-red fluorescence ratio |
| CHM | canopy height model |
| CON | conventional management |
| DOY | day of year |
| DSS | decision support system |
| FLAV | flavonol |
| MFA | multiplex force-A |
| NBI-G | nitrogen balance index on green excitation channel |
| NBI-R | nitrogen balance index on red excitation channel |
| NDVI | normalised difference vegetation index |
| ORG-I | organic annual management |
| ORG-II | organic biennial management |
| PA | precision agriculture |
| PPP | plant protection product |
| SFR-G | sample fluorescence emission ratio on green excitation channel |
| SFR-R | sample fluorescence emission ratio on red excitation channel |
| UAS | unmanned aerial system |

## References


1. Shaheen, A.; Li, Z.; Yang, Y.; Xie, J.; Zhu, L.; Li, C.; Nie, F.; Wang, M.; Wang, Y.; Rasheed, A.; et al. Genetic regulation of wheat plant architecture and future prospects for its improvement. *New Crops* **2025**, *2*, 100048. [CrossRef]
2. Pastrana-Pastrana, Á.J.; Rodríguez-Herrera, R.; Solanilla-Duque, J.F.; Flores-Gallegos, A.C. Plant proteins, insects, edible mushrooms and algae: More sustainable alternatives to conventional animal protein. *J. Future Foods* **2025**, *5*, 248–256. [CrossRef]
3. Ziska, L.H.; Bunce, J.A.; Shimono, H.; Gealy, D.R.; Baker, J.T.; Newton, P.C.D.; Reynolds, M.P.; Jagadish, K.S.V.; Zhu, C.; Howden, M.; et al. Food security and climate change: On the potential to adapt global crop production by active selection to rising atmospheric carbon dioxide. *Proc. R. Soc. Biol. Sci.* **2012**, *279*, 4097–4105. [CrossRef]
4. FAO. Land Cover FAOSTAT, 2017. Available online: https://www.fao.org/faostat/en/#data/QCL. (accessed on 22 October 2024).
5. Lesur-Dumoulin, C.; Malézieux, E.; Ben-Ari, T.; Langlais, C.; Makowski, D. Lower average yields but similar yield variability in organic versus conventional horticulture. A meta-analysis. *Agron. Sustain. Dev.* **2017**, *37*, 45. [CrossRef]
6. Riahi, A.; Hdider, C.; Sanaa, M.; Tarchoun, N.; Ben Kheder, M.; Guezal, I. Effect of conventional and organic production systems on the yield and quality of field tomato cultivars grown in Tunisia. *J. Sci. Food Agric.* **2009**, *89*, 2275–2282. [CrossRef]
7. Grabowska, A.; Caruso, G.; Mehrafarin, A.; Kalisz, A.; Gruszecki, R.; Kunicki, E.; Sękara, A. Application of modern agronomic and biotechnological strategies to valorise worldwide globe artichoke (*Cynara cardunculus* L.) potential—An analytical overview. *Ital. J. Agron.* **2018**, *13*, 1252. [CrossRef]
8. Stea, D.; Sellami, M.H.; Ruta, C.; Pulvento, C.; De Mastro, G. Cropping practices of globe artichoke: A bibliometric study. *Front. Sustain. Food Syst.* **2023**, *7*, 1274714. [CrossRef]
9. Fracchiolla, M.; Lasorella, C.; Cazzato, E.; Renna, M. Living mulch with subterranean clover (*Trifolium subterraneum* L.) is effective for a sustainable weed management in globe artichoke as annual cropping in Puglia (Southern Italy). *Horticulturae* **2022**, *8*, 825. [CrossRef]
10. Pesce, G.R.; Mauromicale, G. *Cynara cardunculus* L.: Historical and economic importance, botanical descriptions, genetic resources and traditional uses. In *The Globe Artichoke Genome*; Portis, E., Acquadro, A., Lanteri, S., Eds.; Series Title: Compendium of Plant Genomes; Springer International Publishing: Cham, Switzerland, 2019; pp. 1–19. [CrossRef]
11. Spezzigu, M. Effect of crop management practices on globe artichoke physiological, nutritional, and nutraceutical quality. Ph.D. Thesis, University of Sassari, Sassari, Italy, 2022.
12. Lombardo, S.; Pandino, G.; Mauromicale, G. Minerals profile of two globe artichoke cultivars as affected by NPK fertilizer regimes. *Food Res. Int.* **2017**, *100*, 95–99. [CrossRef]
13. De Menna, F.; Malagnino, R.A.; Vittuari, M.; Segrè, A.; Molari, G.; Deligios, P.A.; Solinas, S.; Ledda, L. Optimization of agricultural biogas supply chains using artichoke byproducts in existing plants. *Agric. Syst.* **2018**, *165*, 137–146. [CrossRef]
14. Ledda, L.; Deligios, P.A.; Farci, R.; Sulas, L. Biomass supply for energetic purposes from some Cardueae species grown in Mediterranean farming systems. *Ind. Crops Prod.* **2013**, *47*, 218–226. [CrossRef]
15. Rana, R.L.; Bux, C.; Lombardi, M. Carbon footprint of the globe artichoke supply chain in Southern Italy: From agricultural production to industrial processing. *J. Clean. Prod.* **2023**, *391*, 136240. [CrossRef]
16. Deligios, P.A.; Tiloca, M.T.; Sulas, L.; Buffa, M.; Caraffini, S.; Doro, L.; Sanna, G.; Spanu, E.; Spissu, E.; Urracci, G.R.; et al. Stable nutrient flows in sustainable and alternative cropping systems of globe artichoke. *Agron. Sustain. Dev.* **2017**, *37*, 54. [CrossRef]
17. Provost, C.; Pedneault, K. The organic vineyard as a balanced ecosystem: Improved organic grape management and impacts on wine quality. *Sci. Hortic.* **2016**, *208*, 43–56. [CrossRef]
18. Dorais, M.; Alsanius, B.W. Recent advances in organic horticulture technology and management. *Sci. Hortic.* **2016**, *208*, 1–2. [CrossRef]
19. Montemurro, F.; Persiani, A.; Diacono, M. Environmental Sustainability Assessment of Horticultural Systems: A Multi-Criteria Evaluation Approach Applied in a Case Study in Mediterranean Conditions. *Agronomy* **2018**, *8*, 98. [CrossRef]
20. Kristensen, H.L.; Thorup-Kristensen, K. Effects of vertical distribution of soil inorganic nitrogen on root growth and subsequent nitrogen uptake by field vegetable crops. *Soil Use Manag.* **2007**, *23*, 338–347. [CrossRef]
21. Ara, I.; Turner, L.; Harrison, M.T.; Monjardino, M.; deVoil, P.; Rodriguez, D. Application, adoption and opportunities for improving decision support systems in irrigated agriculture: A review. *Agric. Water Manag.* **2021**, *257*, 107161. [CrossRef]
22. Bulacio Fischer, P.T.; Carella, A.; Massenti, R.; Fadhilah, R.; Lo Bianco, R. Advances in Monitoring Crop and Soil Nutrient Status: Proximal and Remote Sensing Techniques. *Horticulturae* **2025**, *11*, 182. [CrossRef]
23. Pallottino, F.; Antonucci, F.; Costa, C.; Bisaglia, C.; Figorilli, S.; Menesatti, P. Optoelectronic proximal sensing vehicle-mounted technologies in precision agriculture: A review. *Comput. Electron. Agric.* **2019**, *162*, 859–873. [CrossRef]
24. Leufen, G.; Noga, G.; Hunsche, M. Fluorescence Indices for the Proximal Sensing of Powdery Mildew, Nitrogen Supply and Water Deficit in Sugar Beet Leaves. *Agriculture* **2014**, *4*, 58–78. [CrossRef]
25. Hoffmann, A.M.; Noga, G.; Hunsche, M. Fluorescence indices for monitoring the ripening of tomatoes in pre- and postharvest phases. *Sci. Hortic.* **2015**, *191*, 74–81. [CrossRef]

26. Yadav, S.K.; Jalink, H.; Groot, S.P.; Van Der Schoor, R.; Yadav, S.; Dadlani, M.; Kodde, J. Quality improvement of aged cabbage (*Brassica oleracea* var. capitata) seeds using chlorophyll fluorescence sensor. *Sci. Hortic.* **2015**, *189*, 81–85. [CrossRef]
27. Ali, S.; Honermeier, B. Post emergence herbicides influence the leaf yield, chlorophyll fluorescence and phenolic compounds of artichoke (*Cynara cardunculus* L.). *Sci. Hortic.* **2016**, *203*, 216–223. [CrossRef]
28. Borgognone, D.; Rouphael, Y.; Cardarelli, M.; Lucini, L.; Colla, G. Changes in Biomass, Mineral Composition, and Quality of Cardoon in Response to NO3−:Cl− Ratio and Nitrate Deprivation from the Nutrient Solution. *Front. Plant Sci.* **2016**, *7* . [CrossRef]
29. Gamon, J.A.; Kovalchuck, O.; Wong, C.Y.S.; Harris, A.; Garrity, S.R. Monitoring seasonal and diurnal changes in photosynthetic pigments with automated PRI and NDVI sensors. *Biogeosciences* **2015**, *12*, 4149–4159. [CrossRef]
30. Wang, J.; Zhang, S.; Lizaga, I.; Zhang, Y.; Ge, X.; Zhang, Z.; Zhang, W.; Huang, Q.; Hu, Z. UAS-based remote sensing for agricultural monitoring: Current status and perspectives. *Comput. Electron. Agric.* **2024**, *227*, 109501. [CrossRef]
31. Squeri, C.; Poni, S.; Di Gennaro, S.F.; Matese, A.; Gatti, M. Comparison and ground truthing of different remote and proximal sensing platforms to characterize variability in a hedgerow-trained vineyard. *Remote Sens.* **2021**, *13*, 2056. [CrossRef]
32. Spanu, E.; Deligios, P.A.; Azara, E.; Delogu, G.; Ledda, L. Effects of alternative cropping systems on globe artichoke qualitative traits. *J. Sci. Food Agric.* **2018**, *98*, 1079–1087. [CrossRef]
33. Deligios, P.A.; Farina, R.; Tiloca, M.T.; Francaviglia, R.; Ledda, L. C-sequestration and resilience to climate change of globe artichoke cropping systems depend on crop residues management. *Agron. Sustain. Dev.* **2021**, *41*, 20. [CrossRef]
34. Piras, F. Codifica fenologica e dinamica dell'assorbimento minerale in *Cynara cardunculus* var. scloymus. Ph.D. Thesis, University of Sassari, Sassari, Italy, 2012.
35. Bürling, K.; Cerovic, Z.G.; Cornic, G.; Ducruet, J.M.; Noga, G.; Hunsche, M. Fluorescence-based sensing of drought-induced stress in the vegetative phase of four contrasting wheat genotypes. *Environ. Exp. Bot.* **2013**, *89*, 51–59. [CrossRef]
36. Ghozlen, N.B.; Cerovic, Z.G.; Germain, C.; Toutain, S.; Latouche, G. Non-destructive optical monitoring of grape maturation by proximal sensing. *Sensors* **2010**, *10*, 10040–10068. [CrossRef]
37. De Souza, R.; Peña-Fleitas, M.T.; Thompson, R.B.; Gallardo, M.; Grasso, R.; Padilla, F.M. Use of fluorescence indices as predictors of crop N status and yield for greenhouse sweet pepper crops. *Precis. Agric.* **2022**, *23*, 278–299. [CrossRef]
38. Siqueira, R.; Mandal, D.; Longchamps, L.; Khosla, R. Assessing Nitrogen Variability at Early Stages of Maize Using Mobile Fluorescence Sensing. *Remote Sens.* **2022**, *14*, 5077. [CrossRef]
39. Sytar, O.; Bruckova, K.; Hunkova, E.; Zivcak, M.; Konate, K.; Brestic, M. The application of multiplex fluorimetric sensor for the analysis of flavonoids content in the medicinal herbs family Asteraceae, Lamiaceae, Rosaceae. *Biol. Res.* **2015**, *48*, 5. [CrossRef] [PubMed]
40. Hazir, M.H.M.; Shariff, A.R.M.; Amiruddin, M.D.; Ramli, A.R.; Iqbal Saripan, M. Oil palm bunch ripeness classification using fluorescence technique. *J. Food Eng.* **2012**, *113*, 534–540. [CrossRef]
41. Mohan, M.; Silva, C.; Klauberg, C.; Jat, P.; Catts, G.; Cardil, A.; Hudak, A.; Dia, M. Individual tree detection from unmanned aerial vehicle (UAV) derived canopy height model in an open canopy mixed conifer forest. *Forests* **2017**, *8*, 340. [CrossRef]
42. Lin, C.; Zhang, J. Detecting nonadditivity in single-facet generalizability theory applications: Tukey's test. *J. Educ. Meas.* **2018**, *55*, 78–89. [CrossRef]
43. Hamann, F.A.; Czaja, S.; Hunsche, M.; Noga, G.; Fiebig, A. Monitoring physiological and biochemical responses of two apple cultivars to water supply regimes with non-destructive fluorescence sensors. *Sci. Hortic.* **2018**, *242*, 51–61. [CrossRef]
44. Kim, K.H.; Shawon, M.R.A.; An, J.H.; Lee, H.J.; Kwon, D.J.; Hwang, I.C.; Bae, J.H.; Choi, K.Y. Effect of shade screen on sap flow, chlorophyll fluorescence, NDVI, plant growth and fruit characteristics of cultivated paprika in greenhouse. *Agriculture* **2022**, *12*, 1405. [CrossRef]
45. Akanbi, O.D.; Bhuvanagiri, D.C.; Barcelos, E.I.; Nihar, A.; Gonzalez Hernandez, B.; Yarus, J.M.; French, R.H. Integrating multiscale geospatial analysis for monitoring crop growth, nutrient distribution, and hydrological dynamics in large-scale agricultural systems. *J. Geovisualization Spat. Anal.* **2024**, *8*, 9. [CrossRef]
46. Sassu, A.; Deidda, A.; Mercenaro, L.; Virgillito, B.; Gambella, F. Multisensor Analysis for Biostimulants Effect Detection in Sustainable Viticulture. *Agriculture* **2024**, *14*, 2221. [CrossRef]

*horticulturae*

MDPI

*Article*

# Optimizing Caraway Growth, Yield and Phytochemical Quality Under Drip Irrigation: Synergistic Effects of Organic Manure and Foliar Application with Vitamins $B_1$ and E and Active Yeast

**Ahmed A. Hassan [1], Amir F.A. Abdel-Rahim [2], Ghadah H. Al Hawas [3,4], Wadha Kh. Alshammari [5], Reda M.Y. Zewail [6,*], Ali A. Badawy [7,*] and Heba S. El-Desouky [6]**

1 Horticulture Department, Faculty of Agriculture, Minia University, Minia 61519, Egypt
2 Agricultural Research and Experiments Center, Minia University, Minia 61519, Egypt
3 Department of Biology, College of Science, Imam Abdulrahman Bin Faisal University, Dammam 31441, Saudi Arabia
4 Basic and Applied Scientific Research Center, Imam Abdulrahman Bin Faisal University, Dammam 31441, Saudi Arabia
5 Biology Department, College of Science, University of Hafr Al Batin, Hafr Al-Batin 31991, Saudi Arabia
6 Botany Department, Faculty of Agriculture (Moshtohor), Benha University, Toukh 13736, Egypt
7 Botany and Microbiology Department, Faculty of Science, Al-Azhar University, Cairo 11884, Egypt
* Correspondence: reda.zewail@fagr.bu.edu.eg (R.M.Y.Z.); ali.abdalhalim@azhar.edu.eg (A.A.B.)

**Abstract**

Despite its value as a culinary, medicinal, and essential oil crop, caraway struggles to grow and develop its biochemical quality in drought-prone sandy soils. To tackle this challenge, we conducted two field trials under drip irrigation, testing four rates of organic manure (0, 5, 10, and 15 ton/hectare (ha) and three foliar biostimulants: vitamin $B_1$ (50 and 100 mg $L^{-1}$), vitamin E (50 and 100 mg $L^{-1}$), and active yeast (100 and 150 mL $L^{-1}$). We used a randomized split-plot design with three replicates, assigning manure rates to main plots and biostimulants to subplots. We measured plant height, stem diameter, branch number, dry biomass, umbels per plant, 1000-seed weight, seed yield (per plant and per ha), essential oil content, chlorophyll a and b, carotenoids, and leaf N, P, and K. All treatments outperformed the unfertilized control. Applying 15 ton/ha of manure alone increased mean plant height by 185.3 cm, stem diameter by 2.93 mm, branch number by 14.5, and herbal weight by 91.97 g across both seasons—a gain of about 11–15%. Foliar application of vitamin $B_1$ at 100 mg $L^{-1}$ (without manure) achieved even larger gains: mean plant height improved by 176.5 cm, stem diameter by 2.6 mm, branches number by 15.1, and herbal biomass by 103.95 g (20–36% growth increases). It also boosted essential oil yield by 1.89 mL per plant (16–50%) and enhanced nutrient uptake. The most pronounced synergy emerged when combining 15 ton/ha of manure with 100 mg $L^{-1}$ vitamin $B_1$, raising seed yield to 1698.8 kg/ha (35%), plant height to 184.7 cm (52%), number of branches to 17.4 per plant (56%), umbels to 38.1 per plant (42%), 1000-seed weight to 16.9 g (48%), and essential oil yield to 2.3 mL per plant (115%), compared to the control. Chlorophyll a increased by 50%, chlorophyll b by 33%, carotenoids by 35%, and leaf N, P, and K by 43%, 90%, and 76%, respectively. Manure combined with vitamin E or yeast delivered moderate improvements. These findings demonstrate that integrating organic manure with targeted foliar biostimulants—especially vitamin $B_1$—under drip irrigation, is a sustainable strategy to maximize caraway yield, oil content, and nutritional quality on marginal sandy soils.

**Keywords:** *Carum carvi* L.; organic fertilization; vitamins; yeast; caraway oil yield; oil quality; sandy soil

## 1. Introduction

The caraway plant (*Carum carvi* L.) is an aromatic plant belonging to the Apiaceae family [1]. Caraway seeds and their essential oil are used medicinally as carminatives, mild stomachics, antispasmodics, and as a tonic in the treatment of digestive disorders. Dried seeds are widely employed for flavoring bread, cake, confectionery, cheese, and various kinds of food products. The odor and flavor of medicinal seed plants are due to their essential oil [2]. Also, the leaves of caraway are edible and are used in salads and as a garnish or to add flavor [3].

Agriculture in reclaimed lands is critical to global food security and can be achieved through engineering interventions and soil amendments. However, agriculture in these areas faces significant distinct challenges with poor soil health due to poor soil structure, low organic matter content (<0.5%), and a lack of beneficial microorganisms, leading to reduced soil fertility and water retention [4,5], as well as nutrient imbalances, particularly nitrogen (N) and phosphorus (P), and nutrient deficiencies [6,7]. To address these problems, key agricultural strategies include soil reclamation with compost (10–20 t/ha) to rebuild organic carbon [8] and the application of water management techniques such as drip irrigation, which has been shown to reduce salinity accumulation by 30% compared to flood irrigation methods [9].

To enhance caraway growth and maximize essential oil production sustainably, researchers have examined the role of organic fertilizers and biostimulants. Organic manure, in particular, enriches soil humus over time and upgrades both its physical and chemical characteristics. Moreover, incorporating organic matter improves soil structure, increases nutrient availability, and stimulates microbial activity—making it central to sustainable agricultural practices [10–12]. Regular fertilization not only maintains soil fertility and meets plant nutritional needs but also reduces environmental stress. Furthermore, organic fertilizers help reduce chemical residues, a growing concern in export markets. Studies have shown that fertilizers are essential for restoring reclaimed sandy soils and improving seed and essential oil production from aromatic plants [13,14].

Thiamine (vitamin $B_1$) plays a role in improving plant metabolism. It is synthesized in plant leaves and can be transported to the plant, where it forms the growth-enhancing enzyme thiamine pyrophosphate (TPP). As TPP, thiamine catalyzes a crucial step in carbohydrate metabolism and supports photosynthesis and respiration [15]. It also contains aromatic content, adding to the value of medicinal and ornamental plants [16]. It also enhances plants' ability to tolerate stresses, such as drought and salinity, by regulating their defense against oxidation, eliminating free radicals, and increasing sweating capacity. This enzyme helps plants grow under adverse conditions [17].

In addition, alpha-tocopherol (Vit. E), the major form of vitamin E in plants, is a low-molecular-weight, lipophilic antioxidant produced exclusively by photosynthetic organisms. It is found in the thylakoid membranes of chloroplasts, where it quenches singlet oxygen and removes lipid peroxide radicals, preventing chain reactions that could damage polyunsaturated fatty acids [18]. In this way, it maintains the fluidity and permeability of membranes, protects photosynthetic proteins from photodamage, and maintains an optimal environment for photosynthesis. In response to drought, salinity, cold, or extreme light stress, plants increase their alpha-tocopherol content, a response that has been associated with stronger growth, increased yield, and enhanced antioxidant defenses [19].

Yeast is a fast-growing unicellular fungus and an excellent environmentally friendly promoter of plant growth and alternative to chemical fertilizers. It contains readily available macronutrients and micronutrients, essential nutrients (including vitamins $B_1$, $B_2$, $B_3$, and $B_{12}$, as well as amino acids, proteins, and carbohydrates), and plant hormones (auxins and cytokinins), which significantly enhance plant production [20]. These elements promote cell division and enlargement, chlorophyll formation, protein and nucleic acid synthesis, carbohydrate accumulation, and growth, ultimately leading to increased yield and seeding [21]. Furthermore, yeast also resists biotic and abiotic stresses, such as drought, by inhibiting pathogens, solubilizing phosphates, producing protective compounds, and reducing stress-induced damage [22]. The beneficial properties of yeast make it a valuable and safe tool for agriculture.

This study aimed to evaluate the synergistic effects of organic manure, foliar treatment with vitamins ($B_1$ and E), and yeast on caraway plants grown under drip irrigation in a poor sandy soil (arid area). The objective was to improve vegetative growth, seed yield, essential oil production, and phytochemical quality (photopigments, nitrogen, phosphorus, and potassium content) of caraway plants while ensuring the sustainability of the cultivation system.

## 2. Materials and Methods

### 2.1. Experimental Design

This study was conducted over two consecutive growing seasons (2023/2024 and 2024/2025) at the Agricultural Researches and Experiments Center, Minia University, Shosha, West Samalout City, Egypt (28.3140° N, 30.7101° E, as observed in Figure 1). A split-plot design within a randomized complete block framework with three replicates was employed. Four rates of organic manure (0, 5, 10, and 15 ton/ha) served as the main-plot factor (A), while foliar treatments of vitamin E (50 and 100 mg $L^{-1}$), vitamin $B_1$ (50 and 100 mg $L^{-1}$), and active yeast (100 and 150 mL $L^{-1}$) comprised the subplot factor (B), resulting in 28 treatment combinations. Soil analyses were performed according to Jackson [23], and the results are presented in Table 1. Based on the physical properties in Table 1, the soil is classified as sandy with low macro- and micronutrient levels, indicating poor fertility. The pH is alkaline, electrical conductivity measures 1.30 dS/m, and the $CaCO_3$ content is 13.96%. Overall, this is a sandy, low-fertility soil [24].

The experimental field was laid out in terraces 70 cm wide and spaced 30 cm apart, each fitted with two drip irrigation lines 40 cm apart. Caraway seeds were sown on 15 October in both seasons at 3–4 seeds per hill, with hills spaced 25 cm along the drip lines (drippers spaced 25 cm apart, 4 l/h discharge). Each 4 m × 1 m unit (4 $m^2$ = 1/2625 ha) contained two irrigation lines and 32 plants (84,000 plants/ha). At 40 days after sowing (24 November), seedlings were thinned to one plant per hill. Metrological data for the experimental area is shown in Table 2. All agronomic practices—planting density, sowing timing, irrigation, and pest, disease, and weed management—followed the Egyptian Ministry of Agriculture's recommendations for caraway cultivation.

### 2.2. Applied Treatments

Organic manure was obtained from a private farm. The organic manure, which is a mixture of cow, sheep, and goat manure, was applied at 0, 5, 10, and 15 ton/ha during the preparation of the soil for cultivation in both seasons. The chemical analysis of organic manure was performed according to Black [25] and is shown in Table 3.

Thiamine (Vit. $B_1$) and α-tocopherol (Vit. E) were obtained from the Algomhoria company for chemical trading in Egypt and applied by hand sprayer three times. The first dose was added 45 days after sowing (30 November), while the second and third ones were applied at three-weeks intervals from the first dose (21 December and 12 January) during both seasons.

**Table 1.** Physical and chemical analysis of the used soil during the two seasons of 2023–2024 and 2024–2025.

| Soil Characteristics | 2023/2024 | 2024/2025 |
|---|---|---|
| | **Physical properties** | |
| Sand (%) | 91.22 | 90.50 |
| Silt (%) | 5.36 | 5.78 |
| Clay (%) | 3.42 | 3.72 |
| Soil type | sandy | Sandy |
| | **Available nutrients** | |
| $Ca^{++}$ (mg $L^{-1}$) | 133.0 | 138.0 |
| $Mg^{++}$ (mg $L^{-1}$) | 60.1 | 62.6 |
| $Na^{+}$ (mg $L^{-1}$) | 79.5 | 86.0 |
| $K^{+}$ (mg $L^{-1}$) | 15.7 | 21.3 |
| $Cl^{-}$ (mg $L^{-1}$) | 10.5 | 11.3 |
| $CO_3^{-2}$ (mg $L^{-1}$) | 12.4 | 13.3 |
| | **Chemical properties** | |
| pH | 8.16 | 8.20 |
| E.C. (dS/m) | 1.38 | 1.23 |
| Organic matter (%) | 0.17 | 0.14 |
| $CaCO_3$ (%) | 13.94 | 13.99 |
| | **DTPA-extractable nutrients** | |
| Fe (mg $L^{-1}$) | 0.85 | 0.94 |
| Cu (mg $L^{-1}$) | 0.40 | 0.44 |
| Zn (mg $L^{-1}$) | 0.31 | 0.34 |
| Mn (mg $L^{-1}$) | 0.54 | 0.60 |

Commercial soft yeast cells (*Saccharomyces cerevisiae*) were cultivated in a nutrient-rich aerobic medium of glucose and casein, which not only supported vigorous growth but also promoted the formation of valuable bio-constituents: proteins, sugars, carbohydrates, fatty acids, hormones, and amino acids. After harvesting, we disrupted the cells using two freeze–thaw cycles, combined with air sparing and precise temperature shifts during incubation, to efficiently release their soluble contents. The glucose–casein medium ensured an optimal balance of carbon, nitrogen, and other essential nutrients throughout this process. The resulting active yeast suspension was then applied as a foliar spray at 100 and 150 mL $L^{-1}$. Each rate was sprayed three times, following the vitamin-application schedule, until the foliage was uniformly coated and runoff began.

### 2.3. Vegetative and Yield Measurements

Vegetative growth was assessed mid-season on 20 February of each year. For each treatment, nine plants per replicate (three replicates) were randomly selected—27 plants per season—to measure plant height (cm), stem diameter (cm), number of branches per plant, and dry herb weight per plant (g).

Figure 1. Location and experimental site of Minya Governorate in Egypt.

Table 2. Metrological data for Minya, Egypt during 2023/2024 and 2024/2025 seasons.

| | 2023/2024 | | | | | | | | | | | | 2024/2025 | | | | | | | | | |
|---|---|---|---|---|---|---|---|---|---|---|---|---|---|---|---|---|---|---|---|---|---|---|
| Months | Oct | Nov | Dec | Jan | Feb | Mar | Apr | May | Jun | Jul | Aug | Sep | Oct | Nov | Dec | Jan | Feb | Mar | Apr | May | Jun | Jul |
| Max. temp. (°C) | 30.24 | 26.0 | 21.4 | 20.3 | 22.1 | 25.6 | 30.6 | 34.7 | 36.8 | 39.0 | 36.9 | 35.4 | 29.12 | 25.0 | 20.1 | 20.4 | 22.1 | 25.6 | 30.6 | 33.5 | 36.8 | 39.0 |
| Mean temp. (°C) | 24.1 | 18.4 | 13.7 | 12.5 | 14.1 | 17.5 | 22.0 | 26.6 | 29.2 | 29.7 | 29.6 | 28.0 | 24.0 | 17.0 | 13.0 | 12.8 | 14.3 | 17.8 | 22.1 | 26.4 | 29.2 | 29.7 |
| Min. temp. (°C) | 17.4 | 11.8 | 7.1 | 5.5 | 6.6 | 9.6 | 13.5 | 18.1 | 21.0 | 22.2 | 22.4 | 20.9 | 17.1 | 12.0 | 7.0 | 5.4 | 6.5 | 10.0 | 13.8 | 18.3 | 21.0 | 22.2 |
| Low temp. (°C) | 10.0 | 3.0 | −0.8 | −3.4 | −0.3 | 1.0 | 4.5 | 9.0 | 14.3 | 16.0 | 16.2 | 13.5 | 10.0 | 3.2 | −0.9 | −3.4 | −0.3 | 1.0 | 3.5 | 7.0 | 14.3 | 16.0 |
| Aver. Preci. (mm) | 0 | 0.21 | 0.18 | 0.68 | 0.75 | 0.83 | 0.03 | 0.14 | 0 | 0 | 0 | 0 | 0 | 0.23 | 0.24 | 0.68 | 0.75 | 0.83 | 0.03 | 0.14 | 0 | 0 |
| Aver. precipitation days (≥1.0 mm) | 0 | 0.08 | 0.08 | 0.23 | 0.38 | 0.27 | 0 | 0.04 | 0 | 0 | 0 | 0 | 0 | 0.08 | 0.1 | 0.23 | 0.38 | 0.27 | 0 | 0.04 | 0 | 0 |
| Average relative humidity (%) | 55 | 61 | 67 | 43 | 37 | 40 | 46 | 51 | 53 | 55 | 42 | 42 | 55 | 54 | 62 | 66 | 42 | 39 | 43 | 53 | 53 | 55 |

Source: NOAA (humidity, dew point, records).

**Table 3.** Chemical analysis of the organic manure used during the 2023/2024 and 2024/2025 growing seasons.

| Manure Characteristics | Values | |
|---|---|---|
| | 2023/2024 | 2024/2025 |
| pH (1:1) | 7.43 | 7.39 |
| E.C. (dS/m) | 1.03 | 1.09 |
| O.M. (%) | 27.43 | 27.75 |
| O.C. (%) | 15.95 | 16.14 |
| C/N ratio | 19.69 | 18.55 |
| Humidity (%) | 8.04 | 7.93 |
| Total N (%) | 0.81 | 0.87 |
| Total P (%) | 0.24 | 0.29 |
| Total K (%) | 1.08 | 1.06 |
| Fe (mg $kg^{-1}$) | 385 | 379 |
| Mn (mg $kg^{-1}$) | 234 | 232 |
| Zn (mg $kg^{-1}$) | 271 | 269 |

Final yield and its components were recorded at 180 days after sowing (last week of April). Ten plants per replicate (30 plants per treatment) were randomly harvested to determine the number of umbels per plant, 1000-seed weight (g), seed yield per plant (g), and seed yield per ha (kg).

### 2.4. Chemical Analysis

Essential oil content was measured by gas chromatography–mass spectrometry (GC–MS) at the Laboratory of Medicinal and Aromatic Plants Research, National Research Centre. A TRACE GC Ultra gas chromatograph (Thermo Scientific, Waltham, MA, USA) was coupled to an ISQ single-quadrupole mass spectrometer. Oil yield was calculated per plant (ml) and per ha (L) according to Pharmacopoeia methods [26]. Three weeks after the final treatment (2 February), photosynthetic pigments were quantified following Moran [27]. Nutrient content (N, P, K) was determined in the dried herb using the standard analytical procedures of Sparks et al. [28] and Ibrahim et al. [29].

### 2.5. Statistical Analysis

The results were tabulated and statistically analyzed using MSTAT-C following Freed et al. [30]. Treatment means were compared by the least significant difference (LSD) test at the 5% level and Duncan letters. Data are presented as a mean of three replicates ± standard deviation.

## 3. Results

### 3.1. Vegetative Growth

The data presented in Tables 4 and 5 show the individual and interactive effects of organic manure (5, 10, and 15 ton/ha), vitamin E (50 and 100 mg $L^{-1}$), vitamin $B_1$ (50 and 100 mg $L^{-1}$), and active yeast (100 and 150 mL $L^{-1}$) treatments on some vegetative growth traits (plant height, stem diameter, number of branches/plant, and herb dry weight/plant) of caraway plants over two growing seasons (2023/24 and 2024/25).

**Table 4.** Effects of organic manure, vitamin E, vitamin B$_1$, and active yeast treatments on plant height (cm) and stem diameter (cm) of caraway plants during the two growing seasons (2023/24 and 2024/25).

| Vitamins and Active Yeast Treatments (B) | Organic Manure (ton/ha) (A) | | | | | | | | | |
|---|---|---|---|---|---|---|---|---|---|---|
| | 0.0 | 5.0 | 10 | 15 | Mean (B) | 0.0 | 5.0 | 10 | 15 | Mean (B) |
| | First Season (2023/2024) | | | | | Second Season (2024/2025) | | | | |
| | Plant Height (cm) | | | | | | | | | |
| Control (Without) | 146.54 ± 4.90 | 155.90 ± 7.21 | 162.45 ± 8.14 | 174.41 ± 7.18 | 159.82 | 149.59 ± 6.88 | 159.06 ± 7.33 | 165.94 ± 8.28 | 178.01 ± 7.26 | 163.15 |
| Vit. E at 50 mg L$^{-1}$ | 157.98 ± 4.93 | 168.06 ± 5.24 | 172.12 ± 6.79 | 182.52 ± 6.55 | 170.17 | 161.36 ± 4.98 | 171.66 ± 5.31 | 175.72 ± 6.93 | 186.34 ± 6.59 | 173.77 |
| Vit. E at 100 mg L$^{-1}$ | 162.34 ± 5.38 | 172.74 ± 5.74 | 179.71 ± 7.22 | 183.77 ± 7.20 | 174.64 | 165.73 ± 5.47 | 176.35 ± 5.84 | 183.42 ± 7.37 | 187.59 ± 7.24 | 178.27 |
| Vit. B$_1$ at 50 mg L$^{-1}$ | 158.39 ± 6.06 | 168.48 ± 6.49 | 172.74 ± 8.19 | 180.44 ± 8.25 | 170.01 | 161.98 ± 6.17 | 171.97 ± 6.57 | 176.45 ± 8.36 | 184.26 ± 8.30 | 173.66 |
| Vit. B$_1$ at 100 mg L$^{-1}$ | 167.23 ± 7.34 | 177.94 ± 7.84 | 188.07 ± 9.51 | 198.08 ± 9.10 | 182.83 | 170.83 ± 7.50 | 181.65 ± 7.97 | 192.00 ± 9.73 | 202.02 ± 9.14 | 186.63 |
| Active yeast at 100 mL L$^{-1}$ | 152.57 ± 5.73 | 162.34 ± 6.48 | 169.21 ± 9.15 | 181.69 ± 9.01 | 166.45 | 155.84 ± 6.03 | 165.73 ± 6.03 | 172.70 ± 5.30 | 185.51 ± 9.01 | 169.94 |
| Active yeast at 150 mL L$^{-1}$ | 160.68 ± 5.55 | 171.50 ± 6.55 | 176.49 ± 5.26 | 183.04 ± 9.15 | 172.93 | 163.85 ± 5.55 | 174.26 ± 6.55 | 180.20 ± 5.26 | 186.96 ± 9.15 | 176.32 |
| Mean (A) | 157.96 | 168.14 | 174.40 | 183.42 | | 161.31 | 171.53 | 178.06 | 187.24 | |
| L.S.D. at 5% | A: 5.95 | | B: 4.11 | | AB: 8.22 | A: 6.01 | | B: 3.55 | | AB: 7.10 |
| | Stem Diameter (cm) | | | | | | | | | |
| Control (Without) | 2.170 ± 0.08 | 2.28 ± 0.09 | 2.38 ± 0.08 | 2.46 ± 0.07 | 2.32 | 2.21 ± 0.05 | 2.32 ± 0.09 | 2.44 ± 0.08 | 2.51 ± 0.08 | 2.37 |
| Vit. E at 50 mg L$^{-1}$ | 2.25 ± 0.07 | 2.39 ± 0.08 | 2.51 ± 0.06 | 2.58 ± 0.06 | 2.43 | 2.29 ± 0.07 | 2.45 ± 0.07 | 2.56 ± 0.06 | 2.63 ± 0.05 | 2.48 |
| Vit. E at 100 mg L$^{-1}$ | 2.31 ± 0.08 | 2.46 ± 0.08 | 2.54 ± 0.07 | 2.61 ± 0.07 | 2.48 | 2.36 ± 0.08 | 2.51 ± 0.08 | 2.59 ± 0.07 | 2.67 ± 0.05 | 2.53 |
| Vit. B$_1$ at 50 mg L$^{-1}$ | 2.26 ± 0.09 | 2.41 ± 0.09 | 2.52 ± 0.08 | 2.60 ± 0.08 | 2.45 | 2.30 ± 0.09 | 2.47 ± 0.09 | 2.57 ± 0.08 | 2.65 ± 0.06 | 2.50 |
| Vit. B$_1$ at 100 mg L$^{-1}$ | 2.43 ± 0.10 | 2.58 ± 0.11 | 2.66 ± 0.09 | 2.70 ± 0.09 | 2.59 | 2.36 ± 0.10 | 2.63 ± 0.10 | 2.72 ± 0.09 | 2.76 ± 0.08 | 2.62 |
| Active yeast at 100 mL L$^{-1}$ | 2.23 ± 0.04 | 2.37 ± 0.05 | 2.49 ± 0.03 | 2.56 ± 0.03 | 2.41 | 2.27 ± 0.04 | 2.43 ± 0.04 | 2.54 ± 0.03 | 2.61 ± 0.01 | 2.46 |
| Active yeast at 150 mL L$^{-1}$ | 2.28 ± 0.05 | 2.44 ± 0.06 | 2.53 ± 0.04 | 2.61 ± 0.04 | 2.46 | 2.32 ± 0.05 | 2.49 ± 0.05 | 2.58 ± 0.02 | 2.67 ± 0.01 | 2.52 |
| Mean (A) | 2.27 | 2.42 | 2.52 | 2.59 | | 2.30 | 2.47 | 2.57 | 2.64 | |
| L.S.D. at 5% | A: 0.06 | | B: 0.01 | | AB: 0.02 | A: 0.05 | | B: 0.01 | | AB: 0.02 |

Data are expressed as means ± standard deviation (SD) of three replicates. Treatment means were compared using the least significant difference (LSD) test at $p = 0.05$.

**Table 5.** Effects of organic manure, vitamin E, vitamin $B_1$, and active yeast treatments on the number of branches/plant and herb dry weight/plant (g) of caraway plant during the two growing seasons (2023/24 and 2024/25).

| Vitamins and Active Yeast Treatments (B) | Organic Manure (ton/ha) (A) | | | | | | | | | |
|---|---|---|---|---|---|---|---|---|---|---|
| | 0.0 | 5.0 | 10 | 15 | Mean (B) | 0.0 | 5.0 | 10 | 15 | Mean (B) |
| | First Season (2023/2024) | | | | | Second Season (2024/2025) | | | | |
| | Number of branches/plant | | | | | | | | | |
| Control (Without) | 11.39 ± 0.72 | 12.11 ± 0.74 | 12.73 ± 0.59 | 13.14 ± 1.33 | 12.34 | 11.6 ± 0.75 | 12.33 ± 0.78 | 12.95 ± 0.62 | 13.36 ± 1.35 | 12.56 |
| Vit. E at 50 mg $L^{-1}$ | 12.11 ± 0.66 | 12.80 ± 0.68 | 13.35 ± 0.50 | 13.87 ± 1.34 | 13.04 | 12.33 ± 0.70 | 13.05 ± 0.72 | 13.68 ± 0.50 | 14.19 ± 1.34 | 13.31 |
| Vit. E at 100 mg $L^{-1}$ | 12.63 ± 0.72 | 13.46 ± 0.73 | 13.97 ± 0.52 | 14.39 ± 1.44 | 13.61 | 12.85 ± 0.76 | 13.78 ± 0.78 | 14.3 ± 0.53 | 14.71 ± 1.45 | 13.91 |
| Vit. $B_1$ at 50 mg $L^{-1}$ | 12.21 ± 0.82 | 13.04 ± 0.83 | 13.46 ± 0.59 | 14.08 ± 1.65 | 13.20 | 12.43 ± 0.76 | 13.26 ± 0.88 | 13.78 ± 0.60 | 14.4 ± 1.66 | 13.47 |
| Vit. $B_1$ at 100 mg $L^{-1}$ | 13.66 ± 0.99 | 14.42 ± 1.01 | 14.59 ± 0.69 | 17.25 ± 1.94 | 14.98 | 13.99 ± 1.05 | 14.75 ± 1.07 | 14.92 ± 0.69 | 17.58 ± 1.94 | 15.31 |
| Active yeast at 100 mL $L^{-1}$ | 11.9 ± 0.08 | 12.63 ± 0.72 | 13.25 ± 0.29 | 13.66 ± 1.32 | 12.86 | 12.12 ± 0.70 | 12.85 ± 0.75 | 13.57 ± 0.30 | 13.99 ± 1.32 | 13.13 |
| Active yeast at 150 mL $L^{-1}$ | 12.010.055 | 12.73 ± 0.75 | 13.66 ± 0.55 | 14.18 ± 1.55 | 13.15 | 12.22 ± 0.75 | 12.95 ± 72 | 13.99 ± 0.35 | 14.5 ± 1.42 | 13.42 |
| Mean (A) | 12.27 | 13.03 | 13.57 | 14.37 | | 12.51 | 13.28 | 13.88 | 14.68 | |
| L.S.D. at 5% | A: 0.38 | | B: 0.42 | | AB: 0.84 | A: 0.41 | | B: 0.52 | | AB: 1.04 |
| | Herb dry weight/plant (g) | | | | | | | | | |
| Control (Without) | 71.13 ± 8.43 | 75.63 ± 8.98 | 78.87 ± 8.35 | 81.48 ± 8.83 | 76.78 | 72.60 ± 8.58 | 77.10 ± 9.19 | 80.45 ± 8.52 | 83.16 ± 8.97 | 78.33 |
| Vit. E at 50 mg $L^{-1}$ | 76.04 ± 8.18 | 80.86 ± 8.72 | 83.78 ± 8.00 | 87.13 ± 8.49 | 81.95 | 77.52 ± 8.34 | 82.43 ± 8.92 | 85.47 ± 8.17 | 88.92 ± 8.63 | 83.58 |
| Vit. E at 100 mg $L^{-1}$ | 84.41 ± 8.65 | 89.85 ± 9.22 | 92.47 ± 8.37 | 94.35 ± 9.00 | 90.27 | 86.09 ± 8.81 | 91.64 ± 9.42 | 94.36 ± 8.54 | 96.24 ± 9.15 | 92.08 |
| Vit. $B_1$ at 50 mg $L^{-1}$ | 78.55 ± 9.92 | 83.58 ± 10.56 | 86.50 ± 9.59 | 87.97 ± 10.38 | 84.15 | 80.13 ± 10.10 | 85.26 ± 10.80 | 88.19 ± 9.79 | 89.76 ± 10.65 | 85.83 |
| Vit. $B_1$ at 100 mg $L^{-1}$ | 96.55 ± 11.81 | 102.72 ± 12.58 | 104.12 ± 11.34 | 108.32 ± 11.86 | 102.93 | 98.44 ± 12.02 | 104.82 ± 12.85 | 106.2 ± 11.56 | 110.42 ± 12.06 | 104.98 |
| Active yeast at 100 mL $L^{-1}$ | 74.06 ± 3.55 | 78.76 ± 3.78 | 82.11 ± 4.44 | 84.94 ± 9.01 | 79.97 | 75.53 ± 3.62 | 80.34 ± 3.85 | 83.79 ± 4.51 | 86.62 ± 9.01 | 81.57 |
| Active yeast at 150 mL $L^{-1}$ | 79.08 ± 9.82 | 84.10 ± 11.15 | 88.39 ± 10.22 | 93.20 ± 8.85 | 86.19 | 80.65 ± 10.25 | 85.78 ± 10.52 | 90.17 ± 10.24 | 95.09 ± 11.12 | 87.92 |
| Mean (A) | 79.97 | 85.07 | 88.03 | 91.06 | | 81.57 | 86.77 | 89.82 | 92.89 | |
| L.S.D. at 5% | A: 2.71 | | B: 1.10 | | AB: 2.20 | A: 2.76 | | B: 1.71 | | AB: 3.42 |

Data are expressed as means ± standard deviation (SD) of three replicates. Treatment means were compared using the least significant difference (LSD) test at $p = 0.05$.

In both seasons, each amendment—organic manure, vitamins, and active yeast—resulted in consistent increases in caraway plant height and stem diameter compared to the untreated control (Table 1). In the first season, applying only 15 ton/ha of manure increased plant height from 146.5 cm to 174.4 cm (+19.0%), and stem diameter from 2.17 cm to 2.46 cm (+13.4%); the second season showed very similar increases. Foliar spraying with 100 mg $L^{-1}$ vitamin $B_1$ alone (without manure) increased plant height by about 14% and diameter by about 11%, while 100 mg $L^{-1}$ vitamin E achieved moderate improvements (12–17% height increase and 9–14% diameter increase). Application of 15 g/L of active yeast increased plant height by 13% and stem diameter by approximately 8%. The most significant synergistic effect came from the combination of 100 mg $L^{-1}$ of vitamin $B_1$ and 15 ton/ha of manure: height increased to 198.1 cm (+35.1%) and stem diameter to 2.70 cm (+24.4%) in the first season (the same increase was also observed in the second season).

In the 2023/24 and 2024/25 seasons (Table 2), each improvement measure increased the number of branches per plant and the dry weight of herb per plant (g), compared to the untreated control group (11.4 branches and 71.1 g dry weight per plant). Applying 15 ton/ha of manure alone increased the number of branches by approximately 15% and the dry weight by 15%. Applying 100 mg $L^{-1}$ of vitamin E (without manure) increased branch number by about 11% and dry weight by about 19%, while applying 100 mg $L^{-1}$ of vitamin $B_1$ alone increased branch number by about 20% and dry weight of herb by about 36%. The increase was lower with 150 mL $L^{-1}$ of active yeast (about a 5% increase in branch number and about an 11% increase in dry weight of herb). The strongest synergistic effect was achieved when combining vitamin $B_1$ (100 mg $L^{-1}$) with 15 ton/ha of manure—branch number increased by about 52% (from 11.4–11.6 g to about 17.3–17.6 g), and dry weight by about 52% (from 71–72 g to about 108–110 g), for both seasons.

### 3.2. Yield and Yield Components

The data presented in Table 6 and Figures 2 and 3 (A and B) show the individual and interactive effects of organic manure (5, 10, and 15 ton/ha), vitamin E (50 and 100 mg $L^{-1}$), vitamin $B_1$ (50 and 100 mg $L^{-1}$), and active yeast (100 and 150 mL $L^{-1}$) treatments on the number of umbels/plant, weight of 1000 seeds, and seed yield/plant and/ha of caraway plants over two growing seasons (2023/24 and 2024/25).

As shown in Table 6, in both growing seasons, different inputs such as organic manure, vitamins, and active yeast resulted in better growth performance of caraway plants compared to the untreated control group. With only 15 ton/ha of manure applied, the number of umbels per plant increased by about 13% in the first season (27.35 vs. 24.13) and by 22% in the second season (30.04 vs. 24.64), and the weight of 1000 seeds increased by about 17% in both growing seasons (13.49 vs. 13.13 g). Spraying 100 mg $L^{-1}$ of vitamin E (without manure) increased the number of umbels by about 24–26% and the seed weight by about 14%, while treatment with 150 mL $L^{-1}$ of active yeast increased the number of umbels by about 24% and the seed weight by about 13%, compared to the control group. With 100 mg $L^{-1}$ vitamin $B_1$ applied to the leaves alone, the number of umbels increased by about 33% and the seed weight by about 17%. The most significant effect was the combination of 100 mg $L^{-1}$ of vitamin $B_1$ with 15 ton/ha. of manure. The increase in umbels was more than 56% (37.67 vs. 24.13 in the first season; 38.45 vs. 24.64 in the second season). The weight of a thousand seeds increased by about 42% in both seasons (16.38 g vs. 11.57 g, and 17.41 g vs. 12.33 g, respectively).

**Table 6.** Effects of organic manure, vitamin E, vitamin $B_1$, and active yeast treatments on the number of umbels/plant, and weight of 1000 seed (g) of caraway plant during the two growing seasons (2023/24 and 2024/25).

| Vitamins and Active Yeast Treatments (B) | Organic Manure (ton/ha) (A) | | | | | | | | | |
|---|---|---|---|---|---|---|---|---|---|---|
| | 0.0 | 5.0 | 10 | 15 | Mean (B) | 0.0 | 5.0 | 10 | 15 | Mean (B) |
| | First Season (2023/2024) | | | | | Second Season (2024/2025) | | | | |
| | Number of umbels/plant | | | | | | | | | |
| Control (Without) | 24.13 ± 2.94 | 25.67 ± 3.12 | 27.25 ± 2.92 | 27.35 ± 3.47 | 26.10 | 24.64 ± 3.00 | 26.22 ± 3.20 | 27.82 ± 2.99 | 30.04 ± 3.01 | 27.18 |
| Vit. E at 50 mg $L^{-1}$ | 25.81 ± 2.68 | 27.45 ± 2.85 | 29.01 ± 2.65 | 32.12 ± 2.59 | 28.60 | 26.35 ± 2.74 | 28.00 ± 2.93 | 29.62 ± 2.72 | 32.79 ± 2.64 | 29.19 |
| Vit. E at 100 mg $L^{-1}$ | 29.99 ± 2.64 | 31.91 ± 2.80 | 33.67 ± 2.62 | 35.74 ± 2.68 | 32.83 | 30.62 ± 2.69 | 32.58 ± 2.87 | 34.38 ± 2.69 | 36.49 ± 2.73 | 33.52 |
| Vit. $B_1$ at 50 mg $L^{-1}$ | 27.65 ± 2.97 | 29.42 ± 3.15 | 30.25 ± 2.84 | 33.15 ± 2.99 | 30.12 | 28.23 ± 3.03 | 30.04 ± 3.23 | 30.88 ± 2.92 | 33.85 ± 3.05 | 30.75 |
| Vit. $B_1$ at 100 mg $L^{-1}$ | 32.04 ± 3.54 | 34.08 ± 3.76 | 34.53 ± 3.28 | 37.67 ± 3.57 | 34.58 | 32.72 ± 3.61 | 34.83 ± 3.85 | 35.30 ± 3.37 | 38.45 ± 3.63 | 35.32 |
| Active yeast at 100 mL $L^{-1}$ | 25.13 ± 3.37 | 26.74 ± 3.59 | 28.39 ± 3.59 | 30.64 ± 3.45 | 27.73 | 25.66 ± 3.44 | 27.30 ± 3.66 | 28.98 ± 3.67 | 31.29 ± 2.56 | 28.31 |
| Active yeast at 150 mL $L^{-1}$ | 29.90 ± 2.42 | 31.81 ± 2.75 | 33.46 ± 2.45 | 35.22 ± 3.22 | 32.60 | 30.53 ± 2.58 | 32.47 ± 2.85 | 34.17 ± 2.55 | 35.96 ± 2.65 | 33.28 |
| Mean (A) | 27.81 | 29.58 | 30.94 | 33.13 | | 28.39 | 30.21 | 31.59 | 34.12 | |
| L.S.D. at 5% | A: 1.18 | | B: 1.24 | | AB: 2.48 | A: 1.22 | | B: 1.25 | | AB: 2.50 |
| | Weight of 1000 seeds (g) | | | | | | | | | |
| Control (Without) | 11.57 ± 0.67 | 12.39 ± 0.61 | 13.09 ± 0.51 | 13.49 ± 0.90 | 12.64 | 12.33 ± 0.70 | 13.19 ± 0.65 | 13.94 ± 0.53 | 14.37 ± 0.95 | 13.46 |
| Vit. E at 50 mg $L^{-1}$ | 12.49 ± 0.53 | 13.29 ± 0.45 | 13.69 ± 0.40 | 14.18 ± 0.85 | 13.41 | 13.30 ± 0.55 | 14.15 ± 0.49 | 14.58 ± 0.41 | 15.11 ± 0.89 | 14.29 |
| Vit. E at 100 mg $L^{-1}$ | 13.19 ± 0.56 | 14.00 ± 0.48 | 14.47 ± 0.41 | 14.67 ± 0.91 | 14.08 | 14.04 ± 0.59 | 14.90 ± 0.52 | 15.44 ± 0.42 | 15.65 ± 0.95 | 15.01 |
| Vit. $B_1$ at 50 mg $L^{-1}$ | 12.59 ± 0.61 | 13.39 ± 50 | 13.79 ± 0.41 | 14.32 ± 1.04 | 13.52 | 13.40 ± 0.64 | 14.26 ± 0.54 | 14.68 ± 0.39 | 15.22 ± 1.10 | 14.39 |
| Vit. $B_1$ at 100 mg $L^{-1}$ | 13.52 ± 0.73 | 14.01 ± 0.60 | 14.50 ± 0.46 | 16.38 ± 1.20 | 14.60 | 14.37 ± 0.77 | 14.98 ± 0.65 | 15.33 ± 0.45 | 17.41 ± 1.26 | 15.52 |
| Active yeast at 100 mL $L^{-1}$ | 12.08 ± 0.67 | 12.89 ± 0.66 | 13.59 ± 0.44 | 14.08 ± 0.82 | 13.16 | 12.86 ± 0.69 | 13.73 ± 0.67 | 14.47 ± 0.45 | 15.01 ± 1.05 | 14.02 |
| Active yeast at 150 mL $L^{-1}$ | 13.03 ± 0.52 | 13.82 ± 0.72 | 14.21 ± 0.35 | 14.61 ± 0.88 | 13.92 | 13.83 ± 0.72 | 14.68 ± 0.62 | 15.11 ± 0.42 | 15.54 ± 0.98 | 14.79 |
| Mean (A) | 12.64 | 13.40 | 13.91 | 14.53 | | 13.45 | 14.27 | 14.79 | 15.47 | |
| L.S.D. at 5% | A: 0.41 | | B: 0.89 | | AB: 1.78 | A: 0.52 | | B: 0.94 | | AB: 1.88 |

Data are expressed as means ± standard deviation (SD) of three replicates. Treatment means were compared using the least significant difference (LSD) test at $p = 0.05$.

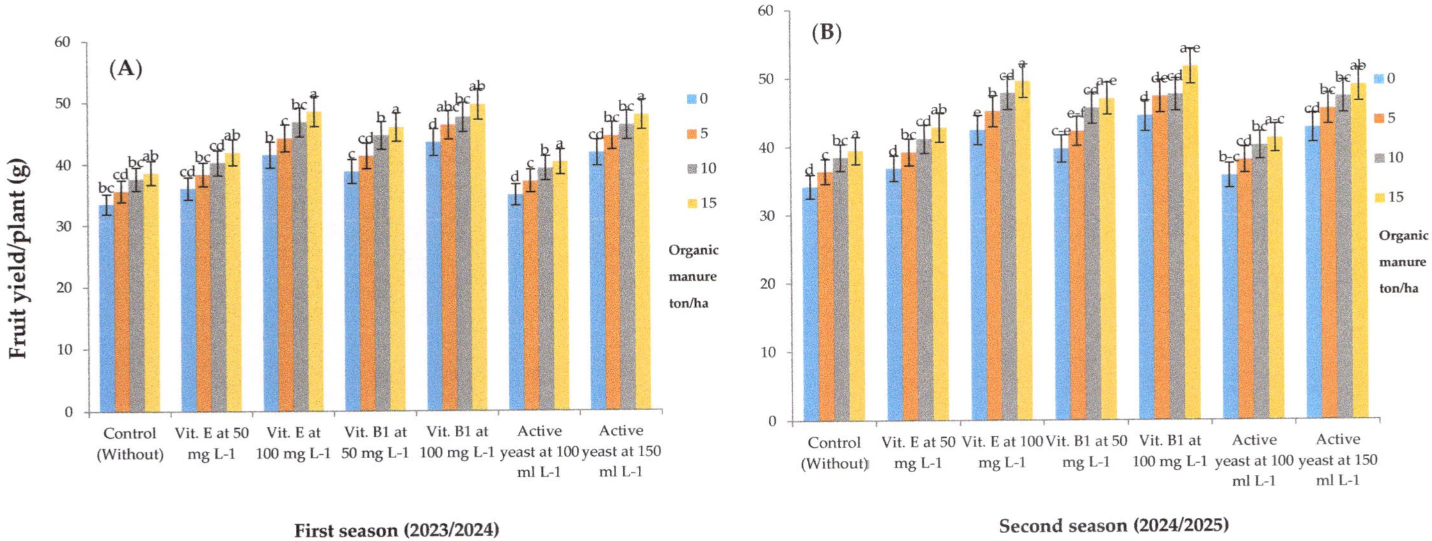


**Figure 2.** (**A**,**B**). Effects of organic manure, vitamin E, vitamin $B_1$, and active yeast treatments on fruit yield/plant (g) of caraway plants during the two growing seasons (2023/24 and 2024/25). The data are statistically different, according to the Duncan Multiple Range Test (DMRT) at $p \leq 0.05$. Interactions are indicated by lowercase letters, unless otherwise mentioned.

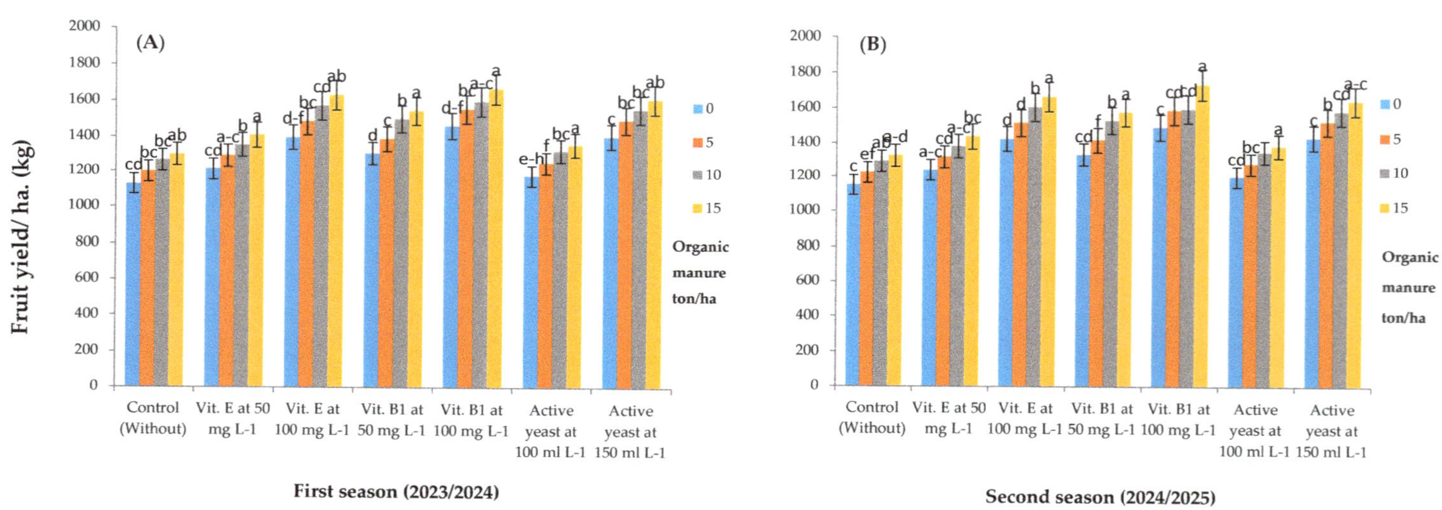


**Figure 3.** (**A**,**B**). Effects of organic manure, vitamin E, vitamin $B_1$, and active yeast treatments on fruit yield/ha (kg) of caraway plants during the two growing seasons (2023/24 and 2024/25). The data are statistically different, according to the Duncan Multiple Range Test (DMRT) at $p \leq 0.05$. Interactions are indicated by lowercase letters, unless otherwise mentioned.

In the 2023/24 and 2024/25 growing seasons, all tested treatments—organic manure, vitamins, and active yeast—produced significantly higher seed yields than the untreated control group, as shown in Figures 2 and 3. Individually, 15 ton/ha of manure were applied, and yield per plant increased in both seasons from 33.53 g to 38.64 g (approximately +15.3%), and yield per ha increased from 1126.6 kg to 1298.3 kg (about +15.3%). Only 100 mg $L^{-1}$ of vitamin E was applied to the leaves, which increased the yield per plant by about 24% (to ~41.5 g) and increased the yield per ha rate by about 24% (to ~1395 kg), while 15 g/l of active yeast also increased the yield per plant (to~41.7 g) and yield per ha (~1401 kg), giving the similar gains of ~24%. Vitamin $B_1$ at 100 mg $L^{-1}$ without manure outperformed both, increasing yield per plant by about 30% (to ~43.5 g) and per ha rate by about 30% (to ~1460 kg). The strongest synergistic effect came from combining 100 mg $L^{-1}$ of vitamin $B_1$ with 15 ton/ha of manure, resulting in 49.54 g yield per plant (≈+47.8%) and 1664.5 kg per ha (≈+47.8%) in the first season, and up to 51.58 g yield per plant (+51.31 kg) and 1733.1 kg yield per ha (+50.7%) in the second season.

### 3.3. Essential Oil Productivity

Figures 4–6 (A and B) show the individual and combined effects of organic manure (5, 10, and 15 ton/ha), vitamin E (50 and 100 mg $L^{-1}$), vitamin $B_1$ (50 and 100 mg $L^{-1}$), and active yeast (100 and 150 mL $L^{-1}$) on the essential oil (%), essential oil yield/plant and essential oil yield/ha of caraway plants during the 2023/2024 and 2024/2025 growing seasons.

Various applications, such as organic manure, vitamins, and active yeast, increased the essential oil content of caraway plants and significantly increased oil production per plant and per ha, compared to the untreated control, in both seasons. Applying only 15 ton/ha of manure increased the essential oil percentage from 3.09% to 3.42% (+10.7%) in the 2023/24 season, and from 3.15% to 3.49% (+10.8%) in the 2024/25 season, while increasing oil production per plant by about 27% (1.04 mL → 1.32 mL; 1.08 mL → 1.38 mL) and oil yield per ha by about 27% (34.94 L → 44.35 l; 36.29 L → 46.37 l). Foliar application of 100 mg $L^{-1}$ vitamin $B_1$ (without manure) increased oil concentration by about 16% (to 3.58–3.66%), increased oil yield per plant by about 50% (to 1.56–1.62 mL), and increased production per ha by about 50% (to 52.42–54.43 mL), compared to the control. A 100 mg $L^{-1}$ application of vitamin E gave moderate increases (oil percentage +9.4–9.5%; yield +35%), and 150 mL $L^{-1}$ of active yeast gave more modest increases (oil % +5.2–5.4%; yield +30%). The synergistic effect was even more pronounced when 100 mg $L^{-1}$ of $B_1$ was combined with 15 ton/ha of manure: essential oil production jumped by about 46% (to 4.52% and 4.66%), doubling oil production per plant (+115%; 2.24 mL and 2.36 mL), and more than doubling oil production per ha (+115%; 75.26 and 79.30).

### 3.4. Biochemical Constituents

#### 3.4.1. Photosynthetic Pigments

As shown in Figures 7–9 (A and B), this study investigated the effects of organic manure (5, 10, and 15 ton/ha), vitamin E and vitamin $B_1$ (50 and 100 mg $L^{-1}$ each), and active yeast (100 and 150 mL $L^{-1}$) on photosynthetic pigments (chlorophyll a, b, and carotenoids) of caraway plants during the 2023/2024 and 2024/2025 seasons.

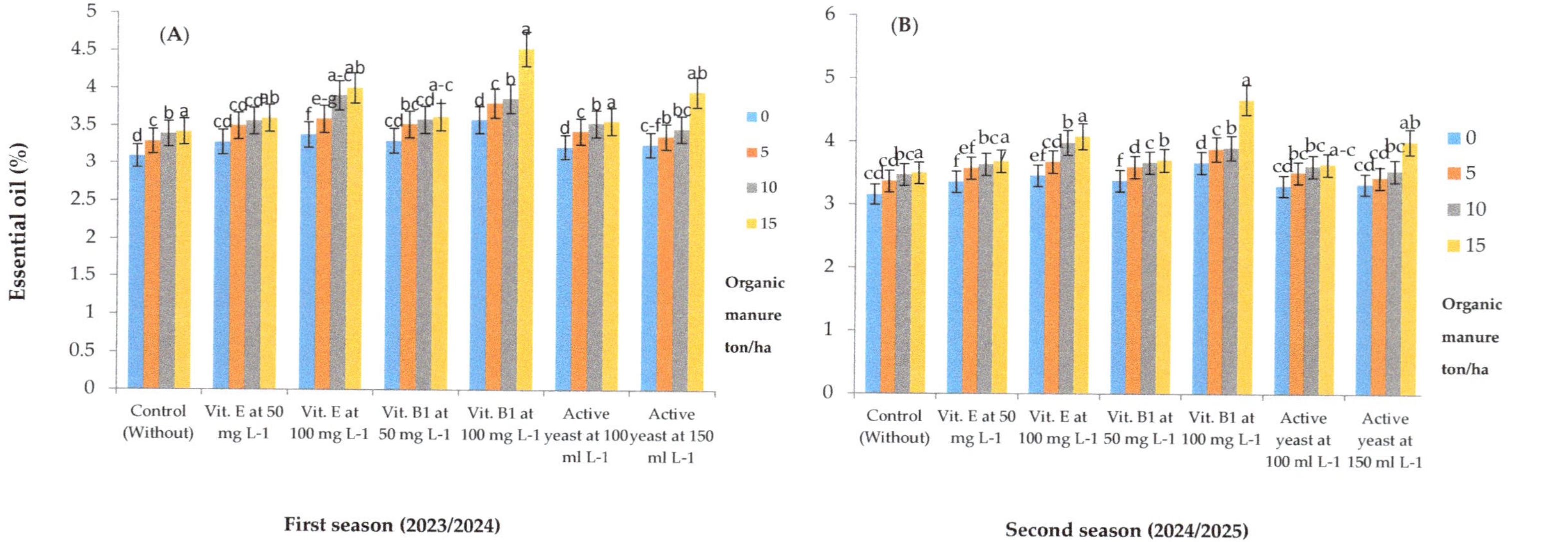


**Figure 4.** (**A**,**B**). Effects of organic manure, vitamin E, vitamin $B_1$, and active yeast treatments on essential oil (%) of caraway plants during the two growing seasons (2023/24 and 2024/25). The data are statistically different, according to the Duncan Multiple Range Test (DMRT) at $p \leq 0.05$. Interactions are indicated by lowercase letters, unless otherwise mentioned.

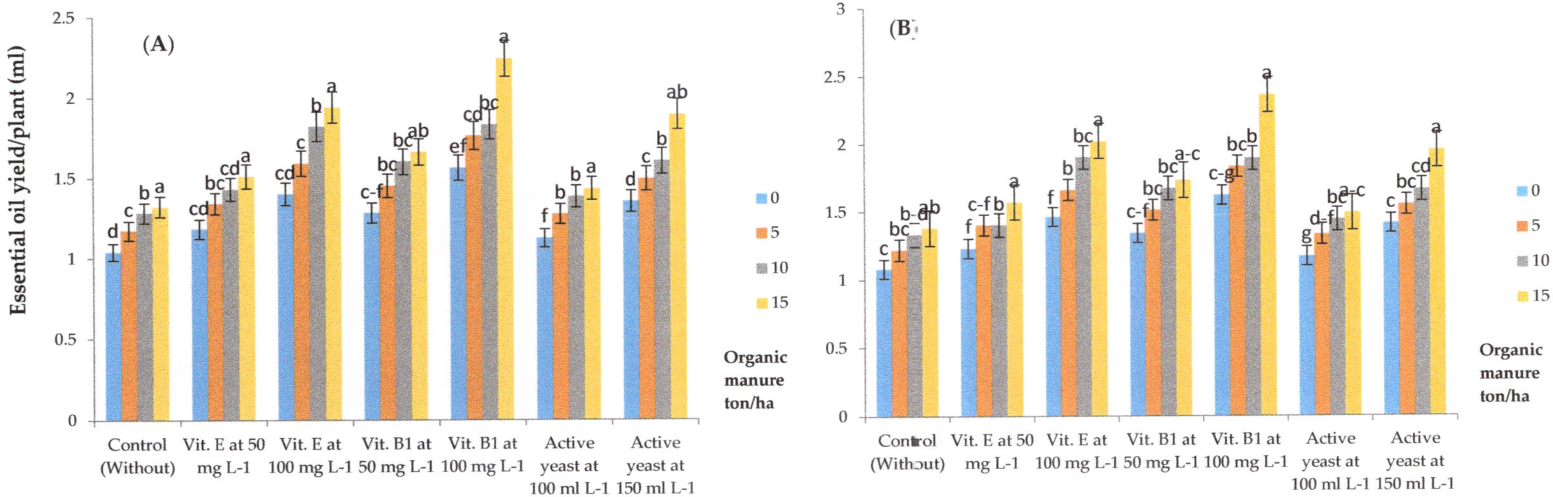


**First season (2023/2024)**

**Second season (2024/2025)**

**Figure 5.** (**A**,**B**). Effects of organic manure, vitamin E, vitamin $B_1$, and active yeast treatments on essential oil yield/plant (mL)of caraway plants during the two growing seasons (2023/24 and 2024/25). The data are statistically different, according to the Duncan Multiple Range Test (DMRT) at $p \leq 0.05$. Interactions are indicated by lowercase letters, unless otherwise mentioned.

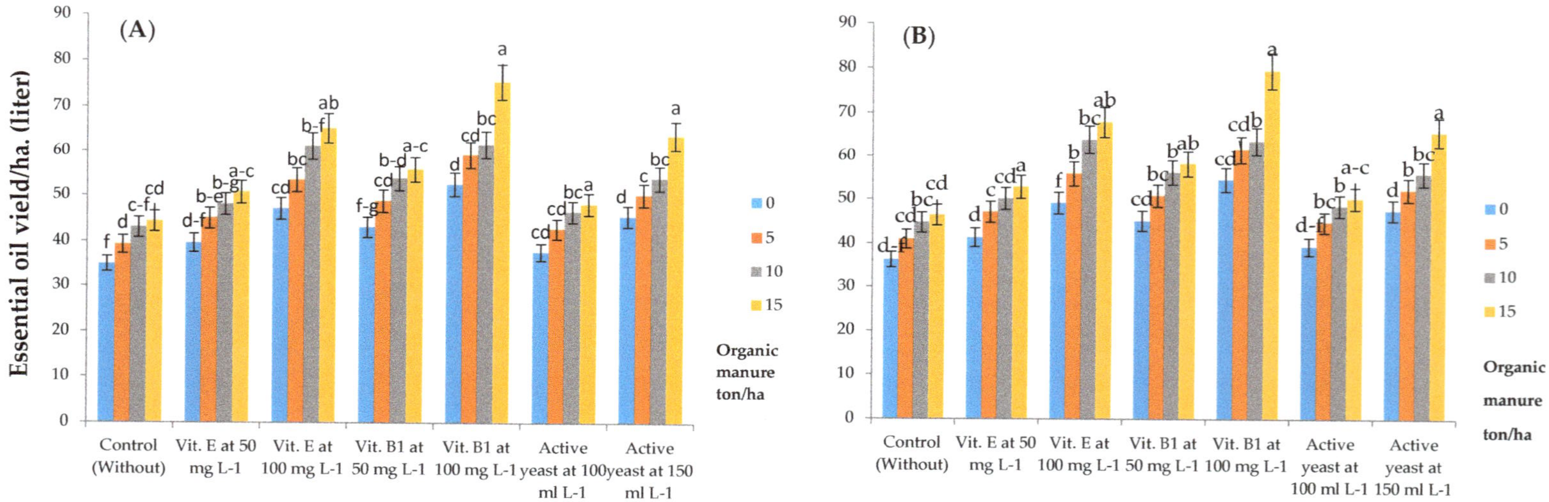


**First season (2023/2024)** **Second season (2024/2025)**

**Figure 6.** (**A**,**B**). Effects of organic manure, vitamin E, vitamin $B_1$, and active yeast treatments on essential oil yield/ha (l) of caraway plants during the two growing seasons (2023/24 and 2024/25). The data are statistically different, according to the Duncan Multiple Range Test (DMRT) at $p \leq 0.05$. Interactions are indicated by lowercase letters, unless otherwise mentioned.

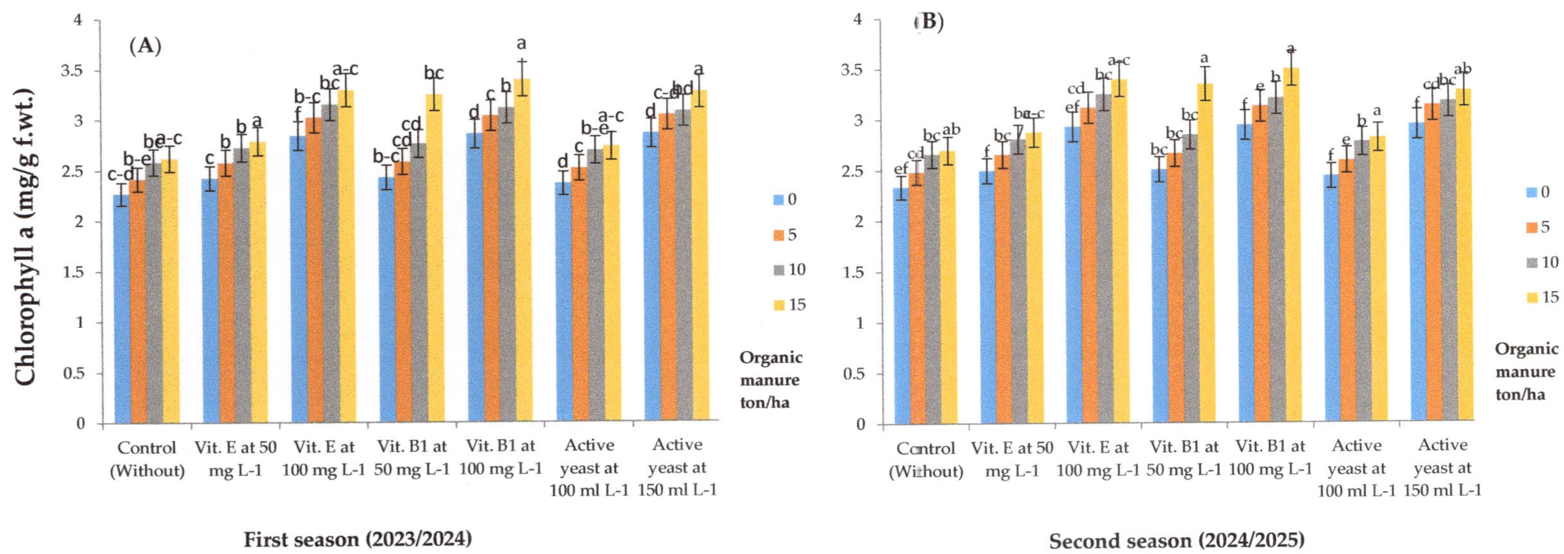


**Figure 7.** (**A**,**B**). Effects of organic manure, vitamin E, vitamin $B_1$, and active yeast treatments on chlorophyll a (mg/g f.w.) of caraway plants during the two growing seasons (2023/24 and 2024/25). The data are statistically different, according to the Duncan Multiple Range Test (DMRT) at $p \leq 0.05$. Interactions are indicated by lowercase letters, unless otherwise mentioned.

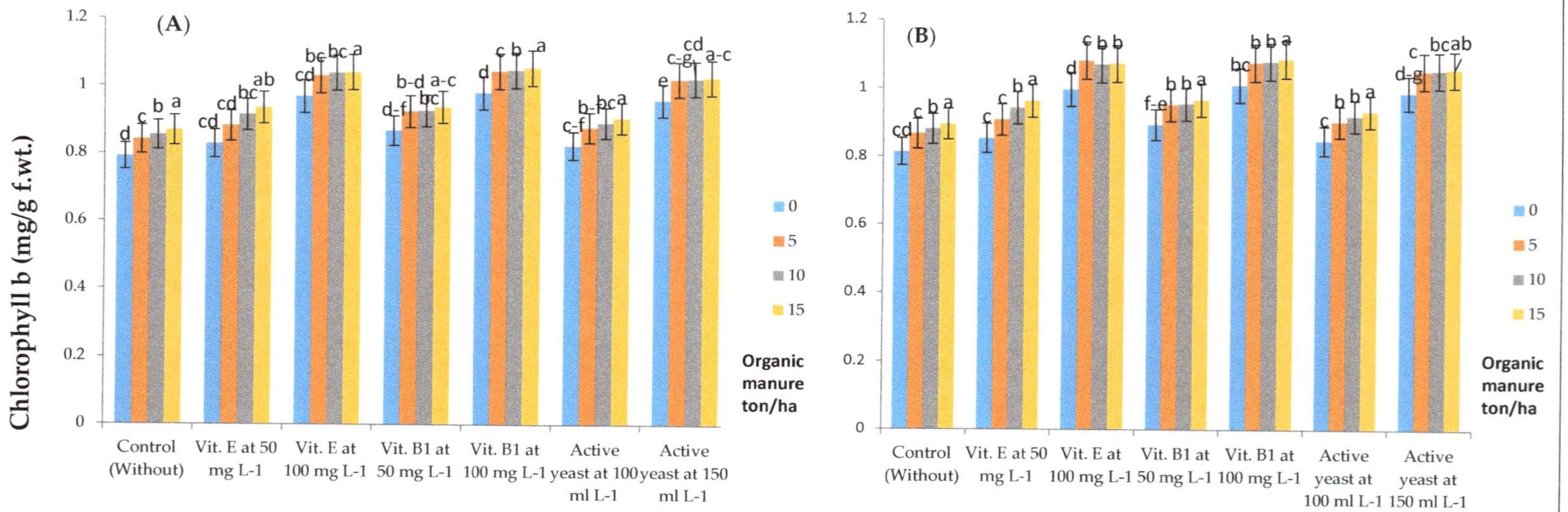


**First season (2023/2024)** **Second season (2024/2025)**

**Figure 8.** (**A**,**B**). Effects of organic manure, vitamin E, vitamin $B_1$, and active yeast treatments on chlorophyll b (mg/g f.wt.) of caraway plants during the two growing seasons (2023/24 and 2024/25). The data are statistically different, according to the Duncan Multiple Range Test (DMRT) at $p \leq 0.05$. Interactions are indicated by lowercase letters, unless otherwise mentioned.

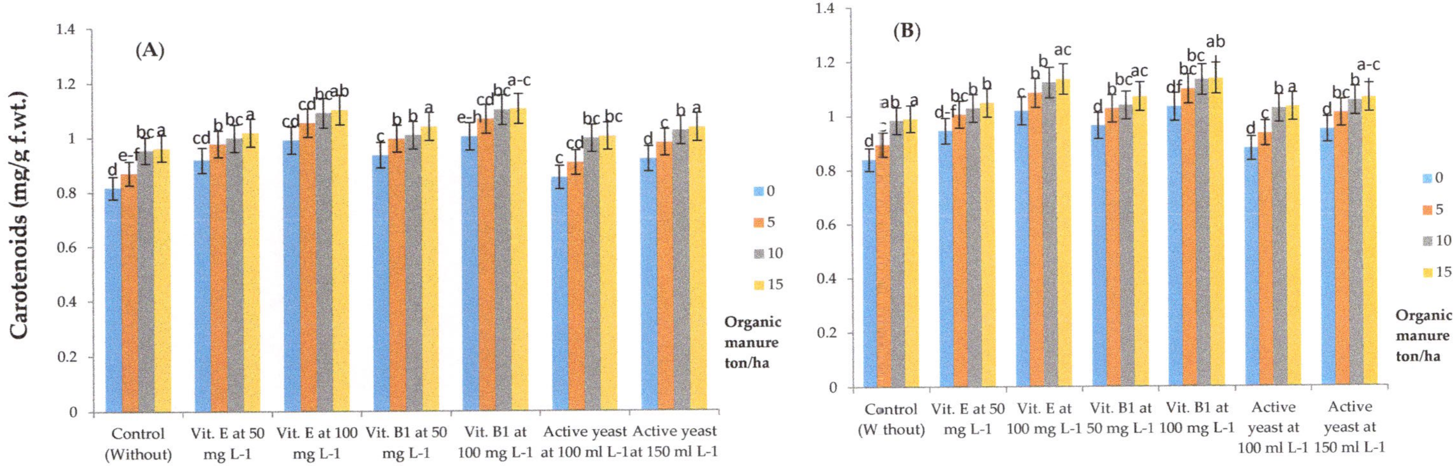


**First season (2023/2024)** **Second season (2024/2025)**

**Figure 9.** (**A**,**B**). Effects of organic manure, vitamin E, vitamin $B_1$, and active yeast treatments on carotenoids (mg/g. f.wt.) of caraway plants during the two growing seasons (2023/24 and 2024/25). The data are statistically different, according to the Duncan Multiple Range Test (DMRT) at $p \leq 0.05$. Interactions are indicated by lowercase letters, unless otherwise mentioned.

Each of the improvements—organic manure, vitamin spray, and active yeast—led to a significant increase in caraway leaf pigment content compared to the untreated control in the 2023/24 and 2024/25 seasons. Using only 15 ton/ha of manure, chlorophyll a content increased from approximately 2.27–2.34 mg/g to 2.62–2.70 mg/g (an increase of approximately 15.5%), chlorophyll b content increased from 0.79–0.81 mg/g to 0.87–0.90 mg/g (an increase of approximately 10%), and carotenoid content increased from 0.82–0.84 mg/g to 0.96–0.99 mg/g (an increase of approximately 17.8%). Foliar application of 100 mg $L^{-1}$ of vitamin $B_1$ (without manure) increased chlorophyll a by about 25.9%, chlorophyll b by about 23.8%, and carotenoids by about 22.6%, while 100 mg $L^{-1}$ of vitamin E achieved almost the same improvements. Applying 150 mL $L^{-1}$ of active yeast alone increased pigments by about 25.9% in chlorophyll a, about 21.0% in chlorophyll b, and about 12.5% in carotenoids. The strongest synergistic effect was achieved when combining 100 mg $L^{-1}$ of vitamin $B_1$ with 15 ton/ha of manure: chlorophyll a increased by about 49.6%, chlorophyll b by about 33.1%, and carotenoids by about 35.0%, compared to the control group.

### 3.4.2. Nitrogen, Phosphorus and Potassium (%)

Table 7 shows the effect of different levels of organic manure (5, 10, and 15 ton/ha), vitamin E and vitamin $B_1$ (50 and 100 mg $L^{-1}$ each), and active yeast (100 and 150 mL $L^{-1}$), alone or in combination, on the percentages of N, P, and K in the dry leaves of caraway plants in the 2023/2024 and 2024/2025 seasons.

Over two growing seasons, untreated caraway leaves contained an average nitrogen content of 2.18–2.24%, phosphorus content of 0.29–0.30%, and potassium content of 0.83–0.85%. Applying only 15 ton/ha of manure increased leaf nitrogen levels to 2.43–2.50% (+11.5% over the control), phosphorus to 0.33–0.34% (+12%), and potassium to 0.92–0.94% (+10.5%). Spraying with 100 mg $L^{-1}$ vitamin E (without manure) increased leaf nitrogen content by about 28% (to 2.72–2.80%), phosphorus by about 76% (to 0.49–0.50%), and potassium by about 55% (to 1.34–1.38%). Applying 150 mL $L^{-1}$ of active yeast alone resulted in moderate increases in nitrogen (19.4%), phosphorus (57.2%), and potassium (40.5%). However, when 100 mg $L^{-1}$ of vitamin $B_1$ was combined with 15 ton/ha of manure, the synergistic effect was most striking: nitrogen increased to 3.11–3.20% (+42–43%), phosphorus to 0.55–0.57% (+90%), and potassium to 1.46–1.50% (+76%). Combining vitamin E with fertilizer resulted in similar, albeit slightly lower increases (N by 38–39%, P by 85–86%, K by 74%).

**Table 7.** Effects of organic manure, vitamin E, vitamin $B_1$, and active yeast treatments on nitrogen, phosphorus and potassium (%) of caraway leaves during the two growing seasons (2023/24 and 2024/25).

| Vitamins and Active Yeast Treatments (B) | Organic Manure (ton/ha) (A) | | | | | | | | | |
|---|---|---|---|---|---|---|---|---|---|---|
| | 0.0 | 5.0 | 10 | 15 | Mean (B) | 0.0 | 5.0 | 10 | 15 | Mean (B) |
| | First Season (2023/2024) | | | | | Second Season (2024/2025) | | | | |
| | Nitrogen (%) | | | | | | | | | |
| Control (Without) | 2.178 ± 0.24 | 2.317 ± 0.25 | 2.385 ± 0.25 | 2.428 ± 0.26 | 2.327 | 2.244 ± 0.24 | 2.388 ± 0.26 | 2.458 ± 0.26 | 2.502 ± 0.27 | 2.398 |
| Vit. E at 50 mg $L^{-1}$ | 2.299 ± 0.21 | 2.445 ± 0.23 | 2.514 ± 0.23 | 2.549 ± 0.24 | 2.452 | 2.369 ± 0.22 | 2.520 ± 0.23 | 2.591 ± 0.24 | 2.627 ± 0.25 | 2.527 |
| Vit. E at 100 mg $L^{-1}$ | 2.718 ± 0.20 | 2.892 ± 0.21 | 2.934 ± 0.22 | 3.016 ± 0.23 | 2.890 | 2.801 ± 0.21 | 2.980 ± 0.22 | 3.023 ± 0.23 | 3.108 ± 0.24 | 2.978 |
| Vit. $B_1$ at 50 mg $L^{-1}$ | 2.530 ± 0.21 | 2.692 ± 0.23 | 2.714 ± 0.24 | 2.749 ± 0.24 | 2.671 | 2.607 ± 0.21 | 2.773 ± 0.23 | 2.797 ± 0.25 | 2.833 ± 0.25 | 2.753 |
| Vit. $B_1$ at 100 mg $L^{-1}$ | 2.781 ± 0.26 | 2.959 ± 0.28 | 3.059 ± 0.29 | 3.108 ± 0.29 | 2.977 | 2.866 ± 0.22 | 3.049 ± 0.28 | 3.153 ± 0.30 | 3.203 ± 0.30 | 3.067 |
| Active yeast at 100 mL $L^{-1}$ | 2.269 ± 0.23 | 2.414 ± 0.25 | 2.484 ± 0.26 | 2.529 ± 0.29 | 2.424 | 2.338 ± 0.24 | 2.488 ± 0.26 | 2.560 ± 0.26 | 2.606 ± 0.30 | 2.498 |
| Active yeast at 150 mL $L^{-1}$ | 2.601 ± 0.21 | 2.768 ± 0.22 | 2.845 ± 0.26 | 2.889 ± 0.41 | 2.776 | 2.690 ± 0.25 | 2.853 ± 0.26 | 2.932 ± 0.28 | 2.977 ± 0.32 | 2.863 |
| Mean (A) | 2.482 | 2.641 | 2.705 | 2.752 | | 2.559 | 2.722 | 2.788 | 2.837 | |
| L.S.D. at 5% | A: 0.055 | | B: 0.046 | | AB: 0.092 | A: 0.042 | | B: 0.037 | | AB: 0.074 |
| | Phosphorus (%) | | | | | | | | | |
| Control (Without) | 0.290 ± 0.09 | 0.308 ± 0.10 | 0.316 ± 0.10 | 0.325 ± 0.10 | 0.310 | 0.300 ± 0.10 | 0.317 ± 0.10 | 0.326 ± 0.11 | 0.335 ± 0.10 | 0.319 |
| Vit. E at 50 mg $L^{-1}$ | 0.317 ± 0.09 | 0.338 ± 0.09 | 0.340 ± 0.10 | 0.348 ± 0.09 | 0.336 | 0.327 ± 0.09 | 0.348 ± 0.10 | 0.350 ± 0.10 | 0.359 ± 0.10 | 0.346 |
| Vit. E at 100 mg $L^{-1}$ | 0.490 ± 0.08 | 0.521 ± 0.09 | 0.537 ± 0.09 | 0.539 ± 0.09 | 0.522 | 0.504 ± 0.08 | 0.537 ± 0.09 | 0.552 ± 0.09 | 0.554 ± 0.09 | 0.537 |
| Vit. $B_1$ at 50 mg $L^{-1}$ | 0.448 ± 0.09 | 0.477 ± 0.09 | 0.483 ± 0.09 | 0.489 ± 0.09 | 0.474 | 0.462 ± 0.09 | 0.491 ± 0.10 | 0.497 ± 0.10 | 0.503 ± 0.10 | 0.488 |
| Vit. $B_1$ at 100 mg $L^{-1}$ | 0.510 ± 0.11 | 0.543 ± 0.12 | 0.547 ± 0.11 | 0.553 ± 0.11 | 0.538 | 0.525 ± 0.11 | 0.558 ± 0.12 | 0.564 ± 0.12 | 0.570 ± 0.12 | 0.554 |
| Active yeast at 100 mL $L^{-1}$ | 0.303 ± 0.10 | 0.321 ± 0.12 | 0.330 ± 0.10 | 0.339 ± 0.11 | 0.323 | 0.312 ± 0.10 | 0.331 ± 0.12 | 0.339 ± 0.10 | 0.349 ± 0.12 | 0.333 |
| Active yeast at 150 mL $L^{-1}$ | 0.456 ± 0.10 | 0.485 ± 0.10 | 0.492 ± 0.11 | 0.504 ± 0.11 | 0.484 | 0.470 ± 0.11 | 0.499 ± 0.12 | 0.508 ± 0.12 | 0.519 ± 0.13 | 0.499 |
| Mean (A) | 0.402 | 0.428 | 0.435 | 0.442 | | 0.414 | 0.440 | 0.448 | 0.456 | |
| L.S.D. at 5% | A: 0.008 | | B: 0.007 | | AB: 0.014 | A: 0.007 | | B: 0.006 | | AB: 0.012 |

**Table 7.** *Cont.*

| Vitamins and Active Yeast Treatments (B) | Organic Manure (ton/ha) (A) | | | | | | | | | |
|---|---|---|---|---|---|---|---|---|---|---|
| | 0.0 | 5.0 | 10 | 15 | Mean (B) | 0.0 | 5.0 | 10 | 15 | Mean (B) |
| | First Season (2023/2024) | | | | | Second Season (2024/2025) | | | | |
| | Potassium (%) | | | | | | | | | |
| Control (Without) | 0.829 ± 0.23 | 0.883 ± 0.24 | 0.891 ± 0.24 | 0.916 ± 0.24 | 0.880 | 0.854 ± 0.23 | 0.909 ± 0.25 | 0.917 ± 0.25 | 0.943 ± 0.25 | 0.906 |
| Vit. E at 50 mg $L^{-1}$ | 0.877 ± 0.21 | 0.932 ± 0.23 | 0.945 ± 0.23 | 0.967 ± 0.23 | 0.930 | 0.903 ± 0.22 | 0.960 ± 0.24 | 0.972 ± 0.24 | 0.996 ± 0.23 | 0.958 |
| Vit. E at 100 mg $L^{-1}$ | 1.343 ± 0.20 | 1.429 ± 0.21 | 1.440 ± 0.21 | 1.444 ± 0.21 | 1.414 | 1.383 ± 0.20 | 1.472 ± 0.22 | 1.483 ± 0.22 | 1.487 ± 0.21 | 1.456 |
| Vit. $B_1$ at 50 mg $L^{-1}$ | 1.146 ± 0.20 | 1.220 ± 0.21 | 1.233 ± 0.22 | 1.248 ± 0.22 | 1.212 | 1.180 ± 0.21 | 1.256 ± 0.22 | 1.270 ± 0.22 | 1.285 ± 0.22 | 1.248 |
| Vit. $B_1$ at 100 mg $L^{-1}$ | 1.353 ± 0.25 | 1.440 ± 0.26 | 1.447 ± 0.22 | 1.460 ± 0.26 | 1.425 | 1.394 ± 0.25 | 1.483 ± 0.27 | 1.490 ± 0.27 | 1.503 ± 0.27 | 1.467 |
| Active yeast at 100 mL $L^{-1}$ | 0.864 ± 0.21 | 0.920 ± 0.22 | 0.928 ± 0.26 | 0.954 ± 0.23 | 0.916 | 0.890 ± 0.22 | 0.947 ± 0.23 | 0.956 ± 0.26 | 0.982 ± 0.25 | 0.944 |
| Active yeast at 150 mL $L^{-1}$ | 1.164 ± 0.21 | 1.238 ± 0.24 | 1.291 ± 0.22 | 1.337 ± 0.25 | 1.257 | 1.199 ± 0.22 | 1.275 ± 0.25 | 1.329 ± 0.22 | 1.377 ± 0.28 | 1.295 |
| Mean (A) | 1.082 | 1.152 | 1.168 | 1.189 | | 1.115 | 1.186 | 1.203 | 1.225 | |
| L.S.D. at 5% | A: 0.018 | | B: 0.016 | | AB: 0.032 | A: 0.019 | | B: 0.017 | | AB: 0.034 |

Data are expressed as means ± standard deviation (SD) of three replicates. Treatment means were compared using the least significant difference (LSD) test at p = 0.05.

## 4. Discussion

Growing caraway in sandy soils (newly reclaimed areas) requires strategies to compensate for their inherently low water-holding capacity, rapid nutrient loss, and low organic matter content. Drip irrigation (which delivers water and dissolved nutrients directly to the root zone) combined with the regular application of organic amendments, such as organic manure, can gradually improve soil texture, increase moisture-retaining pore space, and enhance its cation exchange capacity. Foliar application of biostimulants (such as vitamins $B_1$ and E or active yeast) helps overcome nutrient binding limitations in the soil and delivers key cofactors and antioxidants directly to the leaves, enhancing photosynthesis and stress resistance. Together, these measures transform previously barren sandy soils into a high-yielding environment, enabling rapid root establishment, enhanced water-use efficiency, robust caraway growth, and high essential oil production.

### 4.1. Growth Parameters

This study demonstrated that applying organic manure at 15 ton/ha substantially enhanced sandy soil's organic matter, porosity, and water-holding capacity, translating into 10–15% gains in plant height, branch number, and dry herb weight versus unfertilized controls. These improvements align with evidence that organic amendments boost cation exchange capacity and root permeability, ensuring gradual nutrient release and uptake [11]. Together, these soil enhancements set the stage for foliar biostimulants to deliver their full physiological benefits, reinforcing the proven value of organic manure for sustainable crop productivity [12–14].

In the same way, foliar application of vitamin E (alpha-tocopherol) enhances membrane stability and serves as a signaling molecule in light–heat balance, senescence, abscission, and numerous metabolic pathways [18]. By scavenging lipid peroxide radicals, tocopherol helps preserve thylakoid integrity and supports biomass accumulation. Applied alone, vitamin E boosts nitrogen, phosphorus, and potassium uptake, underscoring its role in alleviating light-induced stress and sustaining growth under drip-irrigated sandy-soil conditions [19]. Prior research has documented its growth-promoting effects on caraway and other species [31–33]. Moreover, combining 15 ton/ha of organic fertilizer with 100 mg $L^{-1}$ vitamin E produces moderate increases in growth, highlighting its value as a secondary biostimulant.

Additionally, foliar application of vitamin $B_1$ (thiamine) markedly enhanced plant nutritional status, reflecting its function both as thiamine pyrophosphate in carbohydrate metabolism and as a trigger for antioxidant defenses under field conditions, in addition to stimulating root respiration [15,16]. The resulting boost in carbon skeletons and cellular energy drove secondary metabolite production, which corresponded to a 20–36% rise in shoot number and dry herb mass. These improvements in caraway performance mirror outcomes reported by other researchers [34,35]. When 15 ton/ha of organic fertilizer was combined with a 100 mg $L^{-1}$ thiamine foliar spray, caraway exhibited substantially stronger physiological activity. Root-zone fertility and leaf metabolic capacity both improved roughly two-fold, in agreement with similar findings in other aromatic species [10].

Moreover, active yeast delivers amino acids, B vitamins ($B_1$, $B_2$, $B_3$, $B_{12}$), minerals, and phytohormones (auxins, cytokinins) that boost cell division, root growth, and protein synthesis [20,22]. Trials report roughly 5% more branching, an 11% increase in herb dry weight, greater plant height and branch number, as well as higher nitrogen and phosphorus in tomato and sugarcane [36]. By stimulating phosphate-solubilizing microbes in the rhizosphere and speeding nutrient cycling, active yeast can be co-applied with fertilizers to enhance microbial interactions and stress resilience [22,37,38].

### 4.2. Yield and Its Components

Integrating 15 ton/ha of organic manure into sandy soil markedly improved its structure and fertility, leading to measurable boosts in caraway's yield components. The elevated soil organic matter and porosity facilitated better moisture retention and root function, while localized acidification during organic matter breakdown enhanced phosphorus and micronutrient availability, reducing nutrient runoff. Enhanced microbial turnover accelerated N, P, and K cycling, reflected in 11.5–12% increases in leaf nutrient content [38]. These results corroborate previous findings on organic amendments' role in optimizing cation exchange and sustained nutrient supply and underline the necessity of pairing soil improvement with foliar biostimulants to maximize physiological responses and overall yield [1,2].

Foliar spraying of vitamin E (alpha-tocopherol) not only reinforces membrane stability but also regulates light–heat balance, senescence, abscission, and a host of other metabolic processes [18]. Applied on its own, vitamin E boosted oil concentration by about 9.5%, demonstrating its ability to alleviate light-induced stress and sustain growth under drip-irrigated sandy-soil conditions [19]. Earlier work has shown similar yield-enhancing effects of vitamin E on caraway and various other crops [31–33]. Furthermore, co-applying 15 tons of organic fertilizer with 100 mg $L^{-1}$ vitamin E delivered moderate but meaningful gains—30% higher oil content and a 70% increase in oil yield—underscoring its value as a secondary biostimulant.

Furthermore, foliar application of thiamine (100 mg $L^{-1}$) boosted caraway's reproductive performance, raising essential oil yield per plant by 50%. By serving as a coenzyme in carbohydrate metabolism and activating antioxidant defenses, thiamine increased carbon skeleton availability, energy supply, nutrient uptake, and secondary metabolite synthesis [16]. These effects mirror reports in other medicinal species, where foliar thiamine alleviates drought and salinity stress through enhanced antioxidant enzyme activity and osmolyte accumulation [17,34]. When combined with 15 tons of organic fertilizer, thiamine spray doubled root-zone fertility and leaf metabolic capacity [10], leading to a 56% increase in umbel yield, a 48% rise in seed yield, and a 115% gain in essential oil content.

Functioning as a biostimulant, active yeast provides plants with essential nutrients—plant-available amino acids, B vitamins ($B_1$, $B_2$, $B_3$, $B_{12}$), and minerals—alongside plant hormones (auxins, cytokinins) that directly stimulate cell division and enlargement [22]. Mechanisms like phosphate solubilization and protective polysaccharide production further support plant growth, potentially leading to modest yield increases. Applied as yeast extracts, it enhances physiological processes such as cell division, root proliferation, and protein synthesis [20]. These effects translate into measurable improvements, notably a 30–35% boost in oil production. Documented benefits include enhanced oil content and yield in caraway plants [31]. Active yeast serves as a valuable, eco-friendly supplement that can be mixed with fertilizers to bolster microbial interactions and improve plant stress tolerance.

### 4.3. Chemical Composition

Our findings confirm that organic manure at 15 ton/ha transforms marginal sandy soil—boosting soil organic matter, porosity, and water-holding capacity—and drives improvements in essential oil output and leaf nutrient levels. The combined effects of enhanced cation exchange capacity, acidification-mediated nutrient solubilization [39], and heightened microbial activity yielded significant increases in leaf N (11.5%), P (12%), and K (10.5%). This soil foundation amplifies the efficacy of subsequent foliar biostimulants, forging a resilient and sustainable cultivation system. These outcomes echo the extensive literature on organic manure's agronomic benefits and support integrated nutrient management as a pathway to improved caraway performance and resource stewardship [40,41].

Foliar application of vitamin E (alpha-tocopherol) likewise fortifies the photosynthetic machinery against oxidative stress, boosting chlorophyll and carotenoid levels. Beyond membrane protection, tocopherol acts as a signaling molecule for light–heat homeostasis [18]. By neutralizing lipid peroxide radicals, it preserves thylakoid membrane integrity, fostering biomass build-up and steady photosynthetic activity; chlorophyll rose by about 25% and carotenoids by roughly 21%, versus the control. Applied alone, vitamin E also increased leaf nitrogen, phosphorus, and potassium by 28–55%, underscoring its capacity to enhance growth performance [19]. Previous studies have documented similar improvements in the chemical profiles of caraway and other crops [31–33]. When combined with 15 ton/ha of organic manure and sprayed at 100 mg $L^{-1}$, vitamin E produced marked gains, highlighting its role as an effective secondary biostimulant.

With respect to thiamine, foliar vitamin $B_1$ markedly enhanced caraway's nutritional and reproductive performance by acting as a coenzyme in carbohydrate metabolism and by activating antioxidant defenses under field conditions [16]. Similar effects have been reported in other medicinal species, where foliar thiamine alleviates drought and salinity stress via upregulated antioxidant enzymes and osmo-protectant accumulation [17,42]. Thiamine alone boosted leaf N, P, and K by 25–28%, indicating improved nutrient transporter activity in infertile soils [34,35]. When paired with 15 t/ha of organic manure and a 100 mg $L^{-1}$ thiamine spray, caraway nutrient uptake rose by 43% (N), 90% (P), and 76% (K), reflecting roughly two-fold gains in root-zone fertility and leaf metabolic capacity—findings consistent with other aromatic crops [10].

Active yeast benefits plants by supplying essential nutrients (amino acids, B vitamins, minerals) and hormones (auxins and cytokinins) that promote cell division and growth [20,22]. It enhances nutrient availability by solubilizing phosphate and producing protective polysaccharides. Applications lead to increased root growth, protein synthesis, and consistent improvements in pigment (13–26%) and leaf nutrient content (19–57%). Yeast also stimulates beneficial soil microbes and nutrient cycling activity [22,37,38], improving plant chemical composition (e.g., N, P, K, Ca, Mg, Fe, Zn, and Mn in caraway [31]; N/P in tomato/sugarcane [36]). Combined with manure (15 mL $L^{-1}$), it increased pigment/nutrients by 35–40%.

The significant synergistic effects between organic fertilizers and biostimulant treatments highlight the importance of integrated soil–plant management in marginal sandy environments. Similar synergistic effects have been recorded in aromatic grasses, where the combination of soil conditioners and foliar antioxidants resulted in higher yield and quality compared to individual inputs [34]. The simultaneous enhancement of plant growth, photosynthetic capacity, nutrient uptake, and lipid biosynthesis suggests that simultaneously enhancing root nutrient supply (via manure) and foliar metabolic regulation (via vitamins) can form a positive feedback loop that improves the overall performance of caraway.

## 5. Conclusions

This study conclusively demonstrated that combining 15 ton/ha of organic manure with 100 mg $L^{-1}$ vitamin $B_1$ foliar spray significantly improved caraway growth in sandy soil under drip irrigation. Key results included increases in vegetative growth (35% increase in plant height, 52% increase in branching, 52% increase in herb dry weight), yield (48% increase in seed yield per ha, 56% increase in umbel yield), oil yield (46% increase in essential oil concentration, 115% increase in oil yield per ha), and biochemical quality (50% increase in chlorophyll a, 43% increase in nitrogen, 90% increase in phosphorus, 76% increase in potassium). Vitamin $B_1$ outperformed vitamin E and active yeast in enhancing metabolic activity and alleviating stress. Therefore, we recommend these combined treatments to maximize yield and oil quality while reducing reliance on synthetic fertilizers, in line with sustainable agricultural practices. Future research should assess the economic feasibility and scalability of this approach across diverse arid regions, explore cost optimization through reduced input rates, elucidate the molecular mechanisms by which vitamin $B_1$ enhances nutrient uptake and oil synthesis, and evaluate these biostimulants on other high-value aromatic crops grown in marginal soils. Overall, our findings demonstrate that integrating organic manure with targeted foliar biostimulants—particularly vitamin $B_1$—under drip irrigation offers a sustainable strategy to optimize caraway yield, essential oil content, and nutritional quality on marginal sandy soils.

**Author Contributions:** Conceptualization, A.A.H., R.M.Y.Z., and A.A.B.; methodology, A.A.H., A.F.A.A.-R., and R.M.Y.Z.; software, A.F.A.A.-R., G.H.A.H., and R.M.Y.Z.; validation, A.A.H., W.K.A., and H.S.E.-D.; formal analysis, A.F.A.A.-R., A.A.B. and H.S.E.-D.; investigation, A.A.H., G.H.A.H., and H.S.E.-D.; resources, A.F.A.A.-R., W.K.A., and R.M.Y.Z.; data curation, A.A.H., A.A.B., and H.S.E.-D.; writing—original draft preparation, A.A.H., A.A.B., and H.S.E.-D.; writing—review and editing, A.F.A.A.-R., G.H.A.H., W.K.A., R.M.Y.Z., and A.A.B.; visualization, A.A.H., A.F.A.A.-R., and R.M.Y.Z.; supervision, A.F.A.A.-R., A.A.B., and H.S.E.-D.; project administration, A.A.H., A.F.A.A.-R., and R.M.Y.Z.; funding acquisition, A.A.H., A.F.A.A.-R., G.H.A.H., W.K.A., R.M.Y.Z., A.A.B., and H.S.E.-D. All authors have read and agreed to the published version of the manuscript.

**Funding:** This research received no external funding.

**Data Availability Statement:** The datasets generated and/or analyzed during the current study are available from the corresponding author upon reasonable request. To clarify, the study did not involve human or animal samples.

**Acknowledgments:** The authors would like to thank the faculty members who helped with advice, materials, or methods in producing this research in its final form.

**Conflicts of Interest:** The authors declare no conflicts of interest.

## References

1. Aćimović, M.G.; Oljača, S.I.; Tešević, V.V.; Todosijević, M.M.; Djisalov, J.N. Evaluation of Caraway Essential Oil from Different Production Areas of Serbia. *Hortic. Sci.* **2014**, *41*, 122–130. [CrossRef]
2. Ibrahim, S.M.; El-labban, H.M.; Mohamed, F.I.; Naga, N.M. Effect of Organic Manures and Chemical Fertilizers on *Foeniculum vulgare* Mill. and *Carum carvi* L. *Bull. Pharm. Sci. Assiut Univ.* **2006**, *29*, 187–201.
3. Hassan, E.A.; El-Gohary, A.E.; Radwan, E.M.A.; Nada, A.S. Response Caraway (*Carum carvi* L.) Plants to Humic Acid, Mycorrhizae Fungi and Azolla Extract. *New Val. J. Agric. Sci.* **2024**, *4*, 22–32.
4. Basak, N.; Rai, A.K.; Barman, A.; Mandal, S.; Sundha, P.; Bedwal, S.; Kumar, S.; Yadav, R.K.; Sharma, P.C. Salt Affected Soils: Global Perspectives. In *Soil Health and Environmental Sustainability: Application of Geospatial Technology*; Springer: Cham, Switzerland, 2022; pp. 107–129.
5. Omer, A.M.; Osman, M.S.; Ibrahim, S.M.; Badawy, A.A. Nanoparticle-Based Strategies for Enhancing Faba Bean (*Vicia faba* L.) Growth and Stress Tolerance in Saline Soils. *Biocatal. Agric. Biotechnol.* **2025**, *67*, 103630. [CrossRef]
6. Zhang, R.; Meng, T.; Sun, Z.; Liu, Z. Effects of Applying Different Organic and Inorganic Soil Amendments to Improve the Late Stage of Reclaimed Soil from Abandoned Homesteads on Soil Nutrients and Maize Yield. *Front. Sustain. Food Syst.* **2025**, *9*, 1542710. [CrossRef]
7. Badawy, A.A.; Alshammari, W.K.; Salem, N.F.G.; Alshammari, W.S.; Hussein, H.A. Arginine and Spermine Ameliorate Water Deficit Stress in Fenugreek (*Trigonella foenum-graecum* L.) by Enhancing Growth and Physio-Biochemical Processes. *Antioxidants* **2025**, *14*, 329. [CrossRef]
8. Sarwar, G.; Ibrahim, M.; Tahir, M.A.; Iftikhar, Y.; Haider, M.S.; Noor-Us-Sabah, N.-U.-S.; Han, K.-H.; Ha, S.-K.; Zhang, Y.-S. Effect of Compost and Gypsum Application on the Chemical Properties and Fertility Status of Saline-Sodic Soil. *Korean J. Soil Sci. Fertil.* **2011**, *44*, 510–516. [CrossRef]
9. Hashem, M.S.; Guo, W.; Qi, X.-B.; Li, P.; She, Y.; Cui, J.; Li, T. Effect of Using Reclaimed Water via Furrow and Subsurface Drip Systems under Alternate Partial Root-Zone Irrigation Mechanism on Crops Growth and Soil Properties. *Irrig. Sci.* **2024**, *42*, 919–936. [CrossRef]

10. Farahat, M.M.; Ibrahim, M.M.S.; Taha, L.S.; El-Quesni, E.M.F. Response of Vegetative Growth and Some Chemical Constituents of *Cupressus sempevirens* L. to Foliar Application of Ascorbic Acid and Zinc at Nubaria. *World J. Agric. Sci.* **2007**, *3*, 496–502.
11. El-Banna, H.; Fouda, K. Effect of Mineral, Organic, Biofertilizers and Humic Acid on Vegetative Growth and Fruit Yield Quality of Caraway Plants (*Carum carvi* L.). *J. Soil Sci. Agric. Eng.* **2018**, *9*, 237–241. [CrossRef]
12. Khater, R.M.; Abd-Allah, W.H.; El Shafay, R. Effect of Organic Fertilization and Spraying *Aloe vera* Extract on the Growth and Productivity of *Carum carvi* L. Plant Under Shalateen Conditions in Egypt. *Plant Arch.* **2020**, *20*, 4959–4971.
13. Sawsan, A.S.E.-Y.; Essam, G.S. Effect of Applied Organic Manure and Foliar Spray with Antioxidants on Growth, Yield and Chemical Constituents of Caraway Plant. *Fayoum J. Agric. Res. Dev.* **2008**, *22*, 42–59. [CrossRef]
14. Zatimeh, A.A.; Al-Fraihat, A.H. Influence of Tryptophan and Glycine Acids on Growth and Productivity of Fennel under Different Organic Fertilizer Levels. *Int. J. Hortic. Sci. Technol.* **2025**, *12*, 207–218. [CrossRef]
15. Hendawy, S.F.; Ezz El-Din, A.A. Growth and Yield of *Foeniculum vulgare* Var. Azoricum as Influenced by Some Vitamins and Amino Acids. *Ozean J. Appl. Sci.* **2010**, *3*, 113–123.
16. Aly, M.; Ahmed, E.; Mohamed, M.; Kasem, M. Response of Anise Plants To Humic Acid, Amino Acids and Thiamine Treatments. *Sci. J. Flowers Ornam. Plants* **2022**, *9*, 153–165. [CrossRef]
17. Tuna, A.L.; Kaya, C.; Altunlu, H.; Ashraf, M. Mitigation Effects of Non-Enzymatic Antioxidants in Maize (*Zea mays* L.) Plants under Salinity Stress. *Aust. J. Crop Sci.* **2013**, *7*, 1181–1188.
18. Arrom, L.; Munné-Bosch, S. Tocopherol Composition in Flower Organs of Lilium and Its Variations during Natural and Artificial Senescence. *Plant Sci.* **2010**, *179*, 289–295. [CrossRef]
19. Orabi, S.A.; Abdelhamid, M.T. Protective Role of α-Tocopherol on Two *Vicia faba* Cultivars against Seawater-Induced Lipid Peroxidation by Enhancing Capacity of Anti-Oxidative System. *J. Saudi Soc. Agric. Sci.* **2016**, *15*, 145–154. [CrossRef]
20. Hammad, S.A.R.R.; Ali, O.A.M.M. Physiological and Biochemical Studies on Drought Tolerance of Wheat Plants by Application of Amino Acids and Yeast Extract. *Ann. Agric. Sci.* **2014**, *59*, 133–145. [CrossRef]
21. El-Serafy, R. Growth and Productivity of Roselle (*Hibiscus sabdariffa* L.) As Affected By Yeast and Humic Acid. *Sci. J. Flowers Ornam. Plants* **2018**, *5*, 195–203. [CrossRef]
22. Kowalska, J.; Krzymińska, J.; Tyburski, J. Yeasts as a Potential Biological Agent in Plant Disease Protection and Yield Improvement—A Short Review. *Agric.* **2022**, *12*, 1404. [CrossRef]
23. Jackson, M.L. *Soil Chemical Analysis: Advanced Course*; UW-Madison Libraries Parallel Press: Madison, WI, USA, 2005.
24. Yost, J.L.; Hartemink, A.E. Soil Organic Carbon in Sandy Soils: A Review. *Adv. Agron.* **2019**, *158*, 217–310.
25. Black, C.A. *Methods of Soil Analysis-Part 2*; ASA Pub: Madison, WI, USA, 1965; pp. 771–1572.
26. *Egyptian Pharmacopoeia*; General Organization for Governmental Printing Office. Ministry of Health, Egypt: Cairo, Egypt, 1984; pp. 31–33.
27. Moran, R. Formulae for Determination of Chlorophyllous Pigments Extracted with N, N-Dimethylformamide. *Plant Physiol.* **1982**, *69*, 1376–1381. [CrossRef]
28. Sparks, D.L.; Page, A.L.; Helmke, P.A.; Loeppert, R.H. *Methods of Soil Analysis, Part 3: Chemical Methods*; ACSESS: Madison, WI, USA, 2020.
29. Ibrahim, E.M.; Badawy, A.A.; Ghanem, N.A.E. The Resilience of *Ephedra alata* Decne in Two Regions of Egypt: Soil Properties, Tissues Anatomy, and Physiological Responses. *Egypt. J. Soil Sci.* **2025**, *65*, 1331–1344. [CrossRef]
30. Freed, R.; Eisensmith, S.P.; Everson, E.; Weber, M.; Paul, E.; Isleib, E. MSTAT-C: A Microcomputer Program for the Design, Management, and Analysis of Agronomic Research Experiments. *East Lansing, MI, USA Michigan State Univ.* **1991**.
31. Medani, R.A.; Taha, R.S. Improving Growth and Yield of Caraway (*Carum carvi* L.), Plants by Decapitation and/or Active Dry Yeast Application. *Int. J. Curr. Microbiol. Appl. Sci.* **2015**, *4*, 47–60.
32. Ayyat, A.; Kenawy, A.G.M.; Aboel-Ainin, M.A.; Abdel-Mola, M.A.M. Improving Growth, Productivity and Oil Yield of *Nigella sativa* L. Plants by Foliar Spraying with Some Stimulants. *J. Plant Prod.* **2021**, *12*, 339–344. [CrossRef]
33. Abdou, M.A.H.; Taha, R.A.; Abdel Gaber, A.F.M. Effect of Alpha Tocopherol and Moringa Leaf Extract on Volatile Oil and Certain Chemical Components of Marjoram Grown under Salinity Stress. *Sci. J. Agric. Sci.* **2024**, *6*, 73–87. [CrossRef]
34. Elgohary, M.A.M.; Dawh, A.K.; Gendy, A.S.H.; Abdelkader, M.A.I. Response of Caraway (*Carium carvi* L.) Growth and Yield Components to NPKMg Fertilizer Rate and Antioxidants Type. *Zagazig J. Agric. Res.* **2020**, *47*, 883–894. [CrossRef]
35. Eisa, E. Effect of Some Different Sourses of Organic Fertilizers and Seaweed Extract on Growth and Essential Oil of Sweet Fennel (*Foeniculum vulgare* Mill.) Plants. *J. Plant Prod.* **2016**, *7*, 575–584. [CrossRef]
36. Lonhienne, T.; Mason, M.G.; Ragan, M.A.; Hugenholtz, P.; Schmidt, S.; Paungfoo-lonhienne, C. Yeast as a Biofertilizer Alters Plant Growth and Morphology. *Crop Sci.* **2014**, *54*, 785–790. [CrossRef]
37. Zewail, R.M.Y.; Khedr, Z.M.A.; Ismael, F.H.M.; Abd Elhameed, E.F. Enhancement of Soya Bean (*Glycine max* L.) Plants Growth, Yield and Seed Quality by Using Putrescine, Benzyladenine and Yeast Extract. *Ann. Agric. Sci. Moshtohor* **2019**, *57*, 493–506. [CrossRef]

38. Taiwo, L.B.; Adediran, J.A.; Sonubi, O.A. Yield and Quality of Tomato Grown with Organic and Synthetic Fertilizers. *Int. J. Veg. Sci.* **2007**, *13*, 5–19. [CrossRef]
39. Aćimović, M.G. The Influence of Fertilization on Yield of Caraway, Anise and Coriander in Organic Agriculture. *J. Agric. Sci. Belgrade* **2013**, *58*, 85–94. [CrossRef]
40. El-Tarawy, M.; Hegazi, M.; Mahmoud, E. Effect of Bio, Organic and Chemical Fertilization on Growth, Productivity and Oil Consituents of Caraway (*Carum carvi* L.). *J. Plant Prod.* **2017**, *8*, 993–997. [CrossRef]
41. Awad, M. Poultry Manure and Humic Acid Foliar Applications Impact on Caraway Plants Grown on a Clay Loam. *J. Soil Sci. Agric. Eng.* **2016**, *7*, 1–10. [CrossRef]
42. Abdou, M.; El-Sayed, A.; Taha, R.; Ahmed, S.; El-Nady, M. Response of Cumin Plant To Some Organic, Biofertilization and Antioxidant Treatments Ii. Essential Oil Production and Chemical Constituents. *Sci. J. Flowers Ornam. Plants* **2019**, *6*, 89–97. [CrossRef]

*horticulturae*

MDPI

*Article*

# Physiological, Productive, and Nutritional Performance of Tomato Plants Treated with Iron and Zinc Nanoparticles via Foliar Application Under Deficit Irrigation

**Erika Caminha Almeida [1], Francisco Hevilásio Freire Pereira [1], Kaiki Nogueira Ferreira [1], Antonio Carlos de Sena Rodrigues [1], Railene Hérica Carlos Rocha Araújo [1], José Ebson Janoca de Souza [1], Carlos Sávio Gomes Ramos [1], Guilherme Lopes [2], Leônidas Canuto dos Santos [2], Francisco Bezerra Neto [3], Francisco Vaniés da Silva Sá [4], José Zilton Lopes Santos [5], Ronaldo do Nascimento [1] and Josinaldo Lopes Araujo Rocha [1,*]**

1 Center for Technology and Natural Resources, Federal University of Campina Grande, Campina Grande 58429-900, PB, Brazil; erikacaminha2017@gmail.com (E.C.A.); francisco.hevilasio@professor.ufcg.edu.br (F.H.F.P.); kaicknf95@gmail.com (K.N.F.); rantoniocarlosdesena@gmail.com (A.C.d.S.R.); railene.herica@professor.ufcg.edu.br (R.H.C.R.A.); ebsonsoares2016@gmail.com (J.E.J.d.S.); savioramos.515@gmail.com (C.S.G.R.); ronaldon453@gmail.com (R.d.N.)

2 Department of Soil Science, Federal University of Lavras, Lavras 37200-900, MG, Brazil; guilherme.lopes@ufla.br (G.L.); leonidas.santos2@estudante.ufla.br (L.C.d.S.)

3 Agricultural Sciences Center, Federal Rural University of Semi-Arid, Mossoró 59625-900, RN, Brazil; bezerra@ufersa.edu.br

4 Agricultural Sciences Center, State University of Paraíba, Catolé do Rocha 58884-000, PB, Brazil; vanies@ufersa.edu.br

5 Faculty of Agricultural Science, University of Amazonas, Manaus 69067-005, AM, Brazil; ziltonlopes@ufam.edu.br

* Correspondence: josinaldo.lopes@professor.ufcg.edu.br

## Abstract

Water deficit in the semi-arid region of Brazil is a critical limiting factor for tomato (*Solanum lycopersicum* Mill.), plant development and productivity. We evaluated whether foliar zinc (ZnO NPs) and iron ($Fe_2O_3$NPs) nano-oxides and their conventional salts ($ZnSO_4 \cdot 7H_2O$ and $FeSO_4 \cdot 7H_2O$) mitigate water deficit effects on tomato (hybrid HM 2798). A split-plot field experiment was conducted with two irrigation levels (50% and 100% ETc) and five foliar treatments: control (no application), $FeSO_4 \cdot 7H_2O$ (T1), $Fe_2O_3$NPs (T2), ZnONPs (T3), $ZnSO_4 \cdot 7H_2O$ (T4), with four replications, totaling 40 experimental plots (2 irrigation levels × 5 foliar treatments × 4 replicates). The water deficit significantly reduced the leaf area index, photosynthetic rate, membrane stability, calcium and boron contents in fruits, and total and marketable yield. Foliar application of iron and zinc nano-oxides and their conventional sources had a limited effect on tomato plant growth but increased the photosynthetic rate under both irrigation levels. Under full irrigation, $ZnSO_4 \cdot 7H_2O$ increased total fruit production by 61% and fruit Zn content by 18.1%. In turn, $Fe_2O_3$ NPs (T2) led increases in fruit iron content by 117.3% under water deficit and 135.2% under full irrigation. Foliar application of Fe as $Fe_2O_3$ NPs is promising to promote the biofortification of tomato fruits with this micronutrient, especially in regions with deficiency problems of this micronutrient.

**Keywords:** *Solanum lycopersicon*; drought stress; fruit quality; gas exchange; mineral nutrients

## 1. Introduction

Tomato (*Solanum lycopersicum*) is one of the most widely cultivated and consumed vegetables globally, notable for its nutritional value, richness in lycopene, vitamins, and antioxidants, which confer human health benefits [1,2]. In Brazil, tomato ranks among the top vegetables in economic value and is widely consumed fresh in salads, as sauces, and processed into pulps and extracts [3]. In the semi-arid Northeast region of Brazil, where edaphoclimatic conditions favor the cultivation of this vegetable, water scarcity is a major factor contributing to reduced yield and commercial quality [4].

Water scarcity is one of the greatest global challenges of the 21st century, exacerbated by climate change, population growth, and unsustainable water resource use, especially in countries like Brazil, where agriculture accounts for about 70% of freshwater consumption [5]. Reduced water availability directly compromises plant productivity, particularly in sensitive crops like tomato [6,7].

Water deficit significantly compromises gas exchange in tomato plants, reducing the photosynthetic rate due to stomatal closure and decreased leaf conductance, which limits $CO_2$ assimilation [8,9]. These physiological alterations inhibit vegetative growth, resulting in leaf area and biomass decreases and also affect the synthesis of photosynthetic pigments such as chlorophylls and carotenoids, essential for plant energy efficiency [10]. In fruits, water stress reduces the translocation of photoassimilates and mineral nutrients (such as calcium, potassium, magnesium, iron, and zinc) to the fruits and negatively impacts productivity, reducing the number and size of fruits per plant [11,12].

The decrease in mineral nutrients in fruits also occurs due to reduced diffusion and mass flow processes, which contribute significantly to the quantity of nutrients reaching the roots and their subsequent redistribution to the fruits [13]. These responses highlight the need for water management strategies to mitigate losses under water scarcity conditions.

Foliar application of zinc oxide nanoparticles (ZnONPs) and iron oxide nanoparticles ($Fe_2O_3$NPs) emerges as a promising strategy to mitigate the effects of water deficit in horticultural crops, as demonstrated by recent advances in agricultural nanotechnology [14–17]. These nanomaterials have been considered better sources of iron and zinc than conventional sources of these elements due to their high specific surface area and unique physicochemical properties [18–21]. On the other hand, advances in knowledge regarding the absorption mode and physiological effects of nanoparticles have been made [21,22]. The primary mode of action of ZnONPs and $Fe_2O_3$NPs would be their stimulatory effect on plant metabolism through interaction with biomolecules, promoting increased enzymatic activity and the synthesis of compounds involved in combating plant oxidative stress [20–23].

In this sense, NPs appear to modify antioxidant enzyme activity, elevating the activity of antioxidant enzymes like superoxide dismutase and catalase [19,20] while simultaneously inducing the synthesis of osmoprotectants that maintain cellular osmotic pressure under drought conditions [23,24]. NPs containing iron or zinc can also act as efficient sources of Fe and Zn for plants, with slower release than conventional sources like zinc and iron sulfate [25], increasing the bioavailability of these elements, which are important plant and human micronutrients [26,27].

Zinc and iron are vital micronutrients in plants [12]. Zinc participates in the synthesis of various hormones, such as gibberellin, auxin, cytokinin, and abscisic acid, and is involved in chlorophyll production, chloroplast development, and cellular membrane integrity, providing protection, stability, and structure [12,28]. These processes, in turn, are associated with water stress tolerance mechanisms and plant growth and development, with implications for fruit productivity and quality [29,30]. Iron, in turn, acts as an enzymatic cofactor in chlorophyll biosynthesis and chloroplast maintenance [12], is a central component of electron transport chains (cytochromes and ferredoxin), crucial for the photochemical phase of photosynthesis [31], and regulates the activity of antioxidant enzymes (like catalase and peroxidase), protecting photosynthetic pigments from oxidative stress [12], with implications for fruit productivity [32]. Previous studies have investigated the use of iron oxide (Fe) and zinc oxide (Zn) nanoparticles to mitigate water deficit in various crops, including fruit vegetables such as tomato. Nevertheless, the comparison between these nanoscale sources and conventional sources remains poorly explored, particularly under field conditions in a semi-arid tropical climate.

The hypothesis of the present study is that foliar application of zinc oxide (ZnO) and iron oxide ($Fe_2O_3$) nanoparticles (NPs) is more effective than conventional sources (salts) in mitigating the negative effects of deficit irrigation on photosynthesis and crop yield. Therefore, this study aims to (i) compare the efficacy of ZnO and $Fe_2O_3$ NPs with their ionic salt counterparts; (ii) quantify their effects on plant physiological parameters and fruit mineral content; and (iii) evaluate potential interactions between the nutrient source and irrigation regime.

## 2. Materials and Methods

### 2.1. Characterization of the Experimental Area

The experiment was conducted under field conditions from October 2023 to February 2024, at the Experimental Farm of the Center for Food Science and Technology (Centro de Ciências e Tecnologia Agroalimentar) of the Federal University of Campina Grande (UFCG), Pombal Campus, Paraíba State (PB). The Experimental Farm is located in the municipality of São Domingos, approximately 30 km from Pombal (PB). The municipality is situated in the Western region of Paraíba State, within the Sertão Paraibano mesoregion and the Sousa microregion. According to the Köppen classification, the predominant climate is type Aw, characterized as hot and humid. Annual precipitation averages around 800 mm, and the thermal amplitude is always less than 5 °C. The Experimental Farm is situated at an altitude of 190 m, at the following coordinates: Latitude −6.83444; longitude −37.88583°. According to the Gaussen classification, the prevailing bioclimatic type is Mediterranean, also known as medium-dry northeastern, featuring a dry season lasting between 4 and 6 months. Area preparation (Figure A1) consisted of one plowing followed by one harrowing using a plow harrow. After this stage, soil sampling was performed in the area, within the 0–20 cm layer, for chemical and physical characterization (Table 1), following the methodology described in the Manual of Soil Analysis Methods [33]. The climatic conditions recorded from 10 October 2023 to 3 February 2024 were as follows: total rainfall of 51.2 mm; monthly average temperatures of 23.0 °C (minimum), 35.8 °C (maximum), and 29.9 °C (mean); and a mean and minimum relative humidity of 55.5% and 28.5%, respectively [34].

**Table 1.** Chemical and physical attributes of the soil used in the experiment.

| Chemical | Value | Physical | Value |
|---|---|---|---|
| pH ($CaCl_2$) | 6.45 | Sand (g kg$^{-1}$) | 389 |
| OM (g kg$^{-1}$) | 11.2 | Silt (g kg$^{-1}$) | 430 |
| P (mg dm$^{-3}$) | 180.1 | Clay (g kg$^{-1}$) | 181 |
| $K^+$ (mg dm$^{-3}$) | 256.8 | BD (g cm$^{-3}$) | 1.30 |
| $Ca^{++}$ (cmol$_c$ dm$^{-3}$) | 7.72 | PD (g cm$^{-3}$) | 2.59 |
| $Mg^{++}$ (cmol$_c$ dm$^{-3}$) | 2.83 | TP (m$^3$ m$^{-3}$) | 0.47 |
| $Na^+$ (cmol$_c$ dm$^{-3}$) | 0.06 | FC (%) | 12.87 |
| $H^+$ $Al^{+++}$ (cmol$_c$ dm$^{-3}$) | 1.35 | PWP (%) | 5.29 |
| Fe (mg dm$^{-3}$) | 18.0 | AWC (%) | 7.58 |
| Zn (mg dm$^{-3}$) | 1.69 | - | - |

P, $K^+$, $Na^+$, Fe, and Zn—Mehlich 1 extractant, $H^+$ + $Al^{+3}$—0.5 mol L$^{-1}$ calcium acetate extractant at pH 7, $Ca^{+2}$, $Mg^{+2}$—1 mol L$^{-1}$ KCl extractant, OM—organic matter, BD—bulk density, PD—soil particle density, TP—total porosity, FC—field capacity, PWP—permanent wilting point, and AWC—available water content.

During the experiment, weather data for temperature (°C), relative humidity (%), and precipitation (mm) were collected via the Agritempo website [34] (Figure 1).

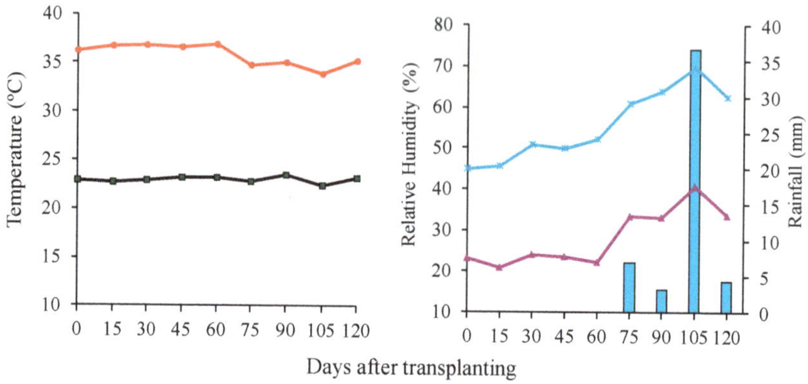


**Figure 1.** Climatological data on maximum (T max) and minimum (T min) temperature, and maximum (RH max) and minimum (RH min) relative humidity and rainfall during the experimental period in the field [34].

## 2.2. *Treatments and Experimental Design*

The experiment was arranged in a randomized block design with split plots in space. The main plots consisted of two irrigation levels, denoted as ID (50% and 100% of crop evapotranspiration—ETc), while the subplots comprised four treatments involving foliar application of iron and zinc sources, termed water deficit alleviators (WDAs). The WDAs treatments were as follows: ferrous sulfate ($FeSO_4{\cdot}7H_2O$ = T1), iron oxide nanoparticles ($Fe_2O_3NP$ = T2), zinc oxide nanoparticles (ZnONP = T3), zinc sulfate ($ZnSO_4{\cdot}7H_2O$ = T4), and a control (C). The design included four blocks, totaling 40 experimental plots.

The total experimental area measured 40.8 m in length by 18.0 m in width. Plants were spaced at 1.2 m between rows and 0.7 m within rows. Each plot contained four rows (3.6 m) with four plants (2.8 m), totaling 16 plants per plot. The net plot area (2.8 m × 1.2 m) in the center of each subplot included six plants for evaluation. Subplots were separated by 1.5 m, and blocks by 2 m, resulting in a total area of 734.4 $m^2$ (Figure S1, Supplementary Materials).

## 2.3. *Seedling Transplanting*

Tomato seedlings (Solanum lycopersicum Mill.), hybrid HM 2798 (Harris Moran Seed Company Co., Ltd., Campinas, Brazil), were produced in expanded polystyrene trays (Isopor®), using a sterilized substrate composed of soil, sand, and cattle manure in a 1:1:1 ratio. Transplanting was carried out when the seedlings had four to six true leaves. The seedlings were planted on ridges measuring 0.4 m in width and 0.30 m in height. Based on the soil analysis (Table 1), basal and top-dressing fertilization was performed according to the recommendations of the Fertilizer Recommendation Manual for the State of Pernambuco [35]. For basal fertilization, the following were applied per planting row (four plants) in each subplot: 134 g (398.8 kg $ha^{-1}$) of NPK fertilizer (10% of N, 10% of $P_2O_5$ and 10% of $K_2O$), 35 g (104.2 kg $ha^{-1}$) of potassium chloride (58% $K_2O$), and 93 g (276.8 kg $ha^{-1}$) of single superphosphate (18% $P_2O_5$). These fertilizers were incorporated into the soil during ridge formation. Additional nitrogen and potassium fertilization was applied via fertigation, consisting of three split applications of 30 g (89.3 kg $ha^{-1}$) urea (45% N) and 12 g (35.7 kg $ha^{-1}$) potassium chloride (58% $K_2O$), applied at 25, 50, and 75 days after transplanting (DAT). All fertilizers used were obtained from Heringer® Fertilizers, Rio Grande, Brazil.

### 2.4. Application of Treatments with Iron and Zinc Sources

As the ZnONP source, zinc oxide (ZnO) nanoparticles (544906-10G, batch number MKCP6471, CAS-No 1314-13-2) from Sigma Aldrich® (Sigma-Aldrich® Co., Ltd., Darmstadt, Germany) were used, with 97% purity, particle sizes < 100 nm, and a specific surface area of 10.8 $m^2$ $g^{-1}$. Suspensions of this product were prepared at a concentration of 0.25 g $L^{-1}$, equivalent to 200 mg of Zn $L^{-1}$. As the $Fe_2O_3$NP source, iron oxide ($Fe_2O_3$) nanoparticles (544884-5G, batch number MKCS3534, CAS-No 1309-37-1) from Sigma-Aldrich® (Sigma-Aldrich® Co., Ltd., Darmstadt, Germany) were used, with 98% purity, particle dimensions < 50 nm, and an average specific surface area of 50–245 $m^2$ $g^{-1}$. Suspensions of this product were prepared at a concentration of 0.30 g $L^{-1}$, equivalent to 200 mg of Fe $L^{-1}$. The definition of the dose of 200 mg $L^{-1}$ of Zn or Fe as NP was based on Martins et al. [11] and Jafari et al. [15]. For conventional zinc and iron sources, the reagents $ZnSO_4 \cdot 7H_2O$ (Vetec Fine chemistry Co., Ltd., Duque de Caxias, Brazil)) and $FeSO_4 \cdot 7H_2O$ (Vetec Fine chemistry Co., Ltd., Duque de Caxias, Brazil)) were employed, respectively, both at a concentration of 4.5 g $L^{-1}$. This concentration was established such that the Fe and Zn concentrations (approximately 1 g $L^{-1}$) from conventional sources were approximately five times greater than their nanoparticle counterparts. This approach accounted for both the anticipated higher efficiency and the greater purchase price of the nanoparticle forms. All products/treatments were applied only once, 20 days after transplanting the seedlings. As a solvent for the preparation of nanoparticle suspensions, only distilled water was used. Solutions were applied at a volume of 300 L $ha^{-1}$, corresponding to approximately 20 mL per plant, using a backpack hand sprayer. To optimize product adhesion to plants, a neutral detergent was used as a spreader–sticker agent at a concentration of 1 mL $L^{-1}$. The suspensions were prepared moments before application, without the need for storage. In each plot, applications were carried out on each plant individually, which were properly isolated with four polystyrene plates, each measuring 1.0 m × 1.0 m. The applications were carried out in the late afternoon (between 4 pm and 6 pm).

### 2.5. Irrigation Depth Application

After transplanting, plants were drip irrigated using emitters spaced at 0.20 m intervals with a nominal flow rate of 1.6 L $h^{-1}$. Following seedling emergence and standardization of plant count and size within subplots, irrigation was applied according to different water depth levels starting 32 days after transplanting (DAT).

The total irrigation requirement (TIR) was calculated using the following expression:

$$\text{TIR} = \frac{(\text{Cc} - \text{Pm}) \times \text{Z} \times \text{BD} \times \text{f}}{10}$$

where

TIR (total irrigation requirement)—Corresponds to the total initial water depth applied, in mm;

FC—Soil moisture at field capacity, %;

PWP—Soil moisture at permanent wilting point, %;

Z—Effective root zone depth (30 cm);

BD—Soil bulk density, g $cm^{-3}$;

f—Water depletion factor (0.5).

Water application uniformity tests were conducted using the Christiansen Uniformity Coefficient (CUC) methodology proposed by [36].

The daily water volume for each irrigation depth was controlled at a standardized time, based on the ratio of emitter flow rate to the time required to meet reference evapotranspiration (ETc) proportions. As the scheduled time interval for each depth was reached, the corresponding drip tapes were sequentially shut off. The 100% ETc irrigation depth was calculated using the following expression [37]:

$$ETc = Kc \times ETo$$

where

ETc—Crop evapotranspiration (mm $day^{-1}$);

ETo—Reference evapotranspiration (mm $day^{-1}$);

Kc—Crop coefficient (dimensionless).

The Kc values for each crop growth stage and daily ETo data were obtained using the FAO Penman–Monteith model [38]. Meteorological data during the experiment were collected from the automated weather station in São Gonçalo, Paraíba (nearest to the experimental site) via the Agritempo platform [34].

Daily irrigation depths were applied based on irrigation time, calculated using system and crop characteristics as per the following expression:

$$Ti = \frac{Eto \times Kc \times A}{Ea \times n \times q}$$

where

Ti—Irrigation time (hours);

ETo—Reference evapotranspiration (mm $day^{-1}$);

Kc—Crop coefficient (dimensionless);

A—Area occupied per plant ($m^2$);

n—Number of emitters per plant;

q—Emitter flow rate (L $h^{-1}$);

Ea—Application efficiency (0.90).

### 2.6. *Cultural Practices and Phytosanitary Control*

Weed control was carried out mechanically, either manually or using hand hoes. Pest and disease control, when necessary, was performed using available natural products, such as neem extract, at a dose of 200 mL $L^{-1}$. For vertical staking, stakes measuring approximately 1.8 m in length were used, alternating between treatments. Following this, string was run from one end to the other of each row, tied horizontally to the stakes to help support the plants during their growth and fruiting.

### 2.7. *Evaluated Variables*

#### 2.7.1. Plant Growth

At the pre-flowering stage, the leaf area index (LAI), plant height, and stem diameter of plants within the useful plot area were evaluated. Plant height and stem diameter were measured 2 cm above the ground. LAI values were obtained using a photosynthetically active radiation (PAR) meter (AccuPAR model LP-80, Decagon Devices, Inc., Pullman, WA, USA). Four readings were taken in each useful plot (net plot area) on newly mature leaves in the central position of the plant.

#### 2.7.2. Gas Exchange and Photosynthetic Pigments

Concurrent with the plant growth evaluation, measurements of gas exchange, pigment concentration, and cell membrane integrity were performed. Gas exchange (net photosynthesis, stomatal conductance, transpiration, and intercellular $CO_2$ concentration) was measured using an infrared gas analyzer (Infra-red Gas Analyzer, Li-Cor 6400, LCpro Analytical Development, Kings Lynn, UK), with a constant light source of 1200 μmol photons $m^{-2}$ $s^{-1}$ and a $CO_2$ concentration of 370 μmol $mol^{-1}$, a leaf temperature of 32 ± 1 °C, and water vapor around 20 ± 1 mmol mol$^{-1}$. Measurements were carried out on newly mature leaves in the central position of the plant.

The evaluations occurred between 07:00 and 9:00 a.m., determining the $CO_2$ assimilation rate—A (μmol $CO_2$ m$^{-2}$ s$^{-1}$), transpiration—E (mmol $H_2O$ m$^{-2}$ s$^{-1}$), stomatal conductance—gs (mol $H_2O$ m$^{-2}$ s$^{-1}$), and internal $CO_2$ concentration—Ci (μmol $CO_2$ mol$^{-1}$). Water use efficiency (WUE) was estimated by the A/E ratio, while the A/Gs ratio was used to estimate intrinsic water use efficiency.

The contents of chlorophyll a (chl a), chlorophyll b (chl b), and carotenoid pigments (carotenoids and chlorophylls) were determined according to the methodology described by Lichtenthaler [39]. Pigments were extracted in pure, chilled acetone, subsequently filtered through 0.45 μm filter paper, and quantified by UV/VIS spectrophotometry (KASVI, K37, KASV Co., Ltd., São José dos Pinhais, Brazil), according to the expressions of Lichtenthaler [39]:

$$\text{chl a}(\text{mg/g}) = \frac{(12.25 \times \text{Abs 663 nm} - 1.79 \times \text{Abs 645 nm}) \times \text{V}}{1000 \times \text{FW}}$$

$$\text{chl b}(\text{mg/g}) = \frac{(21.5 \times \text{Abs 645 nm} - 5.1 \times \text{Abs 643 nm}) \times \text{V}}{1000 \times \text{FW}}$$

$$\text{chl a} + \text{b}(\text{mg/g}) = \frac{(7.15 \times \text{Abs 643 nm} - 18.71 \times \text{Abs 645 nm}) \times \text{V}}{1000 \times \text{FW}}$$

$$\text{Car}(\text{mg/g}) = \frac{(1000 \times \text{Abs 470 nm} - 1.82 \times \text{chl a} - 85.02 \times \text{chl b}) \times \text{V}}{198 \times 1000 \times \text{FW}}$$

where FW and V correspond to the values of leaf fresh mass and extract volume, respectively. Membrane integrity was assessed by estimating electrolyte leakage (Ext) using the methodology suggested by Campos and Thi [40]. For this, eight leaf disks with an area of 113 mm$^2$ each were obtained from plants within the useful plot area of each experimental plot, using a copper hole punch. These disks were washed and placed in Petri dishes containing 20 mL of deionized water. The dishes were sealed and maintained at 25 °C for 90 min. Immediately afterwards, the initial electrical conductivity of the medium (Ci) was measured using a bench conductivity meter (MB11, MS Techonopon®). Subsequently, the dishes containing the leaf disks were subjected to 80 °C for 90 min in a drying oven (SL100/336, SOLAB® Co., Ltd., Piracicaba, Brazil). After cooling, the final electrical conductivity (Cf) was measured. Electrolyte leakage was expressed as a percentage of the initial electrical conductivity relative to the final electrical conductivity, using the expression below, as described in Juzon-Sikora et al. [41].

$$\text{Ext}(\%) = \frac{\text{Ci}}{\text{Cf}} \times 100$$

2.7.3. Fruit Yield

During the fruit production phase, two harvests were carried out throughout the experiment, both performed early in the morning, manually, with the aid of bags and cardboard boxes for better storage and transport of the fruits. All fruits at the ripening stage were harvested from each useful plot. The fruits were placed in bags, all identified with their corresponding plot and treatment.

After harvest, the fruits were immediately transported to the Food Analysis and Food Chemistry and Biochemistry Laboratory of CCTA/UFCG. They were then sorted into marketable and non-marketable fruits. Fruits were considered marketable if they had a diameter equal to or greater than 35 mm and were free of any abnormalities or diseases that could compromise their structure and flavor [42]. The following were evaluated: total number of fruits per plant, number of marketable fruits per plant, longitudinal diameter, yield of non-marketable fruits, yield of marketable fruits, and total yield.

2.7.4. Mineral Nutrient Content in Fruits

Six marketable fruits from each subplot were separated to obtain a composite sample of 100 g of fresh fruit mass. These samples were dried in a forced-air oven (65–70 °C) until constant weight was achieved to subsequently determine dry mass and water content. The dried material was then ground in a Wiley-type mill, followed by sulfuric acid digestion according to the methodology described by Tedesco et al. [43]. Digestion was performed in a Kjeldahl microdigestion block (Marconi Co., Ltd., Londrina, Brazil)using 0.2 g samples of dried plant material in 3 mL of concentrated sulfuric acid and 1 mL of hydrogen peroxide, with an initial temperature of 180 °C and a final temperature of 360 °C. In the sulfuric acid digestate, the contents of the macronutrients calcium (Ca), potassium (K), phosphorus (P), and magnesium (Mg), and the micronutrients boron (B), manganese (Mn), iron (Fe), and zinc (Zn), were determined using ICP-OES (Inductively Coupled Plasma Optical Emission SpectrometryThermo Scientific Co., Ltd., Bremen, Germany) at the Mineral Nutrition Laboratory of the Soil Science Department at the Federal University of Lavras (UFLA). The element contents obtained on a dry mass basis were converted to fresh mass basis contents by correcting the values for the fruit water content.

2.7.5. Radar Chart

A radar chart was created using the variables most impacted by the treatments mitigating water deficit (marketable yield, total yield, photosynthetic pigments, photosynthetic rate, transpiration rate, stomatal conductance, and fruit mineral content) to illustrate the magnitude of the positive (increase) and negative (decrease) effects, expressed as relative values compared to the control treatment, under both irrigation levels. The following expression was used:

$$\text{RVi} = \frac{\text{ViTn}}{\text{ViC}} \times 100$$

where

RVi = Relative value of variable I;

ViTn = Absolute value of variable i obtained in treatment Tn (T1, T2, T3 or T4);

ViC = Absolute value of variable i in the control treatment (C).

### *2.8. Statistical Analyses*

The data were initially subjected to the Shapiro–Wilk normality test [44]. The obtained data were subjected to analysis of variance (see Supplementary Materials in Tables S1–S5), followed by comparison of means (see Supplementary Materials in Tables S6–S10) using Tukey's test ($p \leq 0.05$), as well as a Pearson's linear correlation analysis ($p \leq 0.05$). These statistical analyses were performed using statistical R language version 4.4.1 [45].

## 3. Results

### *3.1. Plant Growth*

Water deficit (50% ETc) reduced the LAI (leaf area index) by 12.6% ($-0.36$ m$^2$ m$^{-2}$) compared to the full irrigation (100% ETc) (Figure 2a). Conversely, plant height (Figure 2b) and stem diameter (Figure 2c) were unaffected by irrigation depths or WDA treatments, except for the stem diameter under 50% ETc, which was influenced by WDA treatments. Under these conditions (50% ETc), foliar application of $ZnSO_4{\cdot}7H_2O$ increased the stem diameter by 25.7% (+5.41 mm) relative to the control (15.65 mm), though with statistically similar effects to other treatments (T1, T2, and T3).

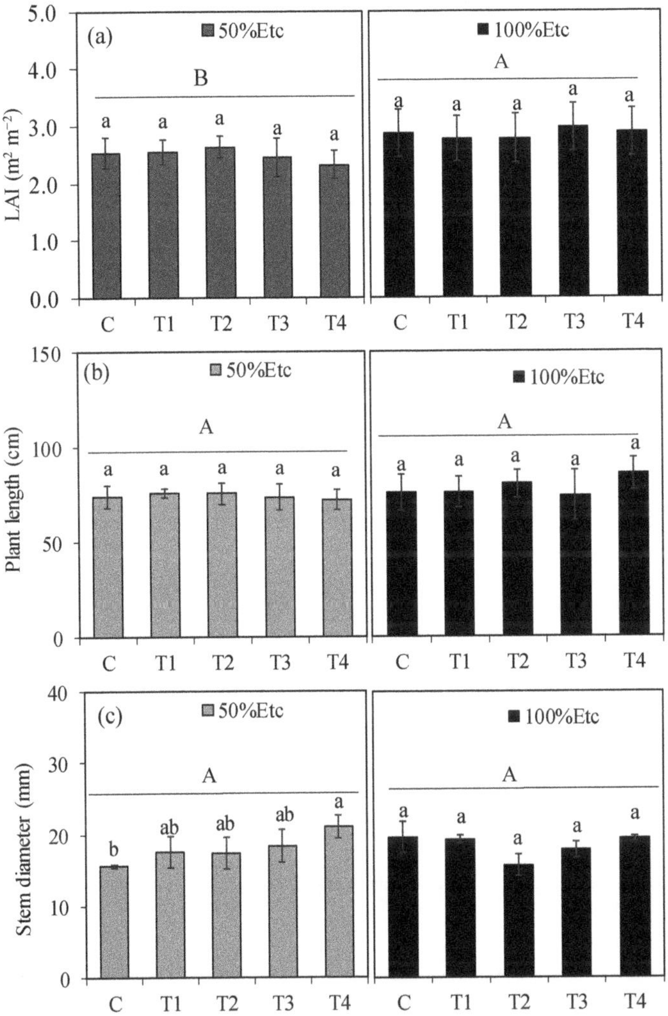


**Figure 2.** Leaf area index-LAI (**a**), plant height (**b**), and stem diameter (**c**) of tomato hybrid 'HM' under different foliar application treatments and irrigation regimes—control (C), $FeSO_4{\cdot}7H_2O$ (T1), $Fe_2O_3$NPs (T2), ZnONPs (T3), and $ZnSO_4{\cdot}7H_2O$ (T4)—subjected to both deficit (50% ETc) and full irrigation (100% ETc). Data represent means ± S.E. Lowercase letters compare treatments (iron and zinc sources), while uppercase letters compare irrigation levels (50% ETc and 100% ETc).

### 3.2. Gas Exchange

Regarding gas exchange, water deficit promoted an average 7.6% decrease (−5.77 µmol $m^{-2}$ $s^{-1}$) in $CO_2$ assimilation (A, Figure 3a) compared to full irrigation (27.03 µmol $m^{-2}$ $s^{-1}$) without significantly affecting transpiration (E, Figure 3b), stomatal conductance (Gs, Figure 3c), internal $CO_2$ concentration (Ci, Figure 3d), or the A/E ratio (Figure 3e). Under deficit irrigation, all nutrient treatments—$Fe_2O_3$NPs (T2), ZnONPs (T3), and their conventional sources ($FeSO_4{\cdot}7H_2O$, $ZnSO_4{\cdot}7H_2O$)—increased $CO_2$ assimilation by 25.4% (+5.38 µmol $m^{-2}$ $s^{-1}$) over the control (20.07 µmol $m^{-2}$ $s^{-1}$), an effect that was even more pronounced under well-watered conditions (100% ETc), where a 27.8% increase (+6.81 µmol $m^{-2}$ $s^{-1}$) was observed.

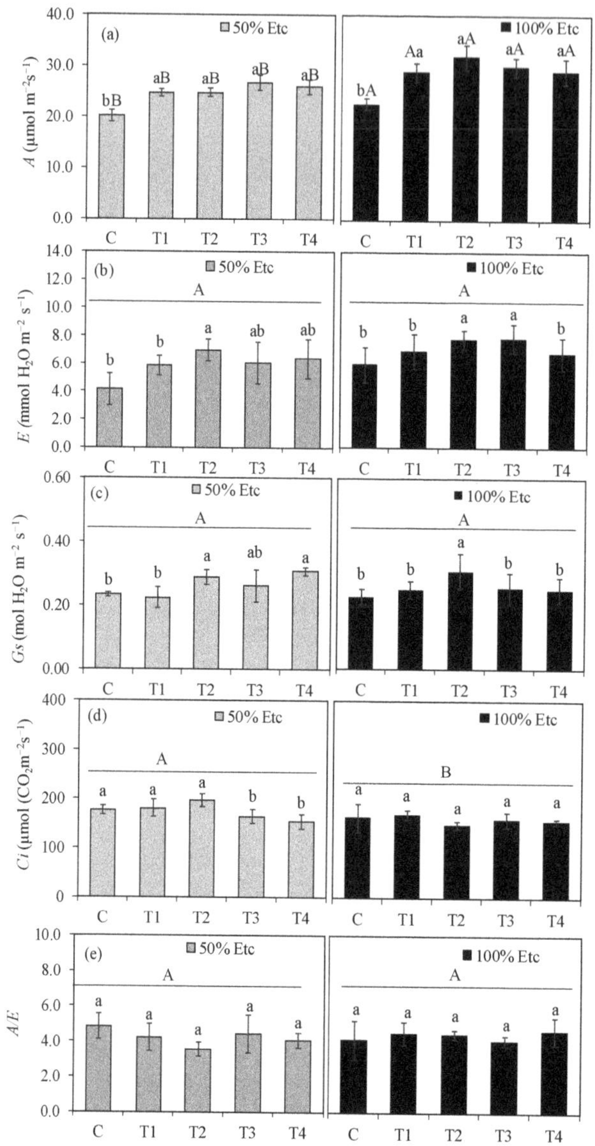


**Figure 3.** $CO_2$ assimilation rate (**a**), transpiration rate (**b**), stomatal conductance (**c**), internal $CO_2$ concentration (**d**), and water use efficiency (**e**) of tomato hybrid 'HM' under different foliar application treatments and irrigation regimes—control (C), $FeSO_4{\cdot}7H_2O$ (T1), $Fe_2O_3$NPs (T2), ZnONPs (T3), and $ZnSO_4{\cdot}7H_2O$ (T4)—subjected to both deficit (50% Etc) and full irrigation (100% Etc). Data represent means ± S.E. Lowercase letters compare treatments (iron and zinc sources), while uppercase letters compare irrigation levels (50% Etc and 100% Etc).

Specifically, foliar application of $Fe_2O_3$NPs (T2) increased the transpiration rate (E) by 34.8% (+2.82 mmol $H_2O$ $m^{-2}$ $s^{-1}$) under 50% ETc, compared to the control treatment (4.14 mmol $H_2O$ $m^{-2}$ $s^{-1}$), though it did not differ from the Zn treatments (T3, T4), while under 100% ETc, the highest E values occurred with either $Fe_2O_3$NPs or ZnONPs, showing an 18% (+1.62 mmol $H_2O$ $m^{-2}$ $s^{-1}$) increase. For stomatal conductance (Gs), treatments T2 and T4 were effective under 50% ETc, whereas only T2 increased Gs under 100% ETc.

Internal $CO_2$ concentration (Ci), when averaged across treatments, was slightly higher under water deficit, and the A/E ratio remained unaffected by either irrigation depth or nutrient application.

### 3.3. Photosynthetic Pigments and Electrolyte Leakage

Water deficit (50% ETc) significantly increased chlorophyll a (Figure 4a), chlorophyll b (Figure 4b), total chlorophyll a + b (Figure 4c), carotenoid content (Figure 4e), and electrolyte leakage (Figure 4f), but did not alter the chlorophyll a/b ratio (Figure 4d). WDA treatments did not affect chlorophyll a, b, total chlorophyll, or carotenoid content at any irrigation level, nor the a/b ratio under 50% ETc. However, under full irrigation (100% ETc), treatments T1, T2, and T4, the chlorophyll a/b ratio showed similar values—62.9% (+1.48 mg $g^{-1}$) higher than the control (2.36 mg $g^{-1}$).

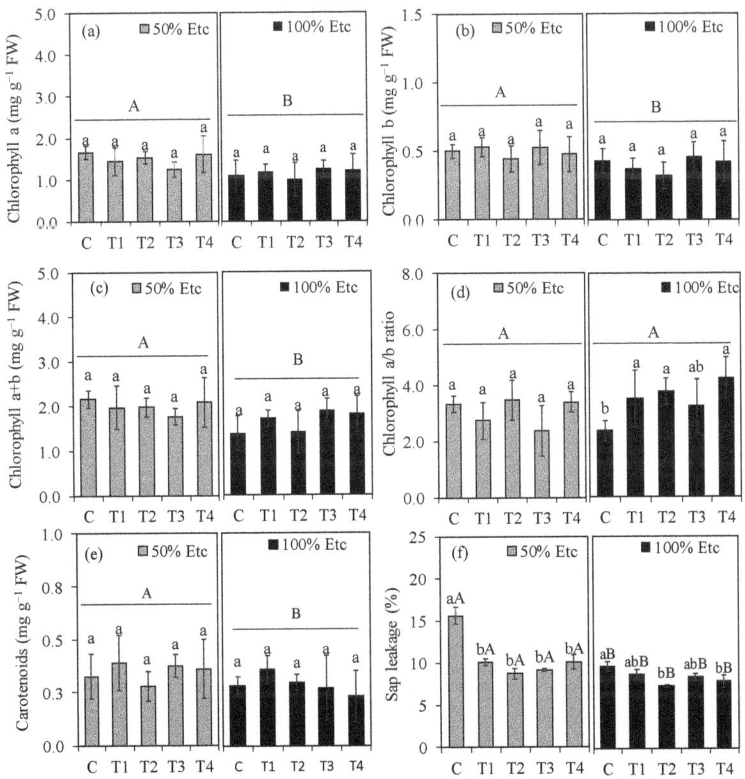

**Figure 4.** Contents of chlorophyll a (**a**), chlorophyll b (**b**), chlorophyll a + chlorophyll b (**c**), chlorophyll a/b ratio (**d**), carotenoids (**e**), and sap leakage (**f**) of tomato hybrid 'HM' under different foliar application treatments and irrigation regimes—control (C), $FeSO_4{\cdot}7H_2O$ (T1), $Fe_2O_3$NPs (T2), ZnONPs (T3), and $ZnSO_4{\cdot}7H_2O$ (T4)—subjected to both deficit (50% ETc) and full irrigation (100% ETc). Data represent means ± S.E. Lowercase letters compare treatments (iron and zinc sources), while uppercase letters compare irrigation levels (50% ETc and 100% ETc).

Regarding electrolyte leakage, under water deficit (50% ETc), all Fe/Zn source treatments (T1–T4) reduced leakage by 39.1% (−6.11%) versus the control (15.62%). Under full irrigation, T2 and T4 reduced leakage by 20.8% (−2.0%) versus the control (9.61%). On average, water deficit increased electrolyte leakage (Ext) by 28.0% (+2.35%) compared to 100% ETc (Figure 4f). Under 50% ETc, treatments T1-T4 all reduced Ext versus control but did not differ from each other. Under 100% ETc, treatments T2-T4 lowered Ext versus control but were statistically similar to T1.

### 3.4. Fruit Productivity

The total number of fruits per plant (Figure 5a) was not significantly affected by the irrigation depths. On the other hand, water deficit reduced the number of marketable fruits per plant (Figure 5b), the transverse diameter (Figure 5c), the longitudinal diameter (Figure 5d), the total yield (Figure 5e), and the marketable fruit yield (Figure 5f). Water restriction decreased total fruit production by 34.9% (−8.00 kg ha$^{-1}$) and marketable production by 71.2% (−10.58 kg ha$^{-1}$), compared to full irrigation (14.87 kg ha$^{-1}$). Under water deficit, treatment T4 increased the total number of fruits by 55.0% (+19.97 fruits per plant), compared to the control (36.33 fruits per plant), but did not alter total fruit yield. Under full irrigation, treatment T4 resulted in a higher number of fruits per plant compared to treatments T1 and T3 (which did not differ from the control and T2 treatments) and increased total yield by 61% (+11.95 kg ha$^{-1}$) compared to the control (19.54 kg ha$^{-1}$).

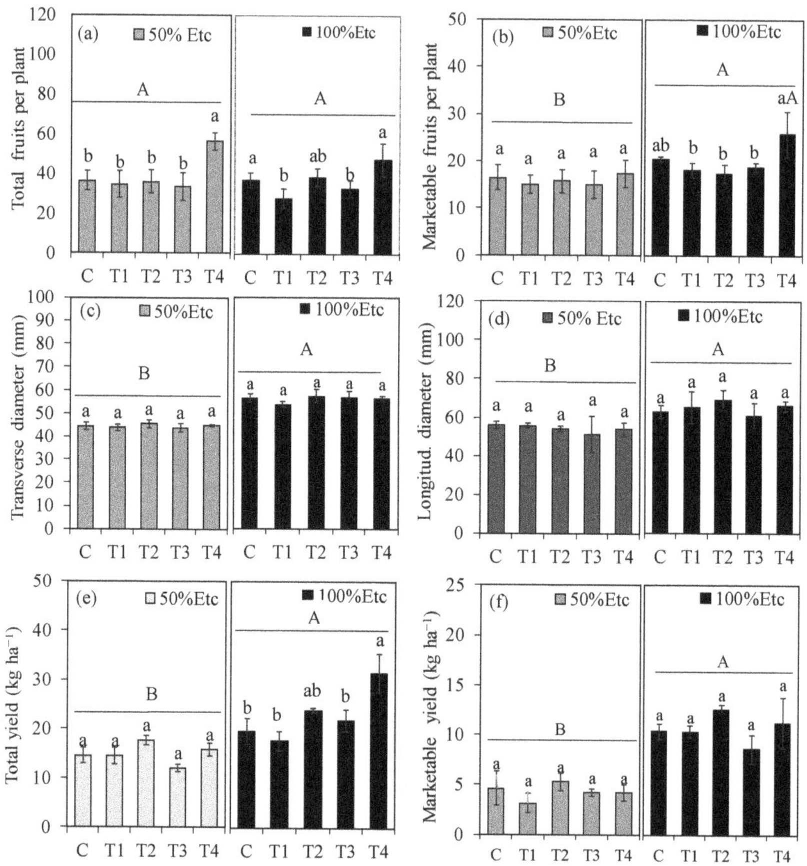


**Figure 5.** Effects of foliar treatments and irrigation regimes on tomato 'HM': (**a**) total fruit number per plant, (**b**) marketable fruit number per plant, (**c**) fruit transverse diameter, (**d**) fruit longitudinal diameter, (**e**) total fruit yield, and (**f**) marketable fruit yield. Treatments: control (C), $FeSO_4{\cdot}7H_2O$ (T1), $Fe_2O_3$ NPs (T2), ZnO NPs (T3), and $ZnSO_4{\cdot}7H_2O$ (T4), under deficit (50% ETc) or full irrigation (100% ETc). Data represent means ± SE. Lowercase letters denote significant differences among foliar treatments (Tukey's test, $p < 0.05$); uppercase letters compare irrigation levels.

The WDA treatments did not modify the quantity of marketable fruits per plant under the 50% ETc irrigation depth (Figure 5b). However, under full irrigation, treatment T4 outperformed treatments T1, T2, and T3, but not the control treatment. Under both irrigation depths, the WDA treatments did not influence the transverse diameter, the longitudinal diameter of the fruits, or the marketable fruit yield.

### 3.5. Mineral Element Content in Fruits

The effect of irrigation depths on boron (B, Figure 6a) content in fruits depended on the WDA treatments. On average, across WDA treatments, calcium (Ca, Figure 6b) content was reduced by 16.5% (−31.63 mg 100 $g^{-1}$) due to water restriction. Water deficit did not affect the content of iron (Fe, Figure 6c), potassium (K, Figure 6d), manganese (Mn, Figure 6e), magnesium (Mg, Figure 6f), phosphorus (P, Figure 6g), or zinc (Zn, Figure 6h), but decreased B levels by 32.9% in the T4 treatment (−0.57 mg 100 $g^{-1}$).

Under water deficit (50% ETc), WDA treatments did not interfere with B levels. Under the 100% ETc irrigation depth, treatment T4 increased boron content by 65.5% (+0.69 mg 100 $g^{-1}$) and calcium (Ca) content by 32.8% (+62.66 mg 100 $g^{-1}$) compared to the control treatment. The WDA treatments particularly affected iron (Fe) content. Under water restriction, treatments T1 ($FeSO_4 \cdot 7H_2O$) and T2 ($Fe_2O_3$NPs) increased the content of this nutrient by 67.5% (+5.73 mg 100 $g^{-1}$) and 117.3% (+9.96 mg 100 $g^{-1}$), respectively, compared to the control (8.50 mg 100 $g^{-1}$). Under full irrigation, treatments T1 and T2 increased iron content by 24.5% (+1.84 mg 100 $g^{-1}$) and 135.1% (+10.15 mg 100 $g^{-1}$), respectively, compared to the control treatment (7.51 mg 100 $g^{-1}$). The WDA treatments influenced zinc (Zn) content only under the 50% ETc irrigation depth, where treatment T4 increased zinc content by 18.5% (+0.26 mg 100 $g^{-1}$) compared to the control treatment.

Figure 7 below presents the relative values (Rv) of the selected variables to illustrate the magnitude of the effects of the WDA treatments under both tested irrigation depths. Iron content was excluded due to the high percentage increase in treatments T1 and T4, which would obscure visualization of effects on other treatments for this variable.

Under water deficit, treatment T1 enhanced the photosynthetic rate (A), transpiration rate (E), carotenoids (Carot), and calcium (Ca) content in fruits (Figure 7a). Under full irrigation, T1 increased values of A, E, total chlorophyll content (Chlt), chlorophyll a, and carotenoids (Figure 7b). Under both irrigation depths, treatment T2 elevated values of A, E, stomatal conductance (gs), Ca content, and total (TYF) and marketable (YMF) fruit yield, in addition to increasing chlorophyll a and carotenoids under 100% ETc. Under water restriction, treatment T3 most markedly increased values of A, E, and Ca content in fruits, while under full irrigation, it elevated A, E, and chlorophylls (a and b). Conversely, treatment T4 promoted increases in Gs, A, E, and Ca and Zn content under water deficit. Under full irrigation, treatment T4 enhanced A, E, chlorophylls (a and b), Ca and B content, and fruit yield (TYF and YMF).

Pearson's linear correlation analysis (Figure 8) revealed significant positive correlations between A and Gs, the number of marketable fruits and B content (r = 0.57), Fe content and physiological parameters A/Ci (r = 0.75), A (r = 0.68), E (r = 0.46), gs (r = 0.62), and Mn content (r = 0.50). Additionally, B content correlated positively with total fruit production (r = 0.55), K content showed direct relationships with Mg (r = 0.52) and P (r = 0.54), and P content strongly correlated with Mn content (r = 0.86). Conversely, electrolyte leakage exhibited negative correlations with E (r = −0.52), A/E (r = −0.76), and A/Ci (r = −0.69).

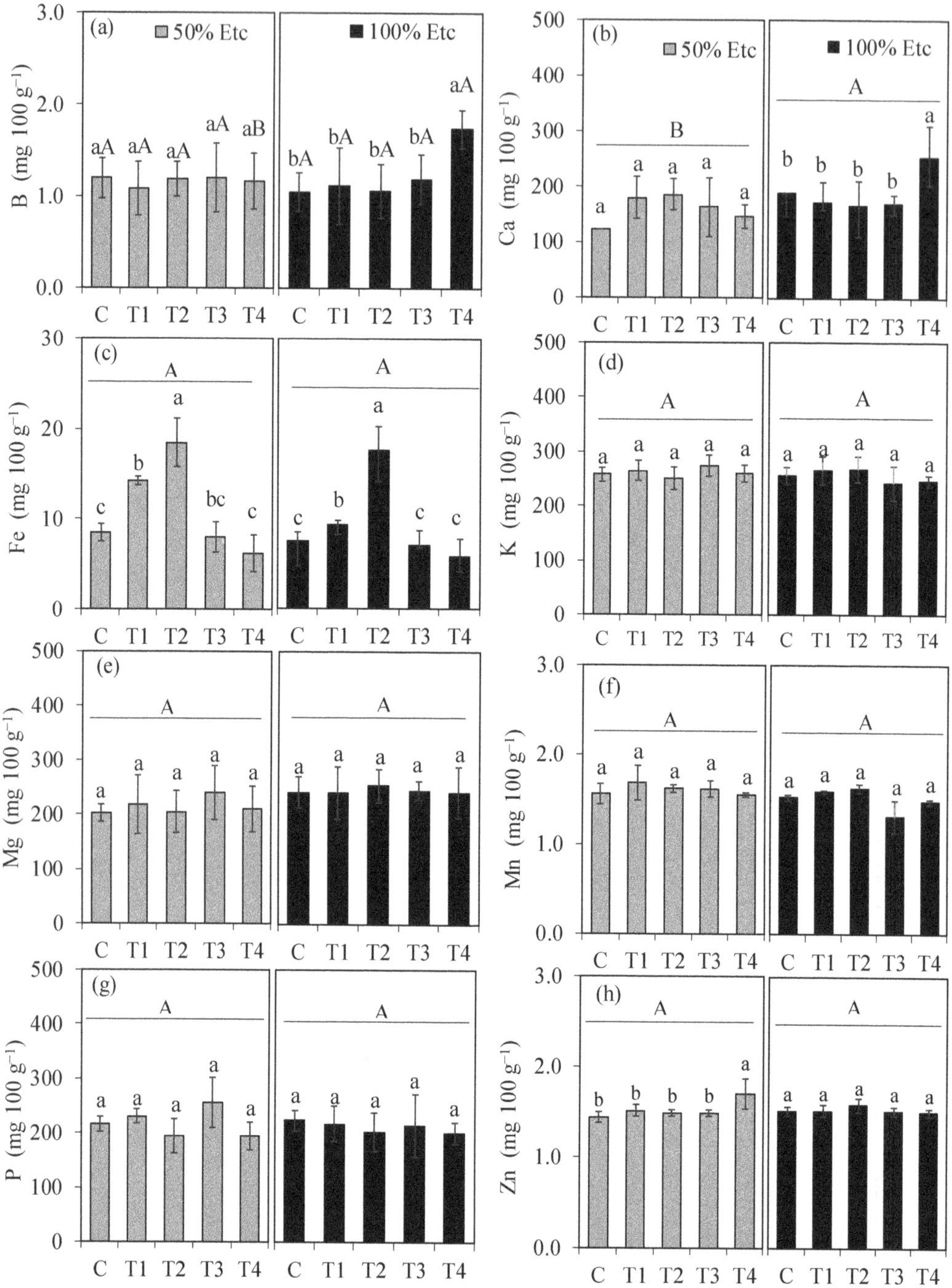


**Figure 6.** Boron (**a**), calcium (**b**), iron (**c**), potassium (**d**), manganese (**e**), magnesium (**f**), phosphorus (**g**), and zinc (**h**) content in fruits of tomato hybrid 'HM' under different foliar application treatments and irrigation regimes—control (C), $FeSO_4{\cdot}7H_2O$ (T1), $Fe_2O_3$NPs (T2), ZnONPs (T3), and $ZnSO_4{\cdot}7H_2O$ (T4)—subjected to both deficit (50% ETc) and full irrigation (100% ETc). Data represent means ± S.E. Lowercase letters compare treatments (iron and zinc sources), while uppercase letters compare irrigation levels (50% ETc and 100% ETc).

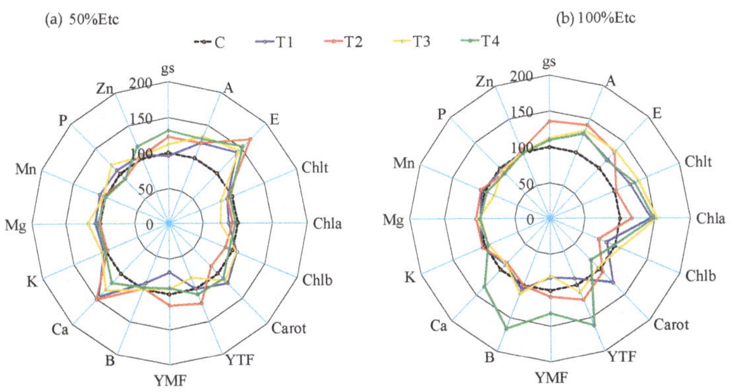


**Figure 7.** Relative values (% of control) of marketable yield (YMF), total yield (TYF), photosynthetic pigments [carotenoids (Carot), chlorophyll (**a**) (chl a), (**b**) (chl b), total chlorophyll (chl t)], physiological parameters [photosynthetic rate (A), transpiration (E), stomatal conductance (gs)], and fruit mineral content [B, Ca, Fe, K, Mg, Mn, P, Zn] in tomato 'HM' under foliar treatments—control (C), $FeSO_4{\cdot}7H_2O$ (T1), $Fe_2O_3$NPs (T2), ZnONPs (T3), and $ZnSO_4{\cdot}7H_2O$ (T4)—and irrigation regimes (50% ETc, 100% ETc).

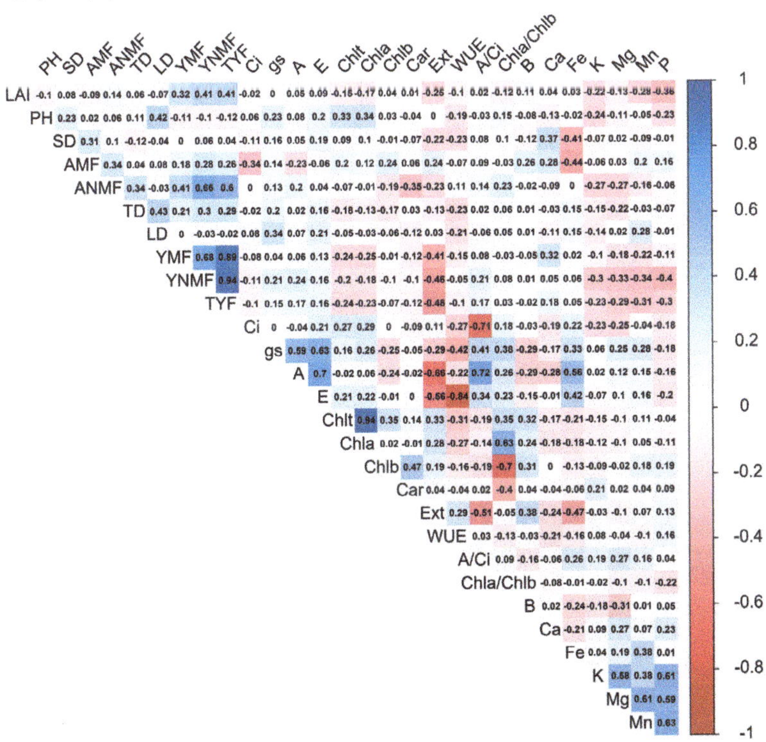

**Figure 8.** Pearson's linear correlation coefficients among the variables leaf area index (LAI), plant height (PH), stem diameter (SD), marketable (AMF) and non-marketable fruit quantity (ANMF), fruit transverse diameter (TD) and longitudinal diameter (LD), marketable (YMF) and non-marketable fruit yield (YNMF), total fruit yield (TYF), internal $CO_2$ concentration (Ci), photosynthetic rate (A), transpiration rate (E), stomatal conductance (gs), total chlorophyll content (chl t), chlorophyll a (chl a), chlorophyll b (chl b), chlorophyll a/b ratio (chl a/chl b), carotenoids (car), electrolyte leakage (Ext), water use efficiency (WUE), A/Ci ratio, and fruit mineral content—boron (B), calcium (Ca), iron (Fe), potassium (K), magnesium (Mg), manganese (Mn), phosphorus (P), and zinc (Zn)—in tomato hybrid 'HM'.

## 4. Discussion

### 4.1. Plant Growth

This study evaluated the potential of iron ($Fe_2O_3$NPs) and zinc ($Fe_2O_3$NPs) nano-oxides, as well as their conventional forms ($FeSO_4{\cdot}7H_2O$ and $ZnSO_4{\cdot}7H_2O$), to mitigate the negative effects of water deficit (50% ETc) on growth traits, gas exchange, photosynthetic pigments, and nutrient content in tomato fruits. Regarding plant growth, the results demonstrate that water deficit did not affect the length of the plant or the diameter of the stem but reduced the leaf area index (LAI). Generally, water stress decreases cell turgor, essential for proper cellular metabolism such as photosynthesis [46], enzymatic activity, and pigment contents [15,46]. The 12.6% reduction in LAI under 50% ETc is attributed to decreased cell expansion and leaf growth [47]. Under drought conditions, plants trigger a series of physiological responses aimed at conserving water [48,49]. A key adaptation is the reduction in leaf area development, which directly limits water loss through evapotranspiration [50]. Foliar application of $Fe_2O_3$NPs, ZnONPs, $FeSO_4{\cdot}7H_2O$, and $ZnSO_4{\cdot}7H_2O$ did not mitigate the negative effect of water deficit on LAI but favored stem diameter. However, the 25.7% increase in the stem diameter with T4 under water deficit reinforces the role of Zn in auxin synthesis and structural integrity [51,52].

### 4.2. Gas Exchange, Photosynthetic Pigments, and Electrolyte Leakage

Water deficit reduced $CO_2$ assimilation (A), but treatments with $Fe_2O_3$NPs and ZnO NPs or their respective conventional sources ($FeSO_4{\cdot}7H_2O$ and $ZnSO_4{\cdot}7H_2O$) increased A by up to 25.4% under 50% ETc and 27.8% under full irrigation, while $Fe_2O_3$NPs increased stomatal conductance (Gs) by 36.9% under full irrigation. In terms of Gs, $Fe_2O_3$NPs and $ZnSO_4{\cdot}7H_2O$ performed better than the control under water deficit, while T2 was more efficient under full irrigation. The increase in A may have been a result of improved chloroplast function [50–52]. These Zn and Fe sources may have enhanced Rubisco activity and electron transport [27]. Iron is crucial for light-dependent reactions, forming components of photosystems and electron transport chains to generate energy [47]. Zinc plays a fundamental role in hormone synthesis and cell division, which contributes to increased root growth and, consequently, greater water absorption; in addition, it also helps stabilize Rubisco itself and thus contributes to increasing the photosynthetic rate [50,51]. The increase in transpiration (E) with $Fe_2O_3$NPs under 50% ETc may link to nanoparticle-mediated stomatal regulation, balancing water loss and photosynthesis [16,27].

Photosynthetic pigments are key cellular components affected by water deficit [31,53]. Water restriction typically reduces pigment content, especially chlorophyll [46,54], due to oxidative damage from reactive oxygen species and increased activity of oxidases (e.g., catalase, peroxidase, SOD) [54]. However, chlorophyll levels may increase or decrease depending on stress severity and plant species [49,54]—e.g., moderate stress elevated chlorophyll in *Lonicera caerulea*. Here, increased chlorophyll/carotenoid content under deficit may indicate an adaptive response to maximize light capture [48], compounded by reduced leaf area concentrating pigments per unit area.

The reduction in electrolyte leakage with T1–T4 indicates that NPs and conventional Fe and Zn sources strengthened cell membranes, likely via antioxidant induction [28]. Zinc activates oxidative stress defense enzymes (e.g., SOD), aids tryptophan (IAA precursor) synthesis, and supports cell division and membrane integrity [28,30].

### 4.3. Yield, Fruit Nutrient Content, and Variable Correlations

Water deficit's negative impact on tomato yield aligns with prior studies [27,32,49]. The 47.9% decrease in marketable yield under 50% ETc reflects stress-induced reductions in fruit number and size [32,55], consistent with a lower LAI, photosynthetic rate, and higher electrolyte leakage. However, T4 ($ZnSO_4 \cdot 7H_2O$) increased total yield by 61% under full irrigation, underscoring zinc's role in cell division and fruit development [12].

Positive correlations between fruit boron (B) content and fruit number (r = 0.57) or marketable yield (r = 0.55) highlight the B importance in cell wall integrity and postharvest quality [28]. Though fruit calcium (Ca) content did not correlate significantly with yield, the deficit reduced Ca. Both B and Ca directly affect cell wall plasticity/elongation [55,56]. These nutrients move via diffusion and mass flow to roots. Water deficit impedes their uptake and phloem-immobile translocation to fruits [12,55].

$Fe_2O_3$ NPs (T2) increased fruit iron content under deficit, confirming Fe oxide NPs' efficacy in nutrient delivery under low water availability, as seen in wheat [56]. Chlorophyll content remained unchanged with Fe treatments, but Fe correlation with photosynthetic parameters (A, gs) confirms its central role in chlorophyll synthesis/electron transport [27,57,58]. Under water stress, ZnO NPs did not alter fruit zinc content, while $ZnSO_4 \cdot 7H_2O$ increased it. Performance differences between NPs or between ZnO NPs and $ZnSO_4 \cdot 7H_2O$ sources may stem from physical properties: $Fe_2O_3$ NPs (<50 nm; 50–245 $m^2\ g^{-1}$ surface area) vs. ZnO NPs (<100 nm; 10.8 $m^2\ g^{-1}$). The mechanisms governing foliar uptake of NPs are complex and not yet fully elucidated. Large-sized nanoparticles (for example: 100 nm) may have difficulty penetrating leaf tissue cells, promoting leaf retention, and thus not producing the expected physiological, nutritional, and fruit yield effects [21]. On the other hand, its foliar retention can promote a slow release of Zn, which may stimulate enzyme activity without, however, meeting the Zn requirements of the tomato plant sufficiently to increase fruit yield [24].

The increased efficacy of $Fe_2O_3$ nanoparticles over conventional $FeSO_4 \cdot 7H_2O$ in boosting tomato fruit iron content may result from their unique properties via direct or indirect mechanisms. The direct mechanism involves a gradual foliar iron release, enhancing bioavailability [21,26], while the indirect mechanism may involve nanoparticle-induced synthesis of organic acids that chelate iron, promoting its translocation to fruits [56,57]. It has been proposed that NP efficacy depends on cellular uptake routes: passive transport (symplastic and apoplastic pathways), endocytosis, or transmembrane protein-mediated transport [59,60]. Internalized $Fe_2O_3$ and ZnO NPs may reach organelles (chloroplasts and mitochondria), releasing Fe and Zn to modulate electron transport chains and associated processes (e.g., oxidative phosphorylation) [15]. However, NP entry is limited by pore size selectivity and accessibility [24,59,60].

The nutritional enhancement of fruits and vegetables is particularly critical in the Brazilian semi-arid region. These areas typically feature soils with inherently low availability of micronutrients such as iron and zinc, a characteristic attributed to the region's soil composition [61]. This scarcity can lead to human malnutrition, notably iron deficiency. According to NIH [62], the daily Fe requirements for men and women over 19 years of age are 8 and 18 mg $day^{-1}$, respectively. Thus, the consumption of 30 g of fresh tomatoes Fe biofortified (+9.96 mg 100 $g^{-1}$ or +117%) could, theoretically, represent up to 37.3% and 17.6% of the daily reference values, for males and females, respectively. The efficient foliar application of these two nutrients presents a highly effective method to increase their concentration in these vegetables, thereby improving the nutritional quality of the harvested produce [11,13,16]. Furthermore, Fe and Zn nanoparticles are environmentally friendly. Due to the low concentrations applied via foliar spray, when they reach the soil through spray drift, these materials do not leave toxic residues, but only Fe, Zn, or sulfate that can be taken up by plant roots or soil microorganisms [63,64].

## 5. Conclusions

Water deficit (50% ETc) significantly reduced the tomato leaf area index (LAI), photosynthetic rate, and total and marketable yield. The foliar application of both nano-oxides ($Fe_2O_3$ and ZnO NPs) and conventional sources ($FeSO_4 \cdot 7H_2O$ and $ZnSO_4 \cdot 7H_2O$) positively influenced several physiological, productivity, and nutritional quality parameters under both irrigation regimes. Notably, $ZnSO_4 \cdot 7H_2O$ was particularly effective under full irrigation, increasing the stem diameter by 25.7% and total yield by 61%. Across both water conditions, all Fe and Zn sources enhanced $CO_2$ assimilation and reduced electrolyte leakage, suggesting a role in protecting cellular membrane integrity. Foliar application of iron and zinc nano-oxides and their conventional sources had a limited effect on tomato plant growth but increased the photosynthetic rate under both irrigation levels. Under full irrigation, $ZnSO_4 \cdot 7H_2O$ increased total fruit production by 61% and fruit Zn content by 18.1%. In turn, $Fe_2O_3$ NPs (T2) led to increases in fruit iron content by 117.3% under water deficit and 135.2% under full irrigation. Conversely, foliar application using $Fe_2O_3$ nanoparticles emerged as a promising strategy for the biofortification of tomato fruits with iron, indicating a specific utility for addressing dietary deficiencies in regions where they are prevalent. However, there is a need to carry out further work in field conditions, under a semi-arid climate with different seasons and tomato cultivars to consolidate these findings.

**Supplementary Materials:** The following supporting information can be downloaded at: https://www.mdpi.com/article/10.3390/horticulturae11101228/s1, Tables S1–S5: Summary of analysis of variance for all analyzed variables; Tables S6–S10: Means ± SE for all analyzed variables; Figure S1: Distribution diagram of the main plots (irrigation depths) and subplots in the field (foliar treatments with Fe and Zn sources; Figure S2: Principal Component Analysis (PCA) for analyzed variables.

**Author Contributions:** E.C.A., J.L.A.R., F.H.F.P., K.N.F., A.C.d.S.R., R.H.C.R.A., and J.E.J.d.S., participated in the experiment setup, acquisition of data collection, data analysis, and article writing. J.L.A.R., C.S.G.R., G.L., and F.B.N. proposed the design and design of the study and writing of the article. F.V.d.S.S., J.Z.L.S., L.C.d.S., and R.d.N. participated in the discussion of the results and contributed to the preparation of the manuscript. All authors have read and agreed to the published version of the manuscript.

**Funding:** This research was funded by FAPESQ (Foundation to support research in the state of Paraíba), Fapesq/PB Edital Universal 09/2021 SEECT/Fapesq. grant number 3071/2021.

**Data Availability Statement:** The original contributions presented in this study are included in the article.

**Acknowledgments:** We sincerely thank the Department of Soil Science at the Federal University of Lavras, lavras-MG for laboratory support for plant material analysis.

**Conflicts of Interest:** The authors declare no conflicts of interest.

## Appendix A

**Figure A1.** Partial view of the experimental area showing the soil tillage (**a**), installation of the drip irrigation system (**b**), seedlings' initial growth (**c**), and the plants in the pre-flowering phase (**d**).

## References

1. Li, M.; Wu, X.; Zhunag, W.; Xia, L.; Chen, Y.; Wu, C.; Rao, Z.; Du, L.; Zhao, R.; Yi, M.; et al. Tomato and lycopene and multiple health outcomes: Umbrella review. *Food Chem.* **2021**, *343*, 128396. [CrossRef]
2. Caseiro, A.; Ascenso, A.; Costa, A.; Creagh-Flynn, J.; Johnson, M.; Simões, S. Lycopene in human health. *LWT Food Sci. Technol.* **2020**, *127*, 109323. [CrossRef]
3. EMBRAPA. *Embrapa Hortaliças. Tomate: Produção e Mercado no Brasil*; Documentos 245; Embrapa: Brasília, Brazil, 2022.
4. Agência Nacional de Águas e Saneamento Básico (ANA). *Conjuntura dos Recursos Hídricos no Brasil 2023: Relatório Pleno*; ANA: Brasília, Brazil, 2023. Available online: https://www.snirh.gov.br/portal/snirh/centrais-de-conteudos/conjuntura-dos-recursos-hidricos/conjuntura_2023.pdf (accessed on 10 June 2024).
5. United Nations. Water (UN-Water). In *The United Nations World Water Development Report 2021*; Valuing Water; UNESCO: Paris, France, 2021.
6. Locatelli, S.; Barrera, W.; Verdi, L.; Nicoletto, C.; Marta, A.D.; Mauricieri, C. Modelling the response of tomato on deficit irrigation under greenhouse conditions. *Sci. Hortic.* **2024**, *326*, 112770. [CrossRef]
7. Wadood, A.; Hameed, A.; Akram, S.; Ghaffar, M. Unraveling the impact of water deficit stress on nutritional quality and defense response of tomato genotypes. *Front. Plant Sci.* **2024**, *15*, 1403895. [CrossRef] [PubMed]
8. Oliveira, E.C.; Carvalho, J.A.; Rezende, F.C.; Freitas, W.A.; Ramos, M.M. Stomatal regulation and photosynthetic performance in tomato under water deficit. *Environ. Exp. Botany* **2021**, *187*, 104465. [CrossRef]
9. Haghpanah, M.; Hashemipetroudi, S.; Arzani, A.; Araniti, F. Drought tolerance in plants: Physiological and molecular responses. *Plants* **2024**, *13*, 2962. [CrossRef]

10. Li, J.; Zafar, S.; Javaid, A.; Perveen, S.; Hasnain, Z.; Ihtisham, M.; Abbas, A.; Usman, M.; El-Sappah, A.H.; Abbas, M.; et al. Zinc nanoparticles (ZnNPs): High-fidelity amelioration in turnip (*Brassica rapa* L.) production under drought stress. *Sustainability* **2023**, *15*, 6512. [CrossRef]
11. Martins, B.L.R.; Ferreira, K.N.; Rocha, J.L.A.; Araujo, R.H.C.R.; Lopes, G.; Santos, L.C.d.; Bezerra Neto, F.; Sá, F.V.d.S.; Silva, T.I.d.; da Silva, W.I.; et al. Nano ZnO and Bioinoculants Mitigate Effects of Deficit Irrigation on Nutritional Quality of Green Peppers. *Horticulturae* **2024**, *10*, 969. [CrossRef]
12. Marschner, H. *Mineral Nutrition of Higher Plants*, 2nd ed.; Academic Press: London, UK, 2012.
13. Uresti-Porras, J.G.; Cabrera-De La Fuente, M.; Benavidez-Mendoza, A.; Olivares-Sáenz, E.; Cabrera, R.I.; Juárez-Maldonado, A. Effect of graft and nano ZnO on nutraceutical and mineral content in bell pepper. *Plants* **2021**, *10*, 2793. [CrossRef]
14. Semida, W.M.; Abdelkhalik, A.; Mohamed, G.F.; Abd El-Mageed, T.A.; Abd El-Mageed, S.A.; Rady, M.M.; Ali, E.F. Foliar application of zinc oxide nanoparticles promotes drought stress tolerance in eggplant (*Solanum melongena* L.). *Plants* **2021**, *10*, 421. [CrossRef]
15. Jafari, A.; Hatami, M. Foliar-applied nanoscale zero-valent iron (nZVI) and iron oxide ($Fe_3O_4$) induce differential responses in growth, physiology, antioxidative defense and biochemical indices in *Leonurus cardiaca* L. *Environ. Res.* **2022**, *215*, 114254. [CrossRef] [PubMed]
16. Zia-ur-Rehman, M.; Mfarrej, M.F.B.; Usman, M.; Anayatullah, S.; Rizwan, M.; Alharby, H.F.; Zeid, I.M.A.; Alabdallah, N.M.; Ali, S. Effect of iron nanoparticles and conventional sources of Fe on growth, physiology and nutrient accumulation in wheat plants grown on normal and salt-affected soils. *J. Hazard. Mater.* **2023**, *458*, 131861. [CrossRef] [PubMed]
17. Golshahi, S.; Ahangar, A.G.; Mir, N.; Ghorbani, M. A comparison of the use of different sources of nanoscale iron particles on the concentration of micronutrients and plasma membrane stability in sorghum. *J. Soil. Sci. Plant Nutr.* **2018**, *18*, 236–252. [CrossRef]
18. Yousaf, N.; Ishfaq, M.; Qureshi, H.A.; Saleem, A.; Yang, H.; Sardar, M.F.; Zou, C. Characterization of Root and Foliar-Applied Iron Oxide Nanoparticles ($\alpha$-$Fe_2O_3$, $\gamma$-$Fe_2O_3$, $Fe_3O_4$, and Bulk $Fe_3O_4$) in Improving Maize (*Zea mays* L.) Performance. *Nanomaterials* **2023**, *13*, 3036. [CrossRef]
19. Kurdekar, A.K.; Desai, B.K.; Koppalakar, B.G.; Sannagoudar, M.S.; Rathod, P.S.; Hiregoudar, S.; Rajesh, N.L.; Almutairi, K.; Elansary, H.; Nazim, M. Comparative evaluation of green synthesized and commercial iron and zinc nanoparticles on germination, growth and productivity of pigeonpea. *Sci. Rep.* **2025**, *15*, 25956. [CrossRef]
20. Kumari, A.; Gupta, A.K.; Sharma, S.; Jadon, V.S.; Sharma, V.; Chun, S.C.; Sivanesan, I. Nanoparticles as a Tool for Alleviating Plant Stress: Mechanisms, Implications, and Challenges. *Plants* **2024**, *13*, 1528. [CrossRef]
21. Li, R.; Zhang, R.; Li, Y.; Liu, C.; Wang, P.; Sun, H.; Wang, L. Foliar Uptake and Distribution of Metallic Oxide Nanoparticles in Maize (*Zea mays* L.) Leaf. *Environ. Sci. Technol.* **2024**, *58*, 16994–17003. [CrossRef]
22. García-López, J.I.; Niño-Medina, G.; Olivares-Sáenz, E.; Lira-Saldivar, R.H.; Barriga-Castro, E.D.; Vázquez-Alvarado, R.; Rodríguez-Salinas, P.A.; Zavala-García, F. Foliar application of zinc oxide nanoparticles and zinc sulfate boosts the content of bioactive compounds in habanero peppers. *Plants* **2019**, *8*, 254. [CrossRef]
23. Azam, M.; Bhatti, H.N.; Khan, A.K.; Zafar, L.; Iqbal, M. Zinc oxide nano-fertilizer application (foliar and soil) effect on the growth, photosynthetic pigments and antioxidant system of maize cultivar. *Biocat Agric. Biotechnol.* **2022**, *42*, 102343. [CrossRef]
24. Yadav, M. Nanoparticle-facilitated targeted nutrient delivery in plants: Breakthroughs and mechanistic insights. *Plant Nano Biol.* **2025**, *12*, 100156. [CrossRef]
25. Amanda-Bandara, R.M.B.; Dissanayaka, D.M.S.B. Agronomic biofortification of vegetables to achieve iron and zinc nutritional security in food systems. *J. Plant Nutr.* **2025**, *48*, 974–994. [CrossRef]
26. Raiesi-Ardali, T.; Ma'mani, L.; Chorom, M.; Moezzi, A. Improved iron use efficiency in tomato using organically coated iron oxide nanoparticles as efficient bioavailable Fe sources. *Chem. Biol. Technol. Agric.* **2022**, *9*, 59. [CrossRef]
27. Zhang, R.; Zhang, W.; Kang, Y.; Shi, M.; Yang, X.; Li, H.; Yu, H.; Wang, Y.; Qin, S. Application of different foliar iron fertilizers for improving the photosynthesis and tuber quality of potato (*Solanum tuberosum* L.) and enhancing iron biofortification. *Chem. Biol. Tech. Agric.* **2022**, *9*, 79. [CrossRef]
28. Elsheery, N.I.; Helaly, M.N.; El-Hoseiny, H.M.; Alam-Eldein, S.M. Zinc oxide and silicone nanoparticles to improve the resistance mechanism and annual productivity of salt-stressed mango trees. *Agronomy* **2020**, *10*, 558. [CrossRef]
29. Ybaez, Q.E.; Sanchez, P.B.; Badayos, R.B. Synthesis and characterization of nano zinc oxide foliar fertilizer and its influence on yield and postharvest quality of tomato. *Philipp. Agric. Sci.* **2020**, *103*, 55–65. [CrossRef]
30. Ahmed, N.; Zhang, B.; Chachar, Z.; Li, J.; Xiao, G.; Wang, Q.; Hayat, F.; Deng, L.; Narejo, M.; Bozdar, B.; et al. Micronutrients and their effects on horticultural crop quality, productivity and sustainability. *Sci. Hortic.* **2024**, *323*, 112512. [CrossRef]
31. Nagao, R.; Kato, K.; Hamaguchi, U.; Ueno, Y.; Tsuboshita, N.; Shimizu, S.; Furutani, S.; Ehira, S.; Nakajima, Y.; Kawakami, K.; et al. Structure of a monomeric photosystem I core associated with iron-stress-induced-A proteins from Anabaena sp. PCC 7120. *Nat. Commun.* **2023**, *14*, 920. [CrossRef]
32. Zahedifar, M.; Moosavi, A.A.; Gavili, E.; Ershadi, A. Tomato fruit quality and nutrient dynamics under water deficit conditions: The influence of an organic fertilizer. *PLoS ONE* **2025**, *20*, e0310916. [CrossRef]

33. EMBRAPA. *Manual of Soil Analysis Methods*, 2nd ed.; Centro Nacional de Pesquisa em Solo: Rio de Janeiro, Brazil, 2011.
34. AGRITEMPO. Agrometeorological Monitoring System: Meteorological Stations for the State of Paraiba. 2023. Available online: https://www.agritempo.gov.br/br/estado/PB/estacoes/ (accessed on 15 October 2023).
35. Cavalcante, F.J.A. *Fertilization Recommendations for the State of Pernambuco (Brazil)*, 2nd ed.; Instituto Agronômico de Pernambuco: Recife, Brazil, 2008.
36. Christiansen, J.E. *Irrigation by Sprinkling*; University of California Agricultural Experiment Station: Berkeley, CA, USA, 1943.
37. Jensen, M.E. Water consumption by agriculture plants. In *Water Deficit and Plant Growth*; Kozlowski, T.T., Ed.; Academic Press: New York, NY, USA, 1968; Volume 2, pp. 1–22.
38. Allen, R.G.; Pereira, L.S.; Raes, D.; Smith, M. *Crop Evapotranspiration—Guidelines for Computing Crop Water Requirements*; FAO Irrigation and Drainage Paper No. 56; FAO: Rome, Italy, 1998.
39. Lichtenthaler, H.K. Chlorophylls and carotenoids: Pigments of photosynthetic biomembranes. *Methods Enzymol.* **1987**, *148*, 350–382. [CrossRef]
40. Campos, P.; Thi, A. Effects of an abscisic acid pretreatment on membrane leakage and lipid composition of Vigna unguiculata leaf discs subjected to osmotic stress. *Plant Sci.* **1997**, *130*, 11–18. [CrossRef]
41. Juzon-Sikora, K.; Lasko's, K.; Warchoł, M.; Czyczyło-Mysza, I.M.; Dziurka, K.; Grzesiak, M.; Skrzypek, E. Water Relations and Physiological Traits Associated with the Yield Components of Winter Wheat (*Triticum aestivum* L.). *Agriculture* **2024**, *14*, 1887. [CrossRef]
42. Gayet, J.P.; Bleinroth, E.W.; Matallo, M.; Garcia, E.E.C.; Garcia, A.E.; Ardito, E.F.G.; Bordin, M.R. *Tomate para Exportação: Procedimentos de Colheita e Pós-colheita*; Série Publicações Técnicas FRUPEX 13; EMBRAPA-SPI: Brasília, Brazil, 1995.
43. Tedesco, M.J.; Gianello, C.; Bissani, C.A.; Bohnen, H.; Volkweiss, S.J. *Analysis of Soil, Plants and Other Materials*, 2nd ed.; Universidade Federal do Rio Grande do Sul: Porto Alegre, Brazil, 1995.
44. Shapiro, S.S.; Wilk, M.B. An analysis of variance test for normality (Complete Samples). *Biometrika* **1965**, *52*, 591–609. [CrossRef]
45. R Core Team. *R*, version 4.4.1. R: A Language and Environment for Statistical Computing [Computer Software]. R Foundation for Statistical Computing: Vienna, Austria, 2025. Available online: https://www.R-project.org/ (accessed on 20 March 2024).
46. Qiao, M.; Hong, C.; Jiao, Y.; Hou, S.; Gao, H. Impacts of drought on photosynthesis in major food crops and the related mechanisms of plant responses to drought. *Plants* **2024**, *13*, 1808. [CrossRef] [PubMed]
47. Taiz, L.; Zeiger, E.; Moller, I.M.; Murphy, A. *Plant Physiology and Development*, 6th ed.; Sinauer Associates: Sunderland, MA, USA, 2017.
48. Suárez-Salazar, J.C.; Guaca-Cruz, L.; Quiceno-Mayo, E.J.; Ortiz-Morea, F.A. Photosynthetic responses and protective mechanisms under prolonged drought stress in cocoa. *Pesqui. Agropecuária Bras.* **2024**, *59*, e03543. [CrossRef]
49. Yan, W.; Lu, Y.; Guo, L.; Liu, Y.; Li, M.; Zhang, B.; Zhang, B.; Zhang, L.; Qin, D.; Huo, J. Effects of drought stress on photosynthesis and chlorophyll fluorescence in blue honeysuckle. *Plants* **2024**, *13*, 2115. [CrossRef]
50. Zhang, J.; Li, M.-H.; Penuelas, J.; Sardans, J.; Du, L.; Yuan, Z.; Luo, Y.; Shen, Y.; Tian, R.; Li, N.; et al. Leaf area modulates the chlorophyll fluorescence of *Leymus chinensis* in response to different drought scenarios. *Environ. Exp. Bot.* **2025**, *237*, 106175. [CrossRef]
51. Cakmak, I.; Marschner, H. Increase in membrane permeability and exudation in roots of zinc deficient plants. *J. Plant Physiol.* **1988**, *132*, 356–361. [CrossRef]
52. Hassan, M.U.; Aamer, M.; Chattha, M.U.; Haiying, T.; Shahzad, B.; Barbanti, L.; Nawaz, M.; Rasheed, A.; Afzal, A.; Liu, Y.; et al. The critical role of zinc in plants facing the drought stress. *Agriculture* **2020**, *10*, 396. [CrossRef]
53. Yang, Y.; Nan, R.; Mi, T.; Song, Y.; Shi, F.; Liu, X.; Wang, Y.; Sun, F.; Xi, Y.; Zhang, C. Rapid and nondestructive evaluation of wheat chlorophyll under drought stress using hyperspectral imaging. *Inter. J. Molec. Sci.* **2023**, *24*, 5825. [CrossRef]
54. Mukherjee, S.; Dash, P.K.; Das, D.; Das, S. Growth, yield and water productivity of tomato as influenced by deficit irrigation water management. *Environ. Process.* **2023**, *10*, 10. [CrossRef]
55. Peroni, E.C.; Messoas, R.S.; Gallo, V.; Bardwosko, J.M.; Souza, E.R.; Avola, L.D.; Bamberg, A.L.; Rambo, C.V. Mineral content and antioxidant compounds in strawberry fruit submitted to drought stress. *Food Sci. Technol.* **2019**, *39*, 245–254. [CrossRef]
56. Nasar, J.; Wang, G.Y.; Ahmad, S.; Muhammad, I.; Zeeshan, M.; Gitari, H.; Adnan, M.; Fahad, S.; Khalid, M.H.B.; Zhou, X.B.; et al. Nitrogen fertilization coupled with iron foliar application improves the photosynthetic characteristics, photosynthetic nitrogen use efficiency, and the related enzymes of maize crops under different planting patterns. *Front. Plant Sci.* **2022**, *13*, 988055. [CrossRef] [PubMed]
57. Gao, D.; Zhao, S.; Huang, R.; Geng, Y.; Guo, L. The effects of exogenous iron on the photosynthetic performance and transcriptome of rice under salt-alkali stress. *Agronomy* **2024**, *14*, 1253. [CrossRef]
58. Wei, L.; Liu, J.; Jiang, G. Nanoparticle-specific transformations dictate nanoparticle effects associated with plants and implications for nanotechnology use in agriculture. *Nat. Commun.* **2024**, *15*, 7389. [CrossRef] [PubMed]
59. Wu, H.; Li, Z. Nano-enabled agriculture: How do nanoparticles cross barriers in plants? *Plant Commun.* **2022**, *3*, 100346. [CrossRef]

60. Alordzinu, K.E.; Appiah, S.A.; AL Aasmi, A.; Darko, R.O.; Li, J.; Lan, Y.; Adjibolosoo, D.; Lian, C.; Wang, H.; Qiao, S. Evaluating the influence of deficit irrigation on fruit yield and quality indices of tomatoes grown in sandy loam and silty loam soils. *Water* **2022**, *14*, 1753. [CrossRef]
61. Oliveira, M.L.J.; Valladares, G.S.; Vieira, J.S.; Coelho, R.M. Availability and spatial variability of copper, iron, manganese and zinc in soils of the State of Ceará, Brazil. *Revista Ciência Agronômica* **2018**, *49*, 371–380. [CrossRef]
62. NIH (National Institute of Health). Iron/Zinc—Health Professional Fact Sheet. 2024. Available online: https://ods.od.nih.gov/factsheets/Zinc-HealthProfessional/ (accessed on 16 July 2024).
63. Bouhadi, M.; Javed, Q.; Jakubus, M.; Elkouali, M.; Fougrach, H.; Ansar, A.; Ban, S.G.; Ban, D.; Heath, D.; Cerne, M. Nanoparticles for Sustainable Agriculture: Assessment of Benefits and Risks. *Agronomy* **2025**, *15*, 1131. [CrossRef]
64. Dokoupil, L.; Mlcek, J. Foliar application of ZnO-NPS influences chlorophyll fluorescence and antioxidants pool in *Capsicum annum* L. under salinity. *Horticulturae* **2022**, *8*, 908. [CrossRef]

*horticulturae*

MDPI

*Article*

# Planting Date and Cultivar Selection Effects on Cauliflower Growth, Physiology, and Yield Performance in North Dakota Growing Conditions

**Ajay Dhukuchhu [1], Ozkan Kaya [1,2] and Harlene Hatterman-Valenti [1,*]**

1 Department of Plant Sciences, North Dakota State University, Fargo, ND 58102, USA; ajay.dhukuchhu@ndsu.edu (A.D.); ozkan.kaya@ndsu.edu or kayaozkan25@hotmail.com (O.K.)
2 Erzincan Horticultural Research Institute, Republic of Türkiye Ministry of Agriculture and Forestry, Erzincan 24060, Türkiye
* Correspondence: h.hatterman.valenti@ndsu.edu

**Abstract**

Investigating the optimal planting strategies for brassica vegetables under variable climatic conditions is essential for developing sustainable production systems in northern agricultural regions. However, comprehensive knowledge about how planting timing modulates growth, physiological responses, and yield parameters across different cultivars remains limited. We investigated vegetative development, root morphology, physiological efficiency, and marketable yield in six cauliflower cultivars ('Amazing', 'Cheddar', 'Clementine', 'Flame Star', 'Snow Crown', and 'Vitaverde') subjected to four planting dates (May 1, May 15, June 1, and June 15) across two growing seasons (2023–2024), followed by detailed morphological and physiological profiling. Planting date, cultivar selection, and seasonal variation significantly influenced all measured parameters ($p < 0.001$), with notable interaction effects observed for fresh root weight, stomatal conductance, water use efficiency, and yield components. Early planted cultivars consistently demonstrated superior performance under variable environmental conditions, maintaining higher growth rates, enhanced root development, and improved physiological efficiency, particularly 'Flame Star', 'Snow Crown', and 'Cheddar', compared to late-planted treatments. Recovery of optimal plant development was most pronounced at May planting dates, with early-established crops showing better maintenance of vegetative growth patterns and enhanced yield potential, including higher curd weights (585.7 g for 'Flame Star') and superior marketable grades. Morphological profiling revealed distinct clustering patterns, with early-planted cultivars forming separate groups characterized by elevated root biomass, enhanced physiological parameters, and superior yield characteristics. In contrast, late-planted crops showed reduced performance, indicative of environmental stress responses. We conclude that strategic early planting significantly enhances cauliflower production resilience through comprehensive optimization of growth, physiological, and yield parameters, particularly under May establishment conditions. The differential performance responses between planting dates provide insights for timing-based management strategies, while the quantitative morphological and physiological profiles offer valuable parameters for assessing crop adaptation and commercial viability potential under variable climatic scenarios in northern agricultural systems.

**Keywords:** *Brassica oleracea*; cultivar performance; planting timing; physiological parameters; yield optimization

## 1. Introduction

Cauliflower (*Brassica oleracea* var. botrytis) is recognized as one of the most important cool-season cruciferous vegetables, serving as a cornerstone of modern agricultural systems due to its economic and nutritional significance worldwide. This versatile member of the Brassicaceae family has a rich evolutionary history, having originated from wild cabbage (*B. oleracea* var. sylvestris), with the Island of Cyprus believed to be its center of origin [1], from where it subsequently spread across diverse geographical regions, including Syria, Turkey, Egypt, Italy, Spain, and eventually North America [2]. The remarkable nutritional profile of cauliflower has positioned it as increasingly vital in sustainable food systems and in promoting human health. It contains an impressive array of bioactive compounds, particularly glucosinolates, carotenoids, and tocopherols, which are widely recognized for their potent anti-cancer properties, including their ability to inhibit malignant cell growth and promote cellular destruction mechanisms [3]. Furthermore, this nutrient-dense vegetable offers exceptional dietary benefits, being naturally low in fat while providing abundant dietary fiber, vitamin C, and vitamin B6, in addition to essential micronutrients including potassium, phosphorus, iron, magnesium, and calcium [4]. These comprehensive health benefits have driven increased global consumption and heightened interest in expanding cauliflower production to new geographical regions.

Despite its agricultural potential and global importance, cauliflower cultivation faces multifaceted challenges that significantly impact yield stability and crop quality across diverse climatic regions. The crop exhibits remarkable sensitivity to environmental conditions, traditionally requiring precise temperature management for optimal growth, with ideal development occurring between 18 and 25 °C and optimal curd formation achieved around 15 °C [5]. However, the increasing prevalence of climate variability has intensified production challenges, as elevated temperatures frequently result in severe quality deterioration, manifesting as loose curds (bolting), unwanted leaf development within heads (buttoning), and complete failure of head development (blindness) [6,7]. Additionally, cauliflower requires carefully balanced moisture management, as consistent hydration is essential for proper growth, yet excess moisture can lead to devastating head rot and crop losses [8]. These environmental sensitivities have necessitated the development of heat-tolerant cultivars designed to adapt to increasingly challenging climatic conditions. The production landscape in the USA dramatically illustrates these regional challenges, where California maintains its dominance by accounting for approximately 90% of national cauliflower output [9–12], largely due to its favorable Mediterranean climate that provides ideal growing conditions. This region benefits from an extended growing season of approximately 270 frost-free days, characterized by mild temperature fluctuations that ensure consistent curd development and high-quality yields, in stark contrast to the significantly shorter and more variable growing period observed in North Dakota. This concentration of production has made California the primary focus for research on environmental stress effects and adaptation strategies, while simultaneously highlighting the vulnerability of the national cauliflower supply to regional climate disruptions. The dominant cauliflower cultivars currently in commercial production across major growing regions include traditional varieties such as 'Snow Crown', 'Cheddar', and 'Amazing', which have been extensively selected for California's Mediterranean climate conditions. However, these established cultivars demonstrate variable performance when introduced to northern production regions with shorter growing seasons and greater temperature extremes. Recent breeding efforts have focused on developing heat-tolerant and early-maturing lines, including newer releases such as 'Flame Star', 'Clementine', and 'Vitaverde', which show promise for adaptation to non-traditional growing regions, though systematic evaluation under northern Great Plains conditions remains limited [13,14].

In contrast to established production regions, North Dakota represents an emerging and strategically important frontier for cauliflower cultivation, presenting both unique opportunities and formidable challenges that demand comprehensive scientific investigation. Currently classified as a minor crop in this region, cauliflower production has experienced notable growth in recent years, driven primarily by increasing consumer demand for its exceptional nutritional benefits and the potential for local food system development [10]. However, the fundamental challenge facing North Dakota producers lies in the region's characteristically short growing season, which spans approximately 130 days between the last and first frost dates, coupled with dramatic temperature fluctuations that occur throughout the critical field season from late spring to early fall [11,12]. These extreme environmental conditions create a complex web of production challenges that extend far beyond simple temperature management. Previous research efforts have been severely limited in scope, focusing predominantly on established production regions with stable climatic conditions, and have consequently failed to address the specific agronomic and physiological challenges inherent to northern plains agriculture. Few researchers have systematically addressed the critical problem of identifying suitable cultivars that can withstand the region's environmental extremes while maintaining acceptable yield and quality standards. Current solutions to cauliflower cultivation in marginal climates remain inadequate and oversimplistic, as most existing studies have concentrated on traditional production areas with more predictable environmental conditions. A fundamental limitation of contemporary research is the conspicuous lack of comprehensive evaluation protocols for new heat-tolerant cultivars under the realistic field conditions that characterize northern Great Plains agriculture. Moreover, there has been remarkably little scientific discussion regarding the physiological mechanisms by which different cauliflower varieties respond to region-specific environmental stresses, particularly concerning critical parameters such as stomatal conductance, water use efficiency, transpiration regulation, and chlorophyll fluorescence as reliable indicators of photosynthetic performance and overall plant vigor.

In light of these significant knowledge gaps and the pressing need for location-specific agricultural solutions in North Dakota, this research aims to accomplish the following objectives to evaluate the potential benefits of commercially available cauliflower cultivars with heat-tolerance characteristics for adaptation to the region's unique growing conditions. The selected cultivars represent both well-established varieties currently in production across various U.S. regions and recently developed heat-tolerant lines that have not yet been systematically evaluated under Northern Great Plains environmental conditions. The specific objectives are: (i) compare various agronomic traits such as yield, head quality, root morphology, and days to maturity; (ii) determine the optimal planting time for these cauliflower cultivars grown as spring crops; (iii) assess the resilience of these cauliflower cultivars to local environmental stresses by examining physiological parameters like stomatal conductance, water use efficiency, transpiration rate, and chlorophyll fluorescence as indicators of net photosynthetic efficiency.

## 2. Materials and Methods

### 2.1. Site Description

The experiment took place at the Horticulture Research Farm near Absaraka, ND (46.986124, −97.355662), during the summer growing seasons of 2023 and 2024 (Figure S1). This location was chosen because it is situated in USDA Plant Hardiness Zone 4A [13], which is characterized by a continental climate with an average annual precipitation ranging from 380 to 760 mm [14]. The experimental site soil type consists of Warsing sandy loam, which is characterized by very gentle slopes ranging from zero to two percent. The soil is scientifically classified as fine-loamy over sandy or sandy-skeletal, mixed, superactive, frigid oxyaquic hapludolls [15]. This particular site was selected due to its representative nature of typical North Dakota agricultural conditions and its suitability for controlled field experimentation.

### 2.2. Experimental Design and Treatments

A randomized complete block design (RCBD) with four replications was utilized in order to minimize experimental error and ensure statistical validity. The experimental design incorporated a two-factorial treatment structure consisting of four different planting dates (1 May, 15 May, 1 June, and 15 June) and six distinct cauliflower cultivars for both experimental years. The cultivars selected for this study were 'Amazing', 'Cheddar', 'Clementine', 'Flame Star', 'Snow Crown', and 'Vitaverde' (Figure S2), all obtained from Johnny's Seeds, all obtained from Johnny's Selected Seeds, a privately held, employee-owned seed producer and merchant headquartered in Winslow, ME, USA (https://www.johnnyseeds.com/about-us/about.html—accessed on 1 October 2025). These specific cultivars were chosen because they represent a diverse range of maturity periods and adaptation characteristics. The maturity information provided for each cultivar was derived from the company's catalog and seed bag data, rather than from the results of the present study. The days to maturity for each cultivar varied considerably: 'Amazing' required 68 days, 'Cheddar' needed 58 days, 'Clementine' matured in 55 days, 'Flame Star' took 62 days, 'Snow Crown' was the earliest at 50 days, and 'Vitaverde' was the latest maturing at 71 days. Each experimental unit was composed of three plants of a single cultivar in order to increase the overall plant population and improve data reliability.

### 2.3. Field Preparation and Soil Analysis

Field preparation was performed in April using a field cultivator and a John Deere 655 rotary hoe to ensure optimal soil conditions for transplanting. In order to characterize the soil conditions, soil samples were collected at depths of 0–15 cm and 15–30 cm using a standard tube auger in a zigzag pattern across the experimental area. The soil nutrient content of the samples was analyzed at the North Dakota State University (NDSU) soil testing laboratory using standard analytical procedures. The experimental field samples exhibited sandy clay loam soil characteristics with pH values of 7.7 and 7.9 at the respective sampling depths. The organic matter content was determined to be 2.3% and 2.1% at depths of 15 and 30 cm, respectively, indicating moderate soil organic matter levels suitable for vegetable production.

### 2.4. Planting and Crop Management

Five-week-old transplants were prepared in a greenhouse environment using standard germination and growth protocols. The transplants were then established in six rows, each measuring 50 m in length and 60 cm in width, at the experimental site. Before transplanting, black plastic mulch with an integrated drip irrigation system was installed in each row to maintain consistent soil moisture and temperature conditions while suppressing weed growth. Plant spacing was set at 45 cm within rows, and row spacing was maintained at 2.5 m to provide adequate space for plant development and equipment access. To maintain proper soil moisture throughout the growing season, irrigation was performed three times per week using the drip irrigation system. Fertilization was accomplished using a 20:20:20 general-purpose water-soluble fertilizer (JR Peters Inc., Allentown, PA, USA) applied manually at 100 ppm concentration. This fertilizer application was performed three times for each planting date, with 20-day intervals after transplanting to ensure adequate nutrient availability during critical growth periods.

### 2.5. Environmental Monitoring

Throughout the growth season, comprehensive environmental monitoring was conducted to track conditions affecting plant development. Hourly air and soil temperatures (°C) were recorded continuously in the experimental field using Watchdog 1000 series micro station dataloggers (Spectrum Technologies, Inc., Aurora, IL, USA). This equipment was chosen because it provides reliable and accurate environmental data collection under field conditions. During the 2023 growing season, air temperatures showed considerable variation, with maximum temperatures reaching 38.3 °C in July and minimum temperatures dropping to 3.2 °C in September. Mean air temperatures ranged from 17.1 °C in September to 22.5 °C in June. Soil temperatures followed similar patterns but with reduced extremes, with maximum soil temperatures reaching 34.9 °C in August and minimum temperatures recorded at 10.6 °C in May. Mean soil temperatures varied from 16.3 °C in May to 24.3 °C in July. The 2024 growing season exhibited different temperature patterns, with air temperatures reaching maximum values of 37.5 °C in June and minimum values of 3.1 °C in May. Mean air temperatures during 2024 ranged from 15.0 °C in May to 22.4 °C in July. Soil temperatures in 2024 showed maximum values of 34.5 °C in June and minimum values of 6.4 °C in May, with mean soil temperatures ranging from 16.0 °C in May to 25.0 °C in July.

### 2.6. Data Collection

Growth parameters were systematically monitored to assess plant development and performance throughout the growing season. Plant height (cm) and number of leaves were observed at 10-day intervals from 20 to 70 days after transplanting to track vegetative growth patterns. After harvesting, comprehensive yield parameters were recorded, including days to harvest, yield per plant (g), and curd diameter (cm). Additionally, marketable grade quality and type of head defects were noted to assess commercial viability. Root parameters were also measured to understand below-ground plant development, including root fresh weight (g), root length (cm), and root width (cm). Furthermore, physiological parameters in the healthy leaves were measured to evaluate plant stress responses and photosynthetic efficiency. These measurements included transpiration rate ($E$), stomatal conductance ($gs$), water use efficiency (WUE), and chlorophyll fluorescence ($Fv'/Fm'$) as an indicator of photosynthetic efficiency. Measurements were taken from one well-grown, healthy, green leaf from each plant, collected between 10:00 AM and 3:00 PM using a CIRAS-3 portable open-flow gas exchange system (PP Systems, Amesbury, MA, USA). The device was configured with specific parameters to ensure consistent and accurate measurements: 100 mL $min^{-1}$ airflow, 1200 μmol $m^{-2}$ $s^{-1}$ photon flux density, 390 ± 5 μmol $mol^{-1}$ $CO_2$ concentration, and 99.9 kPa atmospheric pressure. This equipment was chosen because it provides reliable field-based measurements of plant physiological responses under controlled conditions.

### 2.7. Statistical Analysis

A comprehensive variance analysis (Three-Way ANOVA) was conducted on the quantitative dataset, which combines the responses of cauliflower to treatment combinations, using SAS software version 9.4 (SAS Institute Inc., Cary, NC, USA). The PROC GLIMMIX procedure was used to investigate the impacts of year, planting time, and cauliflower cultivar on several response variables like days to harvest, yield per plant (g), curd diameter (cm), fresh root weight (g), root length (cm), and root width (cm). Year, planting date, cultivar, and interactions were all fixed variables in the model, while replication was a random effect. The significance of the variations among treatment combination means was assessed using Tukey's Honestly Significant Difference (HSD) test, with a significance threshold set at $p \leq 0.05$. The qualitative datasets, such as head defects and grading distribution, were analyzed using a chi-square test and a mosaic heatmap in JMP Student Edition 18.2.0 (SAS Institute Inc., Cary, NC, USA). Regression models for leaf number and plant height were developed using SAS software version 9.4 (SAS Institute Inc., Cary, NC, USA). Non-linear regression models for leaf number and plant height were developed against days after transplanting, averaged across the two years for the four planting dates and each cultivar. The purpose of this modeling approach was to determine the effects of planting dates within the context of North Dakota's short growing season, in order to understand how different sowing times influence vegetative and reproductive growth stages. A hierarchical clustering heatmap was generated using the SRPLOT online platform (https://www.bioinformatics.com.cn/en accessed on 6 July 2025) to visualize the relationships between the primary factors and the studied parameters. The results were depicted clearly using the methodology established by Evgenidis et al. [16].

## 3. Results

### 3.1. Growth Parameters

The regression analysis revealed that leaf development across four planting dates over two growing seasons followed distinct quadratic growth patterns (Figure 1; Table S1). Changes in leaf number were identified using polynomial regression models that demonstrated significant effects of planting timing on vegetative development. The fitted regression equations indicated that all planting dates exhibited quadratic trends, with initial increases in leaf number followed by deceleration as plants matured.

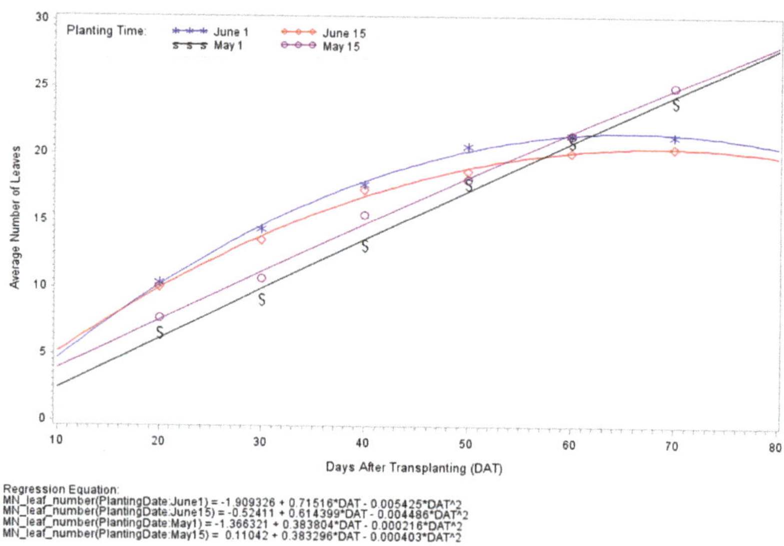


**Figure 1.** Regression models for the average number of cauliflower leaves and days after transplanting under different planting dates averaged over the two years (2023 and 2024) at the Horticulture Research Farm near Absaraka, ND.

The planting date of June 1 showed the highest linear growth rate (0.72 leaves per day) but also demonstrated relatively steep deceleration (−0.005). Similarly, June 15 exhibited a comparable growth pattern with a high initial rate (0.61) and an identical deceleration coefficient. In contrast, May 15 displayed the lowest linear growth rate (0.38) combined with the smallest quadratic deceleration term (−0.0004), indicating slower but more sustained leaf development over the observation period. Although the growth pattern for May 1 appeared nearly linear in Figure 1, the quadratic regression model still demonstrated statistical significance ($p < 0.05$) (Table S1), confirming that a quadratic component better described the overall trend. May 1 demonstrated moderate growth characteristics with intermediate values between the extremes observed in other planting dates. The analysis of cultivar effects on leaf development showed that genetic differences significantly influenced growth patterns across the experimental period (Figure 2). Among the six cultivars evaluated, 'Amazing' exhibited the most vigorous leaf production with the highest linear growth rate (0.57) and uniquely displayed a negative quadratic term (−0.002), indicating a slight deceleration in leaf number until harvest. 'Snow Crown' demonstrated initially high growth rates comparable to 'Cheddar' but subsequently showed the most pronounced deceleration (−0.003), revealing vigorous early development followed by rapid decline. 'Vitaverde' displayed moderate growth rates with similar deceleration patterns (−0.003), while 'Flame Star' and 'Clementine' exhibited the lowest linear growth coefficient (0.47 and 0.44), reflecting a conservative growth pattern (Figure 2). Although 'Vitaverde' appeared phenotypically similar to 'Flame Star' and 'Clementine' in certain growth traits, these cultivars differ in genetic background and curd coloration: 'Vitaverde' (green), 'Cheddar' (yellow), 'Clementine' (orange), and 'Flame Star' (white) are all $F_1$ hybrid seeds (Johnny's Selected Seeds, 2025). Thus, their morphological similarities are more likely due to physiological growth responses than shared parentage.

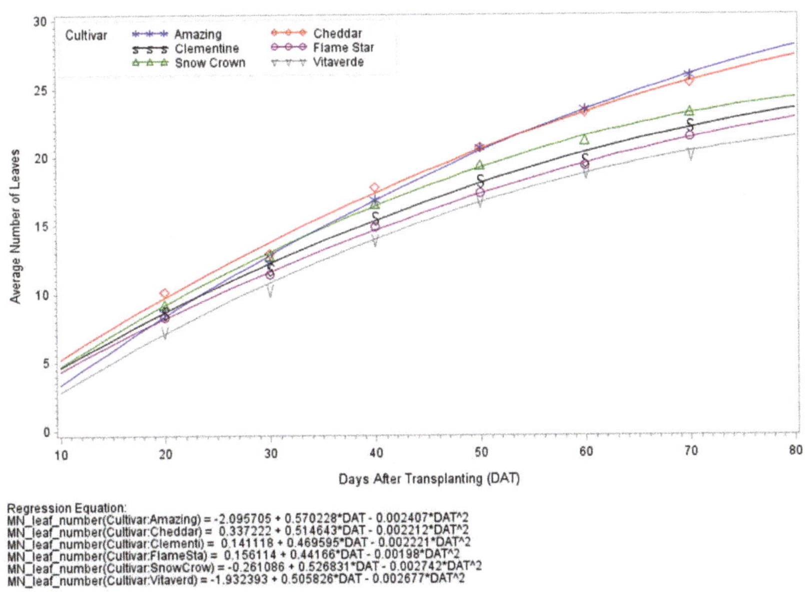


**Figure 2.** Regression models for the average number of cauliflower leaves and days after transplanting under different cultivars averaged over the two years (2023 and 2024) at the Horticulture Research Farm near Absaraka, ND.

In our results, plant height analysis demonstrated that growth patterns varied significantly according to planting date, with all treatments following quadratic trajectories (Figure 3; Table S1). The regression models revealed that later plantings exhibited steeper initial growth slopes compared to earlier plantings. The June 1 and June 15 planting dates had the highest linear growth coefficients (2.51 and 2.45 cm per day, respectively), indicating vigorous early height development (Figure 3). However, these same treatments also demonstrated the most rapid decline in growth rate over time, as evidenced by larger negative quadratic coefficients (−0.022 and −0.020, respectively). Although the growth curve for May 1 planting date appeared nearly linear, the fitted quadratic term was significant ($p < 0.05$) (Table S1), supporting the appropriateness of the quadratic model for describing the growth trajectory. In contrast, plants from the May 1 planting date exhibited moderate growth (1.11) with the smallest deceleration coefficient (−0.001). To contextualize these growth patterns, the two growing seasons (2023 and 2024) exhibited contrasting thermal regimes. The 2023 season experienced higher maximum air temperatures (up to 38.3 °C in July) and greater variability, whereas the 2024 season showed earlier temperature peaks and generally milder conditions (maximum 37.5 °C in June). Plants from May 15 planting showed intermediate characteristics with moderate initial growth (2.02) and moderate deceleration (−0.013). Cultivar analysis for plant height revealed distinct genetic differences in growth architecture and development patterns (Figure 4; Table S1). 'Snow Crown' demonstrated the most vigorous early growth with the highest linear coefficient (2.49) but also exhibited the steepest decline in growth rate (−0.019), indicating rapid early development followed by pronounced maturation. 'Amazing', 'Cheddar', 'Clementine', and 'Flame Star' showed remarkably similar growth patterns with high linear coefficients (1.86, 1.88, 1.94, and 1.92, respectively) and moderate deceleration rates (−0.013, −0.013, −0.013, and −0.012), suggesting robust but controlled height development. "Vitaverde' displayed more moderate growth characteristics with linear coefficients of 2.01 and similar deceleration patterns (−0.014) that indicated stable and predictable height development (Figure 4).

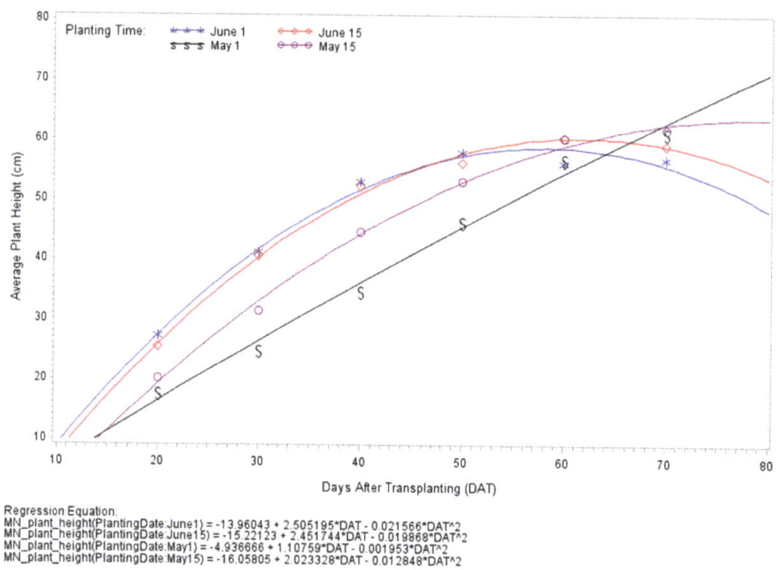


**Figure 3.** Regression models for the average cauliflower plant height and days after transplanting under different planting dates averaged over the two years (2023 and 2024) at the Horticulture Research Farm near Absaraka, ND.

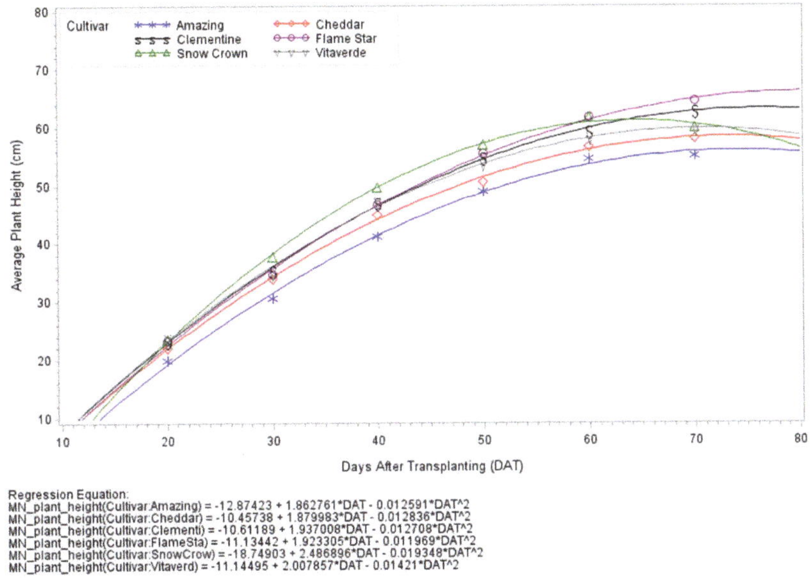


**Figure 4.** Regression models for the average cauliflower plant height and days after transplanting under different cultivars averaged over the two years (2023 and 2024) at the Horticulture Research Farm near Absaraka, ND.

On the other hand, fresh root weight analysis demonstrated significant main effects of cultivar ($p < 0.001$) and planting date ($p < 0.001$), along with significant interactions for year by cultivar ($p < 0.001$), year by planting date ($p < 0.001$), and cultivar by planting date ($p = 0.045$) (Tables S2 and S3). 'Flame Star' consistently demonstrated superior root development in both experimental years, with particularly notable performance in 2024 (158.9 g). In contrast, 'Clementine' exhibited the lowest root biomass across both seasons, with particularly reduced values in 2024 (70.5 g) (Table 1).

**Table 1.** Effect of year by cauliflower cultivar interaction on growth and yield parameters when averaged over four planting dates at the Horticulture Research Farm near Absaraka, ND (Mean ± SE).

| Year | Cultivar | Days to Harvest | Yield (g) | Curd Diameter (cm) | Fresh Root Weight (g) | Root Length (cm) | Foot Width (cm) |
|---|---|---|---|---|---|---|---|
| 2023 | Amazing | 90.9 ± 1.40 b * | 421.9 ± 42.31 a | 13.3 ± 0.52 a | 92.2 ± 7.56 e,f | 21.4 ± 0.84 a | 27.3 ± 1.08 d–f |
| | Cheddar | 89.2 ± 1.52 b | 504.9 ± 35.39 a | 13.9 ± 0.32 a | 112.0 ± 6.23 c,d | 21.8 ± 0.83 a | 27.1 ± 0.93 d–f |
| | Clementine | 84.5 ± 1.52 c | 429.2 ± 37.43 a | 13.5 ± 0.47 a | 80.1 ± 5.88 f,g | 22.0 ± 0.83 a | 26.6 ± 1.02 e,f |
| | Flame Star | 97.0 ± 1.81 a | 466.4 ± 55.49 a | 13.1 ± 0.60 a | 147.2 ± 9.60 a,b | 27.6 ± 1.14 a | 34.5 ± 1.46 b |
| | Snow Crown | 75.9 ± 1.72 d | 501.5 ± 40.99 a | 13.9 ± 0.43 a | 125.9 ± 10.99 c,d | 22.8 ± 1.24 a | 27.0 ± 1.36 e,f |
| | Vitaverde | 99.4 ± 2.06 a | 369.5 ± 31.64 a | 12.5 ± 0.49 a | 104.5 ± 6.81 d,e | 21.7 ± 1.09 a | 29.7 ± 1.24 c–e |
| 2024 | Amazing | 73.7 ± 1.37 d | 612.8 ± 36.26 a | 15.0 ± 0.35 a | 139.3 ± 10.70 a–c | 20.2 ± 0.90 a | 31.4 ± 1.41 b,c |
| | Cheddar | 68.6 ± 0.99 e | 596.3 ± 36.61 a | 14.4 ± 0.33 a | 113.0 ± 7.99 c,d | 19.1 ± 1.16 a | 29.2 ± 1.76 c–e |
| | Clementine | 65.3 ± 1.26 f | 547.5 ± 33.83 a | 14.2 ± 0.43 a | 70.5 ± 5.30 g | 16.4 ± 0.75 a | 24.3 ± 0.97 f |
| | Flame Star | 74.3 ± 0.92 d | 702.5 ± 50.76 a | 15.3 ± 0.50 a | 158.9 ± 8.41 a | 24.4 ± 0.63 a | 38.5 ± 1.09 a |
| | Snow Crown | 60.4 ± 1.00 g | 607.5 ± 33.27 a | 15.0 ± 0.31 a | 105.0 ± 5.17 d | 18.7 ± 0.91 a | 27.0 ± 1.22 d–f |
| | Vitaverde | 68.9 ± 1.08 e | 444.1 ± 27.41 a | 13.5 ± 0.46 a | 114.9 ± 4.32 b–d | 19.5 ± 0.54 a | 29.7 ± 0.86 c,d |
| *p*-value | | <0.001 | 0.154 | 0.206 | 0.001 | 0.055 | 0.039 |

* Means in a column followed by the same letter are not significantly different at $p \leq 0.05$ according to LS Means.

The cultivar by planting date analysis revealed that 'Flame Star' maintained consistently high root weights exceeding 148 g across all planting dates, while 'Clementine' and 'Amazing' showed substantial reductions when planted on June 1 (68.3 g and 63.9 g, respectively) (Table 2). In our results, root length measurements revealed significant effects of all main factors including year, planting date, and cultivar ($p < 0.001$), with a significant year by planting date interaction ($p = 0.039$) (Table S2). The 2023 growing season consistently produced longer root systems across all treatments, with May 1 planting achieving maximum length (25.9 cm). During 2023, May 1 planting produced the highest root biomass (136.4 g), while in 2024, May 15 and June 15 plantings achieved optimal performance with 125.6 g and 117.5 g, respectively (Table 3).

**Table 2.** Effect of cauliflower cultivar by planting date interaction on growth and yield parameters when averaged over two years, 2023 and 2024 at the Horticulture Research Farm near Absaraka, ND (Mean ± SE).

| Cultivar | Planting Date | Days to Harvest | Yield (g) | Curd Diameter (cm) | Fresh Root Weight (g) | Root Length (cm) | Root Width (cm) |
|---|---|---|---|---|---|---|---|
| Amazing | May 1 | 84.8 ± 2.76 c,d * | 735.3 ± 65.35 a | 16.4 ± 0.45 a | 122.9 ± 10.21 a–d | 23.0 ± 1.12 a | 29.4 ± 1.26 a |
| | May 15 | 89.8 ± 1.95 a,b | 486.0 ± 51.01 a | 14.3 ± 0.64 a | 126.1 ± 17.78 a–d | 20.4 ± 1.30 a | 29.1 ± 2.27 a |
| | June 1 | 76.0 ± 2.62 h,i | 426.8 ± 50.51 a | 13.0 ± 0.64 a | 63.9 ± 8.35 h | 17.4 ± 1.10 a | 25.6 ± 1.66 a |
| | June 15 | 78.5 ± 2.29 f–h | 424.1 ± 41.87 a | 13.1 ± 0.58 a | 130.8 ± 10.55 a–c | 22.4 ± 1.09 a | 31.8 ± 1.62 a |
| Cheddar | May 1 | 83.5 ± 2.78 c–e | 685.1 ± 58.71 a | 15.1 ± 0.36 a | 112.6 ± 11.38 c–e | 20.7 ± 1.58 a | 24.9 ± 1.88 a |
| | May 15 | 79.4 ± 3.09 e–h | 652.5 ± 40.37 a | 15.1 ± 0.29 a | 131.8 ± 11.22 a–c | 22.5 ± 1.19 a | 30.4 ± 1.42 a |
| | June 1 | 76.3 ± 2.90 h,i | 489.8 ± 50.21 a | 14.0 ± 0.45 a | 98.6 ± 7.03 c–f | 18.7 ± 1.16 a | 26.4 ± 1.89 a |
| | June 15 | 76.4 ± 2.22 g,h | 398.6 ± 36.42 a | 12.5 ± 0.47 a | 106.3 ± 7.62 b–e | 21.1 ± 1.48 a | 29.8 ± 1.76 a |
| Clementine | May 1 | 81.0 ± 3.23 d–g | 578.5 ± 40.92 a | 14.9 ± 0.58 a | 87.0 ± 10.93 f–h | 21.3 ± 1.47 a | 26.6 ± 1.67 a |
| | May 15 | 78.3 ± 2.00 f–h | 570.1 ± 58.26 a | 15.2 ± 0.55 a | 77.2 ± 8.10 g,h | 20.9 ± 1.19 a | 25.0 ± 1.43 a |
| | June 1 | 71.5 ± 2.92 j,k | 460.5 ± 50.61 a | 13.3 ± 0.63 a | 68.3 ± 4.93 h | 18.1 ± 1.16 a | 25.1 ± 1.33 a |
| | June 15 | 69.0 ± 2.20 j,k | 344.4 ± 42.47 a | 12.0 ± 0.59 a | 69.4 ± 6.60 h | 17.0 ± 1.04 a | 25.3 ± 1.32 a |
| Flame Star | May 1 | 94.3 ± 3.58 a | 755.4 ± 86.91 a | 15.8 ± 0.71 a | 154.3 ± 10.69 a | 27.9 ± 1.51 a | 37.1 ± 1.69 a |
| | May 15 | 87.6 ± 2.93 b,c | 618.5 ± 65.69 a | 14.8 ± 0.70 a | 149.4 ± 14.08 a | 27.5 ± 1.88 a | 36.0 ± 2.34 a |
| | June 1 | 81.4 ± 2.82 d–f | 517.5 ± 65.75 a | 13.5 ± 0.79 a | 148.0 ± 14.37 a,b | 24.1 ± 0.80 a | 34.0 ± 1.95 a |
| | June 15 | 79.4 ± 2.05 e–h | 425.9 ± 81.54 a | 12.4 ± 0.91 a | 158.6 ± 13.21 a | 24.9 ± 0.95 a | 38.3 ± 1.54 a |
| Snow Crown | May 1 | 70.1 ± 2.99 j,k | 639.2 ± 46.20 a | 15.4 ± 0.44 a | 131.8 ± 14.02 a–d | 22.6 ± 2.08 a | 27.2 ± 2.16 a |
| | May 15 | 71.9 ± 2.42 i,j | 732.0 ± 57.19 a | 15.7 ± 0.48 a | 118.5 ± 12.85 b–e | 21.9 ± 1.56 a | 27.9 ± 2.03 a |
| | June 1 | 67.9 ± 2.68 k | 434.7 ± 42.74 a | 13.5 ± 0.55 a | 107.7 ± 14.96 d–g | 19.0 ± 1.42 a | 24.1 ± 1.44 a |
| | June 15 | 62.9 ± 1.62 l | 414.3 ± 35.17 a | 13.3 ± 0.49 a | 104.8 ± 8.73 c–e | 19.8 ± 1.21 a | 28.3 ± 1.41 a |
| Vitaverde | May 1 | 93.0 ± 4.76 ab | 527.2 ± 34.54 a | 14.4 ± 0.51 a | 135.3 ± 7.91 ab | 22.8 ± 1.17 a | 29.8 ± 1.50 a |
| | May 15 | 85.7 ± 3.74 bc | 473.0 ± 41.50 a | 14.4 ± 0.65 a | 100.0 ± 6.69 c–f | 21.0 ± 1.27 a | 30.5 ± 0.93 a |
| | June 1 | 78.1 ± 3.50 f–h | 344.0 ± 35.76 a | 12.2 ± 0.53 a | 94.8 ± 7.73 e–g | 19.4 ± 1.06 a | 27.6 ± 1.22 a |
| | June 15 | 79.8 ± 2.76 e–h | 261.4 ± 28.55 a | 10.5 ± 0.71 a | 111.2 ± 7.62 b–e | 19.2 ± 1.26 a | 30.7 ± 2.05 a |
| *p*-value | | 0.002 | 0.151 | 0.055 | 0.014 | 0.394 | 0.800 |

* Means in a column followed by the same letter are not significantly different at $p \leq 0.05$ according to LS Means.

**Table 3.** Effect of year by planting date interaction on cauliflower growth and yield parameters when averaged over six cultivars at the Horticulture Research Farm near Absaraka, ND (Mean ± SE).

| Year | Planting Date | Days to Harvest | Yield (g) | Curd Diameter (cm) | Fresh Root Weight (g) | Root Length (cm) | Root Width (cm) |
|---|---|---|---|---|---|---|---|
| 2023 | May 1 | 95.4 ± 2.22 a * | 533.0 ± 27.56 b | 14.6 ± 0.32 b,c | 136.4 ± 6.92 a | 25.9 ± 0.92 a | 30.5 ± 1.05 a–c |
| | May 15 | 92.1 ± 1.48 a | 546.6 ± 37.16 b,c | 14.4 ± 0.38 c | 108.3 ± 7.07 c | 23.4 ± 0.98 b | 28.5 ± 1.12 c–e |
| | June 1 | 87.7 ± 1.04 b | 353.7 ± 29.13 d | 12.1 ± 0.37 d | 87.7 ± 6.21 d | 20.4 ± 0.64 b,c | 26.2 ± 0.90 e |
| | June 15 | 82.9 ± 1.22 c | 355.4 ± 34.29 d | 12.4 ± 0.42 d | 109.1 ± 6.77 b,c | 21.94± 0.70 b | 29.6 ± 0.95 a–d |
| 2024 | May 1 | 73.5 ± 1.03 d | 765.3 ± 35.29 a | 16.1 ± 0.29 a | 110.8 ± 6.21 b,c | 20.0 ± 0.73 c,d | 28.0 ± 1.12 d,e |
| | May 15 | 72.2 ± 1.01 d | 631.6 ± 25.37 a | 15.4 ± 0.26 ab | 125.6 ± 7.72 a | 21.1 ± 0.62 b,c | 31.3 ± 1.03 a,b |
| | June 1 | 62.8 ± 0.80 e | 529.5 ± 24.86 b | 14.3 ± 0.29 c | 108.3 ± 7.12 b,c | 18.3 ± 0.74 d | 28.5 ± 1.07 b–e |
| | June 15 | 65.7 ± 0.84 f | 406.8 ± 20.92 c | 12.4 ± 0.33 d | 117.5 ± 5.45 a,b | 19.1 ± 0.74 c,d | 31.8 ± 1.17 a |
| *p*-value | | <0.001 | 0.008 | 0.020 | <0.001 | 0.039 | 0.012 |

* Means in a column followed by the same letter are not significantly different at $p \leq 0.05$ according to LS Means.

The 2024 season demonstrated reduced root elongation across all planting dates, with June 1 planting showing the shortest development (18.3 cm). Other interaction effects including year by cultivar and planting date by cultivar were not statistically significant, although observable trends were noted in the data (Tables 1 and 2). Considering root width, our findings showed significant effects of planting date and cultivar, while year effects were not significant. Significant interactions were detected for year by planting date and year by cultivar ($p < 0.05$) (Table S3). The widest root development was observed in 2024 for June 15 planting (31.8 cm) and in 2023 for May 1 planting (30.5 cm), while the narrowest roots occurred in 2023 for June 1 planting (26.2 cm) (Table 3). 'Flame Star' exhibited the widest root systems in both experimental years, with particularly substantial development in 2024 (38.5 cm), while 'Clementine' demonstrated the narrowest root width in 2024 (24.3 cm) (Table 1). No significant variation was observed for root width among cultivars across different planting dates (Table 2).

### *3.2. Physiological Parameters*

Our results demonstrated that stomatal conductance and associated physiological parameters were predominantly influenced by the experimental year, planting date, and to a lesser extent, by cultivar (Table S2). A highly significant year by planting date interaction was detected ($p < 0.001$). Transpiration rates were markedly affected by year and planting date, whereas cultivar effects were not statistically significant. In contrast, water use efficiency (WUE) was exclusively shaped by planting date and its interaction with year ($p < 0.001$). Regarding chlorophyll fluorescence, both planting date ($p = 0.017$) and cultivar ($p < 0.001$) exerted significant effects, alongside a pronounced year by planting date interaction ($p < 0.001$). The observed physiological variations corresponded with temperature differences between growing seasons, as reported in the Methods section. During 2023, the mean air temperature ranged between 17.1 °C and 22.5 °C, whereas in 2024 it varied between 15.0 °C and 22.4 °C. During the 2023 growing season, stomatal conductance values varied between 584.01 mmol $H_2O$ $m^{-2}$ $s^{-1}$ for May 15 plantings and 689.56 mmol $H_2O$ $m^{-2}$ $s^{-1}$ for June 15 plantings (Table 4). In 2024, overall higher conductance values were recorded, with June 1 plantings achieving the peak value of 855.73 mmol $H_2O$ $m^{-2}$ $s^{-1}$, followed by June 15 at 796.81 mmol $H_2O$ $m^{-2}$ $s^{-1}$.

**Table 4.** Effect of year by planting date interaction on cauliflower physiological parameters when averaged over six cultivars at the Horticulture Research Farm near Absaraka, ND (Mean ± SE).

| Year | Planting Date | *gs* (mmol $H_2O$ $m^{-2}$ $s^{-1}$) | *E* (mmol $H_2O$ $m^{-2}$ $s^{-1}$) | WUE (mmol $CO_2$ $mol^{-1}$ $H_2O$) | *Fv'/Fm'* |
|---|---|---|---|---|---|
| 2023 | 1-May | 675.03 ± 20.45 c * | 7.26 ± 0.28 b | 2.95 ± 0.21 c | 0.61 ± 0.00 a,b |
| | 15-May | 584.01 ± 23.08 d | 6.38 ± 0.23 c | 4.62 ± 0.37 a,b | 0.63 ± 0.01 a |
| | 1-Jun | 590.40 ± 21.03 d | 6.10 ± 0.19 c | 3.06 ± 0.17 c | 0.58 ± 0.01 a,b |
| | 15-Jun | 689.56 ± 19.95 c | 5.99 ± 0.15 c | 4.49 ± 0.66 b | 0.60 ± 0.01 b,c |
| 2024 | 1-May | 608.24 ± 22.36 d | 5.31 ± 0.18 d | 5.36 ± 0.56 a | 0.59 ± 0.01 c,d |
| | 15-May | 720.00 ± 20.51 b,c | 8.29 ± 0.27 a | 3.62 ± 0.26 c | 0.60 ± 0.00 a–c |
| | 1-Jun | 855.73 ± 20.94 a | 8.03 ± 0.15 a | 3.77 ± 0.48 c | 0.61 ± 0.00 a,b |
| | 15-Jun | 796.81 ± 25.28 a,b | 7.23 ± 0.21 a,b | 3.98 ± 0.24 a,b | 0.60 ± 0.01 b,c |
| *p*-value | | <0.001 | <0.001 | <0.001 | <0.001 |

* Means in a column followed by the same letter are not significantly different at $p \leq 0.05$ according to LS Means. *gs*, stomatal conductance; *E*, transpiration; WUE, water use efficiency; *Fv'/Fm'*, chlorophyll fluorescence.

At the cultivar level, 'Snow Crown' exhibited the highest conductance (740.96 mmol $H_2O$ $m^{-2}$ $s^{-1}$), while 'Vitaverde' recorded the lowest (642.43 mmol $H_2O$ $m^{-2}$ $s^{-1}$) (Table 5). Transpiration rates in 2023 ranged from 5.99 ± 0.15 mmol $H_2O$ $m^{-2}$ $s^{-1}$ (June 15 planting) to 7.26 mmol $H_2O$ $m^{-2}$ $s^{-1}$ (May 1 planting) (Table 4). In 2024, rates were generally higher, with May 15 and June 1 plantings reaching maximum values of 8.29 and 8.03 mmol $H_2O$ $m^{-2}$ $s^{-1}$, respectively. No significant cultivar-related differences were detected for transpiration (Table 5). For WUE, in 2023 the highest efficiency was obtained in May 15 plantings (4.62 mmol $CO_2$ $mol^{-1}$ $H_2O$), whereas May 1 and June 1 plantings showed lower values (2.95 and 3.06, respectively) (Table 4). In 2024, an altered trend was observed, with May 1 plantings achieving the maximum WUE (5.36 mmol $CO_2$ $mol^{-1}$ $H_2O$), indicating greater carbon assimilation per unit of water loss under early planting in the second experimental year. Chlorophyll fluorescence values during 2023 ranged between 0.58 (June 1 planting) and 0.63 (May 15 planting) (Table 4). The 2024 season exhibited narrower variation (0.59–0.61), with slightly enhanced efficiencies for May 1 and June 1 plantings. Among cultivars, 'Clementine' demonstrated the highest photosynthetic efficiency (0.62), whereas 'Amazing' consistently recorded the lowest value (0.58) (Table 5).

**Table 5.** Effect of cauliflower cultivar on physiological parameters when averaged over two years, 2023 and 2024 at the Horticulture Research Farm near Absaraka, ND.

| Cultivar | *gs* (mmol $H_2O$ $m^{-2}$ $s^{-1}$) | *E* (mmol $H_2O$ $m^{-2}$ $s^{-1}$) | WUE (mmol $CO_2$ $mol^{-1}$ $H_2O$) | *Fv'/Fm'* |
|---|---|---|---|---|
| Amazing | 674.13 ± 19.69 a,b * | 6.69 ± 0.20 a | 4.58 ± 0.61 a | 0.58 ± 0.01 c |
| Cheddar | 686.47 ± 19.46 a,b | 6.86 ± 0.19 a | 3.72 ± 0.22 a | 0.60 ± 0.01 a–c |
| Clementine | 667.11 ± 22.08 a,b | 6.78 ± 0.20 a | 3.40 ± 0.16 a | 0.62 ± 0.01 a |
| Flame Star | 679.65 ± 19.54 a,b | 6.86 ± 0.21 a | 4.01 ± 0.24 a | 0.60 ± 0.01 a–c |
| Snow Crown | 740.96 ± 20.51 a | 7.15 ± 0.21 a | 4.14 ± 0.42 a | 0.59 ± 0.01 b,c |
| Vitaverde | 642.43 ± 18.39 b | 6.46 ± 0.18 a | 4.03 ± 0.28 a | 0.61 ± 0.01 a,b |
| *p*-value | 0.016 | 0.269 | 0.790 | <0.001 |

* Means in a column followed by the same letter are not significantly different at $p \leq 0.05$ according to LS Means. *gs*, stomatal conductance; *E*, transpiration; WUE, water use efficiency; *Fv'/Fm'*, chlorophyll fluorescence.

### 3.3. Maturity and Yield Parameters

With respect to maturity duration, planting date, cultivar, and year showed significant main effects ($p < 0.001$) with significant interactions observed across all combinations (Table S3). From the cultivar standpoint, 'Vitaverde' and 'Flame Star' displayed the longest maturity periods in 2023 (99.4 and 97.0 days, respectively), while 'Snow Crown' and 'Clementine' exhibited the earliest maturation in 2024 (60.4 and 65.3 days, respectively) (Table 1). In terms of planting date effects, 'Flame Star' consistently showed the longest maturity duration across all planting dates, peaking at 94.3 days for May 1 planting, whereas 'Snow Crown' demonstrated the shortest maturity, particularly at the June 15 planting (62.9 days) (Table 2). These seasonal differences corresponded with variable temperature regimes, as the 2023 growing period exhibited higher diurnal temperature fluctuations than 2024. These differences can be partly explained by the distinct climatic conditions between the two seasons. During 2023, higher summer temperatures and greater diurnal variation delayed curd formation and maturity, whereas in 2024, more stable temperature patterns and earlier seasonal warming accelerated developmental progression. From the yield perspective, year, cultivar, and planting date demonstrated significant main effects ($p < 0.001$) with significant year by planting date interaction ($p = 0.008$) (Table S2). In terms of seasonal performance, the 2024 season produced higher yields for May 1 and May 15 plantings (765.3 g and 631.6 g, respectively) compared to 2023 (533.0 g and 546.6 g, respectively). When considered from the perspective of growing seasons, the 2023 season exhibited longer maturity durations across all planting dates, reaching the highest value at May 1 planting (95.4 days). In contrast, the 2024 season demonstrated markedly reduced maturity durations, particularly for June 1 and June 15 plantings (62.8 and 65.7 days, respectively) (Table 3). When evaluated from the cultivar standpoint, 'Flame Star' exhibited the highest mean yield (585.7 g), followed by 'Snow Crown' and 'Cheddar' (555.7 g and 553.2 g, respectively), while 'Vitaverde' recorded the lowest yield (409.8 g) (Figure 5).

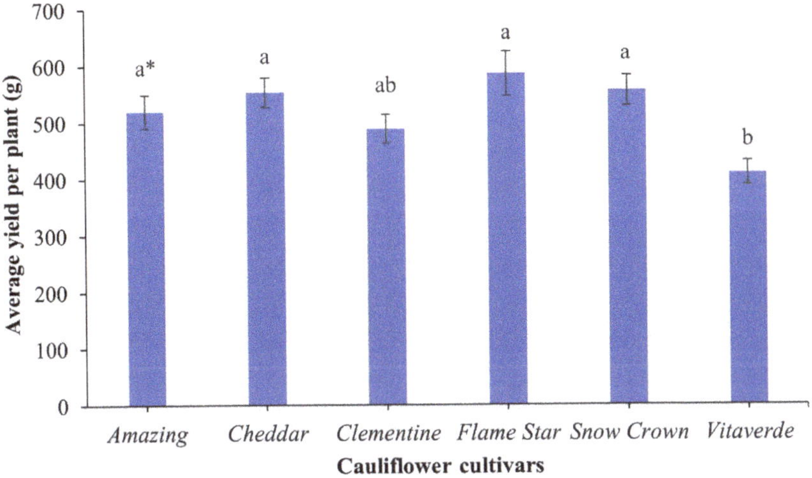


**Figure 5.** Cauliflower cultivar yield averaged over four planting dates and two years. * Means followed by the same letter are not significantly different at $p \leq 0.05$ according to LS Means.

In terms of curd diameter, year and planting date showed significant effects ($p < 0.001$) while cultivar effect was not significant ($p = 0.148$), with a significant year by planting date interaction ($p = 0.020$) (Table S2). From the seasonal perspective, the 2024 season generally produced larger diameters, particularly at the May 1 planting (16.1 cm) compared to 2023 (14.6 cm). The smallest diameters were observed from the June plantings of the 2023 season (12.1 and 12.4 cm for June 1 and June 15, respectively) (Table 3). With respect to head defects, significant associations between cultivar-planting date combinations and defect distributions were determined for both 2023 ($\chi^2 = 275.90$, df = 115, $p < 0.001$) and 2024 ($\chi^2 = 238.63$, df = 115, $p < 0.001$) (Figure 6).

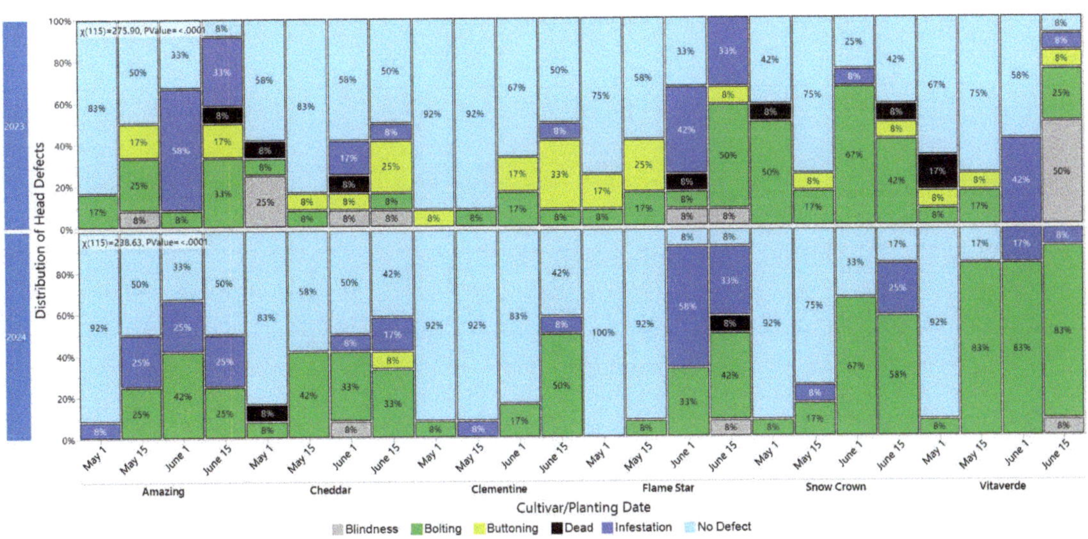


**Figure 6.** Cauliflower defect types influenced by cultivar and planting date for years 2023 and 2024 at the Horticulture Research Farm near Absaraka, ND.

From the temporal standpoint, insect infestation, bolting, and buttoning were prevalent in late plantings during 2023, with 'Amazing' at the June 1 planting date showing 58% infestation and 'Flame Star' at June 15 exhibiting 50% bolting. Early plantings (May 1 and May 15) demonstrated over 50% non-defective heads across cultivars, except for 'Snow Crown' (43% for May 1). The 2024 season showed substantial improvement, with all cultivars achieving over 90% non-defective heads from the May 1 planting. From the cultivar perspective, 'Clementine' consistently produced fewer defective heads across all planting dates in both years. In terms of head quality, significant associations between cultivar-planting date combinations and USDA grading outcomes were established for both 2023 ($\chi^2 = 94.89$, df = 46, $p < 0.001$) and 2024 ($\chi^2 = 190.57$, df = 69, $p < 0.001$) (Figure 7). From the planting date standpoint, early planting dates predominantly achieved 'US No. 1' grade in 2023, while late plantings showed reduced quality, with 'Amazing' at the June 15 planting date producing only 8% 'US No. 1' and 67% non-marketable heads. The 2024 season demonstrated marked improvement, with 'Flame Star' achieving 100% 'US No. 1' grade with the May 1 planting date. When considered from the cultivar perspective, Clementine' consistently achieved the highest proportion of 'US No. 1' graded heads among all cultivars in 2024.

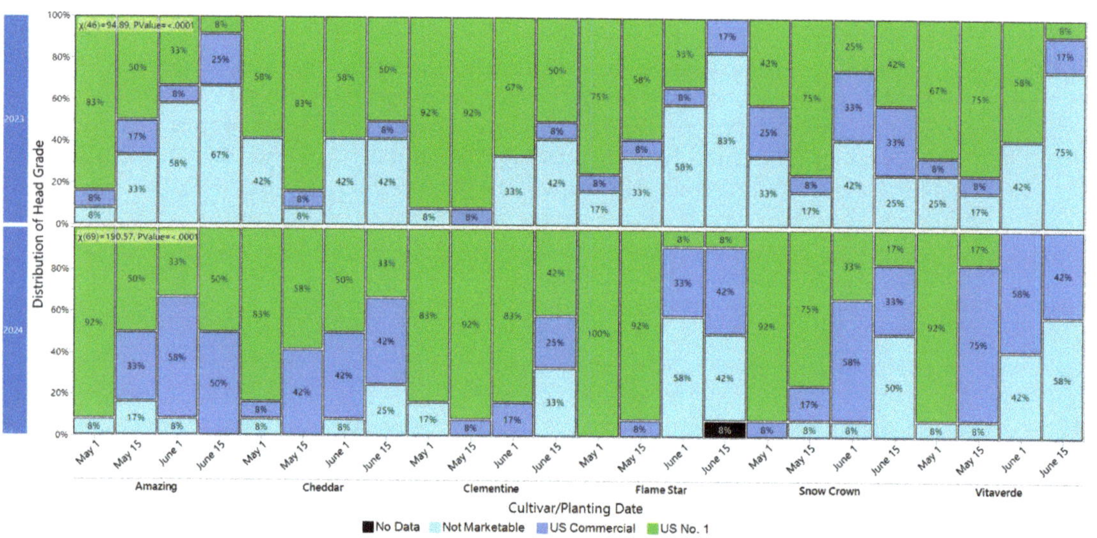


**Figure 7.** Cauliflower grade distribution influenced by cultivar and planting date for years 2023 and 2024 at the Horticulture Research Farm near Absaraka, ND.

### *3.4. General Evaluation of Morphological Parameters*

The heatmap analysis provided a comprehensive overview of the relationships among morphological traits of cauliflower across different years, planting dates, and cultivars (Figure 8). Curd diameter and yield were consistently clustered together, reflecting their strong positive association, while root-related traits (fresh root weight, length, and width) also grouped closely, indicating their interdependence. Vegetative traits such as plant height and number of leaves showed moderate correlation with each other but remained more distinct from yield-related traits. Days to harvest appeared relatively independent, though it showed partial alignment with root parameters rather than curd and yield traits. Across genotype × environment interactions, early plantings in 2024 (e.g., Flame Star and Snow Crown) exhibited superior performance in yield and curd diameter with shorter maturity periods, suggesting favorable adaptation under those conditions. In contrast, late plantings in 2023, particularly with 'Vitaverde' and 'Clementine', were characterized by overall lower performance across most parameters.

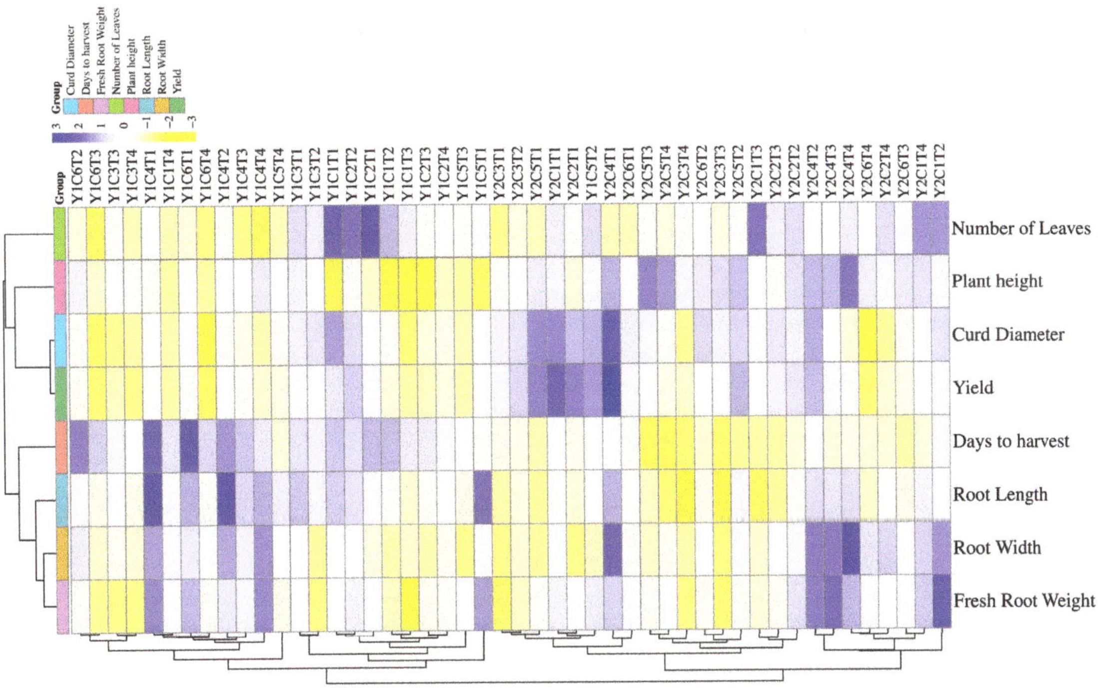


**Figure 8.** Heatmap illustrating the hierarchical clustering of cauliflower growth parameters across year by planting date by cultivar combinations. Colors indicate the relative value of each trait (blue/violet = high, yellow/green = low). Y1: 2023, Y2: 2024, C1: 'Amazing', C2: 'Cheddar', C3: 'Clementine', C4: 'Flame Star', C5: 'Snow Crown', C6: 'Vitaverde', T1: transplanted on May 01, T2: transplanted on May 15, T3: transplanted on June 01, T4: transplanted on June 15.

## 4. Discussion

Although predicting the exact mechanisms underlying vegetative growth responses in brassica crops remains challenging, this study systematically evaluated growth parameters through comprehensive monitoring of plant development across different planting windows and cultivar selections during two consecutive growing seasons (2023 and 2024). In our study, significant variations in leaf number per plant and plant height were determined when cauliflower cultivars were subjected to different planting dates and environmental conditions. The quadratic growth models obtained from each cultivar-planting date combination were accepted as reliable indicators of vegetative development patterns. Previous studies supporting our findings have shown that multiple growth phases of *Brassica* species, such as cabbage, broccoli, cauliflower, and other Cole crops, exhibit distinct responses to planting timing during various developmental stages [17–21]. In some studies, on the other hand, tested for different environmental conditions and planting schedules, the reliability of early season measurements to predict final plant performance has been accepted, although clearly no discernible linear relationship between initial and final growth parameters was seen [22–24].

### 4.1. Vegetative Growth

Many studies examining vegetative growth responses in *Brassica* species have reported that early and late planting effects are not expressed as distinct phases but rather as part of a continuous growth pattern encompassing leaf development, plant height, and root system development [25–28]. Many other researchers, except for Kumar et al. [27], suggested that early planting windows provided optimal conditions for sustained vegetative growth in cauliflower and other brassica crops [20,22,23]. Although vigorous growth is typically observed during early developmental stages, a subsequent decline in growth rate is also common as the season progresses. This pattern has significant implications for production management, suggesting that early-season environmental conditions are critical for establishing the crop's productive potential. Furthermore, growth advantages conferred by early planting cannot be fully compensated for by later favorable conditions.

From a practical perspective, these findings provide a framework for cultivar-specific management decisions based on targeted market timing. For instance, growers seeking high early-season yields may benefit from selecting fast-growing cultivars such as 'Snow Crown', while those targeting mid-season markets may prefer a cultivar like 'Flame Star' that exhibited prolonged vegetative vigor across a wider planting window. Regression-based growth models thus offer valuable predictive tools to align cultivar growth characteristics with optimal harvest timing and market demands. Our findings are consistent with previous reports indicating that cauliflower cultivars within the *B. oleracea* species complex maintain vigorous vegetative growth under favorable early-season conditions [24,25]. Enhanced leaf expansion and canopy development during early planting are likely attributed to the alignment of planting dates with optimal temperature ranges and adequate photoperiods. Once favorable environmental conditions were established, brassica crops tend to sustain growth momentum, with early planting effectively synchronizing developmental processes with the most suitable portion of the growing season [26,28]. Among the cauliflower cultivars evaluated, 'Snow Crown' and 'Cheddar' consistently exhibited superior vegetative growth at early planting dates, and moderate performance under later plantings. Interestingly, all cultivars showed the most vigorous leaf development at early May plantings compared to other planting dates during both 2023 and 2024. In later planting schedules, growth performance varied among cultivars depending on environmental conditions. Current results are compatible with previous studies of Islam et al. [22] and Akhtar et al. [23], who also identified early planting as optimal for maximizing cauliflower vegetative growth. In contrast, later planting schedules resulted in variable growth responses among cultivars, largely influenced by prevailing environmental conditions.

The root morphological characteristics and below-ground development patterns of cauliflower cultivars varied significantly depending on genetic predisposition, planting time, and environmental conditions across both growing seasons. These variations suggest the complex interactions among inherent root development capacity, soil conditions, and seasonal temperature fluctuations. Notably, 'Flame Star' consistently demonstrated superior root system development with the heaviest biomass, longest extension, and widest architecture at optimal planting dates, while 'Clementine' displayed reduced root development under comparable conditions [29]. Such root development differences among cultivars tested under similar soil conditions indicate that genetic factors effectively influence below-ground productivity potential [30,31]. At early planting dates (May 1 and May 15), all cultivars demonstrated enhanced root biomass accumulation and architectural development both years for optimal nutrient and water uptake efficiency. Surprisingly, the root development during the 2023 growing season was generally superior compared to the 2024 season, suggesting that environmental influences on below-ground plant development are not necessarily correlated with high yields [32]. Thus, further research through detailed soil–plant interaction studies is required [33–35]. Based on current evidence, early planting establishes favorable conditions for rapid root system development and efficient soil colonization, conferring significant advantages for overall plant growth and resilience [36,37].

### 4.2. Physiological Response

The physiological responses of cauliflower cultivars, as reflected by *gs*, *E*, WUE, and *Fv′/Fm′*, had significant interactions with planting timing and year. These physiological indicators were more responsive than previously reported in other brassica crops [38–40]. Consistent with prior studies, early- and mid-season plantings significantly enhanced *gs*, *E*, and WUE values compared to late plantings, indicating strong temperature-dependent regulation of gas exchange and photosynthetic efficiency [41–43]. In addition, environmental fluctuations between 2023 and 2024 growing seasons further influenced cultivar performance, as differences in temperature and soil moisture conditions produced measurable variations in physiological traits [44,45]. In both years, early plantings exhibited substantially higher *gs* and *Fv′/Fm′* values compared to delayed planting [46,47]. These findings support the concept of genotype by environment interactions that drive physiological adaptation in *Brassica oleracea*, where optimal environmental conditions enhance photosynthetic capacity and water transport efficiency. Prior to the onset of environmental stress, cultivars maintained near-optimal physiological values, suggesting the importance of early-season conditions for maximizing carbon assimilation and growth potential [34–36]. These observed patterns aligned with environmental monitoring data, confirming anticipated seasonal responses in plant performance.

### 4.3. Yield and Curd Development

Maturity and yield traits varied significantly among cultivars in response to planting date and environmental conditions across both growing seasons. These variations reflect complex interactions among genetic potential, environmental cues, and management practices, and were consistent with previous studies on brassica crops [48–53]. 'Flame Star' and 'Snow Crown' consistently achieved superior yield performance at early planting dates, while other cultivars showed intermediate responses across both growing seasons. Such differences under similar environmental conditions emphasize the influence of cultivar genetics on productivity [51]. Early May plantings consistently resulted in superior curd formation and commercially acceptable yields across all tested cultivars, confirming that proper planting timing is essential for optimal reproductive development. These findings were consistent with prior reports that early planting enhances curd initiation, biomass accumulation, and yield potential [26,52]. Conversely, delayed planting (June 1 and June 15) resulted in reduced curd diameter and lower yield, consistent with earlier findings linking late sowing to suboptimal temperature, shortened growth periods, and reduced head size [54–56]. High-yielding cultivars such as 'Flame Star', 'Snow Crown', and 'Cheddar' produced significantly greater curd weights than lower-performing cultivars like 'Vitaverde', reaffirming the role of cultivar selection in maximizing yield [22]. The enhanced yield response likely reflects favorable temperatures, longer growth periods, and optimal environmental conditions, though this hypothesis requires validation under controlled studies [57]. Current results suggest that early planting promotes optimal cauliflower growth and yield, emphasizing the need for planting schedules aligned with cultivar-requirements and local environments.

Head quality was highly sensitive to planting time and cultivar differences. Early planting dates (May 1 and 15) consistently produced compact, defect-free curds, whereas later plantings (June 1 and 15) were more prone to buttoning, bolting, and bracting, outcomes frequently linked to irregular temperature regimes and elevated heat stress during vegetative and reproductive phases [27,58]. Cooler early-season conditions promoted higher-quality heads with fewer physiological defects [27,59]. Quality assessments indicated that early plantings had a greater proportion of 'US No. 1' marketable curds, while later plantings more often had 'US Commercial' category curds due to discolored and loose curds. This trend was consistent with the USDA quality standards [60] and previous findings in related *Brassica* crops [61–64]. Although favorable weather in 2024 mitigated some late planting negative effects, the consistent advantage of early establishment across cultivars indicates that optimizing planting windows remains the most reliable strategy for achieving high yield and marketable quality in commercial cauliflower production.

### 4.4. Relationship Between Morphological Parameters

Significant correlations were observed among morphological parameters across cultivar and planting date combinations for both seasons. The strongest relationships occurred between curd diameter and yield followed by root morphological traits and then vegetative growth characteristics. Cluster analysis confirmed that yield-related traits were most closely integrated with overall plant performance, followed by root and vegetative parameters. These results indicate that root development acts as a critical intermediary linking vegetative vigor with reproductive success under favorable growing conditions Previous studies have reported similar dynamic interrelationships among morphological traits as environmental conditions shifted across the growing season [65–67]. Our findings support the concept that coordinated morphological development enhances whole-plant performance, with complex interdependencies among growth components rather than simple linear associations [68,69]. Comparative assessment with established growing regions like California suggests that the environmental conditions of North Dakota, characterized by moderate temperatures during early growth stages, are well-suited for cauliflower cultivation. The superior physiological and yield performance observed in early May plantings indicates that cooler early-season conditions favor high-quality curd formation. These results imply that northern regions, such as North Dakota, may gain a comparative advantage for cool-season *Brassica* production as global temperatures rise. Nevertheless, the strong sensitivity of all cultivars to planting time suggests that precise scheduling and region-specific cultivar selection will remain critical for optimizing production in emerging northern growing zones.

## 5. Conclusions

Our study revealed that strategic planting date selection significantly enhances cauliflower production resilience across variable environmental conditions through comprehensive optimization of growth, physiological, and yield parameters. Comparing early and late planting windows across two growing seasons with six distinct cultivars, we found that May planting dates consistently demonstrated superior performance across vegetative development, root morphology, and final yield outcomes. Under optimal early planting conditions, cauliflower plants maintained substantially higher physiological efficiency, with *gs*, WUE, and *Fv'/Fm'* reaching their peak levels compared to delayed planting treatments. The divergent developmental responses between planting dates were particularly evident in root morphological profiles, where early-planted cultivars demonstrated enhanced root system development, particularly 'Flame Star', which achieved maximum fresh root weight, root length, and width parameters under favorable conditions. These differences translated to significant variations in yield performance and quality outcomes, with early-planted crops maintaining higher curd weights and diameters, while enhancing marketable grade classifications such as 'US No. 1' standards across all cultivars. Among cultivars, 'Flame Star', 'Snow Crown', and 'Cheddar' consistently outperformed other genotypes, with 'Flame Star' achieving the highest mean yield and demonstrating exceptional adaptability across different planting windows. These findings have immediate practical applications for *Brassica* vegetable production in regions affected by seasonal temperature fluctuations, where strategic early planting represents a viable management strategy that could significantly improve agricultural sustainability and economic returns.

**Supplementary Materials:** The following supporting information can be downloaded at: https://www.mdpi.com/article/10.3390/horticulturae11111314/s1, Figure S1. Aerial view of experimental site at the NDSU Horticultural Farm near Absaraka, ND in 2023. Figure S2. Example of the cauliflower curds for the six cultivars used in the planting date field studyat the Horticulture Research Farm near Absaraka, ND in 2023 and 2024. Table S1. Parameter estimates and significance of regression lines for the average number of leaves and plant height by date after transplanting (DAT) across planting date and cultivars. Table S2. Analysis of variance (ANOVA) results for physiological parameters of cauliflower cultivars under different planting times across two years. Table S3. Analysis of variance (ANOVA) results for agronomic parameters of cauliflower Cultivars under different planting times across two years.

**Author Contributions:** H.H.-V. and A.D. contributed equally to the experimental design, methodological framework, conceptual development, data interpretation, statistical analysis, figure preparation, and manuscript writing. A.D. and O.K. coordinated the drafting and revision of the text. All authors have read and agreed to the published version of the manuscript.

**Funding:** This research was carried out with financial support from the North Dakota Specialty Crop Block Grant (NOGA 22–224).

**Data Availability Statement:** The original contributions presented in this study are included in the article/Supplementary Material. Further inquiries can be directed to the corresponding author.

**Acknowledgments:** The authors would like to thank Collin Auwarter, the High-value Crops Project Research Specialist and project graduate students for their research assistance.

**Conflicts of Interest:** The authors declare that they have no financial or personal conflicts of interest that could have influenced the objectivity of this research.

## References

1. Swarup, V.; Chatterjee, S.S. Origin and Genetic Improvement of Indian Cauliflower. *Econ. Bot.* **1972**, 381–393. [CrossRef]
2. Sharma, S.R.; Singh, P.K.; Chable, V.; Tripathi, S.K. A Review of Hybrid Cauliflower Development. *J. New Seeds* **2004**, *6*, 151–193. [CrossRef]
3. Picchi, V.; Fibiani, M.; Scalzo, R. Lo Cauliflower. In *Nutritional Composition and Antioxidant Properties of Fruits and Vegetables*; Elsevier: Amsterdam, The Netherlands, 2020; pp. 19–32.
4. Kapusta-Duch, J.; Szeląg-Sikora, A.; Sikora, J.; Niemiec, M.; Gródek-Szostak, Z.; Kuboń, M.; Leszczyńska, T.; Borczak, B. Health-Promoting Properties of Fresh and Processed Purple Cauliflower. *Sustainability* **2019**, *11*, 4008. [CrossRef]
5. Hadley, P.; Pearson, S. Effects of Environmental Factors on Progress to Crop Maturity in Selected Brassica Crops. In Proceedings of the International Symposium Brassica 97, Xth Crucifer Genetics Workshop 459, Rennes, France, 23–27 September 1997; pp. 61–70.
6. Nowbuth, R.D.; Pearson, S. The Effect of Temperature and Shade on Curd Initiation in Temperate and Tropical Cauliflower. In Proceedings of the International Symposium Brassica 97, Xth Crucifer Genetics Workshop 459, Rennes, France, 23–27 September 1997; pp. 79–88.
7. Warland, J.; McKeown, A.W.; McDonald, M.R. Impact of High Air Temperatures on Brassicacae Crops in Southern Ontario. *Can. J. Plant Sci.* **2006**, *86*, 1209–1215. [CrossRef]
8. Kahn, B.A.; Rebek, E.J.; Damicone, J.P. *Cole Crop Production (Broccoli, Cabbage, and Cauliflower)*; Oklahoma State University: Stillwater, OK, USA, 2009.
9. USDA. USDA—National Agricultural Statistics Service—Publications. Available online: https://www.nass.usda.gov/Publications/ (accessed on 16 August 2025).

10. USDA. 2022 Census by State—North Dakota | 2022 Census of Agriculture | USDA/NASS. Available online: https://www.nass.usda.gov/Publications/AgCensus/2022/Full_Report/Census_by_State/North_Dakota/index.php (accessed on 17 August 2025).
11. Farmer Urban North-Dakota-Vegetable-Planting-Calendar. Available online: https://www.ufseeds.com/north-dakota-vegetable-planting-calendar.html?srsltid=AfmBOormxixqGjNBJ4hONI03PBtKVCC9ECXpojS6w4rvMUQUoHRqkqmo (accessed on 17 August 2025).
12. Singh, B.K.; Singh, B.; Singh, P.M. Breeding cauliflower: A review. *Int. J. Veg. Sci.* **2018**, *24*, 58–84. [CrossRef]
13. USDA. USDA Plant Hardiness Zone Map. Available online: https://planthardiness.ars.usda.gov/ (accessed on 17 August 2025).
14. Tollerud, H.; Brown, J.; Loveland, T.; Mahmood, R.; Bliss, N. Drought and Land-Cover Conditions in the Great Plains. *Earth Interact.* **2018**, *22*, 1–25. [CrossRef]
15. USNRCS. Soil Survey of Cass County Area North Dakota: United States. *Natural Resources Conservation Service: Free Download, Borrow, and Streaming: Internet Archive.* Available online: https://archive.org/details/CassND1983 (accessed on 17 August 2025).
16. Evgenidis, G.; Traka-Mavrona, E.; Koutsika-Sotiriou, M. Principal Component and Cluster Analysis as a Tool in the Assessment of Tomato Hybrids and Cultivars. *Int. J. Agron.* **2011**, *2011*, 697879. [CrossRef]
17. Aslam, M.A.; Ahmed, M.; Hassan, F.-U.; Afzal, O.; Mehmood, M.Z.; Qadir, G.; Asif, M.; Komal, S.; Hussain, T. Impact of Temperature Fluctuations on Plant Morphological and Physiological Traits. In *Building Climate Resilience in Agriculture: Theory, Practice and Future Perspective*; Springer: Cham, Switzerland, 2021; pp. 25–52.
18. Cotrina Cabello, G.G.; Ruiz Rodriguez, A.; Husnain Gondal, A.; Areche, F.O.; Flores, D.D.C.; Astete, J.A.Q.; Camayo-Lapa, B.F.; Yapias, R.J.M.; Jabbar, A.; Yovera Saldarriaga, J. Plant Adaptability to Climate Change and Drought Stress for Crop Growth and Production. *CABI Rev.* **2023**. [CrossRef]
19. Quezada-Martinez, D.; Addo Nyarko, C.P.; Schiessl, S.V.; Mason, A.S. Using Wild Relatives and Related Species to Build Climate Resilience in Brassica Crops. *Theor. Appl. Genet.* **2021**, *134*, 1711–1728. [CrossRef]
20. Yu, M.; Luobu, Z.; Zhuoga, D.; Wei, X.; Tang, Y. Advances in Plant Response to Low-Temperature Stress. *Plant Growth Regul.* **2025**, *105*, 167–185. [CrossRef]
21. Wahid, A.; Gelani, S.; Ashraf, M.; Foolad, M.R. Heat Tolerance in Plants: An Overview. *Env. Exp. Bot.* **2007**, *61*, 199–223. [CrossRef]
22. Islam, M.R.; Hoque, T.S.; Khan, R.N.A.; Farzana, S.; Ahmed, M.; Khodabakhshloo, N. Influence of Different Integrated Nutrient Management Strategies on Growth, Yield and Nutritional Qualities of Cauliflower. *Agric. Res.* **2021**, *10*, 656–664. [CrossRef]
23. Akhtar, S.; Naik, A.; Solankey, S.S. Genotypic Selection in Vegetables for Adaptation to Climate Change. In *Advances in Research on Vegetable Production Under a Changing Climate Vol. 1*; Springer: Cham, Switzerland, 2021; pp. 61–89.
24. Suseela, P.; Rangaswami, M.V. Influence of Temperature inside the Greenhouse on Growth Attributes and Yield of Cauliflower. *Karnataka J. Agric. Sci.* **2012**, *24*, 706–708.
25. Tao, J.; Yang, Y.; Wang, Q. Two Growing-Season Warming Partly Promoted Growth but Decreased Reproduction and Ornamental Value of Impatiens Oxyanthera. *Plants* **2024**, *13*, 511. [CrossRef]
26. Rahman, M.A.; Imran, M.; Ikrum, M.; Rahman, M.H.; Rabbani, M.G. Effects of Planting Date and Growth Hormone on Growth and Yield of Cauliflower. *J. Environ. Sci. Nat. Resour.* **2016**, *9*, 143–150. [CrossRef]
27. Kumar, A.; Fatmi, U.; Deepanshu. Effect of Different Planting Dates on Growth, Yield and Quality of Cauliflower (*Brassica oleracea* Var. *botrytis* L.) Hybrids. *Int. J. Res. Agron.* **2024**, *7*, 351–355. [CrossRef]
28. Chatterjee, S.; Aralikatti, O.; Sharma, S.; Mukherjee, D.; Patil, S.; Kanwar, H.S.; Choudhuri, P. Studies of Genetic Variability, Heritability and Genetic Gain for Some Important Horticultural Traits in Cauliflower (*Brassica oleracea* Var. *botrytis* L.). *Int. J. Curr. Microbiol. App. Sci.* **2018**, *7*, 82–92. [CrossRef]
29. Marschner, H. *Marschner's Mineral Nutrition of Higher Plants*; Academic press: Cambridge, MA, USA, 2012; ISBN 0123849055.
30. Giordano, M.; Petropoulos, S.A.; Rouphael, Y. Response and Defence Mechanisms of Vegetable Crops against Drought, Heat and Salinity Stress. *Agriculture* **2021**, *11*, 463. [CrossRef]
31. Biondi, F.; Balducci, F.; Capocasa, F.; Visciglio, M.; Mei, E.; Vagnoni, M.; Mezzetti, B.; Mazzoni, L. Environmental Conditions and Agronomical Factors Influencing the Levels of Phytochemicals in Brassica Vegetables Responsible for Nutritional and Sensorial Properties. *Appl. Sci.* **2021**, *11*, 1927. [CrossRef]
32. Murren, C.J.; Alt, C.H.S.; Kohler, C.; Sancho, G. Natural Variation on Whole-plant Form in the Wild Is Influenced by Multivariate Soil Nutrient Characteristics: Natural Selection Acts on Root Traits. *Am. J. Bot.* **2020**, *107*, 319–328. [CrossRef]
33. Sharma, P.; Meena, P.D.; Chauhan, J.S. First Report of N Igrospora Oryzae (B Erk. & B Roome) Petch Causing Stem Blight on B Rassica Juncea in I Ndia. *J. Phytopathol.* **2013**, *161*, 439–441.
34. Gregory, P. *Plant Roots*; Wiley Online Library: Hoboken, NJ, USA, 2007; ISBN 1405173084.
35. Thakur, P.; Kumar, S.; Malik, J.A.; Berger, J.D.; Nayyar, H. Cold Stress Effects on Reproductive Development in Grain Crops: An Overview. *Env. Exp. Bot.* **2010**, *67*, 429–443. [CrossRef]
36. Balliu, A.; Zheng, Y.; Sallaku, G.; Fernández, J.A.; Gruda, N.S.; Tuzel, Y. Environmental and Cultivation Factors Affect the Morphology, Architecture and Performance of Root Systems in Soilless Grown Plants. *Horticulturae* **2021**, *7*, 243. [CrossRef]

37. Fadilah, L.N.; Lakitan, B. Growth Responses and Yield of Cauliflower (*Brassica oleracea* Var. *botrytis* L.) to the Delayed Transplanting and Drought Stress. In Proceedings of the First International Conference on Assessment and Development of Agricultural Innovation (1st ICADAI 2021), Virtual, 6–7 July 2021; EDP Sciences: Les Ulis, France, 2021; Volume 306, p. 01007.
38. Costa, J.M.; Grant, O.M.; Chaves, M.M. Thermography to Explore Plant–Environment Interactions. *J. Exp. Bot.* **2013**, *64*, 3937–3949. [CrossRef]
39. Kleinhenz, M.D.; Wszelaki, A. Yield and Relationships among Head Traits in Cabbage as Influenced by Planting Date and Cultivar. I. Fresh Market. *HortScience* **2003**, *38*, 1349–1354. [CrossRef]
40. Franks, S.J. Plasticity and Evolution in Drought Avoidance and Escape in the Annual Plant Brassica Rapa. *New Phytol.* **2011**, *190*, 249–257. [CrossRef] [PubMed]
41. Sharma, P. Responses and Tolerance of Brassicas to High Temperature. In *The Plant Family Brassicaceae: Biology and Physiological Responses to Environmental Stresses*; Springer: Singapore, 2020; pp. 277–310.
42. Mohan, V.R.; MacDonald, M.T.; Abbey, L. Impact of Water Deficit Stress on Brassica Crops: Growth and Yield, Physiological and Biochemical Responses. *Plants* **2025**, *14*, 1942. [CrossRef]
43. Levin, G.; Schuster, G. Feeling the Heat: Long-Term Heat Stress Impairs Growth but Not Photosynthesis in a C4 Grass. *Plant Cell* **2024**, *37*, koaf008. [CrossRef]
44. Redondo-Gómez, S. Abiotic and Biotic Stress Tolerance in Plants. In *Molecular Stress Physiology of Plants*; Springer: New Delhi, India, 2013; pp. 1–20.
45. Basu, P.S.; Chaturvedi, S.K.; Gaur, P.M.; Mondal, B.; Meena, S.K.; Das, K.; Kumar, V.; Tewari, K.; Sharma, K. Physiological Mechanisms of Tolerance to Drought and Heat in Major Pulses for Improving Yield under Stress Environments. In *Advances in plant Defense Mechanisms*; IntechOpen: London, UK, 2022; ISBN 1803558024.
46. Uzun, S.; Kar, H. Quantitative Effects of Planting Time on Vegetative Growth of Broccoli (*Brassica oleracea* Var. *italica*). *Pak. J. Bot.* **2004**, *36*, 769–777.
47. Kourani, M.; Mohareb, F.; Rezwan, F.I.; Anastasiadi, M.; Hammond, J.P. Genetic and Physiological Responses to Heat Stress in Brassica Napus. *Front. Plant. Sci.* **2022**, *13*, 832147. [CrossRef]
48. Wien, H.C.; Wurr, D.C.E. Cauliflower, Broccoli, Cabbage and Brussels Sprouts. In *The Physiology of Vegetable Crops*; CABI: Wallingford, UK, 1997.
49. Ara, S.; Afroz, S.; Noman, M.S.; Bhuiyan, M.S.R.; Zia, M.I.K. Variability, Correlation and Path Analysis in F2 Progenies of Inter-Varietal Crosses of Brassica Rapa. *J. Environ. Sci. Nat. Resour.* **2013**, *6*, 217–220. [CrossRef]
50. Lindemann-Zutz, K.; Fricke, A.; Stützel, H. Prediction of Time to Harvest and Its Variability in Broccoli (*Brassica oleracea* Var. *italica*) Part I. Plant Developmental Variation and Forecast of Time to Head Induction. *Sci. Hortic.* **2016**, *198*, 424–433.
51. Weltzien, E.; Rattunde, F.; Christinck, A.; Isaacs, K.; Ashby, J. Gender and Farmer Preferences for Varietal Traits: Evidence and Issues for Crop Improvement. *Plant Breed. Rev.* **2019**, *43*, 243–278.
52. Devi, S.; Kaur, H. Influence of Different Nutrient Management Practices on Growth and Yield of Broccoli (*Brassica oleracea* L. Var. Italica). *Int. J. Sci. Res.* **2022**, *11*, 681–684. [CrossRef]
53. Obaid, M.H.; Saloom, Y.F.; Mejbel, M.H.; Al-Hchami, S.H.J. Effect of Different Planting Dates on Some Plant Growth Traits of Four Hybrids of Broccoli. *Res. Crops* **2019**, *20*, 333–337.
54. Ahmed, I.; Ullah, A.; ur Rahman, M.H.; Ahmad, B.; Wajid, S.A.; Ahmad, A.; Ahmed, S. Climate Change Impacts and Adaptation Strategies for Agronomic Crops. In *Climate Change and Agriculture*; IntechOpen: London, UK, 2019; ISBN 178985668X.
55. Hossain, M.F.; Ara, N.; Uddin, M.R.; Islam, M.R.; Azam, M.G. Effect of Sowing Date and Plant Spacing on Seed Production of Cauliflower. *Bangladesh J. Agric. Res.* **2015**, *40*, 491–500. [CrossRef]
56. Rahman, H.U.; Hadley, P.; Pearson, S. Relationship between Temperature and Cauliflower (*Brassica oleracea* L. Var. *botrytis*) Growth and Development after Curd Initiation. *Plant Growth Regul.* **2007**, *52*, 61–72. [CrossRef]
57. Sarkar, S.; Saha, S.; Ghosh, S.; Paul, S.K.; Dey, S.; Moulick, D.; Santra, S.C.; Brahmachari, K. Abiotic Stress Sensitivity and Adaptation in Field Crops. In *Climate-Resilient Agriculture, Vol 2: Agro-Biotechnological Advancement for Crop Production*; Springer: Cham, Switzerlans, 2023; pp. 319–362.
58. Chen, M.; Zhang, T.-L.; Hu, C.-G.; Zhang, J.-Z. The Role of Drought and Temperature Stress in the Regulation of Flowering Time in Annuals and Perennials. *Agronomy* **2023**, *13*, 3034. [CrossRef]
59. Mareri, L.; Parrotta, L.; Cai, G. Environmental Stress and Plants. *Int. J. Mol. Sci.* **2022**, *23*, 5416. [CrossRef]
60. USDA. Cauliflower Grades and Standards | Agricultural Marketing Service. Available online: https://www.ams.usda.gov/grades-standards/cauliflower-grades-and-standards (accessed on 17 August 2025).
61. Björkman, T.; Pearson, K.J. High Temperature Arrest of Inflorescence Development in Broccoli (*Brassica oleracea* Var. *italica* L.). *J. Exp. Bot.* **1998**, *49*, 101–106. [CrossRef]
62. Wurr, D.C.E.; Fellows, J.R. Leaf Production and Curd Initiation of Winter Cauliflower in Response to Temperature. *J. Hortic. Sci. Biotechnol.* **1998**, *73*, 691–697. [CrossRef]

63. Abed, M.Y.; El-Said, E.M.; Shebl, E.F. Effect of Planting Date and Spacing on Yield and Quality of Cabbage (*Brassica oleracea* Var. *capitata* L.). *J. Plant Prod.* **2015**, *6*, 2093–2102. [CrossRef]
64. Moratagi, R.; Reddy, P.S.S.; Sadarunnisa, S.; Padmaja, V.V.; Reddy, A.R. Effect of Cultivar and Different Plant Spacing on Growth and Yield of Cauliflower under Southern Agro-Climatic Zone of Andhra Pradesh. *Int. J. Curr. Microbiol. App. Sci.* **2021**, *10*, 2965–2973. [CrossRef]
65. Kumar, J.L.; Imtiyaz, M. Yield, Irrigation Production Efficiency and Economic Returns of Broccoli under Variable Drip Irrigation and Lateral Spacing. *J. Sci. Technol.* **2007**, *27*, 111–121. [CrossRef]
66. Lynch, J.P.; Brown, K.M. New Roots for Agriculture: Exploiting the Root Phenome. *Philos. Trans. R. Soc. B Biol. Sci.* **2012**, *367*, 1598–1604. [CrossRef]
67. Pearson, S.; Hadley, P.; Wheldon, A.E. A Model of the Effects of Temperature on the Growth and Development of Cauliflower (*Brassica oleracea* L. *botrytis*). *Sci. Hortic.* **1994**, *59*, 91–106. [CrossRef]
68. Bhat, N.R.; Bhatt, A.; Suleiman, M.K. Growth Stages, Growth Analysis and Growth Modeling. In *Physiology of Growth and Development in Horticultural Plants*; CRC Press: Boca Raton, FL, USA, 2024; pp. 111–120. [CrossRef]
69. Kilbane, E.; Crews, T.; Dehaan, L.; Grillo, M.; Ohsowski, B.; Dybzinski, R. Modeling Carbon Allocation Strategies for High-Yielding Perennial Crops. *Eur. J. Agron.* **2025**, *166*, 127559. [CrossRef]

MDPI AG
Grosspeteranlage 5
4052 Basel
Switzerland
Tel.: +41 61 683 77 34

*Horticulturae* Editorial Office
E-mail: horticulturae@mdpi.com
www.mdpi.com/journal/horticulturae

www.ingramcontent.com/pod-product-compliance
Lightning Source LLC
Chambersburg PA
CBHW061609220926
51850CB00005B/460